Benchmark Papers
in Geology

Series Editor: Rhodes W. Fairbridge
Columbia University

Published Volumes

ENVIRONMENTAL GEOMORPHOLOGY AND LANDSCAPE
    CONSERVATION, VOLUME I: Prior to 1900 / Donald R. Coates
RIVER MORPHOLOGY / Stanley A. Schumm
SPITS AND BARS / Maurice L. Schwartz
TEKTITES / Virgil E. Barnes and Mildred A. Barnes
GEOCHRONOLOGY: Radiometric Dating of Rocks and Minerals / C. T. Harper
SLOPE MORPHOLOGY / Stanley A. Schumm and M. Paul Mosley
MARINE EVAPORITES: Origin, Diagenesis, and Geochemistry / Douglas W.
    Kirkland and Robert Evans
ENVIRONMENTAL GEOMORPHOLOGY AND LANDSCAPE
    CONSERVATION, VOLUME III: Non-Urban Regions / Donald R. Coates
BARRIER ISLANDS / Maurice L. Schwartz
GLACIAL ISOSTASY / John T. Andrews
GEOCHEMISTRY OF GERMANIUM / Jon N. Weber
PHILOSOPHY OF GEOHISTORY: 1785–1970 / Claude C. Albritton, Jr.
ENVIRONMENTAL GEOMORPHOLOGY AND LANDSCAPE
    CONSERVATION, VOLUME II: Urban Areas / Donald R. Coates
GEOCHEMISTRY AND THE ORIGIN OF LIFE / Keith A. Kvenvolden

Additional volumes in preparation

**Benchmark Papers
in Geology**

———— A *BENCHMARK* ® Books Series ————

GEOCHEMISTRY
AND THE
ORIGIN OF LIFE

Edited by
KEITH A. KVENVOLDEN
Ames Research Center
Moffett Field, California

**Dowden, Hutchinson
& Ross, Inc.**
Stroudsburg, Pennsylvania

Distributed by:
HALSTED PRESS
A Division of John Wiley & Sons, Inc.

Acknowledgments
and Permissions

ACKNOWLEDGMENTS

AMERICAN ASSOCIATION FOR THE ADVANCEMENT OF SCIENCE—*Science*
Hydrocarbons of Biological Origin from a One-Billion-Year-Old Sediment
Organic Compound Synthesis on the Primitive Earth
A Production of Amino Acids Under Possible Primitive Earth Conditions

CARNEGIE INSTITUTION OF WASHINGTON
Carnegie Institution of Washington Yearbook 62
The Isolation of Organic Compounds from Precambrian Rocks
Carnegie Institution of Washington Yearbook 64
The Extractable Organic Matter in Precambrian Rocks and the Problem of Contamination
Carnegie Institution of Washington Yearbook 65
Criteria for Suitable Rocks in Precambrian Organic Geochemistry

GEOLOGICAL SOCIETY OF AMERICA
Geological Society of America Bulletin
Geologic History of Sea Water—An Attempt to State the Problem
Geological Society of America Special Paper 62
Geologic Evidence of Chemical Composition of the Precambrian Atmosphere
Petrologic Studies: A Volume to Honor A. F. Buddington
Model for the Evolution of the Earth's Atmosphere

GEOLOGICAL SOCIETY OF SOUTH AFRICA — *Transactions of the Geological Society of South Africa*
A Pre-Cambrian Algal Limestone in Southern Rhodesia

NATIONAL ACADEMY OF SCIENCE — *Proceedings of the National Academy of Science (U.S.)*
Amino Acids in Precambrian Sediments: An Assay
Chemical Events on the Primitive Earth
Nonprotein Amino Acids from Spark Discharges and Their Comparison with the Murchison Meteorite
Amino Acids
On the Early Chemical History of the Earth and the Origin of Life
Prebiotic Synthesis of Hydrophobic and Protein Amino Acids

24th INTERNATIONAL GEOLOGICAL CONGRESS—*24th International Geological Congress, Section 1*
Organic Geochemistry of Early Precambrian Sediments

PERMISSIONS

The following papers have been reprinted with the permission of the authors and copyright holders.

ACADEMIE-VERLAG GmbH, BERLIN — *Abhandlungen der Deutschen der Wissenschaften zu Berlin*
Die Uran- und Goldlagerstätten Witwatersrand—Blind River District—Dominion Reef—Serra de Jacobina: erzmikroskopische Untersuchungen und ein geologischer Vergleich

AMERICAN ASSOCIATION FOR THE ADVANCEMENT OF SCIENCE — *Science*
Alga-like Forms in Onverwacht Series, South Africa: Oldest Recognized Lifelike Forms on Earth
Alga-like Fossils from the Early Precambrian of South Africa
Ammonium Ion Concentration in the Primitive Ocean
Carbon Isotope Fractionation in the Fischer–Tropsch Synthesis and in Meteorites
Carbon Isotopic Studies of Organic Matter in Precambrian Rocks
Endogenous Carbon in Carbonaceous Meteorites
Hydrocarbons of Biological Origin in Sediments About Two Billion Years Old
Microorganisms Three Billion Years Old from the Precambrian of South Africa
Paleobiology of a Precambrian Shale
Precambrian Marine Environment and the Development of Life

AMERICAN CHEMICAL SOCIETY — *Chemical Reviews*
Chemical Evolution

AMERICAN JOURNAL OF SCIENCE (YALE UNIVERSITY) — *American Journal of Science*
A Working Model of the Primitive Earth

AMERICAN METEOROLOGICAL SOCIETY—*Journal of the Atmospheric Sciences*
On the Origin and Rise of Oxygen Concentration in the Earth's Atmosphere

CAMBRIDGE UNIVERSITY PRESS — *Biological Reviews*
Precambrian Micro-Organisms and Evolutionary Events Prior to the Origin of Vascular Plants

ECONOMIC GEOLOGY PUBLISHING CO. — *Economic Geology*
Origin of Precambrian Iron Formations

MACMILLAN JOURNALS LTD.—*Nature*
Evidence for Extraterrestrial Amino-Acids and Hydrocarbons in the Murchison Meteorite
Optical Configuration of Amino-Acids in Pre-Cambrian Fig Tree Chert
Primitive Atmosphere of the Earth

MICROFORMS INTERNATIONAL MARKETING CORPORATION — *Geochimica et Cosmochimica Acta*
Extractable Organic Matter in Precambrian Cherts
Organic Compounds in Meteorites: IV. Gas Chromatographic–Mass Spectrometric Studies on the Isoprenoids and Other Isomeric Alkanes in Carbonaceous Chondrites
Organic Constituents in Meteorites — A Review
The Properties and Theory of Genesis of the Carbonaceous Complex Within the Cold Bokevelt Meteorite

NEW YORK ACADEMY OF SCIENCES — *Annals of the New York Academy of Sciences*
Mass Spectroscopic Analysis of the Orgueil Meteorite: Evidence for Biogenic Hydrocarbons

NORTH-HOLLAND PUBLISHING CO. — *Earth and Planetary Science Letters*
The Oxygen Isotope Chemistry of Ancient Cherts

D. REIDEL PUBLISHING CO.—*Space Life Sciences*
The History of Atmospheric Oxygen

SOCIETY OF ECONOMIC PALEONTOLOGISTS AND MINERALOGISTS—*Journal of Paleontology*
Biogenicity and Significance of the Oldest Known Stromatolites

Series Editor's Preface

The philosophy behind the "Benchmark Papers in Geology" is one of collection, sifting, and rediffusion. Scientific literature today is so vast, so dispersed, and, in the case of old papers, so inaccessible for readers not in the immediate neighborhood of major libraries that much valuable information has been ignored by default. It has become just so difficult, or so time consuming, to search out the key papers in any basic area of research that one can hardly blame a busy man for skimping on some of his "homework."

This series of volumes has been devised, therefore, to make a practical contribution to this critical problem. The geologist, perhaps even more than any other scientist, often suffers from twin difficulties — isolation from central library resources and immensely diffused sources of material. New colleges and industrial libraries simply cannot afford to purchase complete runs of all the world's earth science literature. Specialists simply cannot locate reprints or copies of all their principal reference materials. So it is that we are now making a concerted effort to gather into single volumes the critical material needed to reconstruct the background of any and every major topic of our discipline.

We are interpreting "geology" in its broadest sense: the fundamental science of the planet Earth, its materials, its history, and its dynamics. Because of training and experience in "earthy" materials, we also take in astrogeology, the corresponding aspect of the planetary sciences. Besides the classical core disciplines such as mineralogy, petrology, structure, geomorphology, paleontology, and stratigraphy, we embrace the newer fields of geophysics and geochemistry, applied also to oceanography, geochronology, and paleoecology. We recognize the work of the mining geologists, the petroleum geologists, the hydrologists, the engineering and environmental geologists. Each specialist needs his working library. We are endeavoring to make his task a little easier.

Each volume in the series contains an Introduction prepared by a specialist (the volume editor)—a "state of the art" opening or a summary of the objects and content of the volume. The articles, usually some thirty to fifty reproduced either in their entirety or in significant extracts, are selected in an attempt to cover the field, from the key papers of the last century to fairly recent work. Where the original works are in foreign languages, we have endeavored to locate or commission translations. Geologists, because of their global subject, are often acutely aware of the oneness of our world. The selections cannot, therefore, be restricted to any one country, and whenever possible an attempt is made to scan the world literature.

To each article, or group of kindred articles, some sort of "highlight commentary" is usually supplied by the volume editor. This should serve to bring that article into historical perspective and to emphasize its particular role in the growth of the field. References, or citations, wherever possible, will be reproduced in their entirety—for by this means the observant reader can assess the background material available to that particular author, or, if he wishes, he too can double check the earlier sources.

A "benchmark," in surveyor's terminology, is an established point on the ground, recorded on our maps. It is usually anything that is a vantage point, from a modest hill to a mountain peak. From the historical viewpoint, these benchmarks are the bricks of our scientific edifice.

Rhodes W. Fairbridge

Preface

Geochemical studies of both terrestrial and extraterrestrial materials have provided clues to the presence of early life on earth and to the possible conditions under which life began and evolved. The geologic record of early life is far from complete. Yet during the last twenty years remarkable progress has been made in deciphering that record. From the extensive body of literature that has developed, forty-three papers have been selected to summarize and highlight the work. We have arranged the papers into five parts, such that they tell a story about life on earth. Theoretical concepts and laboratory simulation experiments dealing with the chemical steps that may have led to the origin of life are detailed first. Part II considers the organic chemistry of carbonaceous meteorites. These studies have resulted in observations that support some of the theoretical concepts described in the first part. The third part details the conditions on earth under which life arose and first evolved and describes how life changed the conditions on earth, thereby modifying some geologic processes and creating its own environment. In Part IV an account is given of the organic geochemical record of Precambrian rocks, and the story ends with a description of the evidence of the earliest life that has yet been found in the geologic record. Besides developing a scenario about the origin of life, the book attempts to present the state of our knowledge in 1973, along with a reasonable assessment of the significance of the results that have been obtained.

A book of this type would be impossible without the cooperation of the authors and publishers of the reprints included, and I extend my thanks to all who generously provided permissions to use their work. I am also grateful to Ms. Chris Palmer for her help in typing the text.

Keith A. Kvenvolden

Contents

II. CARBONACEOUS METEORITES—IN SUPPORT OF CHEMICAL EVOLUTION

III. PRIMITIVE ENVIRONMENTS—HYDROSPHERE AND ATMOSPHERE

V. EARLY LIFE ON EARTH—PALEOBOTANY

Contents by Author

Introduction

The origin of life is intrinsically fascinating. Surrounded in mystery, it has been the subject of contemplation and speculation since the beginning of history. Until recently, the study of the origin of life was generally treated metaphysically and philosophically. With increasing scientific knowledge of the cosmos it is now recognized that the subject can be examined by the scientific method and thus falls within the realm of modern scientific inquiry.

The problem in trying to understand the origin of life is enormous. J. D. Bernal, in a statement to the British Physical Society in 1949, aptly said:

> It is probable that even the formulation of this problem is beyond the reach of any one scientist; for such a scientist would have to be at the same time a competent mathematician, physicist, and experimental organic chemist. He should have a very extensive knowledge of geology, geophysics, and geochemistry; and, besides all this be absolutely at home in all biological disciplines. Sooner or later this task will have to be given to groups representing all these faculties and working closely together theoretically as well as experimentally.

This book concentrates on one interdisciplinary approach to trying to understand the origin of life. It is the approach of geologists and chemists—the geochemists—who search the record of the rocks for clues to the origin and evolution of life. Rocks continue to provide a tangible record of the development of life on earth, but when rocks fail to yield informatin about life or processes leading to life, about the only recourse is to theoretical studies and speculation. Fortunately, the samples available to the geochemist are not limited to the rocks of earth. Meteorites, particularly carbonaceous chondrites, provide a source of extraterrestrial information of which the geochemist can avail himself. Also, beginning in 1969, extraterrestrial samples from several localities on the earth's moon became available through the space programs of the United States and the Soviet Union. Thus geochemists have samples spanning almost 5 billion years of time. Meteorites are the most primitive substances and are

1

about 4.6 billion years old. On earth, rocks cover the last 3.7 billion years of earth history. Lunar samples more or less fill the age gap between the oldest rocks on earth and meteorites. Unfortunately, studies of samples from the moon have not provided the kinds of evidence that are readily accepted with regard to the origin of life. For this reason, and because of length constraints, this book will not consider the many papers concerned with the subject of carbon-containing compounds on the moon.

From the evidence, much of it detailed in this book, it is quite certain that life has existed on earth for at least 3 billion years. What exactly preceeded life is still not known, but one concept that provides a reasonable scientific context in which to view the origin of life is the theory of chemical evolution. In this theory it is hypothesized that life arose as a natural consequence of the evolution of matter. The theory was first proposed in modern terms in the 1920s. The starting point was a sterile earth having an atmosphere devoid of oxygen, i.e., a reducing atmosphere. Natural energy sources interacted with this reduced atmosphere, which contained gases rich in hydrogen. As a consequence of this interaction, increasingly complex organic (carbon-rich) substances were formed. They accumulated, concentrated, became organized, and eventually led to living systems. The validity of this theory has not been demonstrated in its entirety. But laboratory simulation experiments have shown that organic molecules can be produced under conditions postulated by the theory, and geochemical measurements have been made which appear to support the theory.

In this book the geochemistry of the origin of life is formulated within the general framework postulated by the theory of chemical evolution. This theory certainly provides a reasonable and convenient model, given our present state of knowledge. Therefore, the first part of the book considers the theory of chemical evolution and a number of laboratory tests that offer support for the theory or at least for concepts envisioned within the theory. The papers, which are more chemical than geological in content, describe attempts to formulate on theoretical grounds models to explain the early chemical history of the earth and the origin of life.

The second part of the book considers carbonaceous meteorites and the information they contain which may relate to the origin of life. Although some researchers have suggested that the organic materials in meteorites represent products of an extraterrestrial biology, most investigators now accept the idea that the organic materials found in meteorites most likely resulted from a nonbiological, chemical synthesis. The important questions at present relate to the cosmochemical mechanisms and conditions that produced the rich collection of molecules found in meteorites. Results from the analyses of meteorites seem to lend strong support to some of the aspects of the theory of chemical evolution.

Having tried to utilize the theory of chemical evolution as a reasonable model in which to consider the origin of life, this book, in its third part, looks into the primitive environment of earth. It is here that inorganic geochemistry plays an important role in understanding the origin and composition of the hydrosphere, atmosphere, and lithosphere. Apparently the principle of uniformitarianism, so useful in many geological considerations, cannot be applied in its strictest sense for interpretations of the Precambrian record. For example, the composition of the atmosphere has apparently changed dramatically over geologic time. Studies of the occurrences of detrital pyrite and uraninite and the distribution of banded iron formations suggest that the content of atmospheric oxygen was much lower in early and middle Precambrian times than measured today. Thus the present is not a direct key to the past, especially in understanding the origin and early evolution of life.

Organic geochemistry, as given in the fourth part of this book, has attempted to unravel some of the mysteries of early life on earth. When the principles of organic geochemistry were first applied to Precambrian rocks, a number of carefully performed analyses revealed the presence of organic compounds which were enthusiastically interpreted as being related to life present at the time the rocks were deposited. With time, many of these interpretations have come to be viewed with skepticism. The results themselves are usually accepted, but it has proved difficult to obtain convincing evidence, especially for the oldest Precambrian rocks, that the compounds found are indeed as old as the rock in which they occur. Nevertheless, the complex organic

GEOLOGIC CLOCK SHOWING EVENTS TAKING PLACE DURING THE LAST FIVE BILLION YEARS OF GEOLOGIC TIME

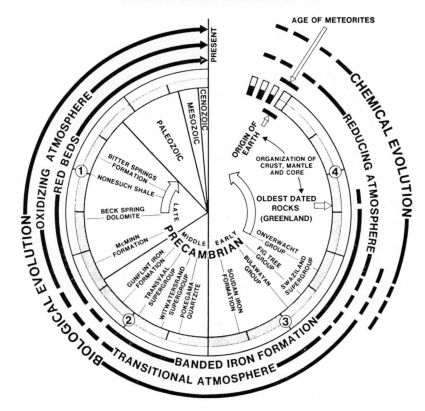

Figure 1.

3

polymers (kerogen) in these old rocks may hold some important clues to early earth history.

The fifth and final part of this book deals with the evidence for earliest life on earth. The work is confined to studies of rocks 3 billion years old and older. Although this section concentrates on paleobotany and the morphological evidence, each paper contains organic geochemical information to support interpretations regarding life. This earliest record of life is not clearly defined, and much controversy still exists with regard to the significance of both paleobotanical and organic geochemical observations. Taken as a whole, however, there appears to be enough evidence to indicate that life quite likely existed on earth at least 3 billion years ago.

Figure 1, developed from an idea originally conceived by J. W. Schopf, attempts to place some of the information regarding geochemistry and the origin of life in a time context consistent with what is known today. The "geologic clock" shows 5 billion years of geologic time; important events and times are indicated. Rock units that have provided chemical and biological evidence for early life are shown. The changing atmospheric compositions and the consequent change in lithologies from banded iron formations to red beds are noted. The clock shows the general temporal relationships of the period of chemical evolution and of biological evolution. The figure should serve as a guide for placing in temporal contexts the geochemical information provided in the papers included in this book.

Our knowledge and understanding of the origin of life has been augmented significantly as a result of geochemical and related studies undertaken during the past 20 years. Particularly important has been the revelation of the great span of time during which life has existed on earth. Although the origins of this life remain a mystery, the conditions on earth during early evolution of life are now better understood. What has been accomplished to date represents only a beginning to potentially rewarding research. As scientific interest in the origin of life increases, geochemistry can expect to play a major part in deciphering the geologic record of life.

Additional Suggested Readings

Brooks, J., and G. Shaw (1973) *Origin and Development of Living Systems*: Academic Press, New York, 412 p.

Calvin, M. (1969) *Chemical Evolution–Molecular Evolution Towards the Origin of Living Systems on Earth and Elsewhere*: Oxford University Press, New York, 278p.

Rutten, M. G. (1971) *The Origin of Life by Natural Causes*: Elsevier, Amsterdam, 420p.

I
Chemical Evolution— Possible Beginnings

Editor's Comments on Papers 1 Through 8

Geochemical studies of the origin of life can be formulated within the context of the theory of chemical evolution. This theory was proposed in modern terms by A. I. Oparin (1924) [see Oparin (1953) for a translation of an expanded version of his early ideas], the Russian biochemist, and by J. B. S. Haldane (1928), the English geneticist. They independently hypothesized that life arose on earth under chemically reducing conditions by means of an evolutionary sequence of events that involved the synthesis and organization of increasingly complex organic substances which over a period of geologic time became living systems. The theory is appealing because parts of it, at least, can be tested not only in the laboratory by simulation experiments but also through geochemical measurements on naturally occurring terrestrial and extraterrestrial materials.

The literature on chemical evolution is extensive, with about 3,000 entries at last count. From this literature eight papers have been selected that provide a framework for considering the geochemistry of the origin of life. The first three papers set the stage for much of the laboratory experimentation in chemical evolution that has been done during the last 20 years. It is only proper that this first paper be one of the very important, early treatises in chemical evolution. In this paper (1) Urey describes in detail his ideas and conclusions relative to the early chemical conditions on the earth and their bearing on the origin of life. He adopts the suggestion of Oparin (1924) that the early atmosphere was reducing in character and builds a model of the history of the early earth. Many elements of this model have survived the numerous tests that have been applied during the last 20 years. One major difference is that the time scale for the evolutionary events incorporated in the model has been extended as a result of increased geochronologic knowledge. In his model Urey stresses the importance of ultraviolet light as an energy source, which, acting upon a reducing atmosphere of methane, ammonia, and water, should produce organic compounds. He suggests that laboratory experiments simulating these conditions would be profitable. Almost as an afterthought he suggests that the "investigation of possible effects of electric discharges on the reactions should also be tried since electric storms in the reducing atmosphere can be postulated reasonably." This last idea was tested in the laboratory the next year by Miller.

Miller's important paper (2) is included here for it represents the first laboratory simulation experiment in chemical evolution which yielded successful results. These historic discoveries have had enormous impact on thinking regarding the origin of life. Most modern books in biology, for example, now cite this work. The experiment involved circulating methane, ammonia, water, and hydrogen past an electric discharge. The following amino acids were identified in the product: glycine, α-alanine, β-alanine, and possibly aspartic acid and γ-amino-n-butyric acid. The experiment clearly demonstrated that compounds of biological significance could be produced under simulated primitive earth conditions.

During the following six years additional experiments were performed and more compounds were identified. A review of this work is given by Miller and Urey (Paper 3), who not only considered their own experiments but also cite the work of numerous other scientists who followed some of their original ideas. By 1959 electric discharge experiments had yielded nine amino acids, three hydroxy acids, three monocarboxylic acids, one dicarboxylic acid, two iminoacetic acids, and two urea compounds. The success of these experiments was viewed with much interest. The results suggested that organic compounds were formed naturally in the environment of the primitive earth; out of this organic milieu life possibly originated.

These classic experiments and their variations have been repeated, modified, and extended by numerous investigators. To provide the reader with a background for this work in chemical evolution up to 1970 the succinct review by Lemmon is included (Paper 4). This review quickly covers most of the experiments since 1952 and illustrates the imagination, vitality, and enthusiasm that characterizes much of the research in this field. Further information on chemical evolution can be found in a review by Ponnamperuma and Gabel (1968) and in the Proceedings of three International Conferences on the Origin of Life (1959; 1965; 1971).

Of course, as in any growing area of scientific inquiry, one cannot always expect unanimity. Noteworthy is the paper by Abelson (5), who challenged the composition of the gas mixture of methane and ammonia used in most of the early chemical evolution simulation experiments. He pointed out that the experiments did not take into account sufficient geological information. Most of the previous work had been done by chemists, who based their ideas on chemical theory. Abelson tried to bring geological considerations into this work by using mixtures of gases which seemed more plausible from the geologists' point of view. For example, ammonia on the primitive earth would have disappeared quickly and, therefore, should not be a major component in primitive atmosphere experiments. Using mixtures of nitrogen, carbon monoxide, and hydrogen in various proportions, he demonstrated that hydrogen cyanide was a major product, along with small amounts of carbon dioxide and water. Hydrogen cyanide

7

appears to be a reactive intermediate which upon further reaction can produce amino acids and at least one of the nucleic acid bases. Abelson's criticisms have influenced the parameters selected for many experiments that followed.

During the last 20 years two models have been developed to account for the generation of the organic materials that eventually may have led to living systems on earth. The primitive-earth atmosphere model, also called the Miller–Urey model, involves the interaction of energy with a reducing atmosphere associated with a planetary body. The second model attempts to simulate conditions in the condensing solar nebula, and these reactions are referred to as Fischer–Tropsch-type syntheses (Studier et al., 1968) because the conditions of the reactions are like those of the Fischer–Tropsch synthesis, which was discovered in organic chemistry in 1923 and has been used industrially to produce gasoline. Two tests of these models have been particularly interesting.

Following the discovery of a complex suite of amino acids in a carbonaceous meteorite (Kvenvolden et al., Paper 14; Kvenvolden et al., 1971) Miller reinvestigated his electric discharge experiment (Ring et al., Paper 6; Wolman et al., Paper 7). Heeding the criticism of Abelson (Paper 5) concerning the excessive ammonia concentration used in previous experiments, Miller successfully synthesized at least 33 amino acids from a mixture of methane, nitrogen, water, and traces of ammonia. For the first time it was shown that amino acids comparable to all those found in carbonaceous meteorites could be generated in a laboratory simulation of the primitive-earth atmosphere. Papers 6 and 7, which represent a single paper published in two different issues of the journal, provide a significant advance in prebiotic chemistry, both in terms of analytical procedures and equipment used and in results obtained.

A positive test of the condensing solar nebula model (Fischer–Tropsch-type synthesis) was provided by Lancet and Anders (Paper 8). In this model the principal active reactants are carbon monoxide and hydrogen, along with various catalysts. Studies of carbon isotopic fractionation in this kind of synthesis showed that the fractionation taking place in this reaction was of the same magnitude and in the same direction as the carbon isotopic fractionation observed in carbonaceous meteorites (Smith and Kaplan, Paper 13). Additional tests of this model have been reviewed by Anders et al. (1973). They conclude that organic compounds in meteorites can be accounted for by involking Fischer–Tropsch-type synthesis of carbon monoxide, hydrogen, and ammonia in the solar nebula. Furthermore, they postulate that these reactions may be a source of prebiotic compounds on the inner planets, including earth.

Tests such as the two described neither prove or disprove the validity of either model. It is quite likely that elements of both models are necessary to simulate the actual processes that took place to produce the organic materials that eventually led to life on earth and perhaps elsewhere.

Numerous individuals have contributed to the development of geochemistry, but Harold C. Urey and Philip H. Abelson are two "giants of geochemistry" whose research and ideas have particularly influenced the field.

Urey's career began in chemistry, but with time his interests have become more directed to the application of chemistry to geological and cosmological problems. Born in Indiana in 1893 and educated in Montana and California, he is well remembered for his discovery of deuterium, for which he received the Nobel Prize in chemistry. His interest in the properties and separation of isotopes led to the use of oxygen isotope geochemistry in paleothermometry. His name is commonly associated with studies of the origin of the earth, planets, meteorites, and the moon. Urey is responsible for the basic ideas incorporated in the two most widely considered models for the origin of organic molecules leading to life. He is now Emeritus University Professor of the University of California.

Abelson is a physical chemist whose interests have included geochemistry, particularly the biological aspects of geochemistry. He was born in the state of Washington in 1913 and was educated there and in California. His early work in nuclear physics, including his discovery of neptunium, was particularly noteworthy. From the chemistry of the nuclear world he turned to the chemistry of the biological world. His work on amino acids played an important role in the development of paleobiochemistry. As the Director of the Geophysical Laboratory of the Carnegie Institution of Washington D.C., he has been responsible for establishing one of the most respected laboratories for organic geochemistry. Professor Abelson is now President of the Carnegie Institution and continues as editor of *Science*, a task that he has carried out with distinction since 1962.

References

Anders, E., R. Hayatsu, and M. H. Studier (1973) Organic compounds in meteorites: *Science*, **182**, 781–790.

Haldane, J. B. S. (1928) The origin of life: *Rationalist Ann.*, **148**, 3–10.

Kvenvolden, K. A., J. G. Lawless, and C. Ponnamperuma (1971) Nonprotein amino acids in the Murchison meteorite: *Proc. Natl. Acad. Sci. (U.S.)*, **68**, 486–490.

Oparin, A. I. (1924) *Proiskhozhdenie zhizni*: Izd. Moskovskii Rabochii, Moscow.

Oparin, A. I. (1953) *The Origin of Life*, 2nd ed.: Dover Publications, New York, 270p.

Ponnamperuma, C., and N. W. Gabel (1968) Current status of chemical studies on the origin of life: *Space Life Sci.*, **1**, 64–96.

Proceedings of the 1st International Symposium on the Origin of Life, Moscow, 1957. The Origin of Life on the Earth, edited by F. Clark and R. L. M. Synge: Pergamon Press, Elmsford, N.Y., 691p. (1959).

Proceedings of the 2nd International Conference on the Origin of Life, Wakulla Springs, Florida, 1963. The Origins of Prebiological Systems and of Their Molecular Matrices, edited by S. Fox: Academic Press, New York, 482p. (1965)

Proceedings of the 3rd International Conference on the Origin of Life, Pont-a-Mousson, France, 1970. Chemical Evolution and the Origin of Life, edited by R. Buvet and C. Ponnamperuma: North-Holland, Amsterdam, 560p. (1971)

Studier, M. H., R. Hayatsu, and E. Anders (1968) Origin of organic matter in early solar system: I. Hydrocarbons: *Geochim. Cosmochim. Acta*, **32**, 151–173.

Additional Suggested Readings

Hayatsu, R., M. H. Studier, A. Oda, K. Fuse, and E. Anders (1968) Origin of organic matter in early solar system: II. Nitrogen compounds: *Geochim. Cosmochim. Acta*, **32**, 175–190.

Hayatsu, R., M. H. Studier, and E. Anders (1971) Origin of organic matter in early solar system: IV. Amino acids: confirmation of catalytic synthesis by mass spectrometry: *Geochim. Cosmochim. Acta*, **35**, 939–951.

Hayatsu, R., M. H. Studier, S. Matsuoka, and E. Anders (1972) Origin of organic matter in early solar system: VI. Catalytic synthesis of nitriles, nitrogen bases and porphyrin-like pigments: *Geochim. Cosmochim. Acta*, **36**, 555–571.

Kimball, A. P., and J. Oró, eds. (1971) *Prebiotic and Biochemical Evolution:* North-Holland, Amsterdam, 296p.

Mamikunian, G., and M. H. Briggs (1965) *Current Aspects of Exobiology*, Jet Propulsion Laboratory, Pasadena, Calif., 420p.

Ponnamperuma, C., ed. (1972) *Exobiology*: North-Holland, Amsterdam, 485p.

Rohlfing, D. L., and A. I. Oparin (1972) *Molecular Evolution–Prebiological and Biological*: Plenum Press, New York, 481p.

Schwartz, A. W., ed. (1971) *Theory and experiment in exobiology*: Wolters-Noordhoff, Groningen, The Netherlands, 160p.

West, M. W., and C. Ponnamperuma (1970) Chemical evolution and the origin of life— a comprehensive bibliography: *Space Life Sci.*, **2**, 225–295.

West M. W., E. D. Gill, and C. Ponnamperuma (1972) Chemical evolution and the origin of life—bibliography supplement 1970: *Space Life Sci.*, **3**, 293–304.

West, M. W., E. D. Gill, and B. R. Sherwood (1973) Chemical evolution and the origin of life—bibliography supplement 1971: *Space Life Sci.*, **4**, 309–328.

Yoshino, D., R. Hayatsu, and E. Anders (1971) Origin of organic matter in early solar system: III. Amino acids: catalytic synthesis: *Geochim. Cosmochim. Acta*, **35**, 927–938.

1

Reprinted from *Proc. Natl. Acad. Sci. (U.S.)*, **38**, 351–363 (1952)

ON THE EARLY CHEMICAL HISTORY OF THE EARTH AND THE ORIGIN OF LIFE

By Harold C. Urey

INSTITUTE FOR NUCLEAR STUDIES, UNIVERSITY OF CHICAGO

Communicated January 26, 1952

In the course of an extended study on the origin of the planets[1] I have come to certain definite conclusions relative to the early chemical conditions on the earth and their bearing on the origin of life. Oparin[2] has presented the arguments for the origin of life under anaerobic conditions which seem to me to be very convincing, but in a recent paper Garrison, Morrison, Hamilton, Benson and Calvin,[3] while referring to Oparin, completely ignore his arguments and describe experiments for the reduction of carbon dioxide by 40 m. e. v. helium particles from the Berkeley 60-inch cyclotron. As I believe these experiments, as well as many previous ones using ultra-violet light to reduce carbon dioxide and water and giving similar results to theirs, are quite irrelevant to the problem of the origin of life, I wish to present my views.

During the past years a number of discussions on the spontaneous origin of life have appeared in addition to that by Oparin. One of the most extensive and also the most exact from the standpoint of physical chemistry is that by Blum.[3a]   It seems to me that his discussion meets its greatest difficulty in accounting for organic compounds from inorganic sources. This problem practically disappears if Oparin's assumptions in regard to the early reducing character of the atmosphere are adopted.

In order to estimate the early conditions of the earth, it is necessary to ask and answer the questions of how the earth originated, and how the primitive earth developed into the present earth.   The common assumption is that the earth and its atmosphere have always been as they are now, but if this is assumed it is necessary to account for the present highly oxidized condition by some processes taking place early in the earth's history. Briefly, the highly oxidized condition is rare in the cosmos and exists in the surface regions of the earth and probably only in the surface regions of Venus and Mars.   Beyond these we know of no highly oxidized regions at all, though undoubtedly other localized regions of this kind exist.   This is essentially the argument of Oparin.

The surface of the moon gives us the most direct evidence relative to the origin of the earth.   Gilbert[4] called attention to the great system of ridges and grooves radiating from Mare Imbrium, and Baldwin[5] has explained this as due to a colliding planetesimal some hundred kilometers in radius. My own studies show that the object contained metallic iron objects and and silicate materials and that water as such or as hydrated silicates arrived with such objects.   This collision occurred during the terminal stage of the moon's formation.   Some five other similar objects left their marks on the moon's surface and they all fell within a time span of some 10^5 years. Their temperatures were not high; probably not appreciably higher than present terrestrial temperatures.   The arguments for these conclusions are long and detailed and cannot be repeated here.

At the time such objects were falling on the moon similar objects fell on the earth.   The conditions were different because of the greater energy of such objects, 22 times as great if they fell from a great distance and 11 times as great if they fell from the circum surface orbit, and because of the presence of a substantial atmosphere on the earth.   The energy was sufficient to completely volatilize the colliding planetesimals and raise the gas to greater than 10,000°K.   An object similar to the Imbrium planetesimal would have distributed its material over a region several thousand kilometers in linear dimensions and the explosion cloud would have risen far above the atmosphere.   Its materials would have fallen through the atmosphere in the form of iron and silicate rains and would have reacted with the atmosphere in the process.   (H. H. Nininger recently showed me spherical iron-nickel objects collected near Meteor Crater, Ariz., which were formed in such a rain.)   The objects contained metallic iron-nickel alloy, silicates, graphite, iron carbide, water or water of crystallization, ammonium salts and nitrides, that is substances which would supply the volatile and nonvolatile constituents of the earth. The temperatures produced in these collisions were very high, but unless the accumulation was very rapid indeed the general temperature of the planet was not excessively high. That such objects fell on the moon and earth at the terminal stage of their formation I regard as certain, and it is difficult in this subject to be certain about anything. But regardless of the detailed arguments, those who postulate oxidizing conditions as the initial state of the earth should present some similar argument to justify their assumption.

The reactions taking place at that time of interest to us here are:

$$FeO + H_2 = Fe + H_2O; \qquad K_{298} = 1.7 \times 10^{-3}, K_{1200} = 0.97$$
$$Fe_3O_4 + H_2 = 3FeO + H_2O; \qquad K_{298} = 2.5 \times 10^{-2}, K_{1200} = 1.0$$
$$C + H_2O = CO + H_2; \qquad K_{298} = 10^{-16}, \qquad K_{1200} = 3.8 \times 10^{-6}$$
$$C + 2H_2 = CH_4; \qquad K_{298} = 7.8 \times 10^{8}, \quad K_{1200} = 1.6 \times 10^{-2}$$
$$Fe_3C + 2H_2 = Fe + CH_4; \qquad K_{298} = 3.2 \times 10^{11}, K_{1200} = 5.9 \times 10^{-3}$$
$$NH_3 = {}^1/_2N_2 + {}^3/_2H_2; \qquad K_{239} = 1.2 \times 10^{-3}.$$

From these equilibrium constants one sees that hydrogen was a prominent constituent of the primitive atmosphere and hence that methane was as well. Nitrogen was present as nitrogen gas at high temperatures but may have been present as ammonia or ammonium salts at low temperatures. At high temperatures hydrogen would escape from the planet very rapidly. but if the temperatures were high objects must have arrived rapidly to replenish the lost hydrogen. If the objects arrived slowly then the temperatures were low and hydrogen did not escape rapidly. Thus it is very difficult to see how the primitive atmosphere of the earth contained more than trace amounts of other compounds of carbon, nitrogen, oxygen and hydrogen than CH_4, H_2O, NH_3 (or N_2) and H_2.

We now consider what could reasonably be expected to convert this atmosphere into the present one existing on the earth.* If there was a large

---

* Poole, J. H. J. (*Proc. Roy. Soc. Dublin*, **22**, 345 (1941)), Harteck, P., and Jensen, J. H. D. (*Z. Naturforschung*, **3a**, 581 (1948)), and Dole, M. (*Science*, **109**, 77 (1949)) have reviewed the modern evidence for the photochemical origin of free oxygen and the evidence against the photosynthetic origin. Their conclusions are accepted here. In a recent paper, Poole, J. H. J. (*Sci. Proc. Roy. Dublin Acad.*, **25**, 201 (1951)) definitely concludes that methane and ammonia were not present in the primitive atmosphere and that it consisted of H_2O, CO_2 and N_2 but contained no oxygen. I cannot accept this conclusion for carbon dioxide and nitrogen are almost as difficult to understand as free oxygen. The interior of the earth and the lavas which reach its surface are highly reducing and are not a likely source of highly oxidized materials. Poole assumes that carbon dioxide was present in large quantities in the primitive atmosphere. How was it produced from methane, graphite or iron carbide? He also assumes that it would remain in the atmosphere, but as is shown in the text and the references cited, carbon dioxide in the presence of water should react rapidly with silicates until its partial pressure reaches values in the neighborhood of those on the earth and Mars.

amount of hydrogen, the outer parts of the atmosphere beyond the convection zone would become highly enriched in hydrogen. Hydrogen would absorb light from the sun in the far ultra-violet[6] and since it does not radiate in the infra-red would become a high-temperature atmosphere just as exists on the earth at present, and hydrogen would be lost very rapidly. As hydrogen was depleted, the atmosphere would become a methane one and since methane and its photochemical disintegration products absorb a wide band of energy in the ultra-violet and radiate in the infra-red, the temperature of the high atmosphere would fall far below the present temperature of 1500° or 2000°C.[7] In fact, it might well approach present terrestrial surface or even lower temperatures. The loss of hydrogen would be decreased, but since an oxidized atmosphere is present on the earth it can be assumed that it escaped at some appropriate rate. The net process was the dissociation of water into hydrogen which escaped, and into oxygen which oxidized reduced carbon compounds to carbon dioxide, ammonia to nitrogen, and reduced iron to more oxidized states. When the methane and ammonia were oxidized free oxygen appeared and the present atmospheric conditions were established. As carbon dioxide was formed it reacted with silicates to form limestones, i.e.,

$$CaSiO_3 + CO_2 = CaCO_3 + SiO_2, \qquad K_{298} = 10^8.$$

Of course the silicates may have been a variety of minerals but the pressure of CO_2 was always kept at a low level by this reaction or similar reactions just as it is now. Plutonic activities reverse the reaction from time to time, but on the average the reaction probably proceeds to the right as carbon compounds come from the earth's interior,[8] and in fact no evidence for the deposition of calcium silicate in sediments seems to exist.†

The histories of Mars and Venus should be similar to that of the earth. Mars has no mountains higher than 750 meters. Thus the initial lunar type mountains were probably eroded by water and no folded mountains have been formed. The oxidation of methane to carbon dioxide and the

† The alternative to this course of events would be the production of carbon dioxide from the earth's interior and there is evidence that at least some oxidized carbon is so produced. Gases escaping from lava lakes of Hawaiian volcanoes are highly oxidized, so much so that it is difficult to account for the high states of oxidation (Day, A. L., and Shepherd, E. S., *J. Wash. Acad. Sci.*, **3**, 457 (1913); Shepherd, E. S., *Bull. Hawaiian Volcano Obs.*, **VII**, 97 (1919); *Ibid.*, **VIII**, 65 (1920).) It seems probable to the writer that atmospheric oxygen directly or indirectly is responsible for the oxidation. Observers report burning gases as escaping from lavas, thus indicating the escape of reduced gases from the lava. Also surface oxidation of the lava pool must occur, as is shown by the high temperatures in the surfaces of such pools. Ferric oxide is probably produced and this in turn would oxidize carbon within the magma. (See *Bull. Hawaiian Volcano Obs.*, **IX**, 113 (1921).) If atmospheric oxygen is the source, then the carbon dioxide does not represent oxidized carbon being supplied to the atmosphere, but on the contrary it represents reduced carbon being so supplied.

formation of limestone proceeded as on earth, but oxygen atoms or water molecules were lost from the planet.   In an atmosphere containing oxygen and nitrogen, a high temperature should have existed on Mars just as exists on the earth, and due to the smaller gravitational field atoms of atomic weight, 16, should escape if atoms of atomic weight, 4, i.e. He, escape from the earth now, as they do   Finally, there results a desert planet with very small amounts of water and a pressure of carbon dioxide in its atmosphere about equal to that on the earth,[9] the excess carbon dioxide having reacted with the silicates to form limestones until its partial pressure was reduced to a low value.   Since mountains are absent, volcanic activity must be small or non-existent and carbon dioxide has not been generated from limestone and silicon dioxide.

Venus started with a reduced atmosphere, which was oxidized to carbon dioxide and limestone by photochemical action.   It cannot lose water, however, and the absence of water means that much less water was present initially than in the case of the earth, probably due to having been formed nearer to the sun and thus at a higher temperature.   Assuming plutonic activity on Venus, the carbon dioxide has been regenerated by processes similar to those of the earth.   In the absence of water the reaction of carbon dioxide with silicates is very slow and hence a dense atmosphere of carbon dioxide is possible.   Thus a reasonable course of events can be postulated for this planet.‡

The Origin of Life.—The problem of the origin of life involves three separate questions in our present discussion: (1) the spontaneous formation of the chemical compounds which form the physical bodies of living organisms; (2) the evolution of the complex chemical reactions which are the dynamic basis of life; and (3) the source of free energy which alone can maintain the chemical reactions and synthesize the chemical compounds. It is only the first and third questions which will be discussed here. At present the source of free energy is sunlight through photosynthesis, but how were primitive living organisms maintained through a long enough period of time for the evolution of photosynthesis to occur?

It is suggested here that life evolved during the period of oxidation of highly reduced compounds to the highly oxidized ones of today.   During

‡ The discussion up to this point, together with the suggestion that life originated during the period of oxidation of reduced carbon compounds to oxidized ones, was presented before the Geological Society of America, Washington, November, 1950.   It was thought that there might be some condition of pressure, temperature and composition such that organic compounds became stable, thus making the synthesis of complex compounds possible.   The general ideas were discussed with Dr. H. E. Suess, who was a Fellow at the Institute for Nuclear Studies.   He made some studies relative to this problem which did not appear to be promising of positive results.   The present paper follows a somewhat different approach to the problem by assuming the synthesis of organic compounds by means of ultra-violet light in the high atmosphere.

15

this period compounds of carbon, oxygen, hydrogen, and nitrogen were present in substantial amounts. It is also suggested that the source of free energy was the absorption of ultra-violet light in the high atmosphere by methane and water and other compounds produced from them by this photochemical action. This process also protected primitive organisms from ultra-violet light in the absence of an ozone layer as exists now.

Organic compounds are generally unstable relative to completely reduced or completely oxidized compounds throughout the entire range of hydrogen and oxygen pressures in chemical equilibrium with water at ordinary temperatures, but photochemical processes should produce such compounds. The exact conditions obtaining in a methane atmosphere can only be roughly estimated. Convection to higher altitudes than now occur on the earth should have been present, since a methane atmosphere radiates in the infra-red while an oxygen-nitrogen atmosphere does not. The atmosphere should have been cooler at high altitudes so that convection extended to higher levels than now, and water vapor should have been carried to high altitudes and photochemical products of the high atmosphere should have been moved rapidly downward.

The photochemical processes in a pure methane atmosphere can be estimated qualitatively. Methane absorbs in the ultra-violet below 1450 A. The total energy of the present solar spectrum below this wavelength is about 5×10^{-6} of the total energy. Methane dissociates into methyl and atomic hydrogen. Methyl probably absorbs at much longer wave-lengths and probably repulsive states exist resulting in the formation of methylene. This compound likewise would be dissociated into CH. Thus the reactions

$$CH_4 + h\nu \ (\lambda < 1500) = CH_3 + H$$
$$CH_3 + h\nu \ (\lambda < 2800) = CH_2 + H$$
$$CH_2 + h\nu \ (\lambda < 2800) = CH + H$$
$$CH \ + h\nu \ (\lambda < 2800) = C + H$$

will occur and with their reversal and the reaction of the primary products with each other and with secondary products a steady state of great complexity will be established, the details of which cannot be estimated because of many unknown factors. The absorption spectra of CH_3 and CH_2 and the kinetics of the back reactions and other reactions are unknown. The fraction of the sun's energy below 2000 A is 3.3×10^{-4} and below 2500 A is 2.2×10^{-3}, so that very appreciable dissociation of CH_3 and CH_2 may be expected. The energies per year cm.$^{-2}$ of the earth's surface for complete absorption of the sun's spectrum at the earth below 1500 A, 2000 A and 2500 A are:

	Cal. yr.$^{-1}$ cm.$^{-2}$
1500	1.6
2000	85
2500	570

Not much methyl would be produced directly but secondary reactions such as $CH + CH_4 = CH_3 + CH_2$ could be expected to produce more of this radical. Altogether a very considerable absorption of solar energy would be expected.

In a pure water atmosphere dissociation of water into hydroxyl and atomic hydrogen would occur by absorption of light below 1900 A. Also dissociation of hydroxyl should occur giving atomic oxygen. Secondary reactions should produce O_2, H_2O_2, HO_2 and O_3 in amounts very difficult to estimate.

A combined methane and water atmosphere quite obviously would give a great variety of compounds of carbon, hydrogen and oxygen. In particular the reaction

$$CH_4 + OH = CH_3 + H_2O$$

would occur, thus producing larger quantities of methyl. These compounds would move by convection to lower levels in some quantities, dissolve in rain water and produce solutions of organic compounds in the oceans. Here ammonium salts should have been present (see below), and and the formation of nitrogen-containing compounds would occur. Given time, some natural catalysts and very slow destruction of organic compounds because of the absence of living organisms, a large number of organic compounds would be expected. If all the present surface carbon were dissolved in the present oceans as organic compounds, the oceans would become approximately a one per cent solution of these compounds. Thus compounds suitable for living organisms were possible and probably abundant.

Though the conditions postulated above are not approximated in any past experiments so far as I have been able to determine, the extensive studies on photochemistry, free radicals produced by various methods and the effects of electrical discharges on chemical substances[10] leave no doubt that many compounds would be formed due to the absorption of ultraviolet light.

A free energy supply for primitive living organisms is necessary, for only in this way can an active metabolism be supported and in the absence of such metabolism only dead and not living organic substance is possible. Rabinowitch[11] estimates that the present annual energy from photosynthesis is 600 cal. cm.$^{-2}$ of the earth's surface. The following table shows the standard free energies for three types of reactions of carbon-hydrogen-oxygen compounds of different oxidation states of the carbon atom.

COMPOUND	DISPROPORTION TO CH₄, CO₂ AND H₂O	OXIDATION TO CO₂ AND H₂O	REDUCTION TO CH₄ AND H₂O
CH_4	0	−195.50	0
CH_3OH	−22.31	−168.94	−30.13
CH_2O	−27.0	−124.75	−42.63
CH_2O_2	−21.83	−70.71	−45.28
CO_2	0	0	−31.26
$^1/_6C_6H_{12}O_6$	−17.23	−114.97	−32.85

It is apparent from this table that substantial quantities of free energy are available from the disproportion and reduction reactions of organic compounds, i.e., reactions possible under reducing conditions, though the free energies of the oxidation reactions possible in the present oxidizing atmosphere are much larger. It is not intended to infer that the reactions listed are necessarily the ones used by primitive life, but only that they indicate the order of magnitude of the free energies available from similar reactions. (Yeast uses a disproportion reaction,

$$^1/_6C_6H_{12}O_6(aq.) = {}^1/_3C_2H_5OH + {}^1/_3CO_2, \qquad \Delta F^\circ_{298} = -8.68$$

as a source of its free energy.) The high energy photochemical reactions of the reducing atmosphere at high altitudes could not be highly effective because of back reactions and because of the high energy used for the elementary processes ($\sim$150,000 cal.). Hence, the free energy supply for the primitive life processes was much less than that of the present time probably not more than 10^{-3} or 10^{-4} as much. However, experimentation on metabolic processes was possible and probably proceeded on a substantial scale. Also, a great advantage accrued to the mutations producing photosynthesis, thus ensuring survival of these processes.

Porphyrins probably appeared during the reducing period as important constituents of enzymes. Also during this time, photosynthesis evolved, and as oxidizing conditions were established green plants became the fundamental, even if they are not the dominant, type of life. In this way the evolution of photosynthesis was possible before the free energy due to it was available to living organisms.

Time and Conditions of Transition Period.—The order in which reduced substances were oxidized depends on the free energies of the oxidation reactions,

$$^1/_2CH_4(g.) + O_2 = {}^1/_2CO_2 + H_2O, \qquad \Delta F^\circ_{298} \text{ kcal.} = -97.75$$
$$^4/_3NH_3(g.) + O_2 = {}^2/_3N_2 + 2H_2O, \qquad \Delta F^\circ_{298} \text{ kcal.} = -110.73$$
$$^1/_2H_2S + O_2 = {}^1/_2H_2SO_4 \text{ (1 molar)}, \qquad \Delta F^\circ_{298} \text{ kcal.} = -84.72$$
$$^1/_2FeS + O_2 = {}^1/_2 FeSO_4 \text{ (aq.)}, \qquad \Delta F^\circ_{298} \text{ kcal.} = -86.7.$$

Thus ammonia should oxidize first, then methane, and hydrogen sulfide third. However, if ammonium salts of organic acids are possible the stability of reduced nitrogen is greatly increased, and thus methane and ammonium ion would be oxidized more or less together. This is shown by the reaction,

$$^4/_3NH_4OOCCH_3 + O_2 = {}^2/_3N_2 + 2H_2O + {}^4/_3CH_3COOH, \quad \Delta F^\circ_{298} = -97.83.$$

The last two reactions show that sulfide sulfur will be oxidized after the methane and ammonium ion have been oxidized. This conclusion is supported by other oxidation-reduction reactions of sulfur and carbon.

The amount of carbon on the earth's surface is about 350 g. atoms cm.$^{-2}$ equivalent to 1.4×10^3 moles cm.$^{-2}$ or 2.5×10^4 g. cm.$^{-2}$ of water, if all carbon was initially present as CH_4 and all had been oxidized to CO_2. But carbon is produced from the earth's interior and its state of reduction is less than that of methane and part of the surface carbon is not now oxidized to carbon dioxide. Hence the total amount of water which has been decomposed in order to oxidize the carbon is more nearly half of the above value, or 700 moles cm.$^{-2}$ of water. Using the energy per year absorbed by the atmosphere in wave-lengths below 1500 A, i.e., 1.6 cal. yr.$^{-1}$ cm.$^{-2}$, and assuming that every quantum produces a hydrogen atom with 200,000 cal. per/gram atom and that the hydrogen atom escaped from the earth, 10^8 years would be required for the oxidation process. If effectively all the energy below 2000 A is utilized in this way, the time would be 2.5×10^6 years. Neither assumption is realistic and no estimate can be made in this way except that the time might be either very long or comparatively short.

The hydrogen must escape in order for an oxidized atmosphere to be established, and if the methane-water atmosphere is cold, escape will be difficult. Interpolating from Spitzer's calculations[12] the escape of the required amount of hydrogen would require about 2×10^9 years, if atomic hydrogen escaped at 325°K. with an effective surface partial pressure of 10^{-3} atmosphere and 2×10^6 years if the surface pressure was one atmosphere. In the latter case the escape formula is not a good approximation but the time would be short nevertheless. If escape was by molecular hydrogen the temperature must be 650°K. for the same times of escape. The temperature of the methane water atmosphere at high altitudes was probably less than 325°K. and hence a long time for the escape of hydrogen from the reducing atmosphere is indicated.

Thode[13] and his coworkers have found that the ratios of the sulfur isotopes in the sulfides, elementary sulfur and sulfates are closely the same as this ratio in meteoritic sulfur until about 8×10^8 years ago, and after this time the sulfur and sulfides contain increasing amounts of S^{32} with time while the sulfates contain less amounts of this isotope. They ascribe this to the action of living organisms in promoting the oxidation and reduction of sulfur compounds, thus leading to a progressive separation of the isotopes, and suggest that life evolved about 8×10^8 years ago. This is a very interesting suggestion and may be a correct conclusion. It does not give a very long time for the evolution of the very comlpex organisms whose remains are found in the Cambrian rocks.

On the other hand, this date might mark the transition from the reducing to the oxidizing atmospheric conditions. The oxidation of sulfur and sulfides to sulfates would probably not occur to any large extent until free oxygen appeared or until photosynthesis was well developed. But limestones were deposited in large quantities early in the earth's history and graphite was not. The two reactions,

$$CaCO_3 + SiO_2 + 4H_2(g.) = CH_4 + 2H_2O + CaSiO_3, \qquad K = 3 \times 10^{16}$$

and

$$C + 2H_2(g.) = CH_4(g.), \qquad K = 8 \times 10^8,$$

make possible an estimate of the pressure of hydrogen and methane that would make these two events possible at the same time. The pressure of methane would only be 5 atmospheres if all the present surface carbon were methane and if part of this carbon were dissolved as organic compounds in the oceans, the partial pressure might well be about one atmosphere. Then, if limestones were deposited, the hydrogen partial pressure was less than 10^{-4} atmosphere if equilibrium existed, but was probably higher since complete equilibrium cannot be expected. The second equation shows that graphite would not be stable under these conditions. As the methane was consumed the hydrogen pressure must have decreased. The partial pressure of carbon dioxide was 10^{-8} atmosphere if calcium carbonate was present, and may have been higher than this as it is today. If the hydrogen pressure fluctuated and for brief periods exceeded some critical pressure, massive deposits of limestones would be possible, but organisms which experimented with calcareous shells would have had great difficulty in preventing the dissolution of their shells during these periods and the extinction of their species, and indeed no certain calcareous fossils have been found in the Precambrian.

It seems just barely possible that reducing conditions were maintained until some 8×10^8 years ago. Limestones could be deposited, graphite need not have been formed, living organisms might find some 10^{-3} atmosphere or even less of hydrogen with photochemically oxidized organic compounds sufficient for their metabolic processes, and the methane pressure could not have been above 5 atmospheres and was maintained at lower pressures by the solubility of the oxidized organic compounds in water. The precipitation of limestones in great quantities presents a difficulty to the hypothesis of a long period during which a reducing atmosphere was present. The presence of highly oxidized iron in "red beds" and hematite (Fe_2O_3) iron ores are justly regarded as evidence for atmospheric oxygen. Red beds apparently are unknown earlier than the late Precambrian.

Most of the great bodies of iron ore were laid down in the late Precambrian (Huronian) or were extensively eroded during this time.   The iron ore of the Vermillion range of Minnesota is much earlier (Keewatin) and thus oxidation of ferrous iron to ferric oxide took place early in the earth's history.[14]   It should be noted that ferrous oxide should be oxidized to magnetic iron oxide by water if the temperature is sufficiently high to make the reaction fast enough with respect to the time available.   However, magnetic iron oxide cannot be oxidized to ferric oxide by water unless the hydrogen is removed.   The relations are shown by the reactions,

$$3FeO + H_2O = Fe_3O_4 + H_2; \quad K_{298} = 10^8, \ K_{500} = 10^4$$
$$2Fe_3O_4 + H_2O = 3Fe_2O_3 + H_2; \quad K_{298} = 10^{-6}, \ K_{500} = 10^{-5}$$

Thus circulating hot water could produce ferric oxide even in the absence of free oxygen, but it would probably be a rare event.   The magnetic iron oxide is formed when iron is corroded by water in boilers.   The conditions of deposition of these ores of the Precambrian are not well understood, though as stated above the presence of highly oxidized iron justifies a strong presumption of an oxidizing atmosphere.   Throughout the calculations it has been assumed that thermodynamic equilibria will be attained except for photochemical effects.   This need not be the case and the presence of living organisms almost certainly would lead to important deviations from such equilibrium.

The red bacteria and some species of algae are able to use hydrogen and carbon dioxide in photosynthesis.   This ability to use hydrogen is especially interesting because they do not find hydrogen available in their natural habitats.   They appear to be living fossils from some former time and would live under conditions outlined above, though they prefer higher pressures of hydrogen than 10^{-3} atmosphere.   Incidentally, modern plants prefer higher concentrations of carbon dioxide than those available in nature.§   If the present atmosphere should slowly change to a reducing one, it is certain that a substantial flora and fauna would survive.   The flora would surely include many green plants and the fauna most of the principal orders of animals with the exception of the mammals and birds, i.e., the warm-blooded animals, for whom the reduced free energy supply would probably be fatal.   A few aerobes would probably survive wherever photosynthesis was very intense.   Aerobic organisms must naturally be most abundant under aerobic conditions, but mutations would surely supply anaerobic ones for life in a reducing atmosphere.[15]

Poole[16] thinks that oxygen may have been absent from the earth's atmosphere for some 10^9 years of the earth's history, but according to the evidence given here his model of the primitive atmosphere is not correct and hence his conclusion does not substantiate the present work.   He shows that Tammann's thermal dissociation is not correct and that photochemical dissociation and the escape of hydrogen are necessary for the formation of free oxygen.   It is contended here that this mechanism is necessary to account for carbon dioxide as well.   Lane[17] has argued that free oxygen did not appear until late Precambrian times because of the reduced condition of the Keewatin Greenstone schists.   MacGregor[18] comes to similar conclusions from the Precambrian rocks of Rhodesia.   He suggests that the Precambrian iron deposits were concentrated from igneous surface rocks by solution of iron in the absence of free oxygen as ferrous carbonate which was precipitated as ferric oxide by the action of green plants in a lake or sea into which the rivers ran.   The plants may have been diatoms and hence the well-known banded structure of hematite and jasper may have been produced.   As indicated above, the origin of these deposits is not well understood and therefore these suggestions, while worthy of consideration, cannot be regarded as conclusive.

The general course of events and the favorable condition for the origin of life outlined in this paper in no way depend on the time of transition from reducing to oxidizing conditions being exactly some 8×10^8 years ago.   However, the evolution from inanimate systems of biochemical compounds, e.g., the proteins, carbohydrates, enzymes and many others, of the intricate systems of reactions characteristic of living organisms, and of the truly remarkable ability of molecules to reproduce themselves seems to those most expert in the field to be almost impossible.   Thus a time from the beginning to photosynthesis of two billion years may help many to accept the hypothesis of the spontaneous generation of life.   On the other hand, our judgment of an approximate time for the origin of life certainly is not so precise that we can say that 2×10^9 years are sufficient but 2×10^8 years are not.

It seems to me that experimentation on the production of organic compounds from water and methane in the presence of ultra-violet light of approximately the spectral distribution estimated for sunlight would be most profitable.   The investigation of possible effects of electric discharges on the reactions should also be tried since electric storms in the reducing atmosphere can be postulated reasonably.

Also theoretical investigations on hydrogen and methane-water atmospheres would be most helpful in estimating the time of transition from the reducing to the oxidizing atmosphere.   Most interesting in this connection would be more experimental data such as those of Dr. Thode and his co-workers on the abundance of the sulfur isotopes.   The time of transition should be recorded in the rocks, and some such indication as that observed, by Thode, or some change in the state of oxidation of some elements should occur and should be detectable providing the time of transition was not too early in the earth's history.

I have profited greatly from discussions of this subject with Professor

James Franck, who has often pointed out to me and others that complex organic compounds, even porphyrins, may have originated under approximately the conditions outlined in this paper.

Note added in proof: Since this paper was written an interesting paper by J. D. Bernal on the Physical Basis of Life (London, Routledge and Kegan Paul, 1951) has come to my attention, in which very similar suggestions have been made, but there are differences in details and the arguments used. His paper is worthy of serious study.

§ I am indebted to Dr. H. Gaffron for information relative to these bacteria and algae.

[1] Urey, H. C., *Geochim. et Cosmochim. Acta*, 1, 209 (1951). This has been revised and extended in *The Planets: Their Origin and Development*, Yale University Press, 1952.

[2] Oparin, A. I., *Origin of Life*, Macmillan Co., New York, 1938.

[3] Garrison, W. M., Morrison, D. C., Hamilton, J. G., Benson, A. A. and Calvin, M., *Science*, 114, 416 (1951).

[3a] Blum, H. F., *Time's Arrow and Evolution*, Princeton University Press, 1951.

[4] Gilbert, G. K., *Bull. Phil. Soc. Wash.*, 12, 241 (1893).

[5] Baldwin, R. B., *The Face of the Moon*, University of Chicago Press, 1949.

[6] See Greenstein, J. L., *Atmospheres of the Earth and Planets*, edited by G. P. Kuiper, University of Chicago Press, 1949, pp. 117 ff.

[7] See Spitzer, L., Jr., *Ibid.*, edited by G. P. Kuiper, University of Chicago Press, 1949.

[8] Rubey, W. W., *Bull. Geol. Soc. Am.*, 62, 1111 (1951).

[9] Kuiper, G. P., *Atmospheres of the Earth and Planets*, University of Chicago Press, 1949, p. 304.

[10] See, for example, Rice, F. O., and Rice, K. K., *The Aliphatic Free Radicals*, The Johns Hopkins Press, 1935; *Free Radical Mechanisms*, Steacie, E. W. R., Reinhold, 1946; Noyes, W. A., Jr., and Leighton, P. A., *The Photochemistry of Gases*, Reinhold, 1941; Rollefson, G. K., and Burton, M., *Photochemistry*, Prentice-Hall, 1939; Glockle G., and Lind. S. C., *Electrochemistry of Gases and Other Dielectrics*, John Wiley, 1939.

[11] Rabinowitch, E. I., *Photosynthesis and Related Processes*, Vol. I, New York, 1945.

[12] Spitzer, Lyman, Jr., *Atmospheres of the Earth and Planets*, edited by G. P. Kuiper, University of Chicago Press, 1949.

[13] Thode, H. G., Am. Chem. Soc. Meeting, New York, September, 1951. See Tudge, A. P., and Thode, H. G., *Can. J. Res.*, 28, 567 (1950).

[14] Leigh, C. K., Lund, R. J., and Leigh, Andrew, *Geol. Survey Professional Paper* 184 (1935).

[15] See Deevey, E. S., Jr., *Scientific American*, 185, No. 4, 68 (1951) for an interesting mention of Chironomius and of other anaerobic animals.

[16] *Loc. cit.*

[17] Lane, A. C., *Am. J. Sci.*, 43, 42 (1917).

[18] MacGregor, A., *South African J. Sci.*, 24, 155 (1927).

2

Reprinted from *Science*, **117**, 528–529 (May 15, 1953)

A Production of Amino Acids Under Possible Primitive Earth Conditions

Stanley L. Miller[1, 2]

G. H. Jones Chemical Laboratory,
University of Chicago, Chicago, Illinois

The idea that the organic compounds that serve as the basis of life were formed when the earth had an atmosphere of methane, ammonia, water, and hydrogen instead of carbon dioxide, nitrogen, oxygen, and water was suggested by Oparin (*1*) and has been given emphasis recently by Urey (*2*) and Bernal (*3*).

In order to test this hypothesis, an apparatus was built to circulate CH_4, NH_3, H_2O, and H_2 past an electric discharge. The resulting mixture has been tested for amino acids by paper chromatography. Electrical discharge was used to form free radicals instead of ultraviolet light, because quartz absorbs wavelengths short enough to cause photo-dissociation of the gases. Electrical discharge may have played a significant role in the formation of compounds in the primitive atmosphere.

The apparatus used is shown in Fig. 1. Water is boiled in the flask, mixes with the gases in the 5-l flask, circulates past the electrodes, condenses and empties back into the boiling flask. The U-tube prevents circulation in the opposite direction. The acids and amino acids formed in the discharge, not being volatile, accumulate in the water phase. The circulation of the gases is quite slow, but this seems to be an asset, because production was less in a different apparatus with an aspirator arrangement to promote circulation. The discharge, a small corona, was provided by an induction coil designed for detection of leaks in vacuum apparatus.

The experimental procedure was to seal off the opening in the boiling flask after adding 200 ml of water, evacuate the air, add 10 cm pressure of H_2, 20 cm of CH_4, and 20 cm of NH_3. The water in the flask was boiled, and the discharge was run continuously for a week.

[1] National Science Foundation Fellow, 1952–53.
[2] Thanks are due Harold C. Urey for many helpful suggestions and guidance in the course of this investigation.

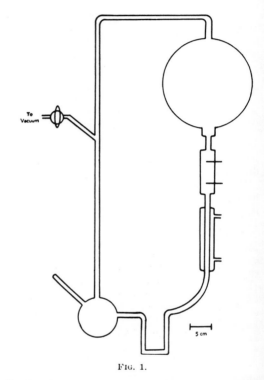

FIG. 1.

During the run the water in the flask became noticeably pink after the first day, and by the end of the week the solution was deep red and turbid. Most of the turbidity was due to colloidal silica from the glass. The red color is due to organic compounds adsorbed on the silica. Also present are yellow organic compounds, of which only a small fraction can be extracted with ether, and which form a continuous streak tapering off at the bottom on a one-dimensional chromatogram run in butanol-acetic acid. These substances are being investigated further.

At the end of the run the solution in the boiling flask was removed and 1 ml of saturated $HgCl_2$ was added to prevent the growth of living organisms. The ampholytes were separated from the rest of the constituents by adding $Ba(OH)_2$ and evaporating *in vacuo* to remove amines, adding H_2SO_4 and evaporat-

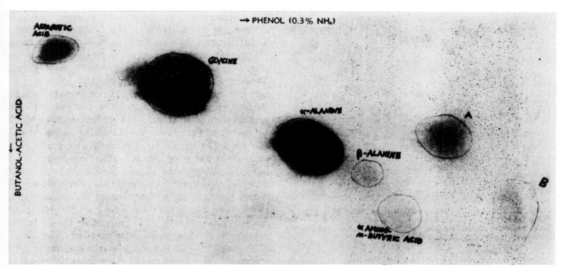

FIG. 2.

ing to remove the acids, neutralizing with $Ba(OH)_2$, filtering and concentrating *in vacuo*.

The amino acids are not due to living organisms because their growth would be prevented by the boiling water during the run, and by the $HgCl_2$, $Ba(OH)_2$, H_2SO_4 during the analysis.

In Fig. 2 is shown a paper chromatogram run in *n*-butanol-acetic acid-water mixture followed by water-saturated phenol, and spraying with ninhydrin. Identification of an amino acid was made when the R_f value (the ratio of the distance traveled by the amino acid to the distance traveled by the solvent front), the shape, and the color of the spot were the same on a known, unknown, and mixture of the known and unknown; and when consistent results were obtained with chromatograms using phenol and 77% ethanol.

On this basis glycine, α-alanine and β-alanine are identified. The identification of the aspartic acid and α-amino-*n*-butyric acid is less certain because the spots are quite weak. The spots marked A and B are unidentified as yet, but may be beta and gamma amino acids. These are the main amino acids present, and others are undoubtedly present but in smaller amounts. It is estimated that the total yield of amino acids was in the milligram range.

In this apparatus an attempt was made to duplicate a primitive atmosphere of the earth, and not to obtain the optimum conditions for the formation of amino acids. Although in this case the total yield was small for the energy expended, it is possible that, with more efficient apparatus (such as mixing of the free radicals in a flow system, use of higher hydrocarbons from natural gas or petroleum, carbon dioxide, etc., and optimum ratios of gases), this type of process would be a way of commercially producing amino acids.

A more complete analysis of the amino acids and other products of the discharge is now being performed and will be reported in detail shortly.

References

1. OPARIN, A. I. The Origin of Life. New York: Macmillan (1938).
2. UREY, H. C. *Proc. Natl. Acad. Sci. U. S.*, **38**, 351 (1952); *The Planets.* New Haven: Yale Univ. Press Chap. 4 (1952).
3. BERNAL, J. D. *Proc. Phys. Soc. (London)*, **62A**, 537 (1949); **62B**, 597 (1949); *Physical Basis of Life.* London: Routledge and Kegan Paul (1951).

Manuscript received February 13, 1953.

3

Reprinted from *Science*, **130**(3770), 245–251 (1959)

CURRENT PROBLEMS IN RESEARCH

Organic Compound Synthesis on the Primitive Earth

Several questions about the origin of life have been answered, but much remains to be studied.

Stanley L. Miller and Harold C. Urey

Since the demonstration by Pasteur that life does not arise spontaneously at the present time, the problem of the origin of life has been one of determining how the first forms of life arose, from which all the present species have evolved. This problem has received considerable attention in recent years, but there is disagreement on many points. We shall discuss the present status of the problem, mainly with respect to the early chemical history of, and the synthesis of organic compounds on, the primitive earth.

Many of our modern ideas on the origin of life stem from Oparin (*1*), who argued that the spontaneous generation of the first living organism might reasonably have taken place if large quantities of organic compounds had been present in the oceans of the primitive earth. Oparin further proposed that the atmosphere was reducing in character and that organic compounds might be synthesized under these conditions. This hypothesis implied that the first organisms were heterotrophic—that is, that they obtained their basic constituents from the environment instead of synthesizing them from carbon dioxide and water. Horowitz (*2*) discussed this point further and outlined how a simple heterotrophic organism could develop the ability to synthesize various cell constituents and thereby evolve into autotrophic organisms.

In spite of the argument by Oparin, numerous attempts were made to synthesize organic compounds under the oxidizing conditions now present on the earth (*3*). Various sources of energy acting on carbon dioxide and water failed to give reduced carbon compounds except when contaminating reducing agents were present. The one exception to this was the synthesis of formic acid and formaldehyde in very small yield (10^{-7} H_2CO molecules per ion pair) by the use of 40-million-electron-volt helium ions from a 60-inch cyclotron (*4*). While the simplest organic compounds were indeed synthesized, the yields were so small that this experiment can best be interpreted to mean that it would not have been possible to synthesize organic compounds nonbiologically as long as oxidizing conditions were present on the earth. This experiment is important in that it induced a reexamination of Oparin's hypothesis of the reducing atmosphere (*5*).

The Primitive Atmosphere

Our discussion is based on the assumption that conditions on the primitive earth were favorable for the production of the organic compounds which make up life as we know it. There are many sets of conditions under which organic compounds could have been produced. All these conditions are more or less reducing. However, before accepting a set of conditions for the primitive earth, one must show that reactions known to take place will not rapidly change the atmosphere to another type. The proposed set of conditions must also be consistent with the known laws for the escape of hydrogen.

Cosmic dust clouds, from which the earth is believed to have been formed, contain a great excess of hydrogen. The planets Jupiter, Saturn, Uranus, and Neptune are known to have atmospheres of methane and ammonia. There has not been sufficient time for hydrogen to escape from these planets, because of their lower temperatures and higher gravitational fields. It is reasonable to expect that the earth and the other minor planets also started out with reducing atmospheres and that these atmospheres became oxidizing, due to the escape of hydrogen.

The meteorites are the closest approximation we have to the solid material from which the earth was formed. They are observed to be highly reduced—the iron mostly as metallic iron with some ferrous sulfide, the carbon as elemental carbon or iron carbide, and the phosphorus as phosphides.

The atmosphere under these reducing conditions would contain some hydrogen, methane, nitrogen, and ammonia; smaller amounts of carbon dioxide and carbon monoxide; and possibly small amounts of other substances such as higher hydrocarbons, hydrogen sulfide, and phosphine. These substances were probably not present in equilibrium concentrations, but compounds which are thermodynamically very unstable in this highly reducing atmosphere —such as oxygen, oxides of nitrogen, and oxides of sulfur—could not have been present in more than a few parts per million. This is true of compounds which are unstable in the present oxidizing atmosphere of the earth, such as hydrogen, ozone, methane, and nitrous oxide.

The over-all chemical change has been the oxidation of the reducing atmosphere to the present oxidizing atmosphere. This

Dr. Miller is a member of the staff of the department of biochemistry, College of Physicians and Surgeons, Columbia University, New York. Dr. Urey is on the staff of the University of California, La Jolla, Calif.

is caused by the loss of hydrogen, which results in the production of nitrogen, nitrate, sulfate, free oxygen, and ferric iron. As is discussed below, many complex organic compounds would have been formed during the course of this over-all change, thereby presenting a favorable environment for the formation of life. Whether the surface carbon of the present earth was all part of the initial atmosphere or whether it has been escaping from the earth's interior in a somewhat reduced condition is not important to the over-all picture.

Escape of Hydrogen

We have learned in recent years that the temperature of the high atmosphere is 2000°K or more, and there is no reason to suppose that the same temperature was not present in the past. One might expect that a reducing atmosphere would be cooler than an oxidizing atmosphere because methane and ammonia can emit infrared radiation while the diatomic molecules, nitrogen and oxygen, cannot. Curtis and Goody (6) have shown that carbon dioxide is ineffective in emitting infrared radiation in the high atmosphere. This is due to the low efficiency of energy transfer from the translational and rotational to the vibrational degrees of freedom, and it seems likely that this would apply to methane as well.

The loss of hydrogen from the earth is now believed to be limited by the diffusion of H_2 to the high atmosphere, since almost all the water is frozen out before it reaches the high atmosphere. Urey (7) has discussed this problem and finds that the loss is entirely due to these effects and not to the Jeans escape formula.

The present rate of escape is 10^7 atoms of hydrogen per square centimeter per second, and it is proportional to the concentration of molecular hydrogen in the atmosphere, which is now 10^{-6} atm at the earth's surface. This rate would result in escape of hydrogen equivalent to 20 g of water per square centimeter in the last 4.5×10^9 years. This rate is not sufficient to account for the oxygen in the atmosphere (230 g/cm²).

In addition, we must account for the oxidation of the carbon, ammonia, and ferrous iron to their present states of oxidation. The oxidation of the 3000 g of surface carbon per square centimeter present on the earth from the 0 to the $+4$ valence state (that is, from C or

H_2CO to CO_2) would require the loss of 1000 g of hydrogen per square centimeter. At the present rate of escape this would require 2.5×10^{12} years. In order for this escape to be accomplished in 2.5×10^9 years (that is, between 4.5×10^9 and 2.0×10^9 years ago), a pressure of hydrogen at the surface of the earth of 0.7×10^{-3} atm would have been required. In order for the nitrogen, sulfur, and iron also to be oxidized, even larger losses and a higher pressure of hydrogen would have been needed. We use a figure of 1.5×10^{-3} atm for the hydrogen pressure in the primitive atmosphere.

These calculations are greatly oversimplified, since methane and other volatile hydrogen compounds would be decomposed in the high atmosphere and therefore a higher concentration of hydrogen might exist in the high atmosphere than is indicated by surface partial pressures. However, the results of the calculation would be qualitatively the same for hydrogen pressures different from the chosen value by an order of magnitude.

Equilibria of Carbon Compounds

The partial pressure of CO_2 in the atmosphere is kept low by two buffer systems. The first system, which is rapid, is the absorption in the sea to form HCO_3^- and H_2CO_3. The second, which is slow, is the reaction of carbon dioxide with silicates; for example

$$CaSiO_3 + CO_2 \rightarrow$$
$$CaCO_3 + SiO_2 \qquad K_{25^\circ} = 10^8$$

The partial pressure of CO_2 at sea level (3.3×10^{-4} atm) is somewhat higher than the equilibrium pressure (10^{-8} atm), but very much lower than would be the case without the formation of limestones ($CaCO_3$).

The equilibrium constant at 25°C in the presence of liquid water for the reaction

$$CO_2 + 4H_2 \rightarrow CH_4 + 2H_2O$$

is 8×10^{22}. Assuming that equilibrium was attained, and using partial pressures $P_{CO_2} = 10^{-8}$ atm and $P_{H_2} = 1.5 \times 10^{-3}$ atm, we find that the pressure of CH_4 would be 4×10^3 atm. In order to have a reasonable pressure of CH_4, the partial pressure of CO_2 would have to be less than 10^{-8} atm, and limestones would not form.

Complete thermodynamic equilibrium could not exist in a reducing atmosphere

because of the dependence of the equilibrium proportions of compounds on pressure and hence on altitude. It is more likely that the steady-state concentrations of CO_2 and CH_4 would be determined not by the equilibrium at sea level but rather by the equilibrium at higher altitude, where the ultraviolet light would provide the activation energy to bring about rapid equilibrium. Under these conditions water would be a gas, and the equilibrium constant would be 10^{20}, so

$$K_{25} = 10^{20} = P_{CH_4}P^2_{H_2O}/P_{CO_2}P^4_{H_2}$$
$$= X_{CH_4}X^2_{H_2O}/X_{CO_2}X^4_{H_2} \quad P^{-2}$$

where the X's are the mole fractions and P is the total pressure. If the surface partial pressures were $P_{CH_4} = 1$, $P_{CO_2} = 3.3 \times 10^{-4}$ (the present value), and $P_{H_2} = 1.5 \times 10^{-3}$, the X's would be equal to these partial pressures. We use $X_{H_2O} = 10^{-6}$, which is the present value for H_2O above the tropopause. Equilibrium will be established under these conditions where $P = 2.5 \times 10^{-9}$ atm—the present atmospheric pressure at about 180 km. It is reasonable to assume that equilibrium was established at some high altitude; therefore, carbon dioxide and hydrogen could both have been present at small partial pressures and methane could have been present at a moderate partial pressure in a reducing atmosphere where the pressure of hydrogen was 1.5×10^{-3} atm.

Carbon monoxide should not have been an important constituent of the atmosphere, as can be seen from the following reaction

$$CO_2 + H_2 \rightarrow CO + H_2O_{(1)} \quad K_{25} = 3.2 \times 10^{-4}$$
$$P_{CO}/P_{CO_2} = 3.2 \times 10^{-4}P_{H_2}$$

Using $P_{H_2} = 1.5 \times 10^{-3}$, we have the ratio $P_{CO}/P_{CO_2} = 5 \times 10^{-7}$, which is independent of pressure. Furthermore, carbon monoxide is a relatively reactive compound, and should any significant quantities appear in the atmosphere, it would react rather rapidly to give organic compounds, carbon dioxide and hydrogen, and formate.

Rubey (8) and Abelson (9) have argued that the surface carbon and nitrogen have come from the outgassing of the interior of the earth instead of from the remaining gases of the cosmic dust cloud from which the earth was formed. The carbon from the outgassing of the earth is a mixture of CO_2, CO, and CH_4, and hydrogen may be present. While outgassing may have been a significant process on the primitive earth, this does not mean that the atmosphere was necessarily composed of CO_2 and CO. The thermodynamic considerations discussed above would still apply. The carbon dioxide would dissolve in the ocean to form bicarbonate, and $CaCO_3$ would be deposited, and the CO would be unstable, as is demonstrated above.

Many writers quote "authorities" in regard to these questions without understanding what is fact and what is opinion. The thermodynamic properties of C, CO, CO_2, CH_4, N_2, NH_3, O_2, H_2O, and other similar substances are all well known, and the equilibrium mixtures can be calculated for any given composition without question. The only point open to argument is whether equilibrium was approximated or whether a nonequilibrium mixture was present. A mixture of hydrogen and carbon monoxide or hydrogen and carbon dioxide is very unstable at 25°C, but does not explode or react detectably in years. But would such mixtures remain in an atmosphere for millions of years subject to energetic radiation in the high atmosphere? We believe the answer is "No." These mixtures would react even without such radiation in geologic times. Hydrogen and oxygen will remain together at low temperatures for long times without detectable reaction by ordinary methods. The use of radioactive tracers shows that a reaction is proceeding at ordinary temperatures nonetheless.

The buffer systems of the ocean and the calcium silicate–calcium carbonate equilibrium were of sufficient capacity to keep the partial pressure of the carbon dioxide in the atmosphere at a low value; hence, the principal species of carbon in the atmosphere would have been methane, even though the fraction of surface carbon in the oxidation state of carbon dioxide was continuously increasing. This would have been true until the pressure of H_2 fell below about 10^{-6} atm. It is likely that shortly after this, significant quantities of molecular oxygen would have appeared in the atmosphere.

Equilibria of Nitrogen Compounds

The equilibrium concentrations of ammonia can be discussed by considering the reaction

$$\frac{1}{2}N_2 + \frac{3}{2}H_2 \rightarrow NH_3 \qquad K_{25} = 7.6 \times 10^2$$

Using $P_{H_2} = 1.5 \times 10^{-3}$, we have P_{NH_3} $P_{N_2}^{1/2} = 0.04$.

Ammonia is very soluble in water and

Table 1. Present sources of energy averaged over the earth.

Source	Energy $(cal\ cm^{-2}\ yr^{-1})$
Total radiation from sun	260,000
Ultraviolet light	
$\lambda < 2500\ A$	570
$\lambda < 2000\ A$	85
$\lambda < 1500\ A$	3.5*
Electric discharges	4 †
Cosmic rays	0.0015
Radioactivity	
(to 1.0 km depth)	0.8‡
Volcanoes	0.13§

* Includes the 1.9 cal cm^{-2} yr^{-1} from the Lyman α at 1216 A (*39*). † Includes 0.9 cal cm^{-2} yr^{-1} from lightning and about 3 cal cm^{-2} yr^{-1} due to corona discharges from pointed objects (*40*). ‡ The value, 4×10^9 years ago, was 2.8 cal cm^{-2} yr^{-1} (*41*). § Calculated on the assumption of an emission of lava of 1 km^3 ($C_p = 0.25$ cal/g, $P = 3.0$ g/cm³) per year at 1000°C.

therefore would displace the above reaction toward the right, giving

$$\frac{1}{2}N_2 + \frac{3}{2}H_2 + H^+ \rightarrow NH_4^+$$

$$(NH_4^+)/P_{N_2}^{1/2}P_{H_2}^{3/2} = 8.0 \times 10^{13}(H^+)$$

which is valid for pH's less than 9. At pH = 8 and $P_{H_2} = 1.5 \times 10^{-3}$, we have

$$(NH_4^+)/P_{N_2}^{1/2} = 47$$

which shows that most of the ammonia would have been in the ocean instead of in the atmosphere. The ammonia in the ocean would have been largely decomposed when the pressure of hydrogen fell below 10^{-5} atm, assuming that the pH of the ocean was 8, its present value. A higher pH would have made the ammonia less stable; the converse is true for a lower pH.

All the oxides of nitrogen would have been unstable and therefore rare. Hydrogen sulfide would have been present in the atmosphere only as a trace constituent because it would have precipitated as ferrous and other sulfides. Sulfur would have been reduced to hydrogen sulfide by the reaction

$$H_2 + S \rightarrow H_2S \qquad K = 6 \times 10^4$$

It is evident that the calculations do not have a quantitative validity because of many uncertainties with respect to temperature, the processes by which equilibrium could be approached, the atmospheric level at which such processes would be effective, and the partial pressure of hydrogen required to provide the necessary rate of escape. In view of these uncertainties, further calculations are unprofitable at the present time. However, we can conclude from this dis-

cussion that a reducing atmosphere containing low partial pressures of hydrogen and ammonia and a moderate pressure of methane and nitrogen constitutes a reasonable atmosphere for the primitive earth. That this was the case is not *proved* by our arguments, but we maintain that atmospheres containing large quantities of carbon monoxide and carbon dioxide are not stable and cannot account for the loss of hydrogen from the earth.

Synthesis of Organic Compounds

At the present time the direct or indirect source of free energy for all living organisms is the sunlight utilized by photosynthetic organisms. But before the evolution of photosynthesis other sources of free energy must have been used. It is of interest to consider the sources of such free energy as well as the origin of the appropriate chemical compounds containing excess free energy which supplied the energy for chemical evolution prior to the existence of what should be called living organisms, and before the evolution of photosynthesis.

Table 1 gives a summary of the sources of energy in the terrestrial surface regions. It is evident that sunlight is the principal source of energy, but only a small fraction of this is in the wavelengths below 2000 A which can be absorbed by CH_4, H_2O, NH_3, CO_2, and so on. If more complex molecules are formed, the absorption can move to the 2500-A region or to longer wavelengths where a substantial amount of energy is available. With the appearance of porphyrins and other pigments, absorption in the visible spectrum becomes possible.

Although it is probable, it is not certain that the large amount of energy from ultraviolet light would have made the principal contribution to the synthesis of organic compounds. Most of the photochemical reactions at these low wavelengths would have taken place in the upper atmosphere. The compounds so formed would have absorbed at longer wavelengths and therefore might have been decomposed by this ultraviolet light before reaching the oceans. The question is whether the rate of decomposition in the atmosphere was greater or less than the rate of transport to the oceans.

Next in importance as a source of energy are electric discharges, such as lightning and corona discharges from pointed objects, which occur closer to the earth's surface and hence would have effected more efficient transfer to the oceans.

Cosmic-ray energy is negligible at present, and there is no reason to assume it was greater in the past. The radioactive disintegration of uranium, thorium, and potassium was more important 4.5×10^9 years ago than it is now, but still the energy was largely expended on the interior of solid materials such as the rocks, and only a very small fraction of the total energy was available in the oceans and atmosphere. Volcanic energy is not only small but its availability is very limited. A continuous source of energy is needed. It contributes little to the evolutionary process to have a lava flow in one part of the earth at one time and to have another flow on the opposite side of the earth years later. For a brief time heat is available at the surface of the lava, but the surface cools and heat flows slowly from the interior for years, making the surface slightly warm. Only a very small contribution to the evolutionary process could be contributed by these energy sources.

Electric Discharges

While ultraviolet light is a greater source of energy than electric discharges, the greatest progress in the synthesis of organic compounds under primitive conditions has been made with electric discharges. The apparatus used by Miller in these experiments was a closed system of glass, except for tungsten electrodes. The water is boiled in a 500-ml flask which mixes the water vapor and gases in a 5-lit. flask where the spark is located. The products of the discharge are condensed and flow through a U-tube back into the 500-ml flask. The first report (*10*) showed that when methane, ammonia, water, and hydrogen were subjected to a high-frequency spark for a week, milligram quantities of glycine, alanine, and α-amino-*n*-butyric acid were produced.

A more complete analysis (*11, 12*) of the products gave the results shown in Table 2. The compounds in the table account for 15 percent of the carbon added as methane, with the yield of glycine alone being 2.1 percent. Indirect evidence indicated that polyhydroxyl compounds (possibly sugars) were synthesized. These compounds were probably formed from condensations of the formaldehyde that was produced by the

electric discharge. The alanine was demonstrated to be racemic, as would be expected in a system which contained no asymmetric reagents. It was shown that the syntheses were not due to bacterial contamination. The addition of ferrous ammonium sulfate did not change the results, and the substitution of N_2 for the NH_3 changed only the relative yields of the compounds produced.

This experiment has been repeated and confirmed by Abelson (*13*), by Pavlovskaya and Passynsky (*14*), and by Heyns, Walter, and Meyer (*15*). Abelson worked with various mixtures of H_2, CH_4, CO, CO_2, NH_3, N_2, H_2O, and O_2. As long as the conditions were reducing conditions—that is, as long as either H_2, CH_4, CO, or NH_3 was present in excess—amino acids were synthesized. The products were the same and the yields as large in many of these mixtures as they were with methane, ammonia, and water. If the conditions were oxidizing, no amino acids were synthesized. These experiments have confirmed the hypothesis that reducing atmospheres are required for the formation of organic compounds in appreciable quantites. However, several of these mixtures of gases are highly unstable. Hence the synthesis of amino acids in these mixtures does not imply that such atmospheres were present on the primitive earth.

Heyns, Walter, and Meyer also performed experiments with different mixtures of gases, with results similar to Abelson's. These workers also used CH_4, NH_3, H_2O, and H_2S. They obtained ammonium thiocyanate, thiourea, and thioacetamide as well as compounds formed when H_2S was absent.

The mechanism of synthesis of the amino acids is of interest if we are to extrapolate the results in these simple systems to the primitive earth. Two alternative proposals were made for the synthesis of the amino and hydroxy acids in the spark discharge system. (i) Aldehydes and hydrogen cyanide are synthesized in the gas phase by the spark. These aldehydes and the hydrogen cyanide react in the aqueous phase of the system to give amino and hydroxy nitriles, which are hydrolyzed to amino and hydroxy acids. This mechanism is essentially a Strecker synthesis. (ii) The amino and hydroxy acids are synthesized in the gas phase from the ions and radicals that are produced in the electric discharge.

It was shown that most, if not all, of the amino acids were synthesized accord-

ing to the first hypothesis, since the rate of production of aldehydes and hydrogen cyanide by the spark and the rate of hydrolysis of the amino nitriles were sufficient to account for the total yield of amino acids (*12*).

This mechanism accounts for the fact that most of the amino acids were α-amino acids, the ones which occur in proteins. The β-alanine was formed not by this mechanism but probably by the addition of ammonia to acrylonitrile (or acrylamide or acrylic acid), followed by hydrolysis to β-alanine.

The experiments on the mechanism of the electric discharge synthesis of amino acids indicate that a special set of conditions or type of electric discharge is not required to obtain amino acids. Any process or combination of processes that yielded both aldehydes and hydrogen cyanide would have contributed to the amount of α-amino acids in the oceans of the primitive earth. Therefore, whether the aldehydes and hydrogen cyanide came from ultraviolet light or from electric discharges is not a fundamental question, since both processes would have contributed to the α-amino acid content. It may be that electric discharges were the principal source of hydrogen cyanide and that ultraviolet light was the principal source of aldehydes, and that the two processes complemented each other.

Ultraviolet Light

It is clear from Table 1 that the greatest source of energy would be ultraviolet light. The effective wavelengths would be $CH_4 < 1450$ A, $H_2O < 1850$ A, $NH_3 < 2250$ A, CO < 1545 A, $CO_2 < 1690$ A, $N_2 < 1100$ A, and $H_2 < 900$ A. It is more difficult to work with ultraviolet light than with electric discharges because of the small wavelengths involved.

The action of the 1849-A Hg line on a mixture of methane, ammonia, water, and hydrogen produced only a very small yield of amino acids (*16*). Only NH_3 and H_2O absorb at this wavelength, but apparently the radical reactions formed active carbon intermediates. The limiting factor seemed to be the synthesis of hydrogen cyanide. Groth (*17*) found that no amino acids were produced by the 1849-A line of mercury with a mixture of methane, ammonia, and water, but that amines and amino acids were formed when the 1470-A and

29

1295-A lines of xenon were used. The 1849-A line produced amines and amino acids with a mixture of ethane, ammonia, and water. The mechanism of this synthesis was not determined. Terenin (*18*) has also obtained amino acids by the action of the xenon lines on methane, ammonia, and water.

We can expect that a considerable amount of ultraviolet light of wavelengths greater than 2000 A would be absorbed in the oceans, even though there would be considerable absorption of this radiation by the small quantities of organic compounds in the atmosphere. Only a few experiments have been performed which simulate these conditions.

In a most promising experiment, Ellenbogen (*19*) used a suspension of ferrous sulfide in aqueous ammonium chloride through which methane was bubbled. The action of ultraviolet light from a mercury lamp gave small quantities of a substance with peptide frequencies in the infrared. Paper chromatography of a hydrolyzate of this substance gave a number of spots with Ninhydrin, of which phenylalanine, methionine, and valine were tentatively identified.

Bahadur (*20*) has reported the synthesis of serine, aspartic acid, asparagine, and several other amino acids by the action of sunlight on paraformaldehyde solutions containing ferric chloride and nitrate or ammonia. Pavlovskaya and Passynsky (*21*) have also synthesized a number of amino acids by the action of ultraviolet light on a 2.5-percent solution of formaldehyde containing ammonium chloride or nitrate. These high concentrations of formaldehyde would not have occurred on the primitive earth. It would be interesting to see if similar results could be obtained with $10^{-4}M$ or $10^{-5}M$ formaldehyde. This type of experiment deserves further investigation.

Radioactivity and Cosmic Rays

Because of the small amount of energy available, it is highly unlikely that high-energy radiation could have been very important in the synthesis of organic compounds on the primitive earth. However, a good deal of work has been done in which this type of energy has been used, and some of it has been interpreted as bearing on the problem of the origin of life.

Dose and Rajewsky (*22*) produced

Table 2. Yields from sparking a mixture of CH_4, NH_3, H_2O, and H_2; 710 mg of carbon was added as CH_4.

Compound	Yield [moles ($\times 10^6$)]
Glycine	63.
Glycolic acid	56.
Sarcosine	5.
Alanine	34.
Lactic acid	31.
N-Methylalanine	1.
α-Amino-*n*-butyric acid	5.
α-Aminoisobutyric acid	0.1
α-Hydroxybutyric acid	5.
β-Alanine	15.
Succinic acid	4.
Aspartic acid	0.4
Glutamic acid	0.6
Iminodiacetic acid	5.5
Iminoacetic-propionic acid	1.5
Formic acid	233.
Acetic acid	15.
Propionic acid	13.
Urea	2.0
N-Methyl urea	1.5

amines and amino acids through the action of x-rays on various mixtures of CH_4, CO_2, NH_3, N_2, H_2O, and H_2. A small yield of amino acids was obtained through the action of 2 Mev electrons on a mixture of CH_4, NH_3, and H_2O (*23*).

The formation of formic acid and formaldehyde from carbon dioxide and water by 40 Mev helium ions was mentioned previously. These experiments were extended by using aqueous formic acid (*24*). The yield per ion pair was only 6×10^{-4} for formaldehyde and 0.03 for oxalic acid. Higher yields of oxalic acid were obtained from $Ca(HCO_3)_2$ and NH_4HCO_3 by Hasselstrom and Henry (*25*). The helium ion irradiation of aqueous acetic acid solutions gave succinic and tricarbolic acid along with some malonic, malic, and citric acids (*26*).

The irradiation of 0.1- and 0.25-percent aqueous ammonium acetate by 2 Mev electrons gave glycine and aspartic acid (*27*). The yields were very small. Massive doses of gamma rays on solid ammonium carbonate yielded formic acid and very small quantities of glycine and possibly some alanine (*28*).

The concentrations of carbon compounds and the dose rates used in these experiments are, in all probability, very much larger than could be expected on the primitive earth, and the products and yields may depend markedly on

these factors, as well as on the effect of radical scavengers such as HS⁻ and Fe^{2+}. It is difficult to exclude high-energy radiations entirely, but if one is to make any interpretations from laboratory work, the experiments should be performed with much lower dose rates and concentrations of carbon sources.

Thermal Energy

The older theories of the formation of the earth involved a molten earth during its formation and early stages. These theories have been largely abandoned, since the available evidence indicates that the solar system was formed from a cold cloud of cosmic dust. The mechanisms for heating the earth are the gravitational energy released during the condensation of the dust to form the earth and the energy released from the decay of the radioactive elements. It is not known whether the earth was molten at any period during its formation, but it is clear that the crust of the earth would not have remained molten for any length of time.

Studies on the concentration of some elements in the crust of the earth indicate that the temperature was less than 150°C during this lengthy fractionation, and that it was probably close to present terrestrial temperatures (*29*).

Fox (*30*) has maintained that organic compounds were synthesized on the earth by heat. When heated to 150°C, malic acid and urea were converted to aspartic acid and ureidosuccinic acid, and some of the aspartic acid was decarboxylated to α- and β-alanine. The difficulty with these experiments is the source of the malic acid and urea on the primitive earth—a question not discussed by Fox. Fox has also synthesized peptides by the well-known reaction (*31*) of heating amino acids at 150° to 180°C, and the yield of peptides has been increased by using an excess of aspartic or glutamic acid (*32*). There is a difficulty connected with heating amino acids and other organic compounds to high temperatures. Geological conditions can heat amino acids to temperatures above 100°C over long periods of time, but it is not likely that this could occur over short periods. Abelson (*33*) has shown that alanine, one of the more stable amino acids, decarboxylates to methylamine and carbon dioxide. The mean life of alanine is 10^{11} years at 25°C but only 30 years at 150°C. Therefore, any extensive heating of amino acids will result in their destruction, and the same is true for most organic compounds. In the light of this, and since the surface of the primitive earth was probably cool, it is difficult to see how the processes advocated by Fox could have been important in the synthesis of organic compounds.

Surface Reactions, Organic Phosphates, and Porphyrins

It is likely that many reactions were catalyzed by adsorption on clay and mineral surfaces. An example is the polymerization of aminoacetonitrile to glycine peptides in the presence of acid clays, by Akabori and his co-workers (*34*). Formaldehyde and acetaldehyde were shown to react with polyglycine adsorbed on kaolinite to give serine and threonine peptides. This field offers many possibilities for research.

Gulick (*35*) has pointed out that the synthesis of organic phosphates presents a difficult problem because phosphate precipitates as calcium and other phosphates under present earth conditions, and that the scarcity of phosphate often limits the growth of plants, especially in the oceans. He proposes that the presence of hypophosphites, which are more soluble, would account for higher concentrations of phosphorus compounds when the atmosphere was reducing. Thermodynamic calculations show that *all* lower oxidation states of phosphorus are unstable under the pressures of hydrogen assumed in this article. It is possible that stronger reducing agents than hydrogen reduced the phosphate or that some process other than reduction solubilized the calcium phosphate. This problem deserves careful attention.

The synthesis of porphyrins is considered by many authors to be a necessary step for the origin of life. Porphyrins are not necessary for living processes if the organism obtains its energy requirements from fermentation of sugars or other energy-yielding organic reactions. According to the heterotrophic theory of the origin of life, the first organisms would derive their energy requirements from fermentations. The metabolism of sulfate, iron, N_2, hydrogen, and oxygen appears to require porphyrins as well as photosynthesis. Therefore, porphyrins probably would have to be synthesized before free energy could be derived from these compounds. While porphyrins may have been present in the environment

before life arose, this is apparently not a necessity, and porphyrins may have arisen during the evolution of primitive organisms.

Intermediate Stages in Chemical Evolution

The major problems remaining for an understanding of the origin of life are (i) the synthesis of peptides, (ii) the synthesis of purines and pyrimidines, (iii) a mechanism by which "high-energy" phosphate or other types of bonds could be synthesized continuously, (iv) the synthesis of nucleotides and polynucleotides, (v) the synthesis of polypeptides with catalytic activity (enzymes), and (vi) the development of polynucleotides and the associated enzymes which are capable of self-duplication.

This list of problems is based on the assumption that the first living organisms were similar in chemical composition and metabolism to the simplest living organisms still on the earth. That this may not be so is obvious, but the hypothesis of similarity allows us to perform experiments to test it. The surprisingly large yields of aliphatic, hydroxy, and amino acids—α-amino acids rather than the other isomers—in the electric-discharge experiments, plus the arguments that such syntheses would have been effective on the primitive earth, offer support for this hypothesis. Further support can be obtained by demonstrating mechanisms by which other types of biologically important compounds could be synthesized.

Oparin (*1*) does not view the first organism as a polynucleotide capable of self-duplication but, rather, as a coacervate colloid which accumulates proteins and other compounds from the environment, grows in size, and then splits into two or more fragments, which repeat the process. The coacervate would presumably develop the ability to split into fragments which are very similar in composition and structure, and eventually a genetic apparatus would be incorporated which would make very accurate duplicates.

These two hypotheses for the steps in the formation of the first living organism differ mainly in whether the duplication first involved the relatively accurate duplication of nucleic acids, followed by the development of cytoplasm duplication, or whether the steps occurred in

the reverse order. Other sequences could be enumerated, but it is far too early to discuss profitably the exact nature of the first living organism.

It was probably necessary for the primitive organisms to concentrate organic and inorganic nutrients from their environment. This could be accomplished by means of a membrane or by absorption on rocks or clays (*36*). The development of optical activity in living organisms is another important problem. This has been discussed by many authors and is not taken up here.

Life on Other Planets

Life as we know it—and we know of no other variety of life than that existing on the earth—requires the presence of water for its chemical processes. We know enough about the chemistry of other systems, such as those of silicon, ammonia, and hydrogen fluoride, to realize that no highly complex system of chemical reactions similar to that which we call "living" would be possible in such media. Also, much living matter exists and grows actively on the earth in the absence of oxygen, so oxygen is *not* necessary for life, although the contrary is often stated. Moreover, the protecting layer of ozone in the earth's atmosphere is not necessary for life, since ultraviolet light does not penetrate deeply into natural waters and also because many carbon compounds capable of absorbing the ultraviolet light would be present in a reducing atmosphere.

It is possible for life to exist on the earth and to grow actively at temperatures ranging from 0°C, or perhaps a little lower, to about 70°C. It seems likely that if hot springs were not so temporary, many plants and possibly animals would evolve which could live in such temperatures. Plants are able to produce and accumulate substances which lower the freezing point of water, and hence they can live at temperatures below 0°C. At much lower temperatures the reactions would probably be too slow to proceed in reasonable periods of time. At temperatures much above 120°C, reaction velocities would probably be so great that the nicely balanced reactions characteristic of living things would be impossible. In addition, it is doubtful whether the organic polymers necessary for living organisms would be stable much above 120°C; this is probably true even when allowance is made for the amazing stability of the enzymes of thermophilic bacteria and algae.

Only Mars, Earth, and Venus conform to the general requirements so far as temperatures are concerned. Mars is known to be very cold and Venus may be too hot. Observations of the black-body emission of radio waves from Venus indicate surface temperatures of 290° to 350°C (37). The clouds of Venus have the polarization of water droplets. Clearing of the clouds occurs, and this indicates that the clouds are composed of some volatile substance, for nonvolatile dust could hardly settle out locally. However, no infrared bands of water have been observed. It is possible that this is due to a very dry, high atmosphere, such as is characteristic of the earth, and to a cloud level that rises to very near the tropopause, so that there is little water vapor above the reflecting layer.

Mars is known to be very cold, with surface temperatures of $+30°C$ to $-60°C$ during the day. The colors of Mars have been observed for many years by many people. The planet exhibits seasonal changes in color—green or bluish in the spring and brown and reddish in the autumn. Sinton (38) has observed an absorption at 3.5 μ in the reflected light of Mars. This corresponds to the C-H stretching frequency of most organic compounds, but many inorganic compounds have absorptions at this wavelength. The changing colors of Mars and the 3.5 μ absorption are the best evidence, however poor it may be, for the existence of life on the planet. One thing that can be stated with confidence is that if life exists there, then liquid water must have been present on the planet in the past, since it is difficult to believe that life could have evolved in its absence. If this was so, water must have escaped from the planet, as very little water remains there now and no liquid water has been observed. Hence, oxygen atoms must escape from the planet. This is possible if the high atmosphere has a temperature of 2000°K, and this may well be the case in view of the high temperatures in the high atmosphere of the earth.

Surely one of the most marvelous feats of 20th-century science would be the firm proof that life exists on another planet. All the projected space flights and the high costs of such developments would be fully justified if they were able to establish the existence of life on either Mars or Venus. In that case, the thesis that life develops spontaneously when the conditions are favorable would be far more firmly established, and our whole view of the problem of the origin of life would be confirmed (42).

References

1. A. I. Oparin, *The Origin of Life* (Macmillan, New York, 1938; Academic Press, New York, ed. 3, 1957).
2. N. H. Horowitz, *Proc. Natl. Acad. Sci. U.S.* 31, 153 (1945).
3. E. I. Rabinowitch, *Photosynthesis* (Interscience, New York, 1945), vol. I, p. 81.
4. W. M. Garrison, D. C. Morrison, J. G. Hamilton, A. A. Benson, M. Calvin, *Science* 114, 416 (1951).
5. H. C. Urey, *Proc. Natl. Acad. Sci. U.S.* 38, 351 (1952); *The Planets* (Yale Univ. Press, New Haven, Conn., 1952), p. 149.
6. A. R. Curtis and R. M. Goody, *Proc. Roy. Soc. (London)* 236A, 193 (1956).
7. H. C. Urey, in *Handbuch der Physik*, S. Flügge, Ed. (Springer, Berlin, 1958), vol. 52.
8. W. W. Rubey, *Geol. Soc. Am. Spec. Paper No. 62* (1955), p. 631.
9. P. H. Abelson, *Carnegie Inst. Wash. Year Book No. 56* (1957), p. 179.
10. S. L. Miller, *Science* 117, 528 (1953).
11. ——, *J. Am. Chem. Soc.* 77, 2351 (1955).
12. ——, *Biochim. et Biophys. Acta* 23, 480 (1957).
13. P. H. Abelson, *Science* 124, 935 (1956); *Carnegie Inst. Wash. Year Book No. 55* (1956), p. 171.
14. T. E. Pavlovskaya and A. G. Passynsky, *Reports of the Moscow Symposium on the Origin of Life* (Aug. 1957), p. 98.
15. K. Heyns, W. Walter, E. Meyer, *Naturwissenschaften* 44, 385 (1957).
16. S. L. Miller, *Ann. N. Y. Acad. Sci.* 69, 260 (1957).
17. W. Groth, *Angew. Chem.* 69, 681 (1957); —— and H. von Weyssenhoff, *Naturwissenschaften* 44, 510 (1957).
18. A. N. Terenin, in *Reports of the Moscow Symposium on the Origin of Life* (Aug. 1957), p. 97.
19. E. Ellenbogen, *Abstr. Am. Chem. Soc. Meeting, Chicago* (1958), p. 47C.
20. K. Bahadur, *Nature* 173, 1141 (1954); in *Reports of the Moscow Symposium on the Origin of Life* (Aug. 1957), p. 86; *Nature* 182, 1668 (1958).
21. T. E. Pavlovskaya and A. G. Passynsky, *Intern. Congr. Biochem. 4th Congr. Abstr. Communs.* (1958), p. 12.
22. K. Dose and B. Rajewsky, *Biochim. et Biophys. Acta* 25, 225 (1957).
23. S. L. Miller, unpublished experiments.
24. W. M. Garrison *et al.*, *J. Am. Chem. Soc.* 74, 4216 (1952).
25. T. Hasselstrom and M. C. Henry, *Science* 123, 1038 (1956).
26. W. M. Garrison *et al.*, *J. Am. Chem. Soc.* 75, 2459 (1953).
27. T. Hasselstrom, M. C. Henry, B. Murr, *Science* 125, 350 (1957).
28. R. Paschke, R. Chang, D. Young, *ibid.* 125, 881 (1957).
29. H. C. Urey, *Proc. Roy. Soc. (London)* 219A, 281 (1953).
30. S. W. Fox, J. E. Johnson, A. Vegotsky, *Science* 124, 923 (1956); *Ann. N.Y. Acad. Sci.* 69, 328 (1957); *J. Chem. Educ.* 34, 472 (1957).
31. For a review, see E. Katchalski, *Advances in Protein Chem.* 6, 123 (1951).
32. S. W. Fox *et al.*, *J. Am. Chem. Soc.* 80, 2694, 3361 (1958); —— and K. Harada, *Science* 128, 1214 (1958).
33. P. H. Abelson, *Ann. N.Y. Acad. Sci.* 69, 276 (1957).
34. S. Akabori, in *Reports of the Moscow Symposium on the Origin of Life* (Aug. 1957), p. 117; *Bull. Chem. Soc. Japan* 29, 506 (1956).
35. A. Gulick, *Am. Scientist* 43, 479 (1955); *Ann. N.Y. Acad. Sci.* 69, 309 (1957).
36. J. D. Bernal, *Proc. Phys. Soc. (London)* 62A, 537 (1949); *ibid.* 62B, 597 (1949); *The Physical Basis of Life* (Routledge and Kegan Paul, London, 1951).
37. G. H. Mayer, T. P. McCullough, R. M. Sloanaker, *Astrophys. J.* 127, 1 (1958).
38. W. Sinton, *ibid.* 126, 231 (1957).
39. W. A. Rense, *Phys. Rev.* 91, 299 (1953).
40. B. Schonland, *Atmospheric Electricity* (Methuen, London, 1953), pp. 42, 63.
41. E. Bullard, in *The Earth as a Planet*, G. P. Kuiper, Ed. (Univ. of Chicago Press, Chicago, 1954), p. 110.
42. One of the authors (S. M.) is supported by a grant from the National Science Foundation.

4

Reprinted from *Chem. Rev.*, **70**, 95–109 (1970)

CHEMICAL EVOLUTION

RICHARD M. LEMMON

Laboratory of Chemical Biodynamics, Lawrence Radiation Laboratory, University of California, Berkeley, California 94720

Received March 21, 1969

Contents

I. Introduction

The term "chemical evolution" has come to mean the chemical events that took place on the primitive, prebiotic Earth (about 4.5–3.5 billion years ago) leading to the appearance of the first living cell. In other words, it is the study of the biologically relevant chemistry that preceded Darwinian evolution.

In this report we shall review the results and implications of chemical evolution experiments performed over the past two decades. We shall concentrate on the results of laboratory experiments simulating the presumed environment of the prebiotic Earth. Less attention will be paid to the other planets because (a) we do not know if any of them are (or ever were) a habitat for life, and (b) we hardly know more about their present (let alone their past) conditions than we know about the Earth of 4×10^9 years ago. Even less can be said about the possibilities of abiogenic synthesis outside our solar system. Our knowledge of this space is limited entirely to the examination of light reaching us from the stars—locales far too hot for the existence of anything like our terrestrial organic chemistry. It is widely assumed that a small fraction, but a huge number, of these stars have planetary systems comparable to our solar system. However likely this may be, we have only indirect evidence for the existence of planets outside our solar system. Research in chemical evolution has little choice but to focus its attention on planet Earth.

There are several reasons for the recent rise of research in chemical evolution. (1) One method of studying the detailed chemistry of the living cell is to try to reconstruct the chemistry of the prebiotic Earth that led to the appearance of life. (2) Laboratory work has shown that the exposure to high energies of samples of the primitive Earth's presumed atmosphere leads to the appearance of many biologically important organic molecules. (3) The immediacy of lunar and planetary exploration has made it important to know what compounds and conditions to seek in determining whether a given locale is now, ever was, or might be in the future an abode of life. (4) Chemical evolution is "interdisciplinary" in the widest and best sense. It is a field of research that promotes the active collaboration of (to name a few) astronomers, geologists, organic chemists, radiation chemists, and biologists.

In the latter half of the 19th century Darwin and Tyndall recognized that the backward extension in time of biological evolution led to a kind of chemical evolution. However, for the most part, this was a neglected area of scientific thought, let alone experiment. There were several reasons for this neglect: lack of any clear idea of the chemical environment of the prebiotic Earth, insufficient development of analytical techniques, and a prevailing opinion that such scientific thought and experiment would be an invasion of the precincts of religion. In addition, Pasteur's famous 1864 experiment appeared to have smashed completely the age-old idea of spontaneous generation. Everyone, most scientists included, agreed that "life can come only from life," and thought little about the chemical events that must have preceded the appearance of life.

The USSR was the locale of the birth of the studies that we call chemical evolution. In the early 1920's the young Soviet biochemist, A. I. Oparin, expressed the idea that life on Earth must have arisen from a preformed "pool" of organic compounds. In brief, he was saying that spontaneous generation was indeed the original route to life, although its "spontaneity" had to be stretched to a billion years or more. Oparin's ideas first appeared in print in 1924 in a booklet entitled "The Origin of Life,"[1] and he has steadily contributed both ideas and experimental data right up to the present. Oparin's ideas were independently reached by the British biologist, J. B. S. Haldane, who suggested in 1928 that large amounts of organic compounds must have accumulated on the prebiotic Earth, particularly in the oceans.[2] It was not until after World War II, however, that direct experimental evi-

(1) A. I. Oparin, "Proiskhozhdenie zhizni," Izd. Moskovskii Rabochii, Moscow, 1924.
(2) J. B. S. Haldane, "Rationalist Annual," 1929; "Science and Human Life," Harper Bros., New York and London, 1933, p 149.

95

dence finally put the Oparin–Haldane ideas on firm ground and established chemical evolution as a serious scientific study.

II. Abiogenic Synthesis on the Primitive Earth

A. TIME SCALE

There is general agreement that the Earth condensed from a dust cloud 4.5–4.8 $\times$ 10[9] years ago.[3] Recent reports of presumed bacterial and algal fossils strongly indicate that life, in the form of unicellular organisms (Protozoa), has been present on the Earth for at least 3.1 $\times$ 10[9] years,[4-6] and that multicellular organisms (Metazoa) may have been here 2–2.5 $\times$ 10[9] years ago.[7] A basic concept of chemical evolution is that it declined at the onset of Darwinian evolution; in our present "biological" era, almost any organic compound produced abiogenetically would be quickly metabolized. The Barghoorn–Schopf 3.1 $\times$ 10[9] year-old algae (which were probably preceded by earlier, anaerobic organisms) seem to leave only about one billion years for the period of chemical evolution, about one-third the time thought available before the Barghoorn and Schopf reports[4,5] appeared. This has been disturbing to some "chemical evolutionists." But it shouldn't be. There is nothing to indicate that 3.0 $\times$ 10[9] years is enough time, but 1.0 $\times$ 10[9] years is not.

B. THE PRIMITIVE EARTH'S ATMOSPHERE

Any discussion of the chemistry of the primitive Earth must begin with its atmosphere. That atmosphere would, in turn, have controlled the chemistry of the oceans and surface rocks.

There is now general agreement that the primitive planet's atmosphere was a hydrogen-dominated, or reduced, one. The realization that the early atmosphere was quite different than it is today began with the discovery, in 1929, that hydrogen was by far the solar system's most common element (about 87% of the mass of the sun[8]). It was, therefore, reasonable to suppose that, as the Earth was forming, most of its carbon, nitrogen, and oxygen would be in the form of methane, ammonia, and water. Oparin, in the 1938 edition of his book, "Origin of Life," concluded that most of the carbon present on the early Earth was in the form of hydrocarbons.[9] Similar conclusions were advanced regarding the chemical nature of the nitrogen and oxygen in the primitive atmosphere.

One problem with respect to this view of the primitive atmosphere is the relatively low percentage of the noble gases in our present atmosphere. Harrison Brown has calculated that the fraction of neon present in the Earth's atmosphere today is only 10[-10] of its cosmic abundance;[10] the other noble gases are similarly conspicuous by their relative absences. At first thought, it seems difficult to imagine the primitive Earth retaining such low molecular weight molecules as methane, ammonia, and water while losing atoms, such as krypton and xenon, with atomic weights around 80 and 130. However, the carbon may have been initially bound as carbides (from whence methane and other hydrocarbons may have come by reaction with water), the water retained as hydrates, and the ammonia as ammonium ion. A similar view is that the Earth's first atmosphere was lost by diffusion, and was replaced by a secondary one (also reduced) that resulted from outgassing from the Earth's interior.[11] Such a secondary atmosphere may have been kept in the reduced form for as long as 10[9] years by the presence of metallic iron in the Earth's upper mantle and crust.[12]

The potent arguments in favor of a reduced primitive atmosphere may be summarized as follows.

(a) The general geochemical arguments of Oparin[9] and Urey[13] that, since our solar system and observable universe are so heavily hydrogen-laden, the Earth's primitive atmosphere must have been highly reduced. This argument also draws attention to the larger and colder planets in our solar system (e.g., Jupiter). The higher gravitational fields and lower temperatures would favor the retention of light molecules, and, indeed, the Jovian planets' atmospheres appear to be rich in methane and ammonia.[13]

(b) The meteorites that reach the Earth are, for the most part, reduced. Most of their carbon appears as elementary carbon, carbides, and hydrocarbons. The iron is mostly metallic or ferrous, and the phosphorus appears as phosphides.[14]

(c) When mixtures of methane, ammonia (or N_2), and water—the principal constituents of the Earth's presumed early atmosphere—are subjected to ultraviolet or ionizing radiation, many biologically important compounds (amino acids, sugars, purines, etc.) are formed. Similar irradiations of samples of the Earth's present atmosphere yield little of biological relevance, e.g., only traces of formic acid and formamide. It appears that the accumulation, on the primitive Earth, of the necessary "building blocks" for the first living cell required a reduced atmosphere.

(d) Molecular oxygen exerts a deleterious effect on many aspects of cell metabolism, a fact difficult to account for if the first living cells had appeared in an oxygenated environment. Chromosomes appear to operate in an anaerobic medium, and cell division takes place during a temporary period of anaerobiosis.[15] These facts seem to point to the early evolution of the living cell in a reduced atmosphere.

(e) The work of Rankama, Randohr, and others (as reported by Rutten[16])—the finding, for example, of increased ferrous iron in the oldest Precambrian sediments—indicates that these deposits were laid down under a reduced atmosphere.

It should be noted that this view of a reduced early atmosphere is not without its detractors. Abelson[17] and Cloud[18] have asserted, mostly because "carbon is not a conspicuous

(3) G. R. Tilton and R. H. Steiger, Science, 150, 1805 (1965).
(4) E. S. Barghoorn and J. W. Schopf, ibid., 152, 758 (1966).
(5) J. W. Schopf and E. S. Barghoorn, ibid., 156, 508 (1967).
(6) A. E. J. Engel, et al., ibid., 161, 1005 (1968).
(7) H. J. Hofman, ibid., 156, 500 (1967).
(8) H. N. Russell, Astrophys. J., 70, 11 (1929).
(9) A. I. Oparin, "Origin of Life," 2nd English ed, 1953, Dover Publications, Inc., New York, N. Y., p 101.
(10) H. Brown, "The Atmospheres of the Earth and Planets," University of Chicago Press, Chicago, Ill., 1952, p 258.

(11) H. D. Holland in "Petrologic Studies: a Volume in Honor of A. F. Buddington," A. E. J. Engel, H. L. James, and B. F. Leonard, Ed., Geological Society of America, New York, N. Y., 1962, p 447.
(12) S. I. Rasool and W. E. McGovern, Nature, 212, 1225 (1966).
(13) H. C. Urey, "The Planets: Their Origin and Development," Yale University Press, New Haven, Conn., 1952.
(14) S. L. Miller and H. C. Urey, Science, 130, 245 (1959).
(15) H. Stern, ibid., 121, 144 (1955).
(16) M. G. Rutten, "The Geological Aspects of the Origin of Life on Earth," Elsevier Publishing Co., Amsterdam, 1962, p 106.
(17) P. H. Abelson, Proc. Natl. Acad. Sci. U. S., 55, 1365 (1966).
(18) P. E. Cloud, Jr., Science, 160, 729 (1968).

component of the oldest sediments," that there was little methane in the primitive atmosphere. However, we have no samples of rocks reliably known to an age greater than about 3.5 billion years. It may be that, by that time, the carbon in the Earth's atmosphere had shifted from predominantly CH_4 to predominantly CO_2. The era of chemical evolution that we are considering may be considered as taking place during the first billion years of the Earth's history—at a time when, it is very probable, the atmosphere was reduced.

The change from the reduced, primitive atmosphere to the present oxidized one is explained by the ultraviolet radiolysis of water in the Earth's upper atmosphere (followed by the preferential escape of hydrogen) and by the development of the process of plant photosynthesis. Along with the change to an oxygenated atmosphere the Earth developed its present "shield" of ozone in the upper atmosphere. Without this shield, which protects our planet from the strong ultraviolet light from the sun, it is difficult to see how the Earth could have become an abode of life.

C. OTHER CONDITIONS ON THE PRIMITIVE EARTH

Other factors on the prebiotic Earth that would have strongly influenced chemical evolution, and that play an important part in the laboratory investigations in this subject, are the energies available, the temperatures, and the oceans and their sediments.

The significant source of energy on our planet, now and in the remote past, is that from the sun. Table I is a compilation of data supplied from several different sources. [14, 19, 20]

Table I

Main Present Sources of Energy at Earth's Surface

Energy source	Energy, cal/year $\times$ 10^{-19}
Total radiation from sun	132,000
Ultraviolet light	
$\lambda < 2500$ Å	300
$\lambda < 2000$ Å	45
$\lambda < 1500$ Å	1.8
Electric discharges	2.1
Radioactivity	0.4[a]
Volcanoes	0.07
Cosmic rays	0.0008

[a] 4×10^9 years ago this value was about 1.4×10^{19} cal/year. [19]

Since methane, ammonia, and water all absorb very little at wavelengths longer than 2000 Å, only shorter wavelengths can be considered as having been effective on the Earth's primitive atmosphere. Since, however, the region below 2000 Å is difficult to employ in the laboratory (*e.g.*, due to uv destruction of the special windows that are needed), researchers have tended to simulate the other primitive Earth-available energies. These include electric sparks and corona discharges (simulating the effects of lightning storms), γ rays and electron beams (simulating cosmic rays and radioactivity

in the rocks), and heat (simulating the thermal effects around volcanoes).

Most laboratory work on the abiogenic synthesis of biologically relevant compounds has been done at ambient temperature. Geologists and geochemists are of the general opinion that by the time the Earth became a solid, cohesive mass it had about the average temperature it has today. The notable exception to ambient-temperature work is that of Professor S. W. Fox and his groups at the Florida State University and the University of Miami; they have emphasized the possible role of local high temperatures, present in areas of volcanic activity, in promoting the initial events of chemical evolution. Using very high temperatures (of the order of 1000°) they have contributed interesting amino acid forming experiments; working at much lower temperatures they have shown the formation of peptides from amino acids. We will come back to this work in later sections of this report.

For over a century scientists have regarded the early oceans as the probable birthplace of life on our planet. Darwinian evolution points to this, and chemical evolution seems to point to it too. As was mentioned earlier, Haldane emphasized the role of the primitive oceans and thought that the early abiogenic synthesis would have accumulated in them a considerable concentration of organic compounds. Sagan calculated, on the basis of (a) average quantum yields for the uv-light conversion of reduced-atmosphere gas mixtures to higher molecular weight compounds and (b) assumed values of the uv photon flux in primitive times, that the abiogenic Earth's oceans could have developed a 1 % solution of organic matter in 3×10^8 years. [21] For this reason, most chemical evolution experiments have been done in dilute aqueous media. Another consideration that strongly favors the oceans as the principal locale for chemical evolution is that organic compounds, once formed, would be protected against the radiolysis that would be caused by the strong uv flux of primitive times. The formed amino acids, sugars, etc., would be expected to be adsorbed on mineral particles (muds, clays, etc.), carried down to the bottoms of lakes and seas, and there be protected from photolysis by the uv light. The ocean would also be an effective medium, or vehicle, for the mixing together of different classes of organic compounds formed at separated points on the Earth's surface as a result of, for example, differing temperatures, cosmic ray fluxes, and available mineral surfaces. The last item brings up the important point of surface catalysis—without doubt, adsorption on mineral surfaces played an important part in chemical evolution; the availability of these surfaces in the transporting and mixing actions of the oceans make the latter even more attractive as promoters of abiogenic syntheses.

III. Simulated Primitive-Earth Experiments. Early Work and Usual Experimental Conditions

A. EARLY EXPERIMENTS

Studies were made, as far back as 1897, on the effects of high-energy sources (electric discharges) on mixtures of carbon dioxide and water. [22] In that experiment, and in subsequent ones throughout the ensuing half-century, claims were made

(19) E. Bullard in "The Earth as a Planet," G. P. Kuiper, Ed., University of Chicago Press, Chicago, Ill., 1964, p 110.
(20) A. J. Swallow, "Radiation Chemistry of Organic Compounds," Pergamon Press, New York, N. Y., 1960, p 244.

(21) I. S. Shklovskii and C. Sagan, "Intelligent Life in the Universe," Holden-Day, Inc., San Francisco, Calif., 1966, p 233.
(22) S. M. Losanitsch and M. Z. Jowitschitsch, Ber., 30, 135 (1897).

that formaldehyde was a detectable product of such irradiations; other workers, on the basis of their own experiments, denied these claims. The possibilities of formaldehyde as a product of CO_2–H_2O interactions intrigued many scientists as a hint of how green-plant photosynthesis might operate. From 1870 until about 1940 it was widely held that formaldehyde was the first product of CO_2 fixation by green plants, and that the formaldehyde was converted to sugars by the well-known polymerization reaction. Although most of the interest in the possible formaldehyde product was from the standpoint of photosynthesis, some workers additionally suggested that this product could have been involved in the formation of organic compounds on the primitive Earth.[23]

In 1950, interest in the possible reduction of CO_2 and its fixation into biologically important compounds through the action of ionizing radiation was rekindled by the experiment of Garrison, Calvin, et al.[24] These workers demonstrated the appearance of formic acid and formaldehyde when CO_2–H_2O–Fe^{2+} solutions were irradiated with an α-particle beam. A far greater interest was engendered by the experiments of Miller, first reported in 1953.[25] Having been convinced by the arguments of Oparin[9] and Urey[13] that the Earth's primitive atmosphere was reducing, Miller reasoned that the really meaningful chemical evolution experiment would be to subject such an assumed atmosphere (mixture of CH_4, NH_3, H_2O, and H_2) to high-energy radiation—in his case, to electrical discharges. In this way, Miller demonstrated the facile appearance of glycine, α-alanine, β-alanine, aspartic acid, and α-aminobutyric acid. This experiment set the pattern for many subsequent ones over the ensuing 16 years. Many investigators, as will be detailed below, used various energy sources and made many alterations on Miller's original gas (primitive atmosphere) mixture, but this work is all basically the same, namely, studies of the effects of ionizing radiation on reduced gas mixtures of the sort that are presumed to be similar to the early terrestrial atmosphere.

B. EXPERIMENTAL CONDITIONS

From the foregoing discussions, the reader can conclude that there is general agreement as to what may constitute a meaningful "chemical evolution" experiment. In order to determine what biologically relevant chemistry may have occurred in the Earth's early atmosphere, one employs a mixture of simple molecules (CH_4, NH_3, H_2O, N_2, CO, H_2, etc.) of an overall reduced character, and with energy sources that were certainly plentiful on the primitive Earth (preferably uv light and electric discharges). If one is interested in the chemistry of the primitive oceans, it is desirable to work at ambient temperature, in dilute ($\sim1\%$) solution, and with reactants that (a) appear as products of the early-atmosphere processes and (b) are reasonably stable toward hydrolysis; the same energy sources are employed. When catalysts are added to such a reaction mixture, they should be of the sort for which there is geological evidence for their existence in the early stages of our planet's development (i.e., around 4×10^9 years ago).

The main disagreement with these experimental conditions comes, as was mentioned earlier, from those who think that

the use of higher temperatures (such as those found in areas of volcanic activity) is also valid. When one considers the problem of the formation of the biopolymers in dilute aqueous solution, one realizes that high temperatures and/or nonaqueous environments may indeed have played important, perhaps indispensable, roles in chemical evolution. However, the dilute solution (ocean) model is preferred because the oceans would have provided a wealth of catalytic surfaces, transportation and mixing of intermediate products, and protection from ultraviolet light. Additional considerations are that the oceans cover such a large fraction of the Earth's surface and that they are the almost certain locale for the emergence of life.

IV. Abiogenic Synthesis of Biomonomers

In this section we shall review the progress that has been made, through 1968, in the laboratory synthesis, under the conditions presented in the previous section, of the biomonomers. By the latter term we refer to the constituent units of the biopolymers (the proteins, nucleic acids, and polysaccharides). To be specific, we shall review the abiogenic synthesis of the amino acids, nucleic acid bases, nucleosides, nucleotides, sugars, fats, and porphyrins.

Since most of the abiogenic syntheses have come from the effects of high energy on methane–ammonia–water mixtures, we shall first consider the chemical-evolution-relevant radiation chemistry of each of these compounds. Most of our knowledge of this chemistry comes from the identifications of reactive fragments in the mass spectrometer, i.e., after electron impact on the compound under observation. However, it is likely that the same fragments appear under the effects of other forms of ionizing radiation, some of which are of greater importance to chemical evolution (e.g., electric discharges or ultraviolet light). To say that "the same fragments appear" does not imply that they necessarily end up in the same product. Different fragment densities, along the tracks of various ionizing rays or particles, may lead to quite different final products.

A. RADIATION CHEMISTRY OF METHANE, AMMONIA, AND WATER

1. Methane

The radiation chemistry of this compound has been extensively studied. Mass spectrometric investigations of the pure gas have shown that the principal carbon-containing ions formed are CH_4^+ (47%), CH_3^+ (40%), CH_2^+ (8%), CH^+ (4%), and C^+ (1%).[26] H^+ is also observed (about 1%), as is a trace of H_2^+. Additional mass spectrometric investigations, at higher pressures, have shown the predominance of the ion–molecule reactions.

$$CH_4^+ + CH_4 \longrightarrow CH_5^+ + \cdot CH_3$$
$$CH_3^+ + CH_4 \longrightarrow C_2H_5^+ + H_2$$

At pressures of about 0.2–2.0 mm the CH_5^+ and $C_2H_5^+$ account for about 70% of the ion current, and ions up to C_7 have been detected.[26, 27] That these ions are branched, as well as straight-chained, is indicated by the identification of isobutane and isopentane among the radiolysis products of

(23) E. I. Rabinowitch, "Photosynthesis," Vol. I, Interscience Publishers, New York, N. Y., 1945, p 82.
(24) W. M. Garrison, et al., Science, 114, 416 (1951).
(25) S. L. Miller, ibid., 117, 528 (1953).

(26) F. H. Field and M. S. B. Munson, J. Am. Chem. Soc., 87, 3289 (1965).
(27) S. Wexler and N. Jesse, ibid., 84, 3425 (1962).

methane.[28] The higher molecular weight ions and radicals arise, at least in part, from such additional reactions as

$$C_2H_5^+ + CH_4 \longrightarrow C_3H_7^+ + H_2$$

$$C_2H_5^+ + e^- \longrightarrow \cdot C_2H_5$$

The G values (molecules produced per 100 eV of absorbed energy) for the principal products observed on methane's radiolysis are[29]

Product	G	Product	G
H_2	6.4	$n\text{-}C_4H_{10}$	0.13
C_2H_4	0.13	$i\text{-}C_4H_{10}$	0.06
C_2H_6	2.1	$i\text{-}C_5H_{10}$	0.05
C_3H_8	0.26		

The prominent ethylene product (C_2H_4) is assumed to arise from

$$C_2H_5^+ + e^- \longrightarrow C_2H_4 + H\cdot$$

and

$$2CH_2 \longrightarrow C_2H_4$$

The ethylene may account for another route to higher molecular weight products

$$R\cdot + C_2H_4 \longrightarrow RCH_2CH_2\cdot \xrightarrow{C_2H_4} R(CH_2)_4\cdot, \text{ etc.}$$

where $R\cdot$ is a radical or a hydrogen atom.

Although acetylene has not been reported as a product of the radiolysis of pure methane, it has been observed when I_2 is present,[29] and $C_2H_2^+$ ions have been observed in the mass spectrum of pure methane.[26] The polymerization of acetylene is presumed to be the route to the benzene ring of the aromatic amino acids, phenylalanine and tyrosine, that have been reported as products of methane–ammonia–water radiolyses. A number of arenes, including benzene, naphthalene, acenaphthalene, phenanthracene, and pyrene, have appeared when CH_4 is passed over silica gel at 1000°.[30]

2. Ammonia

Mass spectrometric studies of gaseous ammonia have shown that NH_2^+, NH_3^+, and NH_4^+ are the principal ions produced.[31,32] The only molecular products are N_2, H_2, and N_2H_4 (hydrazine). The appearance of the hydrazine indicates the production of NH_2 radicals, as does mass spectral observations, indicating that the reaction $NH_3^+ + NH_3 \rightarrow NH_4^+ + NH_2$ occurs with high efficiency.[31] The NH_2 radical may be the source of the amino group of the amino acids that are formed in primitive-Earth atmospheres. That it is important in the formation of the nucleic acid base, adenine, on irradiation of such atmospheres is indicated by a decreased adenine yield with increases of H_2 in the CH_4–NH_3–H_2O–H_2 mixture.[33] The presence of H_2 probably decreases the NH_2 yield by the reaction: $NH_2 + H_2 \rightarrow NH_3 + H$.

3. Water

More radiation chemical studies have been devoted to water than to any other single compound. This is due not only to its importance in radiobiology. It has attracted the attention of the radiation chemists because highly purified samples are easily obtainable and because its relevant physical properties, such as bond angles and lengths, moment of inertia, dielectric constant, etc., are so well known; accurate knowledge of these properties is most useful in formulating radiolysis mechanisms.

Mass spectrometric studies on water have shown that the principal positive ions formed are H_2O^+, H_3O^+, OH^+, and H^+ (the relative intensities are approximately 5:1:1:1). A great variety of studies, of which the principal ones are electron spin resonance spectroscopy, pulse radiolysis accompanied by fast absorption spectroscopy, and addition of radical scavengers, have demonstrated that the principal reactive species in irradiated water are the $\cdot OH$ radical, the $\cdot H$ atom, and the hydrated electron (e_{aq}^-). The principal molecular products are H_2 and H_2O_2, resulting from combinations of $\cdot H$ atoms and $\cdot OH$ radicals. All of the effects of the participation of water in the radiolysis of methane–ammonia–water systems are interpretable on the basis of the participation of one of these reactive species ($\cdot H$, $\cdot OH$, e_{aq}^-, H_2, H_2O_2, or of one of the ions listed above).[34]

B. AMINO ACIDS

It is the great successes in the abiogenic syntheses of the amino acids, beginning with Miller's 1953 experiment, that have made chemical evolution such an attractive area of research. There are several reasons why amino acid syntheses have been so prominent. First, they are the constituent units (monomers) of the proteins, which, with the nucleic acids, are the supremely important biopolymers. Secondly, present data indicate that they form more readily (from CH_4–NH_3–H_2O mixtures) than any of the other biomonomers. Finally, very powerful and sensitive techniques exist for amino acid detection and analysis—commercial "Amino Acid Analyzers," paper chromatography, and very sensitive color tests (Ninhydrin).

For the present discussion, by "amino acids" we are limiting ourselves to the α-amino acids. All of the 20 common amino acids of the natural proteins have their amino group bonded to the α carbon, the carbon that is also bonded to the carboxyl group. It is interesting to note that in all chemical evolution experiments where their yields are reported, α-alanine is formed in much higher yield than is β-alanine.[35-38] The characteristic formation of the amino acids seems also reflected in the preferred formation of α-aminonitriles, as reported by Ponnamperuma and Woeller.[39] It seems very likely that, when life got started on Earth, the α-amino acids were the commonest type available for protein construction. Organisms that developed that could metabolize α-amino acids would be the ones most likely to survive.

At least four mechanisms, or routes, have been proposed to account for the appearance of amino acids in primitive-Earth experiments.

(28) K. Yang and P. J. Manno, *J. Am. Chem. Soc.*, **81**, 3507 (1959).
(29) G. G. Meisels, W. H. Hamill, and R. R. Williams, *J. Phys. Chem.*, **61**, 1456 (1957).
(30) J. Oró and J. Han, *Science*, **153**, 1393 (1966).
(31) L. M. Dorfman and P. C. Noble, *J. Phys. Chem.*, **63**, 980 (1959).
(32) B. P. Burtt and A. B. Zahlan, *J. Chem. Phys.*, **26**, 846 (1957).
(33) C. Ponnamperuma, R. M. Lemmon, R. Mariner, and M. Calvin, *Proc. Natl. Acad. Sci. U. S.*, **49**, 737 (1963).

(34) F. S. Dainton, *et al.*, in "Radiation Research," G. Silini, Ed., North Holland Publishing Co., Amsterdam, 1957 pp 161–262.
(35) S. L. Miller, *J. Am. Chem. Soc.*, **77**, 2351 (1955).
(36) S. L. Miller, *Ann. N. Y. Acad. Sci.*, **69**, 260 (1957).
(37) K. Harada and S. W. Fox, *Nature*, **201**, 335 (1964).
(38) C. U. Lowe, M. W. Rees, and R. Markham, *ibid.*, **199**, 219 (1963).
(39) C. Ponnamperuma and F. H. Woeller, *Currents Modern Biol.*, **1**, 156 (1967).

(a) The cyanohydrin mechanism, invoked by Miller[36] to explain his amino acid products

$$RCHO \xrightarrow{\text{NH}_3,\ \text{HCN}} RCH(NH_2)CN \xrightarrow{\text{H}_2\text{O}} RCH(NH_2)CO_2H$$

This mechanism is reinforced by the fact that aldehydes and HCN are known products in Miller's system. It also explains the predominance of the α-amino acids.

(b) Since electric discharges in anhydrous methane–ammonia mixtures cause the formation of α-aminonitriles,[39] the intermediate aldehyde formation may not be necessary.

$$CH_4 + NH_3 \longrightarrow H_2NCH_2CN,\ H_3CCH(NH_2)CN \xrightarrow{\text{H}_2\text{O}}$$
$$H_2NCH_2CO_2H,\ H_3CCH(NH_2)CO_2H$$

(c) Sanchez, et al.,[40] have suggested a possibly important role for cyanoacetylene (a product of CH_4–N_2 irradiations) in amino acid syntheses. In the presence of NH_3 and HCN, this compound forms considerable asparagine and aspartic acid (ca. 10% yields); the suggested reactions are

$$NCC{\equiv}CH + NH_3 \longrightarrow NCCH{=}CHNH_2 \xrightarrow{\text{HCN}}$$
$$NCCH_2CH(NH_2)CN \xrightarrow{\text{H}_2\text{O}} HO_2CCH_2CH(NH_2)CONH_2 \xrightarrow{\text{H}_2\text{O}}$$
$$\text{asparagine}$$
$$HO_2CCH_2CH(NH_2)CO_2H$$
$$\text{aspartic acid}$$

(d) Abelson,[17] Matthews, Claggett, and Moser,[41–44] and Harada[45] have emphasized a possible key role of HCN oligomers, produced by the base-catalyzed polymerization of HCN. They have found, for example, that the HCN trimer, aminoacetonitrile, and the HCN tetramer, diaminomaleonitrile, give, on heating in water for 24 hr at 100°, as many as 12 of the 20 α-amino acids commonly found in proteins.[43,44] Abelson reports that uv light (2536 Å) speeds HCN polymerization and that, on hydrolysis, glycine, alanine, serine, aspartic acid, and glutamic acid are found.[17]

With the data now available, it is impossible to decide which of the above routes was the most important path of amino acid formation in abiogenic syntheses. Probably all of them contributed to some degree.

Table II[46–54] is a summary of the amino acids whose syntheses have been reported during the past 16 years in "primitive Earth atmosphere" experiments. Yield data are omitted.

However, in all primitive-Earth experiments yields are low, rarely exceeding 5% of the starting carbon (e.g., methane) and usually below 1%. The table reflects some subjective judgment about what is, or isn't, a "primitive Earth" experiment. The table omits reports where "amino acid products" are indicated merely on the basis of ninhydrin-positive tests, or approximate chromatographic positions of the products. We have included only those experiments in which individual amino acid products were firmly established, taking into account the need for control experiments to eliminate bacterial and other contamination. In general, the experiments recorded in Table II were performed on CH_4–NH_3–H_2O–H_2 mixtures; the exceptions are noted. In some cases, the H_2 was omitted; in others, varying amounts of CO, CO_2, or N_2 were added. In all cases the mixture was more reduced than oxidized.

The data of Table II leave no doubt that amino acids, at least those up to six carbons (leucine and lysine), are formed in "reduced atmosphere" experiments. Higher molecular weight amino acids (phenylalanine and tyrosine) have been reported only on heating the CH_4–NH_3–H_2O–H_2 mixtures to about 1000°. Whether this temperature can be considered "primitive Earth" conditions, and how widespread the locale(s) of such temperatures could have been, has been "warmly" debated, and the reader is referred to the written record of one such debate, following the presentation of a paper at the conference held at Wakulla Springs, Fla., in 1963.[55]

In addition to the amino acids whose formation from "primitive-Earth atmospheres" is recorded in Table II, there are many reports in the literature of their formation from the sorts of organic compounds known (e.g., HCHO and HCN) or expected (e.g., N-acetylglycine) to accumulate in the primitive atmosphere and oceans (we shall call these "primitive-Earth compounds"). Many researchers have studied what could be called the secondary formation of amino acids from these compounds. That work is summarized in Table III.[56–66]

C. NUCLEIC ACID COMPONENTS

In reviewing the progress made in the abiogenic syntheses of the nucleic acid monomers (the nucleotides) we shall consider, in turn, synthesis of the purine and pyrimidine bases, the sugars, the nucleosides, and the nucleotides themselves.

1. Purines and Pyrimidines

Compared to the successes in demonstrating amino acid syntheses under primitive-Earth conditions, far less has been achieved in the syntheses of these nucleic acid bases. The more

(40) R. A. Sanchez, J. P. Ferris, and L. E. Orgel, Science, 154, 784 (1966).
(41) C. N. Matthews and R. E. Moser, Proc. Natl. Acad. Sci. U. S., 56, 1087 (1966).
(42) C. N. Matthews and R. E. Moser, Nature, 215, 1230 (1967).
(43) R. E. Moser, A. R. Claggett, and C. N. Matthews, Tetrahedron Lett., 13, 1599 (1968).
(44) R. E. Moser, A. R. Claggett, and C. N. Matthews, ibid., 13, 1605 (1968).
(45) K. Harada, Nature, 214, 479 (1967).
(46) T. E. Pavlovskaya and A. G. Pasynskii in "The Origin of Life on Earth," Proceedings of the 1st International Symposium, Moscow, 1957, Pergamon Press, New York, N. Y., 1959, p 151.
(47) J. Oró, Nature, 197, 862 (1963).
(48) C. Palm and M. Calvin, J. Am. Chem. Soc., 84, 2115 (1962).
(49) K. A. Grossenbacher and C. A. Knight in "The Origins of Pre-Biological Systems," Academic Press, New York, N. Y., 1965, p 173.
(50) W. E. Groth and H. von Weyssenhoff, Planet. Space Sci., 2, 79 (1960).
(51) P. H. Abelson, Carnegie Inst. Wash. Yearbook, 56, 171 (1955/56).
(52) K. Heyns, W. Walter, and E. Meyer, Naturwissenschaften, 44, 385 (1957).
(53) A. S. U. Choughuley and R. M. Lemmon, Nature, 210, 628 (1966).
(54) J. Oró, in ref 49, p 137.

(55) K. Harada and S. W. Fox in ref 49, p 187.
(56) K. Bahadur, Nature, 173, 1141 (1954).
(57) K. Dose and C. Ponnamperuma, Radiation Res., 31, 650 (1967).
(58) T. Hasselstrom, M. C. Henry, and B. Murr, Science, 125, 350 (1957).
(59) C. Reid in ref 46, p 619.
(60) S. W. Fox, Science, 132, 200 (1960),
(61) J. Oró, A. Kimball, R. Fritz, and F. Master, Arch. Biochem. Biophys., 85, 115 (1959).
(62) J. Oró and A. P. Kimball, ibid., 96, 293 (1962).
(63) R. Paschke, R. W. H. Chang, and D. Young, Science, 125, 881 (1957).
(64) K. Heyns and K. Pavel, Naturforsch., 12b, 97 (1957).
(65) N. Getoff and G. O. Schenck in "Radiation Chemistry, Vol. I," Advances in Chemistry Series, No. 81, American Chemical Society, Washington, D. C., 1968, p 337.
(66) G. Steinman, E. A. Smith, and J. J. Silver, Science, 159, 1108 (1968).

Table II

Production of Amino Acids in Simulated "Primitive-Earth Atmosphere" Experiments (1953–1968)

Amino acid formed	Input energy[a]	Product identification[b]	Lit ref	Amino acid formed	Input energy[a]	Product identification[b]	Lit ref
Glycine	A	I	35	α-Aminobutyric	A	I	35
	A	II	46	acid	A	II	46
	A	II	47[e]		B	II	51[f]
	C	I	48		D	II	37
	A	II	49		A	II	49
	B	II	50[c]	N-Methylalanine	A	I	35
	A	II	51[f]	Asparagine	A	II	47[e]
	A	II	52	Aspartic acid	A	I	35
	D	II	37		C	I	48
Alanine	A	I	35		A	II	46
	C	I	48		A	II	47[e]
	A	II	46		A	II	49
	B	II	50[c]		D	II	37
	A	II	52	Glutamic acid	A	I	35
	D	II	37		A	II	46
	A	II	47[e]		A	II	49
	A	II	49		D	II	37
	A	II	51[f]	Valine	D	II	37
β-Alanine	A	I	35	Leucine	D	II	37
	A	II	52		D	II	54
	A	II	46		A	II	49
	A	II	51[f]	Isoleucine	D	II	37
Cysteic acid	A	I	53[d]		D	II	54
N-Methylglycine	A	I	35		A	II	49
(sarcosine)	D	II	37	Alloisoleucine	D	II	37
	A	II	52		D	II	54
	A	II	51	Lysine	A	II	46
Serine	A	II	49		A	II	49
	D	II	37	Phenylalanine	D	II	37
Threonine	A	II	49		D	II	54
	D	II	37	Tyrosine	D	II	37
					D	II	54

[a] A = electric discharge (spark or corona), B = uv radiation, C = ionizing radiation (γ-rays or electrons), D = heat (about 1000°). [b] (I) Product identification very secure; for example, absolute chromatographic coincidence between labeled product (from $^{14}CH_4$) and carrier color or uv absorption, or macro amount of product obtained and physical properties determined. (II) Product identification less secure; usually based on chromatographic R_f values, or elution volumes from ion-exchange columns. [c] Ethane was substituted for methane; with the latter, no amino acid products were detected. [d] H_2S was added to the usual CH_4–NH_3–H_2O mixture. [e] Ethane was added to a CH_4–NH_3–H_2O–H_2 mixture. [f] Experiments on CO_2–N_2–H_2–H_2O mixtures.

complex structures of the latter indicate why this should be expected. The two purines (adenine and guanine) and the

purine   pyrimidine

three pyrimidines (cytosine, thymine, and uracil) of the nucleic acids are formed by the attachments of NH_2, OH, and CH_3 groups to the above basic structures, and the general result is a higher degree of molecular architecture, or specificity, than is exhibited in the amino acid series.

Adenine is the only one of the five nucleic acid bases that has been synthesized in a "primitive-Earth atmosphere" experiment. In 1963, Ponnamperuma, *et al.*, found adenine after irradiating a CH_4–NH_3–H_2O–H_2 mixture with an electron beam.[33] The yield of adenine from starting methane was very small (0.01 %), and none of the other nucleic acid bases were detected. However, considering the complexities of the adenine molecule and the random, radiolytic processes that led to its formation, one can easily believe that this yield may be highly significant for our understanding of chemical evolution. Of further interest was the fact that the adenine yields increased as the H_2 in the starting gas mixture was decreased, a result that may hint at the absence of significant purine synthesis on the primitive Earth until most of the hydrogen was gone. This H_2 effect is not surprising since methane carbon must be oxidized in order to appear finally in a purine. In the adenine synthesis, the principal species effecting the oxidations are probably OH and NH_2 radicals, and these would revert to the starting materials (water and ammonia) on reaction with hydrogen.

Table III

Formation of Amino Acids from Primitive-Earth Compounds (1953–1968)

Reactants	Input energy	Amino acids reported[a]	Lit ref
1. Paraformaldehyde, H_2O. NO_3^-, Fe^{+3}	Sunlight	Gly, Ala, Asp, Val, His, Pro, Lys, Ser, Asp, Arg	56
2. N-Acetylglycine, H_2O, NH_3	γ-Rays	Gly, Asp, Thr	57
3. NH_4Ac, H_2O	e^- beam	Gly, Asp	58
4. H_2NOH, HCHO, CO_2	Uv	Gly, Ala	59
5. HCHO, NH_4Cl, NH_4NO_3	Uv	Ser, Gly, Ala, Glu, Val, Ile, Phe	46
6. HCN (at pH 8–9), followed by hydrolysis	Uv	Gly, Ala, Ser, Asp, Glu	17
7. Formamide	Pyrolysis at 250°, followed by hydrolysis of product	Gly, Ala, Asp, Ser, Thr, Val, Glu, Leu	47
8. Glucose, urea	150–200° hydrolysis of product	Gly	60
9. Malic acid, urea	As above	Asp	60
10. Hydroxyglutamic acid, NH_3	As above	Glu	60
11. HCN, NH_3, H_2O	90°	Asp, Thr, Ser, Glu, Gly, Ala, Ile, Leu, β-Ala, Abu	38
12. HCHO, H_2NOH, H_2O	80–100°	Gly, Ala, β-Ala, Ser, Thr, Asp	61
13. HCN, NH_3, H_2O	27–100°	Gly, Ala, Asp	62
14. $HC\equiv CCN$, NH_3, H_2O, HCN	100°	Asp, $AspNH_2$	40
15. $(NH_4)_2CO_2$ (solid)	γ-Rays	Gly	63
16. Glycine (on quartz)	260–280	Ala, Asp	64
17. $EtNH_2$, HCO_3^-, S^{2-}	γ-Rays	Cys, $(Cys)_2$	65, 66

[a] Ala = alanine, β-Ala = β-alanine, Abu = α-aminobutyric acid, Arg = arginine, Asp = aspartic acid, $AspNH_2$ = asparagine, Cys = cysteine, $(Cys)_2$ = cystine, Gly = glycine, Glu = glutamic acid, His = histidine, Ile = isoleucine, Leu = leucine, Lys = lysine, Pro = proline, Phe = phenylalanine, Ser = serine, Thr = threonine, Val = valine.

$$\cdot OH + H_2 \longrightarrow H_2O + H\cdot$$
$$\cdot NH_2 + H_2 \longrightarrow NH_3 + H\cdot$$

The first reaction is energetically favored since the H–H bond energy is 104.2 kcal and the H–OH bond energy is 119 kcal.[67] The second is slightly unfavored (H–NH_2 bond energy is 103 kcal)[67] and may not occur. Another way in which H_2 may interfere with the production of purines would be through the back reaction $H_2 + \cdot CH_3 \rightarrow CH_4 + \cdot H$. Here, the bond energies are very similar: H–CH_3 (104.0 kcal), H–H (104.2 kcal).[67]

The mechanism of the formation of adenine from the CH_4–NH_3–H_2O mixture is unknown. However, it may be noted that adenine may be considered as a pentamer of HCN, i.e., $(HCN)_5$, and that hydrogen cyanide was a prominent product of Miller's experiment.[85] Furthermore, the formation of adenine from HCN in basic aqueous solution (see below) appears to proceed through a trimer and a tetramer of HCN.

None of the other nucleic acid bases (guanine, cytosine, thymine, or uracil) has been detected after the irradiation of CH_4–NH_3–H_2O mixtures. The apparent preference for adenine synthesis may be related to multiple roles of adenine in biological systems. Not only is it a constituent of both DNA and RNA, but it is also present in many important cofactors, for example, ATP, ADP, DPN, TPN, FAD, and coenzyme A. In addition, molecular orbital calculations have shown that, of all the biologically important purines and pyrimidines,

adenine has the greatest resonance energy.[68,69] This would not only make synthesis of adenine more likely but would, in addition, confer radiation stability upon it.

The first "primitive-Earth" synthesis of adenine was reported by Oró in 1960;[70] this work was elaborated in further publications[62,71] and confirmed by Lowe, et al.[88] Adenine, some amino acids, and a variety of other nitrogen-containing compounds are found when aqueous ammonium cyanide solutions are heated at about 90° for several days. Oró and Orgel and their coworkers have studied the mechanism of this most interesting condensation and have presented evidence that it proceeds through aminomalononitrile (an HCN trimer) and diaminomaleonitrile or diaminofumaronitrile (both HCN tetramers).[71-74] The probable reaction sequence is shown in eq 1. All of the compounds in the above sequence, with the exception of the HCN dimer, have been identified in the aqueous solution chemistry of HCN. Evidence for the existence of the dimer in the (presumably) lower energy form of its radical tautomer (aminocyanocarbene), as well as evi-

(67) J. A. Kerr and A. F. Trotman-Dickenson, "Handbook of Chemistry and Physics," 48th ed, Chemical Rubber Publishing Co., Cleveland, Ohio, 1967–1968, pp F-149–F-154.

(68) B. Pullman and A. Pullman, *Nature*, **196**, 1137 (1962).
(69) B. Pullman and A. Pullman in "Comparative Effects of Radiation," John Wiley and Sons, Inc., New York, N. Y., 1960, pp 111–112.
(70) J. Oró, *Biochem. Biophys. Res. Commun.*, **2**, 407 (1960).
(71) J. Oró and A. P. Kimball, *Arch. Biochem. Biophys.*, **94**, 217 (1961).
(72) J. P. Ferris and L. E. Orgel, *J. Am. Chem. Soc.*, **88**, 3829 (1966).
(73) R. A. Sanchez, J. P. Ferris, and L. E. Orgel, *J. Mol. Biol.*, **30**, 223 (1967).
(74) R. A. Sanchez, J. P. Ferris, and L. E. Orgel, *ibid.*, **38**, 121 (1968).

$$2HCN \longrightarrow [HN{=}CHCN] \xrightarrow{HCN} H_2NCH(CN)_2 \xrightarrow{HCN}$$

dimer   aminomalononitrile (trimer)

$$
\begin{array}{cc}
\underset{NC}{\overset{H_2N}{>}}C{=}C\underset{CN}{\overset{NH_2}{<}} & \underset{NC}{\overset{H_2N}{>}}C{=}C\underset{NH_2}{\overset{CN}{<}} \xrightarrow[\text{formamidine}]{HN=CHNH_2}
\end{array}
$$

diaminomaleonitrile (cis tetramer)   diaminofumaronitrile (trans tetramer)

$$
\underset{\substack{H_2N}}{\overset{NH_2}{HN}}{\diagup}C\diagdown N \xrightarrow[\text{formamidine}]{HN=CHNH_2} \text{adenine} \qquad (1)
$$

4-aminoimidazole-5-carboxamidine   adenine

dence for the spin ground state of the latter, have been presented in a recent publication.[75]

Both adenine and the other nucleic acid purine, guanine, have been synthesized by shining ultraviolet light on dilute solutions of HCN.[76] These two purines have also been reported as products of the reaction of aqueous solutions of HCN or cyanogen with 4-amino-5-cyanoimidazole.[73,74,77]

$$
\underset{H_2N}{\overset{NC}{>}}\diagup\underset{H}{\overset{N}{\diagdown}}N
$$

This compound is also a product that is formed in aqueous solutions of HCN and NH₃.[73] Upon reaction with ammonia it gives the 4-aminoimidazole-5-carboxamidine of the reaction sequence above.

Of the three nucleic acid pyrimidines, cytosine, uracil, and thymine, the first two have been prepared under more or less abiogenic conditions, but thymine has never been reported as a product of such experiments. Uracil has been prepared by heating (130°) malic acid, urea, and polyphosphoric acid.[78] This is a doubtful primitive-Earth experiment because of the temperature, the malic acid (which has never been reported as a product of abiogenic synthesis), and the polyphosphoric acid (a compound whose existence in any watery environment would be transitory). Oró has also recorded the appearance of uracil under conditions he considers relevant to chemical evolution: heating (135°) urea with acrylonitrile (or β-aminopropionitrile, or β-aminopropionamide) in aqueous solution.[54]

The only reported abiogenetic synthesis of cytosine is that of Sanchez, et al.[40,79] They found a 5% yield of this pyrimidine upon heating 0.1 M aqueous cyanoacetylene (which they report as a product of the sparking of a mixture of CH₄ and N₂) with 1.0 M KCNO for 1 day at 100°. These authors also report that cytosine is obtained in a 1% yield when the same solution of the same reactants is allowed to stand at room temperature for 7 days.

Other purines, not normally found in nucleic acids, are also reported to have been synthesized under abiogenic conditions. These include hypoxanthine, diaminopurine, and xanthine.[77]

The abiogenic syntheses of the nucleic acid purines and pyrimidines are summarized in Table IV.

Table IV

Synthesis of Nucleic Acid Purines and Pyrimidines under Primitive-Earth Conditions

Compound	Experimental conditions	Lit ref
Adenine	Aq NH₄CN, HCN, 90°, 1 day	38, 70
	CH₄—NH₃–H₂O; e⁻ beam	33
	Aq HCN, uv	76
	Aq diaminomaleonitrile + HCN, uv, 25°	73, 77
Guanine	Aq HCN, uv	76
	Aq diaminomaleonitrile + C₂N₂, uv, 100°	73, 77
Uracil	Malic acid, urea, H₄P₂O₇, 130°	78
	Acrylonitrile, urea, H₂O, 135°	54
Cytosine	HC≡CCN, KCNO, H₂O, 100°	79

Thymine No recorded "primitive-Earth" synthesis

2. Sugars

To some extent, the problem of the abiogenic synthesis of the common hexoses and pentoses was solved long before the present interest in chemical evolution experiments. Over a century ago, the Russian chemist Butlerov showed that dilute aqueous alkali causes formaldehyde to condense to a complex mixture of sugars.[80] Miller's early work produced evidence, not only for the production of formaldehyde on sparking the CH₄–NH₃–H₂O–H₂ mixture, but also ("possibly") for the appearance of sugars.[14] Consequently, the abiogenic synthesis of sugars on the primitive Earth is easy to visualize, although no one has yet established a specific sugar as a product of CH₄–NH₃–H₂O irradiations. Since we are here focusing our attention on the problem of nucleic acid constituents, it is of considerable interest that the sugars (ribose and deoxyribose) of those biopolymers have been reported as products of the ultraviolet irradiation of formaldehyde.[76] In addition, a 3.8% yield of ribose has been reported to appear on the refluxing of 0.01 M HCHO over kaolinite (hydrated aluminum silicate).[81]

3. Nucleosides

The nucleosides represent a major gap in the synthesis of biologically important intermediates under primitive-Earth conditions. Experiments performed in 1963–1964 indicated the formations of adenosine and deoxyadenosine when 2537-Å light was shown on dilute (about 10⁻³ M in each reactant) solutions of adenosine, ribose (or deoxyribose), and NH₄–H₂PO₄ or NaCN.[82,83] Later work, however, showed that the products were not the natural nucleosides, but closely related (chemically and chromatographically) adenine–sugar adducts.[84] In addition, substitution of adenine by any one of the

(75) R. E. Moser, J. M. Fritsch, T. L. Westman, R. M Kliss, and C. N. Matthews, *J. Am. Chem. Soc.*, **89**, 5673 (1967).
(76) C. Ponnamperuma, in ref 49, p 221.
(77) R. Sanchez, J. Ferris, and L. E. Orgel, *Science*, **153**, 72 (1966).
(78) S. W. Fox and K. Harada, *ibid.*, **133**, 1923 (1961).
(79) J. P. Ferris, R. A. Sanchez, and L. E. Orgel, *J. Mol. Biol.*, **33**, 693 (1968).

(80) A. Butlerov, *Ann.*, **120**, 296 (1861).
(81) N. W. Gabel and C. Ponnamperuma, *Nature*, **216**, 453 (1967).
(82) C. Ponnamperuma, C. Sagan, and R. Mariner, *ibid.*, **199**, 222 (1963).
(83) C. Ponnamperuma and P. Kirk, *ibid.*, **203**, 400 (1964).
(84) C. Reid, L. E. Orgel, and C. Ponnamperuma, *ibid.*, **216**, 936 (1967)

other four nucleic acid bases led to no detectable nucleoside formation.[85]

4. Nucleotides

In considering the formation of the nucleotides (the base–sugar–phosphate, or monomeric, unit of the nucleic acids), we should first consider the state of the element phosphorus on the primitive Earth. In any watery environment the predominant chemical form of phosphorus is phosphoric acid; the lower oxidation states of phosphorus are unstable except under higher H_2 pressures than are reasonable for a primitive-Earth atmosphere.[86] With respect to chemical evolution, the main problem is not the chemical form of the phosphorus, but rather the very low solubility of the alkaline earth phosphates. There would have been, just as today, very little phosphate dissolved in the ocean. However, the phosphate-incorporating reactions may have taken place at the surface of phosphate minerals. The adsorption of dissolved organic compounds onto the surface of such minerals, and their ensuing phosphorylation at those sites, may have been key steps in chemical evolution.

In the work quoted above regarding adenosine formation, the same workers found no detectable adenosine phosphate (adenylic acid) when adenosine was treated with phosphoric acid and ultraviolet light.[82] They did show that when "polyphosphate ester" (a complex mixture of compounds formed by reacting P_2O_5 with ethyl ether[87]) is substituted for the phosphoric acid, adenylic acid was formed. They also detected the synthesis of adenosine diphosphate (ADP) and adenosine triphosphate (ATP); the latter compound is the chief storage depot for energy, and chief supplier of energy, for biological processes. Unfortunately, the "polyphosphate ester" seems a very unlikely primitive-Earth compound (for one thing, it is very quickly hydrolyzed), but arguments have been advanced in favor of its possible existence on the early Earth.[87]

In subsequent work, Ponnamperuma and Mack heated separately (in the absence of water) the five nucleic acid nucleosides (adenosine, guanosine, cytidine, uridine, and thymidine) with sodium dihydrogen phosphate (NaH_2PO_4) at 160° for 2 hr.[88] They demonstrated the formation of the phosphates (nucleotides) of each of the nucleosides, many of the individual monophosphates (2'-, 3'-, 2',3'-cyclic, and 5'-phosphates), a dinucleoside phosphate (UpU), and a dinucleotide (UpUp). They also showed appreciable yields at temperatures as low as 80°, and that the reactions would take place in the presence of small amounts of water. Evidence has been presented that these phosphorylations are effected by a prior formation of pyrophosphate (and higher condensed phosphates).[89]

Waehneldt and Fox have recently demonstrated the phosphorylation (on the hydroxyl group of the pentose moiety) of adenosine, cytidine, deoxycytidine, guanosine, uridine, and thymidine.[90] They employed temperatures from 0 to 22°, and their phosphorylating reagent was polyphosphoric acid; the latter compound may well have existed on the primitive Earth, both for the reason given above and because its principal constituent, pyrophosphoric acid, is formed in dilute aqueous solution by the action of cyanamide (a compound that results from CH_4–NH_3–H_2O irradiations[91]) on phosphoric acid.[92] Pyrophosphate has also been prepared by Miller and Parris by the reaction of potassium cyanate, KCNO, with hydroxyl apatite, $Ca_{10}(PO_4)_6(OH)_2$ (this is the Earth's commonest phosphorus-containing mineral).[93] From the standpoint of chemical evolution, the work of Waehneldt and Fox, with its low temperatures and high yields (25–45%) of nucleoside phosphates, is particularly attractive.

Orgel, et al., have recently reported the formation of uridine 5'-phosphate, uridine 2'(3')-phosphate, and some uridine diphosphate on heating (65–85°) uridine with inorganic phosphates for 9 months.[94] The most effective inorganic phosphates in promoting this reaction were $Ca(H_2PO_4)_2$ and $(NH_4)_2HPO_4$. The authors suggest that of the two, the ammonium hydrogen phosphate is the more likely to have played a part in chemical evolution. $Ca(H_2PO_4)_2$ is precipitated only from acid solution, whereas in an ocean containing substantial amounts of ammonia the $(NH_4)_2HPO_4$ may have formed on the evaporation of shallow pools. Further work from the same laboratory has shown that approximately 1–4% yields of uridine 5'-phosphate are achieved by heating aqueous uridine (0.1 M) with $H_2PO_4^-$ (~1 M, pH 5–8) at 95° for 5 hr in the presence of a variety of cyanide derivatives (~1 M). Of the latter compounds, results of this work led the authors to believe that cyanogen (and its partial-hydrolysis products, cyanformamide and cyanate), cyanamide, and cyanamide dimer (dicyandiamide) were the most effective in producing these condensations.[95] (The latter two compounds are discussed later in this report under "Abiogenic Synthesis of Biopolymers.") The presence of cyanamide dimer (0.01 M, aqueous solution) has been found to produce a small (<1%) yield of adenosine 5'-phosphate when adenosine (0.01 M) and H_3PO_4 (0.01 M) are allowed to stand at room temperature for 4.5 hr.[96] The only difficulty about this work, with respect to chemical evolution, is the necessity of a low pH (2–3); at pH 7 the nucleotide formation is not observed.

D. FATS

The fatty (i.e., aliphatic or carboxylic) acids are another class of compounds whose appearance on the primitive Earth is easily imagined. Experiments unrelated to the present interest in chemical evolution have demonstrated the formation of a wide variety of hydrocarbons, with molecular weights in the hundreds, on the passage of ionizing radiation through methane.[97,98] In addition, it has been shown that ionizing radiation effects the direct addition of CO_2 to a hydrocarbon to form the corresponding fatty acid ($RH + CO_2 \rightsquigarrow RCO_2$-

(85) C. Ponnamperuma, private communication.
(86) N. H. Horowitz and S. L. Miller, Progr. Chem. Org. Nat. Prod., 20, 453 (1962).
(87) G. Schramm, H. Grötsch, and W. Pollmann, Angew. Chem. Intern. Ed. Engl., 1, 1 (1962).
(88) C. Ponnamperuma and R. Mack, Science, 148, 1221 (1965).
(89) J. Rabinowitz, S. Chang, and C. Ponnamperuma, Nature, 218, 442 (1968).
(90) T. V. Waehneldt and S. W. Fox, Biochim. Biophys. Acta, 134, 1 (1967).

(91) A. Schimpl, R. M. Lemmon, and M. Calvin, Science, 147, 149 (1964).
(92) G. Steinman, R. M. Lemmon, and M. Calvin, Proc. Natl. Acad. Sci. U. S., 52, 27 (1964).
(93) S. L. Miller and M. Parris, Nature, 204, 1248 (1964).
(94) A. Beck, R. Lohrmann, and L. E. Orgel, Science, 157, 952 (1967).
(95) R. Lohrmann and L. E. Orgel, ibid., 161, 64 (1968).
(96) G. Steinman, R. M. Lemmon, and M. Calvin, ibid., 147, 1574 (1965).
(97) S. C. Lind and D. C. Bardwell, J. Am. Chem. Soc., 48, 2335 (1926).
(98) W. Mund and W. Koch, Bull. Soc. Chim. Belges, 34, 119 (1925).

H)[99] and to amines to form amino acids.[100] And it should be remarked here that, even in a predominantly reduced primitive atmosphere, the presence of some CO_2 would be expected.

Allen and Ponnamperuma have recently shown that exposure of methane and water to a semicorona discharge results in the formation of monocarboxylic acids from C_2 to C_{12}.[101] The authors identified acetic, propionic, isobutyric, butyric, isovaleric, valeric, and isocaproic acids, and they presented mass spectrographic evidence that their C_6–C_{12} acids were predominantly branched-chain.

Since the fats are esters of glycerol, it remains to be remarked that no one has yet reported glycerol as a product of primitive-Earth experiments. The compound is not as easily detected as most of the other biologically relevant compounds that we have been discussing, and this is probably the reason it has not yet been reported. It is a likely product of some future abiogenic (laboratory) synthesis that will be similar to the modern industrial synthetic route, starting with the C_3 hydrocarbon propylene, and proceeding through a series of hydrations and dehydrations.

E. PORPHYRINS

Another biologically very important class of compounds that need be considered are the porphyrins. Chlorophyll and heme are, respectively, magnesium and iron complexes (chelates) of substituted porphyrins. The porphyrins (or similar visible-light absorbing pigments) had to preexist the first alga. The established biosynthetic route to the porphyrins utilizes glycine and succinic acid (both known primitive-Earth compounds) and proceeds through a C_5 compound, δ-aminolevulinic acid.[102] The latter has been sought, but not found, as a product of CH_4–NH_3–H_2O irradiations.[103] In the biosynthetic pathway two molecules of δ-aminolevulinic acid are condensed to form a pyrrole, four of which, in turn, condense to form a porphyrin. Szutka has demonstrated the appearance of pyrrolic compounds on the uv irradiation of aqueous solutions of δ-aminolevulinic acid.[104] Hodgson and Baker have reported[105] the formation of metal porphyrin complexes on heating (84°) formaldehyde and pyrrole in the presence of cations; the yields, however, were very low (10^{-3}–10^{-5}%). A very recent report asserts that porphyrins are detectable (no yields are given) after the passage of electric discharges through CH_4–NH_3–H_2O mixtures.[106]

It has been pointed out that during the Earth's transition from a reduced to an oxidized atmosphere, the presence of the porphyrins may have been critical for the further progress of chemical evolution. The gradual building up of O_2 (brought about, before the appearance of plant photosynthesis, by the radiolysis of water in the Earth's upper atmosphere) would have led to the appearance of hydrogen peroxide; the latter, in turn, would have produced widespread oxidation (i.e., destruction) of organic compounds. However, Calvin has

pointed out that the incorporation of ferric ion into the porphyrin chelate heme increases, by a factor of 1000, the catalytic effectiveness of the iron's ability to destroy peroxide.[107]

F. SUMMARY

The most important organic molecules (biomonomers) in living systems have been enumerated as the 20 amino acids of the natural proteins, the 5 nucleic acid bases, glucose, ribose, and deoxyribose.[108] Of these, laboratory experiments under conditions clearly relevant to probable conditions on the primitive Earth have resulted in the appearance of at least 15 of the 20 amino acids, 4 of the 5 nucleic acid bases, and 2 of the 3 sugars. In addition, representatives of the biologically important nucleosides, nucleotides, fatty acids, and porphyrins have been observed. This research has made it clear that these compounds would have accumulated on the primitive (prebiotic) Earth—that their formation is the inevitable result of the action of available high energies on the Earth's early atmosphere.

V. Abiogenic Synthesis of Biopolymers

Granted a primitive ocean with a lot of biomonomers in solution, and adsorbed on mineral surfaces, there now arises the problem of how these compounds may have gotten condensed to form the biopolymers. In all cases the monomers are condensed with a concomitant elimination of water (dehydration condensation). These attachments are shown in Scheme I for the three major classes of biopolymers: proteins, nucleic acids, and polysaccharides. In all cases the elimination of water is indicated by the dotted rectangles.

Research in chemical evolution has used two methods, both with some success, in trying to accomplish these condensations under assumed primitive-Earth conditions. The first is to use the high-temperature, relatively anhydrous conditions (the volcanic-areas rationale). The second is to search for simple primitive-Earth compounds whose free energy, and selectivity of hydration, is such that they could promote the dehydration condensations, even in dilute aqueous solution. We shall review the progress toward a picture of the abiogenic synthesis of the biopolymers by separately considering the proteins and the nucleic acids. As yet, there is very little to report on polysaccharide abiogenic synthesis.

A. PROTEINS

As was mentioned earlier under "Other Conditions on the Primitive Earth," Fox and his coworkers have emphasized the possible role of high temperatures (around volcanoes) in promoting chemical evolution. They have done considerable research which has shown that dry mixtures of amino acids are condensed to protein-like material ("protenoids") by simple heating to temperatures of 150–200°; in the presence of polyphosphoric acid, temperatures below 100° are effective.[109–112] The presence of sufficient excess of one of

(99) B. C. McKusick, W. E. Mochel, and F. W. Stacey, *J. Am. Chem. Soc.*, **82**, 723 (1960).
(100) N. Getoff and G. O. Schenk, *Radiation Res.*, **31**, 486 (1967).
(101) W. V. Allen and C. Ponnamperuma, *Currents Mod. Biol.*, **1**, 24 (1967).
(102) D. Shemin, Proceedings of the 3rd International Congress on Biochemistry, Brussels, 1955, p 197.
(103) R. M. Lemmon, unpublished results.
(104) A. Szutka, *Nature*, **212**, 401 (1966).
(105) G. W. Hodgson and B. L. Baker, *ibid.*, **216**, 29 (1967).
(106) G. W. Hodgson and C. Ponnamperuma, *Proc. Natl. Acad. Sci. U. S.*, **59**, 22 (1968).

(107) M. Calvin, *Science*, **130**, 1170 (1960).
(108) G. Wald, *Proc. Natl. Acad. Sci. U. S.*, **52**, 595 (1964).
(109) S. W. Fox and K. Harada, *J. Am. Chem. Soc.*, **82**, 3745 (1960).
(110) S. W. Fox and S. Yuyama, *Ann. N. Y. Acad. Sci.*, **108**, 487 (1963).
(111) K. Harada and S. W. Fox in ref 49, p 289.
(112) S. W. Fox and T. V. Waehneldt, *Biochim. Biophys. Acta*, **160**, 246 (1968).

Scheme I

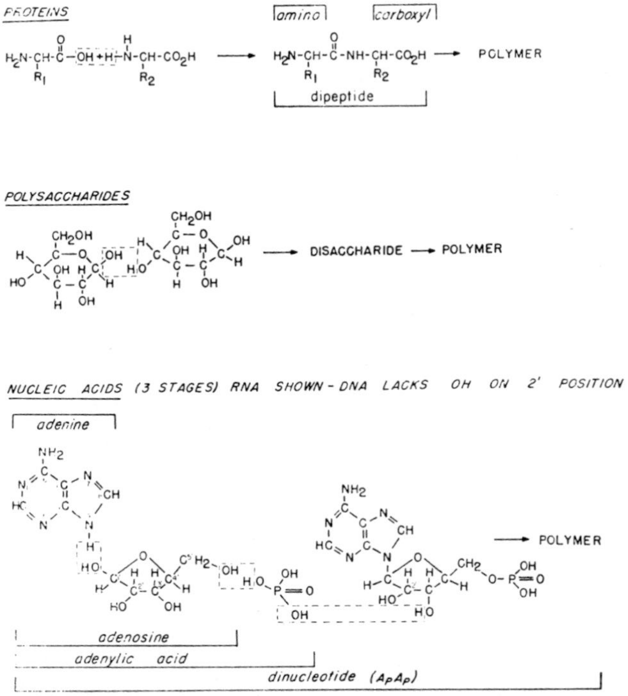

the dicarboxylic amino acids, glutamic acid or aspartic acid, is necessary to achieve these condensations. In promoting these condensations, a special, but unexplained, role for lysine is also claimed.[113] When glutamic and aspartic acids are in excess in mixtures containing essentially all the natural-protein amino acids, the resultant copolymer is found to incorporate all the starting amino acids. Glutamic acid is particularly effective in promoting these condensations, and it is presumed that its effectiveness lies in the thermal production of its lactam, 2-pyrrolidone-5-carboxylic acid, a five-membered cyclic anhydride

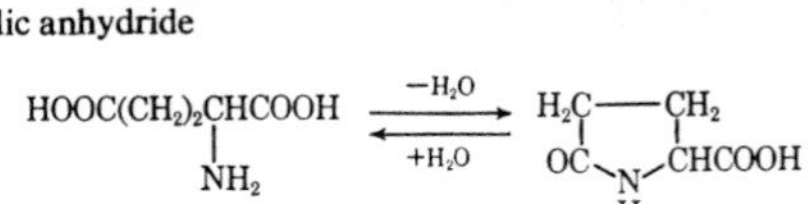

followed by the condensation of this lactam with an amino acid to give a mixed dipeptide.[114]

The amino acid thermal condensations are not completely random; aside from glutamic and aspartic acids, the incorporation of glycine, alanine, lysine, and methionine are slightly favored.[109] The amino acids lose their optical activity during the process of being built into the "protenoid."

Fox and coworkers call their products protein-like because (a) their infrared spectra show the typical bands of peptide bonds, (b) they have molecular weights in the thousands, (c) they give positive biuret reactions (color tests for peptide bonds), (d) they can be hydrolyzed to amino acids, (e) they are susceptible to attack by proteolytic enzymes, and (f) they show some weak catalytic activity. Also intriguing, but somewhat outside the scope of this report, is the behavior of this "protenoid" material when it is dissolved in warm water and the solution allowed to cool. When the solution (suspension) is viewed under the microscope, it is seen to contain a great number of small globules, or "microspheres." [110,115] These globules have some properties that have led Fox and coworkers to call them proto- or precellular. They are about 2 μ in diameter (in the size range of many living cells), they are not broken up by centrifugation at 3000 rpm, they can be sectioned and stained, they have pronounced outer membranes, and they show ATP-splitting (hydrolyzing) ability. When one imagines the control over subsequent chemistry that these membranes would confer upon the proteinoid globules, it seems reasonable to refer to them as possible "protocells." [115]

Peptides have also been reported as products from heating glycine in aqueous NH_4OH,[116] products of the γ-irradiation of N-acetylglycine,[57] and indicated as products of the thermal (90°) treatment of ammonium cyanide solutions.[38] Matthews

(113) D. L. Rohlfing, Nature, 216, 657 (1967).
(114) K. Harada and S. W. Fox, J. Am. Chem. Soc., 80, 2694 (1958).

(115) S. W. Fox in ref 49, p 361.
(116) J. Oró and C. L. Guidry, Arch. Biochem. Biophys., 93, 166 (1961).

and Moser found peptide-like material upon hydrolysis of the products obtained from (a) sparking CH_4–NH_3 mixtures[41] and (b) heating HCN–NH_3 mixtures;[42] these experiments the authors advance as support for the idea that polypeptides may have resulted on the primitive Earth directly from the hydrolysis of HCN polymers.[117] Others have suggested a key role for hydrolyzed aminoacetonitrile polymers in the chemical evolution of the polypeptides (proteins), and this reaction has been used successfully to produce di- and triglycine.[118–120]

Some successes have been reported in forming peptides in dilute aqueous solutions at room temperature. This has been done by using simple analogs of the carbodiimides (RN=C= NR'), reagents that have been used for over a decade in synthetic organic chemistry to effect dehydration condensations (in nonaqueous media).[121] Cyanamide, $H_2NC{\equiv}N$, a tautomer of the parent (and unknown) member of the carbodiimide series (HN=C=NH) has been used in the form of its dimer (dicyandiamide, $H_2NC(=NH)N=C=NH$, a known primitive-Earth compound[91]) to prepare simple peptides (glycylleucine, leucylglycine, and alanylalanine)[92,96,122]. A closely related compound, dicyanamide (NC–NH–CN), has also been used, again in dilute aqueous solution at room temperature, to convert up to 30% of starting glycine into diglycine; considerable yields of triglycine (about 15%) and tetraglycine (about 7%) were also observed.[123]

The carbodiimide analogs referred to here, and the related cyanide derivatives (such as the cyanogen hydrolysis products referred to earlier under "Nucleotides"), have shown a wide usefulness in chemical evolution research. They not only promote the dehydration condensations, in dilute aqueous solution at room temperature, of amino acids to peptides, but under similarly mild conditions they produce glucose 6-phosphate from glucose and H_3PO_4, adenosine 5'-phosphate (AMP) from adenosine and H_3PO_4, and $H_4P_2O_7$ (pyrophosphate) from H_3PO_4.[96] These compounds thus show a preference to remove (react faster with) the elements of water from a variety of molecules (amino acids, sugars, H_3PO_4) in aqueous solution, rather than to react with the surrounding water molecules. The condensing agent itself is hydrolyzed, for example

$$H_2N{-}CN \xrightarrow{H_2O} H_2NCONH_2$$
cyanamide — urea

$$NC{-}NH{-}CN \xrightarrow{H_2O} NC{-}NH{-}CONH_2 \xrightarrow{H_2O} H_2NCO{-}NH{-}CONH_2$$
dicyanamide — cyanurea — biuret

The mechanisms of the reaction of dicyanamide with aqueous glycine have been extensively investigated and are summarized in Scheme II.[123] The kinetic data indicate that the dipeptide, diglycine, is formed through the unstable N-cyano-O-aminoacetylurea (route to the right). The route to the left shows an overall catalytic effect of the glycine in promot-

ing dimerization of the dicyanamide to N,N,N'-tricyanoguanidine; this route is, of course, useless in promoting peptide formation.

All that the ambient-temperature, dilute-solution "chemical evolution" experiments have so far achieved is some low-yield preparations of peptides—at most, the tetrapeptides. However, even if we do not wish to invoke the high-temperature "protenoid" route to proteins (but in no way denying that this route may have been the key one), we can still easily imagine protein syntheses in the prebiotic oceans. Although the coupling of amino acids to peptides is accompanied by a positive free-energy change, there probably existed on the primitive Earth far better combinations of conditions for protein synthesis (catalysts, coupling agents, etc.) than have yet been found in the laboratory "chemical evolution" experiments. Adsorption of oligopeptides on appropriate surfaces may have increased coupling rates and diminished hydrolysis rates, leading to net synthesis of polypeptides. The molar free energy ($-\Delta G$) of adsorption of glycine peptides to the clay mineral montmorillonite increases with increasing length of the peptide.[124] This same mineral has been reported to increase the rate of dicyanamide-promoted peptide synthesis.[125]

We have good reason to expect that polypeptides, once formed, would have tended to aggregate themselves into the conformations (secondary and tertiary structures) that we find in present-day proteins. This tendency is shown by the facile laboratory denaturation and renaturation of proteins. Either a raising of the temperature, or a lowering of the pH, disrupts the helical structure and produces a random coil. On subsequent reversal of the temperature or pH, the helical structure reappears. These shifts in conformation are revealed by the hypochromism of the helix; it shows a lesser optical absorptivity (at about 190 nm) than does the random coil.[126,127] Similar observations have been made with respect to the restoration of, or natural tendency to form, the tertiary structure of proteins. There is good reason to expect the accumulation, on the prebiotic Earth, of proteins similar in most, if not all, respects to the present-day biologically produced proteins.

B. NUCLEIC ACIDS

Just as some modest progress has been recorded in chemical evolution experiments in producing peptides and protein-like material, so also has progress been made in producing polynucleotides (on the way to the nucleic acids). Although it cannot be called a chemical evolution experiment (enzymes were used) the *in vitro* synthesis of nucleic acids by Ochoa, Kornberg, *et al.*, is of great relevance. They demonstrated the following reaction.[128,129]

$$\text{nucleoside} + \text{triphosphate} \xrightarrow[\text{Mg}^{2+}]{\text{enzyme (e.g., polymerase)}} \text{newly synthesized}$$
"primer" nucleic acid — nucleic acid

(117) R. M. Kliss and C. N. Matthews, *Proc. Natl. Acad. Sci. U. S.*, **48**, 1300 (1962).
(118) S. Akabori in ref 46, p 189.
(119) H. Hanfusa and S. Akabori, *Bull. Chem. Soc. Japan*, **32**, 626 (1959).
(120) J. H. Reuter, "On the Synthesis of Peptides under Primitive Earth Conditions," 3rd International Meeting on Organic Geochemistry, London, 1966.
(121) J. C. Sheehan and G. P. Hess, *J. Am. Chem. Soc.*, **77**, 1067 (1955).
(122) C. Ponnamperuma and E. Peterson, *Science*, **147**, 1572 (1965).
(123) G. Steinman, D. H. Kenyon, and M. Calvin, *Biochim. Biophys. Acta*, **124**, 339 (1966).

(124) D. J. Greenland, R. H. Laby, and J. P. Quirk, *Trans. Faraday Soc.*, **58**, 829 (1962).
(125) G. Steinman and M. N. Cole, *Proc. Natl. Acad. Sci. U. S.*, **58**, 735 (1967).
(126) I. Tinoco, Jr., A. Halpern, and W. T. Simpson, "Polyamino Acids, Polypeptides, and Proteins," University of Wisconsin Press, Madison, Wis., 1962, p 147.
(127) M. Calvin, *Proc. Royal Soc.* (London), **A288**, 441 (1965).
(128) A. Kornberg, "Enzymatic Synthesis of Deoxyribonucleic Acid," Academic Press, New York, N. Y., 1961, pp 83–112.
(129) M. Goulian, A. Kornberg, and R. L. Sinsheimer, *Proc. Natl. Acad. Sci. U. S.*, **58**, 2321 (1967).

Scheme

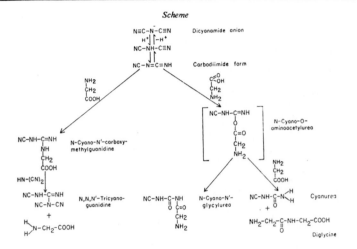

If the "primer" is left out of the reaction mixture, nucleic acid synthesis still occurs, only at a slower rate. The nucleoside phosphates, we have already seen, are synthesized under more or less primitive-Earth conditions (see section IV.C.4), and magnesium constitutes about 2 atom % of the Earth's crust.[130] The above reaction is not detected in the absence of enzyme. However, from the chemical-evolutionary viewpoint, for a given reaction, time can always be substituted for a catalyst. The demonstration of the above reaction in the laboratory is suggestive that nucleic acids could have accumulated on the prebiotic Earth. One can also visualize the concomitant accumulation, at that time, of both the polynucleotides and the polypeptides, each tending to catalyze the formation of the other.

In a more typical chemical evolution experiment it has been shown that cytidine phosphate (cytidylic acid) can be condensed to oligonucleotides (average number of monomers per chain = 5.6), at 65°, by the action of polyphosphoric acid.[131] The oligonucleotide was found to contain the 3'–5' phosphate linkages (to the ribose) that are characteristic of the natural nucleic acids. The difficulty with this work is, of course, the use of the polyphosphoric acid. However, Fox and others have suggested the possibility of localized concentrations of acid on the primitive Earth.[128,132] Such concentrations, plus volcanic heat, might have made it possible for polyphosphoric acid to play a role in chemical evolution. The formation of oligonucleotides has also been reported on the heating (160°) of uridine phosphate and uridine.[133]

It was mentioned earlier (section IV.C.4) that Schramm, *et al.*, have prepared a "polyphosphate ester" that has been used in dehydration condensation work.[87] The same authors also reported the preparation of polynucleotides (from nucleotides) with this reagent and have claimed some chemical-evolutionary relevance for their results.[87] However, the exis-

tence of their reagent on the primitive Earth would have required a very unlikely set of circumstances.

Although they did not investigate nucleotide condensations, Steinman, *et al.*, found that the previously mentioned cyanamide compounds did form (from glucose and H_3PO_4, in dilute aqueous solution), *via* dehydration condensation, the sugar–phosphate ester bond characteristic of the nucleic acids.[92]

VI. Concluding Remarks

This review has concentrated on the "synthesis" aspects of chemical evolution. Outside of its scope is the subject of how the biomacromolecules progressed toward the degree of organization that we now see in the living cell. The principal ideas of how this took place are the "coacervate" theory of Oparin[134] and the "microsphere" concept of Fox.[115] Both of these ideas are but halting steps on the long road toward an understanding of the genesis of the tremendously complex organization of the living cell. However, they both are based on the notion that the emergence of life is the inevitable outcome of associational and organizational forces inherent in the macromolecules' chemistry.

An outstanding, and very often discussed, problem in chemical evolution is that of the origin of optical activity in biological compounds. We still do not know why, to pick the main example, the natural proteins are composed of L- instead of D-amino acids. Little further understanding of this problem has developed during the past decade, and the reader is referred to excellent discussions of the subject by Wald, in 1957,[135] and by Horowitz and Miller, in 1962.[86] A new possibility in understanding this problem, however, has been suggested in a recent paper by Garay.[136] This work suggests that β-ray emission results in circularly polarized bremsstrahlung, and that such γ-radiation might radiation-decompose one enantiomorph faster than the other. Indeed, this report is ac-

(130) B. Mason, "Principles of Geochemistry," John Wiley and Sons, New York, N. Y., 1952, p 41.
(131) A. W. Schwartz and S. W. Fox, *Biochim. Biophys. Acta*, **134**, 9 (1967).
(132) A. P. Vinogradov in ref 46, p 23
(133) J. Morávek, *Tetrahedron Lett.*, **18**, 1707 (1967).

(134) A. I. Oparin, "The Chemical Origin of Life," translation by A. Synge, Charles Thomas Publishers, Chicago, Ill., 1964.
(135) G. Wald, *Ann. N. Y. Acad. Sci.*, **69**, 352 (1957).
(136) A. S. Garay, *Nature*, **219**, 338 (1968).

companied by experimental data that indicate that ^{90}Sr β-rays decompose D-tyrosine faster than L-tyrosine. Any evaluation of the importance of this process in chemical evolution must await a quantitative estimate of its effect, and a determination of how general the effect is.

The relatively new research "area" of chemical evolution is extremely broad and has excited the imaginations of all sorts of scientists, from astronomers to physiologists. Its steady progress, and resultant enrichment of our knowledge of our origins, is assured.

VII. Addenda

Since the original writing of this manuscript, two monographs on chemical evolution have appeared, and the reader is referred to these for elaboration of many of the topics discussed in this review.[137,138]

The work of Degens and Matheja[139] has reinforced the notion that the minerals available in the primitive oceans would have provided excellent templates for protein synthesis. They have also presented evidence that certain amino acid adsorptions may change the subsequent peptide formation from thermodynamically unfavored to favored.

A series of recent papers by Sulston, *et al.*,[140–142] have shown that a water-soluble carbodiimide [1-ethyl-3-(dimethylaminopropyl)carbodiimide hydrochloride] induces in dilute water solution (about 0.01 M in all reactants) the formation of di- and trinucleotides from mixtures of nucleosides and nucleotides. The reaction is performed at ambient tempera-

tures for several days, and the yields are good (a few per cent). Furthermore, these reactions are aided by the presence of synthetic polynucleotides, and the Watson–Crick pairing rules seem to apply. That is, the adenosine–adenylic acid condensations (but not those between adenylic acid and guanosine, cytidine, or uridine) are greatly aided by the presence of polyuridylic acid, and a polycytidylate template aids in the condensations of guanosine with guanylic acid. Although unnatural condensates are also formed (*e.g.*, 2′–5′ linkages), this work demonstrates a likely route for polynucleotide synthesis on the primitive Earth.

Friedmann and Miller have recently shown that phenylacetylene is synthesized in good yield from various hydrocarbons (of which ethane is the best) by high temperatures (1300°), electric discharge, and uv light.[143] They have also shown that the phenylacetylene can be hydrolyzed under primitive-Earth conditions to phenylacetaldehyde, and that the latter may be converted to phenylalanine on treatment with NH$_3$ and HCN, followed by hydrolysis. If the hydration step of phenylacetylene to acetaldehyde is carried out in the presence of H$_2$S and uv light, a better yield of phenylalanine (about 20% from the starting phenylacetylene) is obtained, accompanied by a trace of tyrosine. These results make it easier to envisage the appearance of the aromatic amino acids on the primitive Earth.

Krampitz and Fox have demonstrated that a carbodiimide (dicyclohexylcarbodiimide) causes the adenylates of the natural amino acids to condense easily in water solution at room temperature. The product is a high molecular weight, peptide-bond-containing material, similar in many respects to the "proteinoids" obtained in the anhydrous, high-temperature conditions. This work is a notable addition to our understanding of possible routes of protein synthesis on the prebiotic Earth.[144]

(137) D. H. Kenyon and G. Steinman, "Biochemical Predestination," McGraw-Hill Book Co., Inc., New York, N. Y., 1969.
(138) M. Calvin, "Chemical Evolution," Oxford University Press, Oxford, 1969.
(139) E. T. Degens and J. Matheja, *J. Brit. Interplanet. Soc.*, 21, 52 (1968).
(140) J. Sulston, *et al.*, *Proc. Natl. Acad. Sci. U. S.*, 59, 726 (1968).
(141) J. Sulston, *et al.*, *ibid.*, 60, 409 (1968).
(142) J. Sulston, *et al.*, *J. Mol. Biol.*, 40, 227 (1969).

(143) N. Friedmann and S. L. Miller, *Science*, 166, 766 (1969).
(144) G. Krampitz and S. W. Fox, *Proc. Natl. Acad. Sci. U. S.*, 62, 399 (1969).

5

Reprinted from *Proc. Natl. Acad. Sci. (U.S.)*, **55**(6), 1365–1372 (1966)

CHEMICAL EVENTS ON THE PRIMITIVE EARTH

By Philip H. Abelson

GEOPHYSICAL LABORATORY, CARNEGIE INSTITUTION OF WASHINGTON, WASHINGTON, D. C.

Communicated April 28, 1966

During the past 15 years, many workers employing a variety of energy sources have demonstrated the abiologic production of a large number of biologically interesting substances from many simple starting materials. Most of the experiments, however, have had a curious deficiency. While designed to elucidate origin of life on earth, they do not take into account a body of geologic information.

In this paper the nature of the primitive atmosphere and ocean is considered in the light of geologic and geophysical information. The hypothesis of an early methane-ammonia atmosphere is found to be without solid foundation and indeed is contraindicated. Geologists favor an alternative view—that genesis of air and oceans is a result of planetary outgassing. Some consequences of this view are examined. Volatiles from outgassing interacted with the alkaline crust to form an ocean having a pH 8–9 and to produce an atmosphere consisting of CO, CO_2, N_2, and H_2. Radiation interacting with such a mixture yields HCN as a principal product. Ultraviolet irradiation of HCN solutions at pH 8–9 yields amino acids and other important substances of biologic interest.

The nature of the earth's environment limited the kinds of compounds that might have accumulated in a soup. Arguments concerning feasible components support the view that amino acids and proteins preceded sugars and nucleic acids.

If the methane-ammonia hypothesis were correct, there should be geochemical evidence supporting it. What is the evidence for a primitive methane-ammonia atmosphere on earth? The answer is that there is *no* evidence for it, but much against it. The methane-ammonia hypothesis is in major trouble with respect to the ammonia component, for ammonia on the primitive earth would have quickly disappeared.

The effective threshold for degradation by ultraviolet radiation is 2,250 Å. A quantity of ammonia equivalent to present atmospheric nitrogen would be destroyed in ∼30,000 years. Small amounts of ammonia would be reformed, but this process is unimportant in comparison to the destruction.

If large amounts of methane had ever been present in the earth's atmosphere, geologic evidence for it should also be available. Laboratory experiments show that one consequence of irradiating a dense, highly reducing atmosphere is the production of hydrophobic organic molecules which are adsorbed by sedimenting clays. The earliest rocks should contain an unusually large proportion of carbon or organic chemicals. This is not the case.

The composition of the present atmosphere with respect to the gases neon, argon, krypton, and xenon is crucial. Neon is present on earth to an extent about 10^{-10} that of cosmic abundance,[1] and similarly argon, krypton, and xenon are relatively absent. It seems likely that if xenon of atomic weight 130 could not accumulate, other volatile light constituents such as hydrogen, nitrogen, methane, and carbon monoxide would also be lost at the same time. The concept that the earth had a dense methane-ammonia atmosphere is not supported by geochemistry, and it is contraindicated by the scarcity of xenon and krypton in our present atmosphere.

Understanding the evolution of our present atmosphere does not require *ad hoc* assumptions. Geological evidence suggests that the atmosphere evolved as a result of outgassing of the earth. This process has been discussed by a number of investigators, including Rubey,[2, 3] Holland,[4] and Berkner and Marshall.[5]

The geologic record shows that volcanism and associated outgassing have been going on for more than 3,000 million years. Rubey[2] has emphasized that the composition of volcanic gases is similar to that of the volatile substances that must be accounted for at or near the surface of the earth. These major volatile substances are H_2O, $16,700 \times 10^{20}$ gm; C as CO_2, 921×10^{20} gm; and N_2, 43×10^{20} gm.

Studies of the composition of volcanic gases by Shepherd and others have shown that water and CO_2 are the major volatiles produced by outgassing. These gases are accompanied by a significant amount of reducing potential in the form of hydrogen. One can estimate the amount of reducing power brought to the surface of the earth through outgassing by making a balance sheet of the oxidized and reduced chemicals in the atmosphere, biosphere, and sedimentary rocks. This estimate does not take into account gain or loss of hydrogen from the top of the atmosphere.

Most of the carbon is present in sedimentary rocks as carbonate. Part is also present as reduced carbon. Rubey estimates that the organic carbon has a composition 68×10^{20} gm C, 25×10^{20} gm O, and 9.6×10^{20} gm H. To burn this material to $CO_2 + H_2O$ would require 235×10^{20} gm O_2.

Most of the organic carbon present in the sediments appears to have been derived from photosynthetic organisms, presumably forming organic matter from $CO_2 + H_2O$. In the process, 235×10^{20} gm O_2 would be liberated. Hutchinson[6] points out that the quantity of atmospheric and fossilized oxygen that can be recognized is much less than this. Assuming that all the sulfate of the sea and the sediment represents oxidized sulfide and that oxygen has been fossilized in the production of ferric from ferrous iron, he calculates the total free and fossil oxygen to be

Free in atmosphere	12×10^{20} gm
Sulfate in sea and sediments	47×10^{20} gm
As ferric iron derived from ferrous	14×10^{20} gm
Total	73×10^{20} gm

This leaves 162×10^{20} gm O_2 to be accounted for. Rubey[2] suggested that the most probable explanation for the discrepancy is that an appreciable amount of the carbon released from volcanoes was in the form of CO. Hutchinson points out that other reducing gases such as H_2 are also possible, and cites Cotton.[7]

One can estimate the nature of the reducing gases. The major amount of magma accompanying volcanism is basaltic in composition and contains ferrous silicates. Holland[4] has provided an estimate of the equilibrium relations between H_2O and basalt, and CO_2 and basalt, using a simplified system which approaches basalt in composition. His calculation is based on the quaternary invariant point in the system $MgO\text{-}FeO\text{-}Fe_2O_3\text{-}SiO_2$ at $1255°C$. At this temperature and an estimated P_{O_2} of $10^{-7.1}$ atm, $P_{H_2O}/P_{H_2} = 105$ and $P_{CO_2}/P_{CO} = 37$. These ratios correspond to values observed at Kilauea by Eaton and Murata,[8] which were $P_{H_2O}/P_{H_2} = 137$ and $P_{CO_2}/P_{CO} = 31$. Assuming that the discrepancy in the balance sheet for oxygen can be accounted for by H_2 and CO emitted from volcanoes, and using Holland's estimate of the ratios of P_{H_2O}/P_{H_2} and P_{CO_2}/P_{CO}, one can obtain values for the amounts of these gases which have reached the surface during outgassing. The values are H_2, 19×10^{20} gm, and CO, 17×10^{20} gm.

One can obtain another estimate for the amounts of H_2 and CO using the known amounts of volatiles now at the surface and the ratios of H_2O/H_2 and CO_2/CO quoted by Holland. The values obtained by these different routes are H_2, 16×10^{20} gm, and CO, 14×10^{20} gm. Thus the amounts of H_2 and CO accompanying the water and CO_2 are of the right magnitude to account for the reduced carbonaceous matter found in the sedimentary rocks.

The nature of the primitive atmosphere was to a large degree determined by the ocean-atmosphere interaction, and the pH of the primitive ocean is crucial. Most of the present components of the crust probably reached the surface in the form of basalt. The igneous rocks gradually disintegrate in the presence of water to form partially ionized products. In effect, silicate minerals are salts of strong bases and weak acids, and in dissolving form mildly alkaline solutions. If pulverized basalt is added to water, the pH immediately rises to about 9.6.[9]

Within a short time after outgassing began, there would be sufficient water to form extensive lakes. Rainstorms would begin to occur, accompanied by a weathering process which would bring alkaline waters back to the lakes. These rivers would carry Na^+, Ca^{++}, and clay minerals much as do the rivers of today. These waters would have a pH somewhat less than 9.6, for their pH would be lowered toward 8 by CO_2 and other acid gases. During the initial phases of weathering, the pH of the ocean may have been higher than it is today. At present only about 28 per cent of the surface is available for weathering and most of the continental exposures are sedimentary rocks. These are not so alkaline as are basalts. Sillén[10] has emphasized the buffering capacity of the huge amounts of silicates that have been weathered. The buffering system of weathered silicates and ocean has tended to maintain the pH of the ocean at about 8–9 since outgassing began.

The major products of outgassing were quickly removed from the atmosphere. Water condensed, CO_2 was dissolved in water and converted to carbonate, and other acid gases were converted to nonvolatile salts. The residual gases to be accounted for are CO, N_2, and H_2. The oceanic buffering system affected the distribution of CO. Carbon monoxide is slightly soluble in water and slowly reacts to produce formic acid.[11]

The equilibrium between carbon monoxide in the atmosphere and in the ocean is governed by

$$CO_{(g)} + \overline{OH} \rightleftarrows \overline{HCO_{2}}_{(aq)} \qquad \Delta F = -9.6 \text{ kcal.}$$

Most CO would be converted to formate.

Assuming that the proportions of volatiles have not changed with time and that $16{,}600 \times 10^{20}$ gm H_2O and 17×10^{20} gm CO have reached the surface, the concentration of formate would amount to $0.035\ M$. If the reducing capacity represented by the H_2 were employed in producing CO, the initial concentration of formate might have been as great as $0.6\ M$. Correspondingly, the partial pressure of CO could have been as high as 0.06 atm.

Water vapor was another important component of the atmosphere. Its partial pressure at low altitudes was determined largely by the atmospheric temperature at the surface of the earth. The amount of water vapor at high altitudes was governed by the temperature profile above the earth. It is likely that the temperature of the primitive atmosphere decreased rapidly with height to reach a minimum. In the present atmosphere the temperature distribution is complicated by the effects of oxygen and ozone. The temperature drops from 289°K at the surface to 214°K at 14 km, then rises due to effects of ozone before dropping again to 178°K at 86 km.[12] In the primitive atmosphere with little oxygen present, there was only one temperature minimum. Very low temperatures might have been reached at altitudes not much greater than 20 km. This low-temperature region would limit the fraction of moisture in the region above the temperature minimum.

Most of the nitrogen that has reached the surface of earth is now in the atmosphere. From a small value the nitrogen content has gradually increased to its present amount.

Hydrogen is an important component of volcanic gases. Its abundance in the atmosphere would be governed in part by hydrogen escape, which is sharply dependent on the temperature at the top of the atmosphere. The abundance of hydrogen would also be governed by chemistry occurring at the top of the atmosphere. This too is influenced by temperature. A well-considered calculation of the temperature of the ionosphere of the primitive earth would be an important contribution. However, that is beyond the scope of this paper. Crude guidance is obtained from measurements of the present atmosphere. Above the minimum at 86 km, temperatures rise to 800°K at 140 km, 1200°K at 180 km, and to higher temperatures, e.g., ~1500°K, above that. In a primitive atmosphere possessing a different composition, the temperature would not be the same. However, at the top of the atmosphere where energetic radiations from the sun impinge and where chemical events of interest occur, the temperature would have been elevated.

The reaction $CO_2 + H_2 \rightleftharpoons CO + H_2O$ is of importance. At 25°C this reaction has a $\Delta F = +6.831$ kcal.[13] However, at such temperatures the equilibrium time is about 10^{20} years. At 1200°K the reaction proceeds rapidly[14] and $\Delta F = \sim 0$.

The composition of gases at the top of the atmosphere was governed by thermodynamic relationships existing at the higher temperatures. It was also influenced by the action of the cold trap in limiting the partial pressure of H_2O. These conditions combined to favor production of CO from $CO_2 + H_2$. As a result, carbon monoxide was at one time a major component of the earth's atmosphere.

Above the earth the principal agent causing radiation-induced transformations is ultraviolet radiation and this acts at the top of the atmosphere where temperatures are high. An electric discharge producing high temperatures and short ultraviolet radiation provides a situation qualitatively similar to the terrestrial circumstances. By conducting the discharge in a vessel with a cold trap, one can improve the similarity to nature. The cold trap maintains a low partial pressure of H_2O and serves to remove complex molecules produced by radiation.

To obtain an indication of the kinds of chemicals that might be synthesized, Dr. T. C. Hoering and I have conducted experiments under such conditions. Starting mixtures, with pressures measured in centimeters of Hg, were (1) N_2, 8; CO, 8; H_2, 2; (2) N_2, 4; CO, 4; H_2, 4; (3) N_2, 2; CO, 4; H_2, 6; (4) N_2, 2; CO, 4; H_2, 24. The principal product formed in the last three mixtures was HCN and H_2O. Small amounts of CH_4 and CO_2 were also made. In the first mixture, CO_2 was the major product, with HCN and H₂O second. The products were analyzed by T. C. Hoering in a mass spectrometer. Thus it was simple to look for formaldehyde and for the unexpected. No formaldehyde was detected, which meant that if made, it was present in amounts no more than about 10^{-3} those of HCN. Similarly, other products such as nitriles, acids, and hydrocarbons would have been seen. When hydrogen is not present, the principal products from $N_2 + CO$ or CO alone are CO_2 and C_3O_2. This reaction was studied by Harteck, Groth, and Faltings.[15] We also studied other mixtures. If the cold finger is not refrigerated, water is present in the gas phase. Yields of interesting products such as HCN are small. In summary, irradiation of a variety of mixtures of CO, N_2, H_2 produces HCN as the major product and little else except C_3O_2, CO_2, and H_2O.

In the natural situation any formaldehyde would tend to be destroyed. CH_2O is unstable and it decomposes to $CO + H_2$. The reaction is rapid at 500°C.[16] Formaldehyde if produced at the top of the atmosphere would not survive the temperatures or radiation there; CH_2O is decomposed by quanta of wavelengths as long as 3650 Å.[17]

Any surviving formaldehyde which reached the alkaline ocean would be subject to further attenuation. Formaldehyde undergoes disproportionation to methyl alcohol plus formic acid. In addition, CH_2O reacts rapidly with HCN, amines, and amino acids at room temperatures at pH 8–9. Experiments made in the last century showed that formaldehyde in lime water yields sugarlike substances. Suppose that a small amount of a carbohydrate were formed. This product would also be subject to rapid degradation and attenuation. First there is the well-known browning reaction. In addition, carbohydrates such as glucose combine readily with amino acids to form nonbiologic products. This reaction proceeds at room temperature, and even at 0°C there is a noticeable reaction in a week. At pH 8–9, amino acids and free carbohydrates are simply incompatible. Reaction between them leaves whatever is present in excess. As will be seen, synthesis of amino acids is relatively much more favored than synthesis of carbohydrates. In addition, substances such as glycine and alanine are very stable and hence could accumulate. Thus it is unlikely that the primitive ocean ever contained more than traces of free glucose, free ribose, or deoxyribose.

This discussion of limitations on the production and preservation is illustrative of similar arguments which can be made with respect to important constituents of any primitive "thick soup." For instance, arginine is adsorbed by sedimentary clay as are chlorophyll and porphyrins. Fatty acids form insoluble salts with magnesium and calcium, and hence would be removed from the soup. At least five major factors limit the kinds of compounds that might have accumulated in the primitive ocean.

First, there are limitations on what can be made by inorganic means; second, all organic matter degrades spontaneously with time; third, some substances are readily destroyed by radiation; fourth, many compounds would have been removed from the ocean by precipitation or adsorption; fifth, there are serious chemical incompatibilities among the constituents of living matter, and some of the components of the soup would react to form nonbiologic substances. In view of these limitations, one is challenged to seek a series of steps toward life that are compatible with the environment.

What kind of prebiologic chemistry is there that can occur at ordinary temperatures, in dilute solutions at pH 8.0? How can one form carbon-carbon bonds under these circumstances? Among simple substances, three methods are notably feasible. One is condensation of aldehydes. A second is condensation of aldehydes with HCN. A third is polymerization of HCN. I have indicated that formaldehyde would not be produced in quantity and it is easily destroyed. In contrast HCN can be readily produced and is stable at high temperatures. In slightly alkaline solution, HCN combines with itself to yield a number of interesting compounds.

The reactions involved are importantly sensitive to pH. Thus a 1 M solution of HCN having a pH 4.6 will not polymerize. Polymerization does not occur at very high pH. However, the reaction proceeds well at pH 8–9. Thus the range of pH likely in the primitive ocean is also a favorable one for polymerization of HCN. With solutions 0.1 M in HCN + cyanide, the reaction occurs rapidly at 100°C, more slowly at 25°C. A major product is the tetramer:

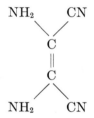

This reaction has long been known, and it has further been found that on hydrolysis this tetramer gives small yields of glycine. Oró and Kamat[18] have also noted production of small amounts of alanine and aspartic acid when 2.2 M cyanide solutions were heated at 70°C for 25 days.

Production of glycine from HCN has been noted in this laboratory. However, yields can be strikingly improved by radiation, and polymerization of dilute cyanide solutions can be made to occur at ambient temperatures. When solutions containing HCN at pH 8–9 are irradiated with 2,536-Å radiation and the product mixture is hydrolyzed, glycine, alanine, serine, aspartic acid, and glutamic acid result. The relative yields of the acids depend on time of irradiation. With short irradiations, yields are low and glycine is prominent. With longer times, yields are better and the larger amino acids are more prominent. At very long times, yields decrease. The ultraviolet radiation speeds the polymerization of HCN and, in view of the products obtained, must cause internal rearrangements including disproportionation. Production of amino acids from solutions 0.002–0.10 M in cyanide have been noted. After irradiation the solutions were hydrolyzed with 6 M HCl. The resultant amino acids were determined by E. Hare using an amino acid analyzer. In one instance, 1 mmole HCN gave rise to the following amounts of amino acids (μmole): 30.0 gly, 3.5 ser, 0.9 ala, 0.8 asp.

Others have studied the production of interesting substances from cyanide. These include Oró,[19] who produced adenine. Additional important roles involving cyanide and its products have been discovered. From irradiation of aqueous HCN, cyanamide is produced.[20] Cyanamide in turn has the important facility of bringing about peptide synthesis from amino acids in dilute solution under the influence of ultraviolet radiation.[21, 22]   Other dehydrating reactions of great biologic interest are also facilitated in a similar way.[23]

One difficulty that must be considered is that HCN is subject to hydrolysis.   The rate of reaction is such that HCN decomposes to $NH_4OH + HCOOH$ in about 10 years at 25°C and about 70 years at 15°C.[24]   At the same time, the production of HCN is limited.   Assuming formation of cyanide requires photolysis of nitrogen, and assuming an over-all efficiency of 10 per cent of the quanta available, one obtains a production rate of the order of 4×10^{14} gm/year.   Even in the early stages of the ocean when the volume was less than, e.g., 10^{22} ml, the resultant solution would be quite dilute ($\sim 10^{-4}$ M in HCN).

A dehydrating mechanism for resynthesis of HCN might have been available. We have already seen one example.   It is possible that silicates could do the job even better.   The energy involved in dehydration reactions is in the range 0 to perhaps 10 kcal/mole.   On the other hand, typical quanta of visible light represent ~ 50 kcal/mole.   If visible light were harnessed to reform HCN from NH_4COOH, a large amount of HCN would be available for synthetic activities.

Formate would be a consequence of volcanic outgassing.   The ammonia produced by hydrolysis of HCN would be available for recycling.   As an example, let us examine the production of serine.   The net chemical reaction is

$$5HCN + 7H_2O \rightarrow C_3H_7O_3N + 4NH_3 + 2CO_2.$$
$$\text{Serine}$$

If the dehydration system were working, the product ammonia would be combined with more formate.

Given a dehydration mechanism utilizing cyanamide or some other agent, the step to an enzyme is relatively short.   Eck and Dayhoff[25] have developed some interesting arguments about ferredoxin which are relevant.

Ferredoxin is an unusually simple protein containing only 55 amino acid residues. It occurs in primitive anaerobic organisms, both photosynthetic and nonphotosynthetic.   The functions of ferredoxin are basic to cell chemistry.   The enzyme participates in oxidation, reduction, energy transfer, fixation of nitrogen, formation of ATP, and synthesis of pyruvate.[26]

Eck and Dayhoff have made an analysis of amino acid content and structure of the enzyme.   They find that it contains an unusual proportion of glycine, alanine, serine, aspartic acid, and cysteine.   From a study of the sequence of amino acids in ferredoxin, the authors are led to the conclusion that the original molecule was based on a repeating sequence of alanine, serine, aspartic acid, and glycine.   These amino acids are those which are produced most readily from HCN.

When one examines the processes of biosynthesis, he is impressed by how few mechanisms and basic building blocks are involved.   Pyruvate, acetate, and carbonate are key chemicals in the synthesis of amino acids.   In some microorganisms these three sources can furnish almost all the carbon in

Serine, glycine, cysteine, alanine, valine, leucine, isoleucine, lysine, aspartic acid, threonine, methionine, glutamic acid, proline, arginine.

All three of these carbon sources were present in the primitive ocean.   Pyruvate and acetate result from degradation of serine.   Snell[27] points out that the reaction

$$\text{CH}_2\text{OHCHNH}_2\text{COOH} + \text{H}_2\text{O} \rightarrow \text{CH}_3\text{COCOOH} + \text{NH}_3$$
$$\text{Serine} \qquad\qquad\qquad\qquad\qquad \text{Pyruvate}$$

proceeds to the right. From pyruvate and from malonic acid, acetate can be derived.

The principal processes required to synthesize the 14 amino acids are condensations (such as those involving pyruvate, carbonate, and acetate in the Krebs cycle), hydrogenation, and transfer of NH_3. Furthermore, given condensation and hydrogenation, one has a mechanism for producing fatty acids. Formate would be a convenient source of hydrogen. Simple receptors for solar radiation might have helped speed the condensation reactions. Thus one can visualize that natural conditions might have favored synthesis of increasingly complex molecules from the simple but versatile substances available.

[1] Brown, Harrison, in *The Atmospheres of the Earth and Planets* (Chicago: The University of Chicago Press, 1952), p. 258.

[2] Rubey, W. W., *Bull. Geol. Soc. Am.*, **62**, 1111 (1951).

[3] Rubey, W. W., *Geol. Soc. Am. Spec. Paper*, **62**, 631 (1955).

[4] Holland, H. D., in *Petrologic Studies: A Volume to Honor A. F. Buddington*, ed. A. E. J. Engel, H. L. James, and B. F. Leonard (New York: Geological Society of America, 1962), p. 447.

[5] Berkner, L. V., and L. C. Marshall, *J. Atm. Sci.*, **22**, 225 (1965).

[6] Hutchinson, G. E., in *The Earth as a Planet*, ed. G. P. Kuiper (Chicago: The University of Chicago Press, 1954), p. 371.

[7] Cotton, C. A., *Nature*, **154**, 399 (1944).

[8] Eaton, J. P., and K. J. Murata, *Science*, **132**, 925 (1960).

[9] For a discussion of the pH of water in contact with silicate minerals, see Stevens, R. E., and M. K. Carron, *Am. Mineralogist*, **33**, 31 (1948).

[10] Sillén, L. G., in *Oceanography*, ed. Mary Sears (Washington, D. C.: American Association for the Advancement of Science Publication No. 67, 1961), p. 549.

[11] Branch, G. E. K., *J. Am. Chem. Soc.*, **37**, 2316 (1915).

[12] International Council of Scientific Unions, Committee on Space Research, *Cospar International Reference Atmosphere*, 1961, compiled by H. Kallmann-Bijl *et al.* (Amsterdam: North-Holland Publishing Co., 1961).

[13] *JANAF* (Joint Army-Navy-Air Force) *Interim Thermochemical Tables* (Midland, Michigan: Thermal Laboratory, The Dow Chemical Co., 1960), vols. 1 and 2.

[14] Graven, W. M., and F. J. Long, *J. Am. Chem. Soc.*, **76**, 2602 (1954).

[15] Harteck, P., W. Groth, and K. Faltings, *Z. Elektrochem.*, **44**, 621 (1938).

[16] Fletcher, C. J. M., *Proc. Roy. Soc. (London)*, Ser. A, **146**, 357 (1934).

[17] Ellis, Carleton, and A. A. Wells, *The Chemical Action of Ultraviolet Rays*, revised and enlarged edition by F. F. Heyroth (New York: Reinhold Publishing Corp., 1941), p. 417.

[18] Oró, J., and S. S. Kamat, *Nature*, **190**, 442 (1961).

[19] Oró, J., *Nature*, **191**, 1193 (1961).

[20] Schimpl, A., R. M. Lemmon, and M. Calvin, *Science*, **147**, 149 (1965).

[21] Steinman, G., R. M. Lemmon, and M. Calvin, *Science*, **147**, 1574 (1965).

[22] Ponnamperuma, C., and E. Peterson, *Science*, **147**, 1572 (1965).

[23] Steinman, G., R. M. Lemmon, and M. Calvin, these Proceedings, **52**, 27 (1964).

[24] Peiker, A. L., thesis presented to the Graduate Committee, Trinity College, June, 1927, cited in Krieble, V. K., F. C. Duennebier, and E. Colton, *J. Am. Chem. Soc.*, **65**, 1479 (1943).

[25] Eck, R. V., and M. O. Dayhoff, *Science*, **152**, 363 (1966).

[26] Arnon, D. I., *Science*, **149**, 1460 (1965).

[27] Snell, E. E., in *The Origins of Prebiological Systems and of Their Molecular Matrices*, ed. Sidney W. Fox (New York: Academic Press, 1965), p. 203.

6

Reprinted from *Proc. Natl. Acad. Sci. (U.S.)*, **69**(3), 765–768 (1972)

Prebiotic Synthesis of Hydrophobic and Protein Amino Acids

(electric-discharge synthesis/gas chromatography–mass spectrometry/amino-acid analyzer)

DAVID RING*, YECHESKEL WOLMAN†, NADAV FRIEDMANN‡, AND STANLEY L. MILLER§

Department of Chemistry, University of California, San Diego, La Jolla, Calif. 92037

Communicated by Harold C. Urey, December 20, 1971

ABSTRACT   The formation of amino acids by the action of electric discharges on a mixture of methane, nitrogen, and water with traces of ammonia was studied in detail. The presence of glycine, alanine, α-amino-n-butyric acid, α-aminoisobutyric acid, valine, norvaline, isovaline, leucine, isoleucine, alloisoleucine, norleucine, proline, aspartic acid, glutamic acid, serine, threonine, allothreonine, α-hydroxy-γ-aminobutyric acid, and α,γ-diaminobutyric acid was confirmed by ion-exchange chromatography and gas chromatography–mass spectrometry. All of the primary α-amino acids found in the Murchison Meteorite have been synthesized by this electric discharge experiment.

Most prebiotic syntheses that start with the primitive atmospheric constituents give substantial yields of glycine, alanine, and α-amino-n-butyric acid (1). Prebiotic syntheses of the higher aliphatic amino acids have been claimed, for example, by the action of electric discharges on $CH_4 + NH_3 + H_2O$ (2–6), by heating $CH_4 + NH_3 + H_2O$ to 900–1200° (7, 8), and by the action of shock waves on CH_4, C_2H_6, NH_3, and H_2O (9). The amino acids were identified only by an amino-acid analyzer (2–4, 6, 7, 9), only by paper electrophoresis (8), or only by gas chromatography (5). However, these techniques are not sufficient by themselves to identify an amino acid.¶

In the original synthesis of amino acids by electric discharges (10–12), only glycine, alanine, α-amino-n-butyric acid, α-aminoisobutyric acid, and β-alanine, of the simple aliphatic amino acids, were synthesized in sufficient yield to obtain identification by a melting point of a derivative. Recently developed techniques permit the identification of compounds found in lower yield by this synthesis.

The synthesis under prebiotic conditions of aspartic and glutamic acid (2–8, 12–14), serine (2, 5, 6–8, 13), threonine

(2, 3, 5–7), and proline (3, 4, 7, 8, 13) have been reported but they have not been properly identified [except for aspartic acid (14)]. The synthesis of these amino acids (except proline) has also been reported from the polymerization of HCN (1), but again without proper identification. A prebiotic synthesis of threonine should also yield allothreonine, but this amino acid has never been reported. In addition, several investigators have reported the appearance of a large peak at the isoleucine position on the amino-acid analyzer (2, 4–6, 13). The identification of this peak as isoleucine has been questioned (4, 13). It is evident that this compound cannot be isoleucine, since a corresponding peak for alloisoleucine is not observed.

Most electric-discharge experiments have been done with a large amount of ammonia present. Use of nitrogen instead of ammonia in such experiments does not change the major products, although the yield of amino acids is lower (12). The use of a higher concentration of ammonia in such experiments has been criticized (15), and it is now thought that the ammonia concentration in the prebiotic atmosphere was not likely to have been greater than 10^{-5} atm (16, 17). Although this is a small percent of the atmosphere, this value corresponds to a significant concentration of NH_4^+ and NH_3 in the ocean, where NH_3 would have played an important role in prebiotic synthesis of organic compounds.

We have synthesized a wide variety of amino acids by the action of electric discharges on a mixture of methane, nitrogen, and water, with traces of ammonia. The compound chromatographing as isoleucine on the amino-acid analyzer is shown to be α-hydroxy-γ-aminobutyric acid. These amino acids were all positively identified by gas chromatography–mass spectrometry.

MATERIALS AND METHODS

To avoid contamination from reagents, the HCl was distilled, constant boiling. NH_4OH was prepared by adding gaseous NH_3 to water redistilled from alkaline permanganate. NH_4Cl was recrystallized from water. Authentic samples of the amino acids were obtained from Calbiochem, except for α-hydroxy-γ-aminobutyric acid, which was prepared by selective deamination of α,γ-diaminobutyric acid (18). *Sec*-Butanol was 90% of the (+) butanol isomer (Norse Laboratories, Santa, Barbara, Calif.).

To a 3-liter flask with two tungsten-electrodes (4) was added 100 ml of 0.05 M NH_4 Cl. The flask was evacuated, and sufficient NH_3 was added to bring the pH to 8.7. Methane (200 mm) and N_2 (200 mm) were then added, and the spark discharge was run for 48 hr. The tesla coil was the same kind

* Pomona College, Claremont, Calif. 91711; † Department of Organic Chemistry, The Hebrew University, Jerusalem; ‡ Albert Einstein College of Medicine, Bronx, N.Y. 10461.
§ To whom reprint requests should be addressed.

¶ The correct elution time on the amino-acid analyzer is obviously insufficient by itself to identify an amino acid. Many nonprotein amino acids have peaks that coincide with the protein amino acids. This is clearly shown in Hamilton, P. B. (1963) *Anal. Chem.* **35**, 2055–2064. The same limitation is true for paper chromatography or electrophoresis, even with several different solvents. Coincidence of the radioactivity of an unknown with the color of a known sample on paper chromatography has led to several errors. The only methods that seem reliable for amino acids are identification by the melting point and mixed melting point of a suitable derivative, or analysis by gas chromatography–mass spectrometry.

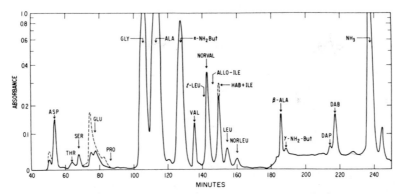

FIG. 1. Chromatogram from an amino-acid analyzer of the desalted amino acids after the electric discharge synthesis. The *arrows* show the elution time of the indicated amino acids. The *dashed line* shows the increase in color on heating the ninhydrin–buffer mixture for 30 min instead of 8 min. (*solid line*). *DAP* is α,β-diaminopropionic acid. *Norleu* is norleucine; *Norval* is norvaline; *DAB* is α,γ-diaminobutyric acid; *HAB* is α-hydroxy-γ-aminobutyric acid.

as previously used (11). The temperature of the flask remained between 20° and 25°. The aqueous solution, presumably containing the amino nitriles rather than the amino acids (12), was hydrolyzed with 3 M HCl for 24 hr, desalted, and evaporated to dryness. The dried sample was hydrolyzed again with 3 M HCl in order to open the rings of glutamic acid, α,γ-diaminobutyric acid, and α-hydroxy-γ-aminobutyric acid that may have been formed during the desalting. Seven similar runs were combined. A sample of the desalted amino acids was then run on the amino-acid analyzer (19).

In order to separate the various amino acids, the combined runs were chromatographed on Dowex 50(H⁺) (38.5 × 2.2 cm) and eluted with HCl (20) (400 ml of 1.5 M HCl, 700 ml of 2.5 M HCl, 400 ml of 4.0 M HCl, and 600 ml of 6.0 M HCl).

TABLE 1. *Yields and mole ratios of amino acids from sparking 336 mmol of* CH_4

	μmol	Mole ratio		μmol	Mole ratio
Glycine	440	100	Norleucine	6.0	1.4
Alanine	790	180	*tert*-Leucine	<0.02	—
α-Amino-*n*-butyric acid	270	61	Proline	1.5	0.3
α-Aminoiso-butyric acid	~30	~7	Aspartic acid	34	7.7
Valine	19.5	4.4	Glutamic acid	7.7	1.7
Norvaline	61	14	Serine	5.0	1.1
Isovaline	~5	~1	Threonine	~0.8	~0.2
Leucine	11.3	2.6	Allothreonine	~0.8	~0.2
Isoleucine	4.8	1.1	α,γ-Diaminobutyric acid	33	7.6
Alloisoleucine	5.1	1.2	α-Hydroxy-γ-aminobutyric acid	74	17

The yields of glycine and alanine, based on the carbon, are 0.26% and 0.71%, respectively. The total yield of amino acids listed in the table is 1.55%.

18 Fractions were collected and evaporated to dryness in a vacuum desiccator; each fraction was quantitated on the amino-acid analyzer. α-Aminoisobutyric acid and isovaline were not separated completely on the amino-acid analyzer. Therefore, these amino acids were estimated by the areas of the peaks found on gas chromatography. The amino-acid analyzer gave the sum of threonine and allothreonine; the gas chromatography peaks, which were completely separated, indicated about equal amounts of these compounds.

The identity of the amino acids was confirmed by gas chromatography–mass spectrometry of the *N*-trifluoroacetyl-amino acid-(+)-2-butyl esters (21). All the amino acids except α,γ-diaminobutyric acid were separated on a capillary column (45.7 meters) containing OV 225 as a liquid phase (we thank G. Pollock for this column). This column separated the four stereoisomers of threonine–allothreonine; the isoleucines separated into three peaks, D-alloisoleucine, L-alloisoleucine + D-isoleucine, L-isoluecine. The α,γ-diaminobutyric acid was separated on a 1% OV-1 on chromosorb W column (0.6 × 180 cm). The instrument was a LKB 9000 (ionization potential, 70 V; mass range, 8–400).

RESULTS

The chromatogram from the amino-acid analyzer is shown in Fig. 1. The peaks labeled valine, norvaline, alloisoleucine, leucine, and norleucine contain between 50 and 80% of the indicated compound. Most of the peak labeled α-hydroxy-γ-aminobutyric acid (HAB) + Ile is α-hydroxy-γ-aminobutyric acid. The *dashed line* in Fig. 1 shows the increase in color yield on heating the column eluent and ninhydrin for 30 min instead of the usual 8 min. The color yield of protein amino acids is not changed by the additional heating, but *N*-substituted amino acids, and some β-amino acids, give substantial increases in color yield (22). This is particularly noticeable in the glutamic-acid region of the chromatogram. The chromatography on Dowex 50(H⁺) cleanly separated aspartic acid, threonine, serine, and glutamic acid from interfering compounds, and allowed quantitation on the amino-acid analyzer.

Quantitation of the amino acids after elution from Dowex 50(H⁺) is given in Table 1. The identity of each amino acid was confirmed by the gas chromatography–mass spectrometry. The elution time and the mass spectrum of an unknown and of known standards confirmed the identification based on HCl elution from Dowex 50, and elution time from an amino-acid analyzer.

In addition to confirming the identity of the unknown, the gas-chromatographic analysis showed that each of the amino acids (except for isovaline, α-hydroxy-γ-aminobutyric acid, α,γ-diaminobutyric acid, and aspartic acid that do not form two peaks on the columns used) were racemic within the experimental error (45–55% D-isomer). This result shows that there was no significant contamination from reagents or dust during the separation process. This conclusion applies particularly to the proline, where the yield was sufficiently low that contamination was a reasonable possibility. The mass spectrum of α-hydroxy-γ-aminobutyric acid is shown in Fig. 2.

DISCUSSION

The results in Table 1 show that there is no selective synthesis of the branched-chain amino acids that occur in proteins. Indeed, the yield of norvaline is three times that of valine, although the yield of norleucine is 50% that of leucine and that of (isoleucine + alloisoleucine). Therefore, the occurrence of glycine, alanine, valine, isoleucine, and leucine in proteins, but the absence of α-amino-n-butyric acid, norvaline, alloisoleucine, and norleucine, cannot be understood on the basis of the yields from this type of synthesis.

The absence of *tert*-leucine in this synthesis may be due to the instability of its amino nitrile or its precursor aldehyde (pivalaldehyde). Two aliphatic amino acids with both α hydrogens substituted, α-aminoisobutyric acid and isovaline (α-amino-α-methyl-n-butyric acid) were found. The six-carbon amino acids of this class were not looked for. The relatively low yield of α-aminoisobutyric acid and isovaline may be due to the instability of the corresponding aminonitrile, as has been discussed elsewhere (12).

The yield of proline is quite low—seven times lower than leucine. A yield this low suggests that an electric-discharge synthesis of this type was not the only source of proline on the primitive earth.

The yield of 5- and 6-carbon amino acids is substantially lower than the glycine, alanine, or even the α-amino-n-butyric acid. The mole ratios of glycine:alanine:α-amino butyric acid:(valine + norvaline):6-carbon amino acids are 100:180:60:18:6. There is some variation in these ratios in different experiments, with the same conditions and spark source. The reason for the lower yields of the 5- and 6-carbon amino acids is not clear. If we lowered the temperature to 0° during the sparking, the yields of the 5- and 6-carbon amino acids were not increased, nor did the use of ethane instead of methane increase the yields.

It seems unlikely that the amino acids that were important in prebiotic polypeptides were present in the primitive ocean in about equal concentrations. Several mechanisms would have been available to concentrate certain amino acids in prebiotic polypeptides. One possible mechanism would have concentrated the sea water in a lagoon by evaporation until the amino acids were partially precipitated, and then synthesized peptides with the precipitated amino acids. The process would have concentrated the 5- and 6-carbon amino acids, since they are less soluble than glycine, alanine, and

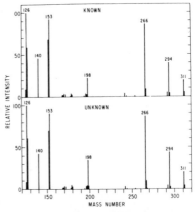

FIG. 2.   The mass spectrum of the known and unknown *N,O*-ditrifluoroacetyl-2-butyl ester derivative of α-hydroxy-γ-aminobutyric acid.

α-amino-n-butyric acid. The concentration of the higher aliphatic amino acids could also have occurred by adsorption on suitable mineral surfaces.

Peptide bonds of the 5- and 6-carbon amino acids are more stable to hydrolysis than those of glycine and alanine. This stability has been correlated with the "rule of 6" (23), and it has been suggested that the relative rates of hydrolysis generated sequences of peptides in the primitive ocean that were hydrolytically stable (24). The same considerations would predict that the higher aliphatic amino acids would concentrate in the peptides of a primitive ocean.

These considerations make it plausible that the yields of the 5- and 6-carbon amino acids obtained in these experiments would have been adequate for prebiotic peptide synthesis.

All of the aliphatic α-amino acids reported in the Murchison Meteorite (25–28), as well as aspartic acid, glutamic acid, β-alanine, and proline, were obtained in this electric-discharge experiment. We have also found *N*-alkylated amino acids, as well as other nonprotein amino acids, in these experiments. The similarities of these amino acids with those found in the Murchison Meteorite, and their significance, will be discussed in a forthcoming paper in this journal.

We thank Drs. Tom Bond and John Wright for helpful discussion, and Mrs. Rosiland Smith for running the amino-acid analyzer. This research was supported by National Science Foundation Grant GB 25048. The gas chromatograph–mass spectrometer was obtained, in part, with NSF Grant GP 18245. D. R. was supported by a undergraduate summer fellowship (NIH 5T01-GM-00945-08).

1. Lemmon, R. H. (1970) *Chem. Rev.* **70**, 95–109.
2. Grossenbacher, K. A. & C. A. Knight (1965) in *The Origins of Prebiological Systems*, ed. Fox, S. W. (Academic Press, New York), pp. 173–183.
3. Czuchajowski, L. & Zawadzki, W. (1968) *Rocz. Chem.* **42**, 697–702.
4. Oró, J. (1963) *Nature* **197**, 862–867.
5. Ponnamperuma, C., Woeller, F., Flores, J., Romiez, M. & Allen, W. (1969) in *Chemical Reactions in Electric Discharges*, ed. Gould, R. F. (Advances in Chemistry series no. 80, A.C.S., Washington), pp. 280–288.
6. Matthews, C. N. & Moser, R. E. (1966) *Proc. Nat. Acad. Sci. USA* **56**, 1087–1094.

7. Harada, K. & Fox, S. W. (1964) *Nature* **201**, 335–336.
8. Taube, M., Zdrojewski, S. Z., Samochocka, K. & Jezierska, K. (1967) *Angew. Chem.* **79**, 239.
9. Bar-Nun, A., Bar-Nun, N., Bauer, S. H. & Sagan, C. (1970) *Science* **168**, 470–473.
10. Miller, S. L. (1953) *Science* **117**, 528–529.
11. Miller, S. L. (1955) *J. Amer. Chem. Soc.* **77**, 2351–2361.
12. Miller, S. L. (1957) *Biochim. Biophys. Acta* **23**, 480–489.
13. Fox, S. W. & Windsor, C. R. (1970) *Science* **170**, 984–986.
14. Oró, J. (1968) *J. Brit. Interplanet Soc.* **21**, 12–25.
15. Abelson, P. H. (1966) *Proc. Nat. Acad. Sci. USA* **55**, 1365–1372.
16. Sillén, L. G. (1967) *Science* **156**, 1189–1197.
17. Bada, J. L. & Miller, S. L. (1968) *Science* **159**, 423–425.
18. Dinnill, P. A. & Fowden, L. (1965) *Phytochemistry* **4**, 445–451.

19. Dus, K., Lindroth, S., Pabst, R. & Smith, R. A. (1967) *Anal. Biochem.* **18**, 532–549.
20. Wall, J. S. (1953) *Anal. Chem.* **25**, 950–953.
21. Pollock, G. E. & Oyama, V. I. (1966) *J. Gas Chromatogr.* **4**, 126–131.
22. Coggins, J. R. & Benoiton, N. L. (1970) *J. Chromatogr.* **52**, 251–256.
23. Whitfield, R. E. (1963) *Science* **142**, 577–579.
24. Nicholson, I. (1970) *J. Macrmol. Sci., Chem.* **A4**, 1619–1625.
25. Kvenvolden, K., Lawless, J., Pering, K., Peterson, E., Flores, J., Ponnamperuma, C., Kaplan, I. R. & Moore, C. (1970) *Nature* **228**, 923–926.
26. Kvenvolden, K. A., Lawless, J. G. & Ponnamperuma, C. (1971) *Proc. Nat. Acad. Sci. USA* **68**, 486–490.
27. Oró, J., Gibert, J., Lichtenstein, H., Wikstrom, S. & Flory, D. A. (1971) *Nature* **230**, 105–106.
28. Cronin, J. R. & Moore, C. B. (1971) *Science* **172**, 1327–1329.

Reprinted from *Proc. Natl. Acad. Sci. (U.S.)*, **69**(4), 809–811 (1972)

Nonprotein Amino Acids from Spark Discharges and Their Comparison with the Murchison Meteorite Amino Acids

(gas chromatography–mass spectrometry/prebiotic synthesis/amino-acid analyzer)

YECHESKEL WOLMAN*, WILLIAM J. HAVERLAND, AND STANLEY L. MILLER†

Department of Chemistry, University of California, San Diego, La Jolla, Calif. 92037

Communicated by Harold C. Urey, December 20, 1971

ABSTRACT   All the nonprotein amino acids found in the Murchison meteorite are products of the action of electric discharge on a mixture of methane, nitrogen, and water with traces of ammonia. These amino acids include α-amino-n-butyric acid, α-aminoisobutyric acid, norvaline, isovaline, pipecolic acid, β-alanine, β-amino-n-butyric acid, β-aminoisobutyric acid, γ-aminoisobutyric acid, sarcosine, N-ethylglycine, and N-methylalanine. In addition, norleucine, alloisoleucine, N-propylglycine, N-isopropylglycine, N-methyl-β-alanine, N-ethyl-β-alanine α,β-diaminopropionic acid, isoserine, α,γ-diaminobutyric acid, and α-hydroxy-γ-aminobutyric acid are produced by the electric discharge, but have not been found in the meteorite.

A number of amino acids have been found recently in the Murchison meteorite (1–4). The evidence is strong that the meteorite was not contaminated with amino acids after it fell on the earth, since the amino acids are racemic and since many of them do not occur or are rare in terrestrial biology. Several of these meteorite amino acids have been properly demonstrated to be synthesized by the action of electric discharges on CH₄, NH₃, H₂O, and H₂ (5–7). These include glycine, alanine, sarcosine, β-alanine, N-methylalanine, α-amino-n-butyric acid, and α-aminoisobutyric acid. We have recently repeated the electric discharge experiment (8) and have demonstrated, by modern techniques, the presence of aspartic acid, glutamic acid, valine, isovaline, norvaline, and proline, all of which are found in the meteorite. We have also demonstrated the electric discharge synthesis of leucine, isoleucine, alloisoleucine, norleucine, serine, threonine, allothreonine, α-hydroxy-γ-aminobutyric acid, and α,γ-diaminobutyric acid, which have not yet been found in the meteorite, although some of them may be present [e.g., the leucines (3)].

The strong resemblance between the amino acids from the electric discharge and in the meteorite induced us to look for the other nonprotein amino acids found in the meteorite. We have found in the electric discharge reaction all the amino acids so far reported to be present in the meteorite.

MATERIALS AND METHODS

The authentic samples of the N-alkylated amino acids were prepared by reaction of the amine with the corresponding haloacid, and recrystallization from alcohol.

The same sample of amino acids produced by electric discharge that was used for the identification of hydrophobic amino acids (8) was used in the present analysis. The 18 frac-

tions from the Dowex 50(H⁺) chromatography were quantitated with an amino-acid analyzer. The amino acids were sufficiently separated to quantitate the β-alanine, γ-aminobutyric acid, isoserine, and α,β-diaminopropionic acid. Sarcosine and N-ethylglycine were quantitated on the amino-acid analyzer with a pH 2.80 buffer; the column eluent and ninhydrin were heated for 30 min instead of the usual 8-min heating period (9). The other amino acids were either present in amounts too small or the color yield was too low to allow quantitation on the amino-acid analyzer. They were, therefore, estimated by comparison of the area of the peaks of the unknown on gas chromatography with the area of an amino acid, on the same gas chromatogram that had been quantitated on the amino-acid analyzer.

The unknown amino acids were identified by gas chromatography–mass spectrometry (8). α,β-Diaminopropionic acid was separated on an OV-1 column, while all the other amino acids were separated on a capillary column of OV-225. Of the compounds in Table 1, only the diastereoisomers of β-aminoisobutyric acid were resolved on these columns.

Identification of the unknown amino acids was based on the elution time from Dowex 50 (H⁺), retention time on the gas chromatography column, and identity with the mass spectrum

TABLE 1.   *Yields and mole ratios from sparking CH₄ (336 mmol), N₂, and H₂O with traces of NH₃*

Amino acid	μmol	Mole ratio
Sarcosine	55	12.5
N-Ethylglycine	30	6.8
N-Propylglycine	~2	~0.5
N-Isopropylglycine	~2	~0.5
N-Methylalanine	~15	~3.4
N-Ethylalanine	<0.2	—
β-Alanine	18.8	4.3
β-Amino-n-butyric acid	~0.3	~0.1
β-Aminoisobutyric acid	~0.3	~0.1
γ-Aminobutyric acid	2.4	0.5
N-Methyl-β-alanine	~5	~1.0
N-Ethyl-β-alanine	~2	~0.5
Pipecolic acid	~0.05	~0.01
α,β-Diaminopropionic acid	6.4	1.5
Isoserine	5.5	1.2

The mole ratios are relative to glycine as 100 (8). The amino acids in this table account for 0.35% of the carbon added as methane.

---

* Present address: Department of Organic Chemistry, The Hebrew University, Jerusalem.
† To whom reprint requests should be addressed.

of the authentic compound. The mass spectra of β-aminoisobutyric acid and β-amino-*n*-butyric acid contained a few extraneous peaks; therefore, the identification of these compounds is tentative.

RESULTS

The yields of the amino acids are shown in Table 1. Several other amino acids were found in the gas chromatography-mass spectrometry analysis, among them no less than eight compounds with a molecular mass number (M^+) of 269. These are isomers of valine, but we were unable to demonstrate their identity with any of the known compounds in our possession. There were also two compounds of M^+ 281 (isomers of pipecolic acid). Two unidentified compounds of M^+ 269 and two of M^+ 281 were found in the meteorite (2).

DISCUSSION

There is a striking similarity between the products and relative abundances of the amino acids produced by electric discharge and the meteorite amino acids. Unfortunately, there are only a few quantitative values (4) for the meteorite amino acids, but we have estimated their relative abundances from the published gas chromatography data (2). Table 2 shows a comparison, including the aliphatic amino acids previously reported from the same sample (8).

The most notable difference between the meteorite and the electric-discharge amino acids is the pipecolic acid, the yield being extremely low in the electric discharge. Proline is also present in relatively low yield from the electric discharge. The amount of α-aminoisobutyric acid is greater than α-amino-*n*-butyric acid in the meteorite, but the reverse is the case in the electric discharge. The amounts of aspartic and glutamic acids in the meteorite are comparable, but there is

TABLE 2. *Relative abundances of amino acids in the Murchison meteorite and an electric-discharge synthesis*

Amino acid	Murchison meteorite	Electric discharge
Glycine	4	4
Alanine	4	4
α-Amino-*n*-butyric acid	3	4
α-Aminoisobutyric acid	4	2
Valine	3	2
Norvaline	3	3
Isovaline	2	2
Proline	3	1
Pipecolic acid	1	<1
Aspartic acid	3	3
Glutamic acid	3	2
β-Alanine	2	2
β-Amino-*n*-butyric acid	1	1
β-Aminoisobutyric acid	1	1
γ-Aminobutyric acid	1	2
Sarcosine	2	3
N-Ethylglycine	2	3
N-Methylalanine	2	2

Mole ratio to glycine ($= 100$): 0.05–0.5, 1; 0.5–5, 2; 5–50, 3; >50, 4. The meteorite abundances are estimated from the published gas chromatogram (2); these estimates are approximate.

five times as much aspartic acid as glutamic acid in the electric discharge.

We do not believe that reasonable differences in ratios of amino acids detract from the overall picture. Indeed, the ratio of α-aminoisobutyric acid to glycine is quite different in two meteorites of the same type, being 0.4 in Murchison and 3.8 in Murray (4).

One would expect quantitative differences between the meteorite composition and the electric-discharge products, even if the mechanism of formation in the two cases were identical. Thus cyanoacetylene, which is synthesized by sparking CH_4 and N_2, is probably the major precursor of aspartic acid, but the yield of cyanoacetylene is decreased by the addition of small amounts of NH_3 to the $CH_4 + N_2$ mixture (10). Therefore, local differences in the NH_3 partial pressures on the parent body of the carbonaceous chondrites would result in substantial differences in the aspartic acid concentration if the amino acids were not completely mixed on the parent body. Temperature differences can also affect the yields. For example, the stability of α-aminoisobutyronitrile is quite sensitive to temperature and cyanide concentration, while α-amino-*n*-butyronitrile, being more stable, is not particularly sensitive to these factors.

The electric-discharge amino acids reported here appear to have been synthesized through the aminonitriles for three reasons: there was considerable hydrogen cyanide produced by the electric discharge, the HCN concentration in the aqueous phase being 0.03 M. Secondly, previous electric-discharge experiments (7) have shown that most, if not all, of the amino acids were produced from nitriles. Thirdly, the presence of α-hydroxy-γ-aminobutyric acid and α,γ-diaminobutyric acid, as well as isoserine and α,β-diaminopropionic acid, is consistent with nitrile precursors. Under the conditions of the aqueous phase (pH 8.7, 0.05 M NH_4Cl) the following equilibrium is established rapidly.

$$\underset{+NH_3 \quad\quad OH}{CH_2\text{—}CH_2\text{—}CH\text{—}CN} + NH_3 \rightleftarrows$$
$$\underset{+NH_3 \quad\quad NH_2}{CH_2\text{—}CH_2\text{—}CH_2\text{—}CN} + H_2O \quad [1]$$

On acid hydrolysis in these experiments, or slower hydrolysis under primitive earth or meteorite conditions, both α-hydroxy-γ-aminobutyric acid and α,γ-diaminobutyric acid would be obtained. A similar equation can be written for isoserine and α,β-diaminopropionic acid.

Although we were able to find substantial yields of sarcosine, *N*-ethyl-, *N*-propyl-, and *N*-isopropylglycine, we could find *N*-methylalanine, but not *N*-ethylalanine. This result can be understood if aminonitriles are the precursors of these amino acids. The following equilibria would have been attained under the conditions of the experiment:

$$\underset{NH_2}{R\text{—}CH\text{—}CN} + CH_3\text{—}NH_2 \overset{K_m}{\rightleftarrows} \underset{NHCH_3}{R\text{—}CH\text{—}CN} + NH_3 \quad [2]$$

$$\underset{NH_2}{R\text{—}CH\text{—}CN} + C_2H_5NH_2 \overset{K_e}{\rightleftarrows} \underset{NHC_2H_5}{R\text{—}CH\text{—}CN} + NH_3 \quad [3]$$

and similar equilibria for propylamine and isopropylamine. There is also a similar equilibrium for the aminonitrile and hydroxynitrile (Eq. [1]).

The values of K_m and K_e are not known, but one would expect K_m and K_e for formaldehyde to be about the same. However, in the case of acetaldehyde, steric hindrance between the alkyl group on the nitrogen and the R group on the α-carbon could make K_e substantially smaller than K_m.

These considerations are valid whether the aminonitriles are formed from the aldehyde, ammonia, or amine, and HCN, or whether the nitrile is formed directly in the electric discharge (11). The equilibria [2] and [3] are rapidly reversible in solution under the pH conditions of the experiment.

The substantial and comparable yields of β-alanine, N-methyl-β-alanine, and N-ethyl-β-alanine can be accounted for by the addition of ammonia, methylamine, and ethylamine to acrylonitrile, which has been postulated (7) to be a product of an electric discharge in $CH_4 + NH_3$. The rate constants for the addition of NH_3 and the amines would be expected to be comparable (12); the concentrations of the amines, by Eq. [2] and [3] and the yields of sarcosine and N-ethylglycine, should also be comparable. The synthesis of β-alanine and the N-alkylated derivatives could also be accounted for by the direct synthesis of the nitriles in an electric discharge (11).

The yields of amino acids from the electric-discharge experiments allow a simple calculation, similar to Urey's (13, 14), of the concentration of amino acids in the oceans of the primitive earth. If we assume that all the surface carbon on the earth (3000 g/cm²) passed through the atmosphere as CH_4, and that decomposition of amino acids after synthesis was minimal, then a 1.9% yield (0.35% from Table 1 + 1.55% in Table 1, ref. 8) would give 57 g of carbon in amino acids per cm², or about 0.6 mol/cm² of amino acids. If the primitive oceans were the present size (300 liters/cm²), the concentration of amino acids would have been about 2 mM.

The yields from these experiments can also be used to calculate the expected amino-acid concentration in the carbonaceous chondrites. The Murchison meteorite contains a total of 2% organic carbon (1) and about 30 µg/gram of amino acids, so the amino acids are 0.15% of the organic carbon. If all the methane on the parent body was converted to organic compounds, the 0.15% of amino acids is a 10-times less than the yields from this electric-discharge experiment. These yields are in reasonable agreement, if we consider that synthesis on the parent body may not have been as efficient as the spark discharge in the laboratory and that various losses of amino acids would have occurred on the parent body.

The results of these experiments with an electric discharge do not exclude other sources of energy, such as ultraviolet light, high-energy radiation, shock waves, etc. from having contributed to the amino acids and other organic compounds in the carbonaceous chondrites. Indeed, these other sources of energy must have played some role, even if small. It is not possible to make a realistic calculation of the relative importance of the various sources of energy, since the conditions on the parent body are not known. These considerations also apply to the recent synthesis of amino acids and other organic compounds from carbon monoxide, hydrogen, and ammonia by a Fischer–Tropsch type of synthesis (15, 16). The yields of amino acids are 10- to 100-times lower than the electric-discharge yields. This synthesis, which is asserted to have been the sole source of organic compounds in the carbonaceous chondrites, presents several problems, such as the source of the CO and the complicated temperature sequence. In any case, the low yields from the Fischer–Tropsch type of synthesis make this hypothesis less attractive from the standpoint of amino-acid production than electric discharges, although the Fischer–Tropsch process might have been the major source of other organic compounds, such as straight-chain hydrocarbons.

We thank Mrs. Rosiland Smith and Mr. Fred Castillo for running the samples on the amino-acid analyzer. This research was supported by National Science Foundation Grant GB 25048. The gas chromatograph–mass spectrometer was obtained, in part, with NSF Grant GP 18245.

1. Kvenvolden, K., Lawless, J., Pering, K., Peterson, E., Flores, J., Ponnamperuma, C., Kaplan, I. R. & Moore, C. (1970) *Nature* **228**, 923–926.
2. Kvenvolden, K. A., Lawless, J. G. & Ponnamperuma, C. (1971) *Proc. Nat. Acad. Sci. USA* **68**, 486–490.
3. Oró, J., Gibert, J., Lichtenstein, H., Wickstrom, S. & Flory, D. A. (1971) *Nature* **230**, 105–106.
4. Cronin, J. R. & Moore, C. B. (1971) *Science* **172**, 1327–1329.
5. Miller, S. L. (1953) *Science* **117**, 528–529.
6. Miller, S. L. (1955) *J. Amer. Chem. Soc.* **77**, 2351–2361.
7. Miller, S. L. (1957) *Biochim. Biophys. Acta* **23**, 480–489.
8. Ring, D., Wolman, Y., Friedmann, N. & Miller, S. L. (1972) *Proc. Nat. Acad. Sci. USA* **69**, 765–768.
9. Coggins, J. R. & Benoiton, N. L. (1970) *J. Chromatogr.* **52**, 251–256.
10. Sanchez, R. A., Ferris, J. P. & Orgel, L. E. (1966) *Science* **154**, 784–785.
11. Ponnamperuma, C. & Woeller, F. (1967) *Curr. Mod. Biol.* **1**, 156–158.
12. Friedman, M. & Wall, J. S. (1964) *J. Amer. Chem. Soc.* **86**, 3735–3741.
13. Urey, H. C. (1952) *Proc. Nat. Acad. Sci. USA* **38**, 351–363.
14. Urey, H. C. (1952) in *The Planets* (Yale Univ. Press, New Haven), pp. 152–153.
15. Yoshino, D., Hayatsu, R. & Anders, E. (1971) *Geochim. Cosmochim. Acta* **35**, 927–938.
16. Hayatsu, R., Studier M. H. & Anders, E. (1971) *Geochim. Cosmochim. Acta* **35**, 939–951.

8

Reprinted from *Science*, **170**, 980–982 (Nov. 27, 1970)

Carbon Isotope Fractionation in the Fischer-Tropsch Synthesis and in Meteorites

MICHAEL S. LANCET and EDWARD ANDERS

Abstract. *Carbon dioxide and organic compounds made by a Fischer-Tropsch reaction at 400°K show a kinetic isotope fractionation of 50 to 100 per mil, similar to that observed in carbonaceous chondrites. This result supports the view that organic compounds in meteorites were produced by catalytic reactions between carbon monoxide and hydrogen in the solar nebula.*

Most of the carbon in carbonaceous chondrites exists in reduced form, largely as an aromatic polymer and, to a lesser extent, as extractable organic compounds (*1*). A minor part (~ 3 to 5 percent) exists in oxidized form, as the carbonates breunnerite [$(Mg,Fe)CO_3$] and dolomite [$MgCa(CO_3)_2$] (*2*).

These two types of reduced carbon differ greatly in their isotopic composition. The polymeric carbon is light: $\delta C^{13} = -15$ to -17 per mil in both type I and type II carbonaceous chondrites (*3*). The carbonate carbon is much heavier: $\delta C^{13} = +60$ to $+70$ per mil in type I and $+40$ to $+50$ per mil in type II carbonaceous chondrites (*3*, *4*), for a total difference of 83 to 87 and 57 to 67 per mil. For comparison, terrestrial organic matter and carbonates (including that formed cogenetically, for example, by shellfish) generally differ by only about 25 per mil, typical values being -25 to -30 per mil for organic carbon and ~ 0 per mil for carbonates. The meteoritic carbonates fall well outside the range of terrestrial carbon, -80 to $+25$ per mil (*5*).

Clayton (*4*) has considered four possible origins of the meteoritic carbon isotope fractionation. The first of these possibilities, single-stage equilibrium fractionation, can, in principle, give an effect of the right sign and magnitude. Data for organic molecules of appropriate complexity are unavailable, but if calculations for CH_4 and CO_2 (*6*)

are used as a first-order approximation, temperatures of 265° to 335°K would seem to be required for the observed fractionations in type I and type II carbonaceous chondrites. However, no mechanisms on earth are known that would establish equilibrium at the lower of these temperatures. Kinetic isotope effects are a second alternative, but in the reactions studied thus far, the fractionation was too small (20 to 40 per mil) and in the wrong direction, the oxidized carbon becoming lighter rather than heavier (*7*). The third possibility, biological activity, produces fractionations of the right sign, but usually of insufficient magnitude [algae, 10 to 12 per mil; vascular plants, 20 to 30 per mil (*8*)]. Finally, it is conceivable that the differences reflect nuclear processes, the two types of carbon having had different histories of nucleosynthesis. This alternative was favored by Urey (*9*). However, all work to date has failed to bring to light any isotopic differences that might be attributed to incomplete mixing in the solar nebula.

The origin of the organic compounds is, of course, pertinent to this problem. Studier *et al.* (*10*) have shown that nearly all compounds identified in meteorites can be produced by a Fischer-Tropsch type of reaction between CO, H_2, and NH_3 in the presence of a nickel-iron or magnetite catalyst. They proposed that the organic compounds in meteorites were made by such reactions, catalyzed by

dust grains in the solar nebula.

The Fischer-Tropsch reaction produces both oxidized and reduced carbon. Much of the oxygen from CO usually is converted into H_2O (Eq. 1), but a variable proportion is converted to CO_2 (Eq. 2):

$$n\,CO + (2n + 1)\,H_2 \rightarrow C_nH_{2n+2} + n\,H_2O \tag{1}$$

$$2n\,CO + (n + 1)\,H_2 \rightarrow C_nH_{2n+2} + n\,CO_2 \tag{2}$$

Nothing is known about carbon isotope fractionation in this reaction, however. We, therefore, decide to investigate it.

A difficulty encountered in our experiments was the tendency of any CO_2 formed to disappear by secondary reactions. [Hydrogenation of CO_2 to hydrocarbons is thermodynamically feasible (*11*).] We circumvented this problem by collecting CO_2 as fast as it was formed, using a $Ba(OH)_2$ absorbent.

For our syntheses, we heated 0.8 liter of an equimolar mixture of CO and H_2, initially at 1 atm, to $400° \pm 10°K$ in a Vycor flask in the presence of 1.5 g of cobalt catalyst (*12*). (Cobalt was chosen because it is effective at lower temperatures than iron catalysts; otherwise there is little difference between the two.) As the reaction progressed CO_2 was removed continuously by absorption in 5 ml of a saturated $Ba(OH)_2$ solution in a 35-ml sampling tube, connected to the reaction vessel by a side arm. The sampling tube was replaced from time to time. In order to ensure complete recovery of all CO_2 formed during the sampling interval, the tube was cooled to $-196°C$ just before removal. All condensable gases were thus frozen out. After closing the stopcock between the sampling tube and the reaction flask, we warmed the tube to room temperature and shook it to promote absorption of CO_2. The stopcock was opened briefly to equalize pressure, and the sampling tube was replaced by a new one.

The gas phase was separated into three fractions: CO, CH_4, and heavier hydrocarbons (C_{2+}). Those components that remained gaseous at $-196°C$ (CO, H_2, and CH_4) were passed over CuO at 150°C. This reaction converted CO to CO_2 and H_2 to H_2O, which were frozen out. The CH_4 remaining was oxidized to CO_2 by passage over CuO at 850°C. Trial runs showed that this procedure separated CO from CH_4 with 99.1 to 99.5 percent efficiency.

The fraction condensed at $-196°C$ (the "C_{2+} fraction"), which consisted of hydrocarbons heavier than CH_4, was oxidized to CO_2 by passage over CuO at 850°C. Finally, the original CO_2 fraction was recovered from the $BaCO_3$-$Ba(OH)_2$ suspension by the addition of 85 percent H_3PO_4 to the sampling tube. The volume of each fraction was determined on a mercury manometer.

Because of the nature of our sampling procedure, the CO_2 fraction was not quite comparable to the other three fractions. It was removed continuously from the entire reaction volume, and thus represents the *incremental* yield between successive samplings. The other three fractions were isolated from small portions of the gas phase and thus represent *cumulative* yields. We cannot completely rule out the possibility that the CO_2 was slightly fractionated as a result of incomplete freezing out and absorption. This result would tend to make the recovered CO_2 too light. A control experiment showed, however, that CO_2 recovery was ≥ 89 percent complete in 4 hours, our shortest sampling interval. Moreover, the CO-CO_2 fractionation remained constant throughout the experiment as sampling times increased.

Isotopic compositions were measured on a 15-cm, 60°-sector double-collecting mass spectrometer (*13*). All results are given relative to the Pee Dee belemnite (PDB) standard. The isotopic measurement itself is usually accurate to ± 0.1 per mil. However, samples smaller than 40 μmole had to be diluted with standard CO_2 before measurement, and, since our initial procedure for measuring small gas volumes was not accurate enough, uncertainties in the dilution factor led to rather large errors in isotopic composition.

The results reported here are based on two experiments under nominally identical conditions, comprising a total of eight fractions (Table 1). The

Table 1. Carbon isotope fractionation in Fischer-Tropsch synthesis; equimolar ratio of CO to H_2; temperature = 400°K; pressure = 1 atm; cobalt catalyst.

Time (hr)	Percentage of total*				δC^{13} (‰)†			
	CO	CH_4	C_{2+}	CO_2	CO	CH_4	C_{2+}	CO_2
				Experiment 1				
7	90.2	8.2	0.5	0.05	-37.6	-81.2	-30. ± 30	0 ± 9
31	30.5	68.6	0	0.05	-46.8	-38.5		-8 ± 6
58	15.8	80.7	1.5	0.10	-47.7	-35.1	-61 ± 3	-11.5 ± 3
320	8.1	87.9	3.3	0.68	-48.0	-35.3	-62.2	-11.1
				Experiment 2				
4	95.6	1.9	1.5	0.05	-37.3	-100 ± 4	-62 ± 3	-1 ± 5
15	93.8	3.5	1.7	0.05	-37.4	-94 ± 3	-55 ± 2	
30	64.7	29.4	2.3	0.18	-33.1	-50.2	-61.2	-5.3
217	24.1	69.0	3.0	0.20	-47.2	-37.1	-63.1	-11.1

* Values for CO, CH_4, and C_{2+} were determined on a 1/20th aliquot and reflect the composition of the gas phase at the time of sampling. Carbon dioxide, on the other hand, was removed continuously from the entire reaction mixture. The amounts given in the table are 1/20th of the amount actually collected and refer to the CO_2 formed between consecutive sampling intervals.   † Initial CO: -38.6 per mil. Errors are ± 0.1 per mil, unless otherwise indicated. A "wax" fraction was recovered from the catalyst in experiment 1 at the end of the reaction. It comprised 1.5 percent of the total carbon and had a δC^{13} of -72.2 per mil.

catalyst in experiment 2 seems to have been slightly less active, judging from the slower rate of reaction. In combining the results on a single graph, we have therefore used degree of conversion rather than time as the abscissa (Fig. 1).

A strikingly large fractionation (~ 100 per mil) between CH_4 and CO_2 appears in the early stages of the reaction. The equilibrium fractionation at 400°K is only 45 per mil (*6*), and so this is clearly a kinetic isotope effect.

As the reaction progresses, the methane becomes isotopically heavier. This is due in part to material balance: as CH_4 becomes the dominant product, it must,

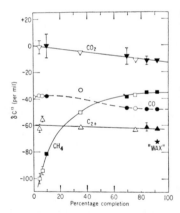

Fig. 1. Carbon isotope fractionation in the Fischer-Tropsch reaction at 400°K. Solid symbols, experiment 1; open symbols, experiment 2. In the early stages of the reaction CH_4 shows a fractionation as large as 100 per mil relative to CO_2. Hydrocarbons of higher molecular weight (C_{2+}, "wax") show smaller fractionations (50 to 60 per mil) which, however, persist throughout the reaction.

of necessity, approach the isotopic composition of the starting material. This trend seems to be further enhanced by isotopic equilibrium. The CH_4-CO fractionation approaches the equilibrium value of 15 per mil (*14*) in the latter stages of the reaction. Apparently the CH_4-CO exchange is fairly rapid under our experimental conditions.

The C_{2+} fraction is perhaps more pertinent to the meteorite problem, because the bulk of the reduced carbon in meteorites consists of polymeric material and molecules with ten or more carbon atoms. This fraction remains ~ 50 per mil lighter than the CO_2 throughout the reaction, as its amount increases from ~ 0.5 to ~ 3 percent. Apparently, its rate of exchange with CO is much slower than that of CH_4.

Because of the nature of our experiment, the C_{2+} fraction was not truly representative of all heavy hydrocarbons produced. Only those that were volatile at 400°K found their way into the sampling tube. Compounds of high molecular weight remained on the catalyst in the reaction vessel and were recovered only after the end of the experiment. The fractionation between this "wax" fraction and CO_2 was 61 per mil in the first experiment, some 10 per mil greater than the C_{2+}-CO_2 fractionation. Unfortunately the wax fraction in experiment 2 was too small for isotopic analysis.

It thus appears that the Fischer-Tropsch reaction gives a carbon isotope fractionation of the same sign and magnitude as that observed in meteorites. Although our results thus strengthen the case for catalytic synthesis of the organic compounds in meteorites, they also place additional constraints on the model.

On the basis of the observed fractionation between "wax" and CO_2 at 400°K, the organic compounds in type II carbonaceous chondrites may have formed at ~ 400°K and those in type I carbonaceous chondrites at a slightly lower or higher temperature, with the exact temperature dependent on the temperature coefficient of the kinetic isotope effect. This value is in good agreement with previous estimates (15), based on trace element content and the formation temperatures of certain minerals in carbonaceous chondrites (Fe_3O_4, 400°K; hydrated silicates, ~ 350°K). However, in a cooling solar nebula it may be difficult to keep most of the carbon in the form of CO down to such low temperatures. Although CO is the stable form of carbon at high temperatures, it becomes unstable with respect to hydrogenation (Fischer-Tropsch) reactions at lower temperatures. Thermodynamically, the most favored product is CH_4; the next most favored product is ethane, then benzene, and other hydrocarbons. Thus, if equilibrium were maintained, most of the CO would be converted to CH_4 before formation of heavier hydrocarbons became possible. (Their formation temperatures lie some 100° to 150° below those of methane.) At total pressures of 10^{-4} and 10^{-5} atm, only 1 percent of the carbon would remain in the form of CO by the time the temperature had dropped to 570° and 520°K. Essentially no CO would be left at 400°K.

This dilemma does not seem to be insoluble. Hydrogenation of CO is known to be very slow in the absence of a suitable catalyst. The fact that carbon compounds are abundant only in meteorites containing magnetite and hydrated silicates may mean that these compounds themselves were the required catalysts. Above the formation thresholds of these minerals (400°K and ~ 350°K) lack of a catalyst may have permitted the survival of CO. In support of this suggestion, there is some evidence that the solar nebula cooled rapidly from high temperatures down to the accretion range of meteorites, and then much more gradually (16). For the ordinary chondrites, the accretion range was a 100-degree interval centered on 510^{+80}_{-60}°K (17). For the carbonaceous chondrites, this range seems to have been ~ 300° to ~ 400°K (15). No carbonaceous chondrites without magnetite are known, which suggests that accretion in the source region of carbonaceous chon-

drites commenced only below 400°K. Presumably the range above 400°K was traversed too rapidly for any significant accretion to take place.

Formation of hydrated silicates may also have provided the cations needed for the carbonate. These silicates, although mineralogically ill-defined (18), have a distinctly lower cation-to-silicon ratio than the olivine from which they seem to be derived (2, 15). In a gas of cosmic composition at a total pressure of 10^{-4} atm, formation of hydrated silicates is expected to commence below about 350°K (15). The following equations schematically represent the process:

$$12(Mg,Fe)_2SiO_4 + 14H_2O \rightarrow$$
$$2Fe_3O_4 + 2H_2 + 3(Mg,Fe)_6(OH)_8Si_4O_{10}$$
$$(3)$$

$$4(Mg,Fe)_2SiO_4 + 4H_2O + 2CO_2 \rightarrow$$
$$2(Mg,Fe)CO_3 + (Mg,Fe)_6(OH)_8Si_4O_{10}$$
$$(4)$$

Just as in terrestrial serpentinization reactions (19), some basic cations are left over. The excess cations can appear either as magnetite (Eq. 3) or carbonate (Eq. 4). Basic oxides react rapidly with CO_2 and would thus remove it from the gas phase fast enough to prevent secondary reactions or isotopic equilibration.

Alternatively, the organic polymer in meteorites may have a formation temperature close to that of methane. Thermodynamic calculations (20) have shown that aromatic polymers may become the favored product under a variety of conditions. Moreover, an essential first step in the Fischer-Tropsch synthesis is adsorption of CO on the surface of the catalyst. At the low partial pressures of CO expected in the solar nebula (10^{-7} to 10^{-8} atm), it is conceivable that effective adsorption may not commence until temperatures have fallen low enough to permit the formation of products other than methane.

Our experiments also provide a very tentative clue to the initial isotopic composition of the carbon in the nebula. Carbonaceous chondrites contain only about 6 percent of their cosmic complement of carbon; the remainder was presumably left behind in the nebula as compounds of low molecular weight. Assuming that CO was the sole starting compound, and comparing the CO_2-CO-wax fractionation in our experiments with the polymer-carbonate fractionation in type II carbonaceous chondrites, we would estimate an initial carbon composition of +15 per mil on

the PDB scale. This is remarkably close to the isotopic composition of carbon in lunar soil (21), which is a mixture of indigenous, meteoritic, and solar-wind carbon.

References and Notes

1. J. M. Hayes, *Geochim. Cosmochim. Acta* **31**, 1395 (1967).
2. E. R. DuFresne and E. Anders, *ibid.* **26**, 1085 (1962).
3. J. W. Smith and I. R. Kaplan, *Science* **167**, 1376 (1970). For the extractable organic matter, values from −5 to −27 per mil have been reported [see also H. R. Krouse and V. E. Modzeleski, *Geochim. Cosmochim. Acta* **34**, 459 (1970) and unpublished work by T. C. Hoering and E. Anders], but at least some of these results may have been affected by contamination. All δC^{13} values are given relative to the PDB standard [H. Craig, *Geochim. Cosmochim. Acta* **12**, 133 (1957)].
4. R. N. Clayton, *Science* **140**, 192 (1963).
5. P. Deines, *Geochim. Cosmochim. Acta* **32**, 613 (1968).
6. Y. Bottinga, *ibid.* **33**, 49 (1969).
7. P. E. Yankwich and R. L. Bedford, *J. Amer. Chem. Soc.* **75**, 4178 (1953); M. Zielinski, *J. Chem. Phys.* **47**, 3686 (1967).
8. P. H. Abelson and T. C. Hoering, *Proc. Nat. Acad. Sci. U.S.* **47**, 623 (1961); S. Park and S. Epstein, *Plant Physiol.* **36**, 133 (1961). A larger fractionation (90 per mil) has been seen in the bacterial degradation of methanol in the laboratory [W. D. Rosenfeld and S. R. Silverman, *Science* **130**, 1658 (1959)] but seems to have few, if any, parallels in nature.
9. H. C. Urey, *Quart. J. Roy. Astron. Soc.* **8**, 23 (1967).
10. M. H. Studier, R. Hayatsu, E. Anders, *Geochim. Cosmochim. Acta* **32**, 151 (1968); R. Hayatsu, M. H. Studier, A. Oda, K. Fuse, E. Anders, *ibid.*, p. 175.
11. P. H. Emmett, Ed., *Catalysis* (Reinhold, New York, 1956), vol. 4, p. 9.
12. A Fischer-Tropsch catalyst containing 38 percent Co was prepared by precipitating cobalt carbonate on Filter-Cel diatomaceous earth (Johns-Manville), according to the procedure of Emmett (*11*, p. 46). The catalyst was reduced in H_2 at 300°C and at a pressure of 1 atm before reaction. The carbon monoxide used in our reactions was purified from traces of CO_2 by overnight exposure to a cold trap at −196°C.
13. A. O. C. Nier, *Rev. Sci. Instrum.* **18**, 398 (1947); C. R. McKinney, J. M. McCrea, S. Epstein, H. A. Allen, H. C. Urey, *ibid.* **21**, 724 (1950).
14. H. Craig, *Geochim. Cosmochim. Acta* **3**, 53 (1953).
15. J. W. Larimer and E. Anders, *ibid.* **31**, 1239 (1967); E. Anders, *Accounts Chem. Res.* **1**, 289 (1968).
16. A. G. W. Cameron, *Icarus* **1**, 13 (1962); ———, in *Meteorite Research*, A. Oda, Ed. (Reidel, Dordrecht, 1969), pp. 7–15; S. P. Clark, K. K. Turekian, L. Grossman, in preparation.
17. R. R. Keays, R. Ganapathy, E. Anders, *Geochim. Cosmochim. Acta*, in press.
18. J. F. Kerridge, in *Mantles of the Earth and Terrestrial Planets*, S. K. Runcorn, Ed. (Interscience, London, 1967), pp. 35–47.
19. I. Barnes, V. C. LaMarche, Jr., G. Himmelberg, *Science* **156**, 830 (1967).
20. M. O. Dayhoff, E. R. Lippincott, R. V. Eck, *ibid.* **146**, 1461 (1964); R. V. Eck, E. R. Lippincott, M. O. Dayhoff, Y. T. Pratt, *ibid.* **153**, 628 (1966).
21. S. Epstein and H. P. Taylor, Jr., *ibid.* **167**, 533 (1970); I. R. Kaplan and J. W. Smith, *ibid.*, p. 541.
22. We are indebted to Prof. R. N. Clayton for the use of his mass spectrometer and for valuable criticism. Mrs. T. Mayeda kindly provided advice and assistance. Work supported in part by NASA grant NGL 14-001-010.

29 June 1970; revised 30 September 1970

II
Carbonaceous Meteorites—
In Support of
Chemical Evolution

Editor's Comments on Papers 9 Through 14

9   **Mueller:** *The Properties and Theory of Genesis of the Carbonaceous Complex Within the Cold Bokevelt Meteorite*

10   **Nagy, Meinschein, and Hennessy:** *Mass Spectroscopic Analysis of the Orgueil Meteorite: Evidence for Biogenic Hydrocarbons*

11   **Hayes:** *Organic Constituents in Meteorites—A Review*

12   **Gelpi and Oró:** *Organic Compounds in Meteorites: IV. Gas Chromatographic–Mass Spectrometric Studies on the Isoprenoids and Other Isometric Alkanes in Carbonaceous Chondrites*

13   **Smith and Kaplan:** *Endogenous Carbon in Carbonaceous Meteorites*

14   **Kvenvolden, Lawless, Pering, Peterson, Flores, Ponnamperuma, Kaplan, and Moore:** *Evidence for Extraterrestrial Amino-Acids and Hydrocarbons in the Murchison Meteorite*

At one time carbonaceous meteorites were thought to represent some of the well-mixed rubble from the inner solar system; now they are generally believed to be relatively unaltered condensates from the early solar nebula (Anders, 1971). They contain a unique record of the chemistry that took place perhaps 4.6 billion years ago and seem to provide evidence in support of the theory of chemical evolution. Because the organic chemistry of meteorites may reflect some of the early chemistry that led to life, six representative papers are included here.

That carbonaceous meteorites contain organic substances has been known for more than a century. Berzelius (1834) extracted complex organic substances from the Alais meteorite, which fell in France in 1806, and he wondered about the significance of his findings and their relationships to the possibility of extraterrestrial life. Between 1834 and 1953 studies of the organic chemistry of meteorites were undertaken only sporadically; not a single paper appeared between 1899 and 1953. But starting with the work of Mueller (Paper 9) and Nagy et al. (Paper 10), the field has gained momentum, and during the last 12 years significant advances have been made.

Mueller's paper (9) is included here for historical reasons. His analytical approaches were unique for their time but hardly sophisticated by modern standards. The important point is that this paper helped reignite interest in the organic chemisty of meteorites and set the stage for the application of modern analytical techniques. Mueller made a number of significant observations, which have been supported by more recent investigations. For example, he found that besides extractable organic substances, which he thought to be mainly organic acids, the Cold Bokkeveld meteorite, which fell in South Africa in 1838, contained a high proportion of an insoluble, carbonaceous complex. Because his extracts were optically inactive, he suggested that the organic material was formed through nonbiological processes. Also he concluded that the meteorite likely condensed under low-temperature conditions.

Eight years later much excitment was created when Nagy et al. (Paper 10) applied modern analytical techniques, particularly mass spectrometry, to distillates of the Orgueil meteorite, which had fallen in 1864 in France. They compared mass-spectral fragmentation patterns for normal, monocyclo-, bicyclo-, and tetracycloalkanes in butter, recent sediments, crude oils, and the Orgueil meteorite. From this comparison they discovered that the "hydrocarbons in the Orgueil meteorite resemble in many important aspects the hydrocarbons in the products of living things and sediments on earth." This observation led to the suggestion that the meteorite hydrocarbons provide evidence for biologic activity; hence evidence for extraterrestrial life could be inferred. This optimistic interpretation directly contrasts with the conclusions reached by Mueller (Paper 9) based on the optical inactivity of his extracts.

Following Nagy et al. there was a flood of scientific activity directed toward trying to ascertain if, indeed, the organic materials in meteorites signaled the existence of life outside the earth. Whether or not the observed organic compounds were biogenic or abiogenic became a highly controversial subject. Numerous studies were made and numerous papers were written. Hayes (Paper 11) made a particularly comprehensive review which covered this period of controversy, and his paper is included here because it serves as the most complete summary of the organic chemistry of meteorites prior to 1967. He concluded that contamination was a crucial factor in meteorite analyses and that most of the indigenous compounds that had been found could be accounted for by abiotic processes. He believed that, in the future, studies of organic substances in meteorites should not be directed to a search for extraterrestrial life but should be focused on the study of organic cosmochemistry. Other reviews and summaries are Briggs and Mamikunian (1963), Mason (1963), Urey (1966), Nagy (1966, 1968), Baker (1971), Ponnamperuma (1972), Lawless et al. (1972), and Oró (1972).

Detailed investigations continued. An example of the sophisticated application of modern techniques to a number of meteorite specimens can be found in the paper by Gelpi and Oró (Paper 12). In a comprehensive study, utilizing gas chromatography and mass spectrometry, of the saturated hydrocarbons in six different carbonaceous meteorites, they found 10 homologous series of alkanes, including the regular isophrenoid hydrocarbons, which, when found in terrestrial samples, are usually interpreted as being of biological origin. The authors did not choose to speculate about their results in the context of extraterrestrial life, but they did point out that "the contribution of our terrestrial environment to the aliphatic hydrocarbon patterns of carbonaceous chondrites may be significant enough as to mask the presence of any indigenous component."

Smith and Kaplan (Paper 13) also noted that terrestrial contamination was common in seven carbonaceous meteorites, but they did suggest that carbonate carbon, residual carbon, and part of the extractable organic material which they found was indigenous and of extraterrestrial origin. This paper is important because it reports the effective use of carbon isotopic abundance measurements on various meteorite fractions to interpret possible chemical processes that generated the organic materials in meteorites. For the first time nonrandom isotopic distribution of carbon in meteorites was demonstrated. In addition, carbon isotopic measurements permitted terrestrial contamination to be separated from indigenous, meteoritic organic carbon.

The fall of the Murchison meteorite in September 1969 in Australia has provided extraterrestrial material that has characteristics consistent with an abiotic synthesis

(Kvenvolden et al. Paper 14). Specimens from this fall were examined almost immediately and the problem of terrestrial contamination was minimized. One class of compounds of interest in this meteorite is the amino acids. The compounds had been found previously in other meteorites (Kaplan et al., 1963), but most, if not all, of the compounds observed could be accounted for as terrestrial contamination. The suite of amino acids recovered from the Murchison meteorite was different in composition from terrestrial contamination, and the amino acids with asymmetric centers were present as racemic mixtures. The presence of amino acids in the Murchison meteorite has been confirmed independently in three laboratories (Kvenvolden et al., 1971; Cronin and Moore, 1971; Oró et al., 1971). These results "represent, taken together, the first solid evidence for the presence of amino acids in meteorites which are probably indigenous and abiogenic" (Oró, 1972).

Other classes of organic compounds that have been found in the Murchison meteorite are hydrocarbons (Kvenvolden et al., Paper 14; Oro et al., 1971; Pering and Ponnamperuma, 1971; Studier et al., 1972), heterocyclics (Folsome et al., 1971, 1973; Anders et al., 1973), and fatty acids (Yuen and Kvenvolden, 1973). There is a lack of agreement concerning interpretation of some of the results on hydrocarbons and heterocycles. Sufficient time has not elapsed for an independent determination of the fatty acids to be made.

The rich collection of molecules of biological significance indigenous to carbonaceous meteorites was likely generated by nonbiological, cosmogenic processes. The nature of these processes is not yet well known, but certainly the molecules must have resulted from reactions involving simpler substances; that is, the molecules must have evolved by processes similar to those predicted by the theory of chemical evolution. Therefore, the finding of these molecules in meteorites is thought to lend support to this theory.

Anders et al. (1973) suggested that carbonaceous meteorites, falling on the primitive earth, may have made significant contributions to the earth's initial endowment of organic matter. It is intriguing to think that the organic molecules in this prebiotic organic matter may have served as precursors for life on earth, an idea developed by Breger et al. (1972). Thus the chemical evolution of organic molecules used in the first living systems on earth may not necessarily have originated in the primitive atmosphere of that planet, but may have been synthesized in meteoritic materials during the condensation of the solar nebula. Indeed, carbonaceous meteorites may be the "Rosetta stone" for understanding the origin of the solar system and also the origin of life.

Two investigators who were among the first to apply the techniques of organic geochemistry to carbonaceous meteorites were Bartholomew Nagy and Juan Oró.

Nagy was born in 1927 in Budapest, Hungary. He came to the United States and attended Columbia University and Pennsylvania State University. As a geochemist his early work was conducted in the petroleum industry, and his early scientific contributions were in petroleum geochemistry. His academic career includes appointments at Fordam, University of California at San Diego, and University of Arizona, where he is presently Professor of Geosciences. His interest have concentrated on organic geochemistry and on the organic chemistry of meteorites.

Juan Oró began his career as a baker in Barcelona, Spain. He switched careers early in life to follow interests in biochemistry; he is now Professor in Biophysical Sciences at the University of Houston and the University of Barcelona. He has been concerned with organic cosmochemistry and molecular and biological evolution. These concerns have led to his important contributions in organic geochemistry and in the application of organic geochemistry to the study of meteorites. He was one of the first to apply the sophisticated techniques of gas chromatography coupled to mass spectrometry to problems in organic geo- and cosmochemistry.

References

Anders, E. (1971) Meteorites and the early solar system: *Ann. Rev. Astron. Astrophy.*, **9**, 1–34.

Anders, E., R. Hayatsu, and M. H. Studier (1973) Organic compounds in meteorites: *Science*, **182**, 781–790.

Baker, B. L. (1971) Review of organic matter in the Orgueil meteorite: *Space Life Sci.*, **2**, 472–497.

Berzelius, J. J. (1834) Uber Meteorsteine: *Ann. Phy. Chem.*, **33**, 113.

Breger, I. A., P. Zubovic, J. C. Chandler, and R. S. Clarke, Jr. (1972) Occurrence and significance of formaldehyde in the Allende carbonaceous chondrite: *Nature*, **236**, 155–158.

Briggs, M. H., and G. Mamikunian (1963) Organic constituents of carbonaceous chondrites: *Space Sci. Rev.*, **1**, 647–682.

Cronin, J. R., and C. B. Moore (1971) Amino acid analysis of the Murchison, Murray, and Allende carbonaceous chondrites: *Science*, **172**, 1327–1329.

Folsome, C. E., J. Lawless, M. Romiez, and C. Ponnamperuma (1971) Heterocyclic compounds indigenous to the Murchison meteorite: *Nature*, **232**, 108–109.

Folsome, C. E., J. G. Lawless, M. Romiez, and C. Ponnamperuma (1973) Heterocyclic compounds recovered from carbonaceous chondrites: *Geochim. Cosmochim. Acta*, **37**, 455–465.

Kaplan, I. R., E. T. Degens, and J. H. Reuter (1963) Organic compounds in stony meteorites: *Geochim. Cosmochim. Acta*, **27**, 805–834.

Kvenvolden, K. A., J. G. Lawless, and C. Ponnamperuma (1971) Nonprotein amino acids in the Murchison meteorite: *Proc. Natl. Acad. Sci. (U.S.)*, **68**, 486–490.

Lawless, J. G., C. E. Folsome, and K. A. Kvenvolden (1972) Organic matter in meteorites: *Sci. Am.*, **226**, 38–46.

Mason, B. (1962-1963) The carbonaceous chondrites: *Space Sci. Rev.*, **1**, 621–646.

Nagy, B. (1966) Investigations of the Orgueil carbonaceous meteorite: *Geol. Foren. Stockholm. Forhand.*, **88**, 235–272.

Nagy, B. (1968) Carbonaceous meteorites: *Endeavour*, **27**, 81–86.

Oró, J. (1972) Extraterrestrial organic analysis: *Space Life Sci.*, **3**, 507–550.

Oró, J., J. G. Gilbert, H. Lichtenstein, S. Wikström, and D. A. Flory (1971) Amino acids, aliphatic and aromatic hydrocarbons in the Murchison meteorite: *Nature*, **230**, 105–106.

Pering, K. L., and C. Ponnamperuma (1971) Aromatic hydrocarbons in Murchison meteorite: *Science*, **173**, 237–239.

Ponnamperuma, C. (1972) Organic compounds in the Murchison meteorite: *Ann. N.Y. Acad. Sci.*, **194**, 56–70.

Studier, M. H., R. Hayatsu, and E. Anders (1972) Origin of organic matter in early solar system: V. Further studies of meteoritic hydrocarbons and a discussion of their origin: *Geochim. Cosmochim. Acta*, **36**, 189–215.

Urey, H. C. (1966) Biological materials in meteorites: review: *Science*, **151**, 157–166.

Yuen, G. U., and K. A. Kvenvolden (1973) Monocarboxylic acids in Murray and Murchison carbonaceous meteorites: *Nature*, **246**, 301–303.

Additional Suggested Readings

Gibson, E. K., C. B. Moore, and C. F. Lewis (1971) Total nitrogen and carbon abundances in carbonaceous chondrites: *Geochim. Cosmochim. Acta*, **35**, 599–604.

Han, J., B. R. Simoneit, A. L. Burlingame, and M. Calvin (1969) Organic analysis on the Pueblito de Allende meteorite: *Nature*, **222**, 364–365.

Krouse, H. R. and V. E. Modzeleski (1970) C^{13}/C^{12} abundances in components of carbonaceous chondrites and terrestrial samples: *Geochim. Cosmochim. Acta*, **34**, 459–474.

Lawless, J. G., K. A. Kvenvolden, E. Peterson, C. Ponnamperuma, and C. Moore (1971) Amino acids indigenous to the Murray meteorite: *Science*, **173**, 626–627.

Lawless, J. G., K. A. Kvenvolden, E. Peterson, C. Ponnamperuma, and E. Jarosewich (1972) Evidence for amino acids of extraterrestrial origin in the Orgueil meteorite: *Nature*, **236**, 66–67.

Levy, R. L., C. J. Wolf, M. A. Grayson, J. Gilbert, E. Gelpi, W. S. Updegrove, A. Zlatkis, and J. Oró (1970) Organic analysis of the Pueblito de Allende meteorite: *Nature*, **227**, 148–150.

Levy, R. L., M. A. Grayson, and C. J. Wolf (1973) The organic analysis of the Murchison meteorite: *Geochim. Cosmochim. Acta*, **37**, 467–483.

Nooner, D. W., and J. Oró (1967) Organic compounds in meteorites: I Aliphatic hydrocarbons: *Geochim. Cosmochim. Acta*, **31**, 1359–1394.

Olson, R. J., J. Oró, and A. Zlatkis (1967) Organic compounds in meteorites: II Aromatic hydrocarbons: *Geochim. Cosmochim. Acta*, **31**, 1935–1948.

Oró, J., S. Nakaparksin, H. Lichtenstein, and E. Gil-Av (1971) Configuration of amino acids in carbonaceous chondrites and a Precambrian chert: *Nature*, **230**, 107–108.

Schildknecht, V. H., R. Keller, and K. Penzien (1971) Über die Kohlenstoffverbindungen des Meteoriten von Essebi: *Chem. Zeit.*, **19**, 807–819.

9

Reprinted from *Geochim. Cosmochim. Acta*, **4**, 1–10 (1953) with permission of Microforms International Marketing Corporation as exclusive copyright licensee of Pergamon Press journal back files.

The properties and theory of genesis of the carbonaceous complex within the cold bokevelt meteorite

G. Mueller, B.Sc., Ph.D.

Dept. of Geology, University College, London, W.C.1

(*Received* 4 *December* 1952)

ABSTRACT

The carbonaceous substances within the Cold Bokevelt Meteorite and other similar types of stones show uniform distribution. The Cold Bokevelt carbonaceous stone proved to contain over 1% extractable organic substances. The results of analyses, de-ashing experiments, etc., indicate that in addition the stone may bear a considerably higher proportion of carbonaceous matter. The extracts consist mainly of complex organic acids, they contain some organic chlorine, an element which has not been so far detected within any of the terrestrial bitumens. The fact that the extracts are optically neutral indicates that the organic matter may have been condensed from the atmosphere of the parent body through non-biological processes of polymerisation, etc. The low decomposition temperature of the finely divided carbonaceous substances seems to point to the hypothesis that the meteorite in question must have condensed under low-temperature conditions.

INTRODUCTION

Organic substances in meteorites were reported long ago, especially from 1835 to 1885; abstracts of most of the old references are given in the book of E. Cohen (1894). Claims were made that substances had been extracted with ether and alcohol from the following stones: Allais, Carcote, Cold Bokevelt, Collescipoli, Kaba, Meghei and Nagaya. Orgueil and Goalpara were recorded as yielding small quantities of carbonaceous residues after the elimination of silicates with acids. In addition, there are eleven other meteorites, in which the presence of organic substances was claimed on less positive grounds, such as dark grey coloration, odour on heating, etc. The old literature does not refer to the composition, structure or problems of origin of the carbonaceous matter within the meteorites, and the present research was undertaken in order to investigate these interesting questions.

EXPERIMENTAL WORK

In the collections of the British Museum there are samples from each of the twenty "carbonaceous" meteorites, referred to above. The stones are all chondrites with very little or no metal. Compared with the average chondrite, the majority are unusually loose in texture, particularly those stones, all without metal, in which the presence of organic substances was established by previous workers using reasonably reliable methods.

Under the binocular microscope, all the stones show uniform, light to dark grey coloration beneath a somewhat darker surface crust, the shade being the same throughout any specimen and even in different stones of the same shower. Thin sections invariably reveal "bituminous" substances, concentrated within the turbid, amorphous ground mass cementing the light coloured chondrules and the occasional phenocrysts. No discrete carbonaceous particles can be resolved

1

in the matrix even under the highest powers of magnification. The stones show as a rule no fluorescence in ultra-violet radiation. This fact is held to indicate the absence of free hydrocarbons, with a hydrogen-carbon ratio over 1·4, as such bitumens from terrestrial sources invariably fluoresce. The only exception is a faint brown luminescence in the surface crust of the Kaba meteorite, and indeed this cannot be resolved into discrete fluorescent particles with the microscope. It is interesting to note that this was the only stone from which ether extracts of vaselinous consistency were recorded (F. WÖHLER (1859) and W. HARDINGER (1859)). The extracts from the other meteorites were said to be more or less resinous.

In the course of research work during 1951–52 the hydrocarbons within the Cold Bokevelt meteorite were investigated in some detail, although the alcohol extracts of this stone had already been briefly described by F. Wohler and W. Hardinger (1859).

For purposes of organic microanalysis a small sample of the meteorite without surface crust was ground and treated with dilute hydrochloric acid in order to expel CO_2 from any carbonates, but no trace of carbonate could be detected. The sample was dried for 48 hours over calcium chloride at ordinary pressure, and on analysis was found to contain

$$
\begin{array}{rl}
C: & 2 \cdot 20\% \\
H: & 1 \cdot 53\% \\
S: & 1 \cdot 99\% \\
O, \text{ etc.:} & 4 \cdot 84\% \\
\text{Residue:} & 89 \cdot 44\%
\end{array}
$$

An X-ray examination of the partly de-ashed substance of the meteorite produced no evidence of graphite (see below), so it is likely that the whole of the carbon found originates from the hydrocarbon. Part of the H and O on the other hand might have been produced from the decomposition of hydrated silicates. The results of extraction indicate that a small proportion of the S was present in the elemental state; iron sulphide was not detected with the microscope.

In order to separate the carbonaceous substances, 12 g of finely-ground powder of the meteorite was treated with a mixture consisting of equal parts of concentrated HCl, HF, and water. After two weeks' digestion at room temperature the residue still contained over 60% of silicates. Subsequently the same sample was placed in a similar acid mixture for three months and then heated on a water bath for 100 hours. The insoluble residue was reduced by this further treatment to approximately 30%. It should be stressed that, under similar treatment, the insoluble residue of shales would have been reduced well below 2%. The difficulty experienced in the "de-ashing" of the meteorite might be due to the fine grains of silicates being coated with a protecting sheath of carbonaceous matter (see p. 7).

It was this impure, partially "de-ashed" residue, whose X-ray powder diagram showed no lines of the graphite pattern.

A 20·0 g sample of the powdered meteorite was distilled and the products of distillation condensed at −50°C. The temperatures given on the left in the table

2

were measured at the neck of the flask, and it seems from the thermal decomposition of another small sample (see below), that the substance overheats to a considerable extent during distillation. The following fractions were separated:

95–105°C: 0·783 g (3·93%) Water with a slight trace of finely divided sulphur.

105–200°C: 1·584 g (7·42%) Water with 0·0086 g (0·043%) sulphur, coloured dark brown by a trace of carbonaceous substances.

200–500°C: 0·521 g (2·60%) Water with 0·0042 g (0·21%) sulphur, coloured black with traces of carbonaceous substances.

Loss: 0·461 g (2·30%) Inflammable gases, not analyzed.
Residue: 16·638 g (83·190%)

The watery fractions had an aromatic, rather spicy odour, different from that of the distillation products of terrestrial bitumens. The meteoritic powder became considerably darker—nearly black—during heating.

After the distillation the residue was placed in a vessel over water in order to determine the percentage of moisture which could be reabsorbed, with the following results:

	Residue: wt. in g		%
After heating to 500°C:	16·638	Loss:	16·810
After 1 day standing:	16·775	Regained:	0·685
After 2 days standing:	16·889	Regained:	1·265
After 5 days standing:	16·953	Regained:	1·575
After 10 days standing:	17·061	Regained:	2·115
After 20 days standing:	17·064	Regained:	2·130
After 30 days standing:	17·066	Regained:	2·140
After 40 days standing:	17·067	Regained:	2·145
After 50 days standing:	17·067	Regained:	2·145
After 100 days standing:	17·067	Regained:	2·145
After 250 days standing:	17·067	Regained:	2·145

The above results show that of the total of 13·95% water lost during heating of the meteorite powder to 500°C, only 2·15% were re-absorbed at room temperature in a saturated atmosphere. Hence it seems that the bulk of the water was chemically attached to the carbonaceous complex and the hydrated silicates within the meteorite. It is most unlikely, therefore, that the water content of the meteorite would be of terrestrial origin, especially as the introduction of over 10% moisture into the molecules constituting the stone would have certainly caused some visible signs of chemical alteration, and most likely a complete decomposition to powder.

3

The thermal decomposition of a small powdered sample of the meteorite was determined by maintaining it for 5 hours at a time at temperatures ranging from 50 to 350°C.

Temp. °C	Weight g	Loss g	Loss per cent
—	0·1382	—	—
50	0·1382	0·0000	0·00
75	0·1380	0·0002	0·15
100	0·1379	0·0003	0·22
125	0·1376	0·0006	0·43
150	0·1371	0·0011	0·80
175	0·1369	0·0013	0·94
200	0·1366	0·0016	1·16
225	0·1361	0·0021	1·52
250	0·1357	0·0024	1·73
300	0·1350	0·0032	2·32
350	0·1268	0·0114	8·25

After heating to red heat the residue weighed 0·1160 g, the total loss being 16·05%. This is slightly less than the loss of 16·81% by the distillation of the larger sample in a closed vessel, possibly because, during the prolonged heating of the small sample in an open crucible, some oxygen may have combined with the ferrous iron.

12·5 g of finely-powdered meteorite was extracted for 8 hours each at the boiling point, with the following solvents; methyl alcohol, ethyl alcohol, chloroform, benzene and carbon disulphide. It was found that a more prolonged extraction with a given solvent removed only slight additional traces of organic substances. The extractions for 8 hours at the boiling points of solvents yielded:

Methyl alcohol: 0·508%
Ethyl alcohol: 0·204%
Chloroform: 0·160%
Benzene: 0·242%
Carbon disulphide: 0·002%

Total: 1·116%

Another sample of 10·0 g, was extracted under identical conditions with the same solvents except carbon disulphide in the reverse order:

Benzene: 0·655%
Chloroform: 0·017%
Ethyl alcohol: 0·424%
Methyl alcohol: 0·029%

Total: 1·125%

From the results ethyl alcohol and benzene were evidently the best extracting agents.

4

The extracts were insoluble in water, but soluble in all the organic solvents. All had an identical, rather aromatic, spicy smell. There were slight differences however in colour and fluorescence: the benzene extracts had a dark brown colour and brown fluorescence, the extracts with other solvents were of light brown colour and greenish-yellow fluorescence. The consistency was soft and resinous, extracts with chloroform being particularly soft, almost semi-liquid. None of the extracts was attacked by acids; with alkalies they formed brown solutions, with precipitation of some iron oxide.

The extracts were subsequently mixed and re-precipitated from alcohol. Some sulphur crystals, representing 10 to 12% of the total weight were eliminated by hand picking. Analyses of the remainder yielded:

C:	19·84%	24·26%	excluding ash
H:	6·64%	8·12%	,, ,,
N:	3·18%	4·00%	,, ,,
S:	7·18%	8·78%	,, ,,
Cl:	4·81%	5·89%	,, ,,
O, etc.:	40·02%	48·95%	,, ,,
Ash:	18·33%		
	100·00%	100·00%	

The halogen found was calculated as chlorine, because this would be presumably the dominant element present from the Cl-Br-I group. The presence of halogen is of considerable theoretical interest, since no elements of this group have so far been recorded from any of the terrestrial bitumens. A 0·02 g sample of the extracts was treated with an acid solution of silver nitrate, in order to determine any halogen which might have been extracted by the solvents in the form of inorganic salts. At room temperature no trace of precipitate could be detected, and after 15 minutes boiling there was only a very slight turbidity in the silver nitrate solution, which could have been caused by some of the organic halogen derivatives, decomposing spontaneously. Thus the above experiment indicates that certainly the bulk of the halogen in the extracts was organic.

The thermal decomposition of the extracts was measured by heating a small sample for 5 hours to temperatures ranging from 100 to 350°C:

Temp. °C	Weight g	Loss g	Loss per cent
—	0·00510	—	—
100	0·00475	0·00035	6·9
125	0·00440	0·00070	13·7
150	0·00420	0·00090	17·6
175	0·00385	0·00125	24·5
200	0·00315	0·00195	38·2
225	0·00260	0·00250	49·0
250	0·00245	0·00265	52·0
300	0·00240	0·00270	52·9
350	0·00230	0·00280	54·9

5

The residue was black, evidently containing carbon, and after ignition it left a brown ash. It is interesting to note that the extracts of the meteorite lost weight at a faster rate, at relatively low temperatures, than the powdered stone, possibly because the decomposition temperature of the hydrated silicates was higher than that of the carbonaceous complex, particularly its extractable portions.

From another portion of the extracts a benzene solution was made in order to determine the optical rotation. The optical rotation of a 5-cm column of 5% solution proved to be below the sensitivity of the instrument, that is less than 0·02°.

The residue of the second extraction was treated with a large excess of concentrated $H_2SO_4 + HNO_3$ mixture in order to nitrate any highly polymerized aromatic bodies in the complex carbonaceous substances, and make these extractable. The residue was filtered after dilution and neutralization and extracted with benzene and nitrobenzene. The solvents extracted 0·006 g and 0·011 g respectively. The extracts were dark brown, and insoluble in alcohol. Further investigation was precluded by the small quantity of the substances.

Theoretical Considerations

1. Structure of the carbonaceous complex

The finely divided carbonaceous complex within the ground mass of the Cold Bokevelt meteorite differs in two fundamental aspects from bituminous substances of the terrestrial sedimentary rocks. (1) On distillation, whereas the meteorite yields only water, most of the shales, etc., produce traces at least of liquid hydrocarbons, mainly paraffinic. (2) The carbonaceous rocks of terrestrial origin are not extractable as a rule with solvents, whereas the meteorite yields well over 1% of chlorinated resinous extracts. For the latter reason the term "retigen" is suggested for the carbonaceous complex of the Cold Bokevelt meteorite in contrast to the "kerogen" of shales. The extracts of the carbonaceous complex might be termed "chlorobitumens."

The high over-all percentage of oxygen in the extracts, and presumably in the whole carbonaceous complex, seems to account for the absence of liquid hydrocarbon fractions, for these would tend to decompose to water and carbon.

The good alkali solubility and the bulk composition of the extracts suggest that these are essentially complex organic acids with some substituted N, S and halogen; and possibly with small quantities of other elements that have not been so far detected. The terrestrial hydrocarbon mineral which the extracts most closely resembles is dopplerite, a calcium salt of humic acids. It is true that no water-extractable inorganic salts could be detected in the extracts of the meteoric hydrocarbon; a high percentage of the acid radicals seems to form salts with the iron and possibly other metals, and these are left behind in the ashes after ignition.

The original carbonaceous complex may be of a structure mainly similar to that of its extractable portions, though richer in carbon and possibly containing some free amorphous C as well.

The extracts by the various solvents all seem to be heterogeneous; that is, they contain substances of differing solubility, some of which fall out as flakes from relatively concentrated solutions, whereas others leave a film after the complete

6

evaporation of the solvent. There can be little doubt, particularly if we consider experience with terrestrial hydrocarbons, that both the original carbonaceous complex and its extracts contain a considerable variety of molecules of complicated structure, though the bulk of these belong to the above mentioned acidic type. It is unlikely that sufficient material would be available from any of the carbonaceous meteorites for an attempt to separate and identify some of the substances, but it is conceivable that infra-red and ultra-violet spectroscopy, X-ray analysis, determination of saponification value, and application of other organic chemical methods on somewhat larger amounts of the organic substance, might yield most interesting information concerning the proportion of radicals and the main types of molecules present. Complete analyses of the carbonaceous and inorganic phases of meteorites would give us the proportions of each element between these.

The organic constituents are likely to contain the residues of the more volatile elements (with exception of the inert gases), the bulk of which would escape during the condensation of the meteoritic substance. Hence their compositions would give us valuable data for the computation of the relative abundance of elements within the universe. Thus, for example, the high proportions of S, Cl, and particularly N, in the extracts of the Cold Bokevelt meteorite suggest that these elements must have been present in still higher proportions in the original substance from which the meteorite condensed. It should be stressed, however, that the affinity for carbon of a volatile element would also help to determine its proportion in the organic complex.

2. *Problem of genesis of the carbonaceous complex*

Regarding the problem of genesis of the carbonaceous substances in meteorites, only brief and tentative suggestions can be found in the literature.

P. E. SPIELMANN (1924) put forward a theory of formation of the hydrocarbons through hydrolysis of carbides, originally in the meteorites, by terrestrial water. This hypothesis should be discarded for the same reasons as the suggestion of a terrestrial origin for the water content (see p. 2). In order to produce over 1% extractable hydrocarbons, i.e. possibly from 3 to 5% of carbonaceous matter, a minimum of 15 to 25% of carbides would be necessary. The decomposition of such a high proportion of the stone would most certainly cause crumbling to powder, or at least the appearance of iron oxide stains, and these cannot be seen anywhere in the stone, notwithstanding its general uniformity.

It is often implied in popular works that the presence of organic substances in meteorites would indicate the existence of life on the celestial body of their origin. However, the extracts from the Cold Bokevelt meteorite proved to have no measurable optical rotation, and this fact must be held to indicate a non-biological origin. The earth's biologically formed carbonaceous minerals, such as oils, or retorting products of coals, etc., invariably show optical rotation.

In the light of modern organic chemical experience, there seems to be no great difficulty in accounting for a non-biological origin of the organic substances. The existence of CH, CN, and similar radicals in atmospheres of comets has been proved spectroscopically. It is reasonable to conjecture that under conditions of varying illumination and temperature a proportion of the constituents of such an

7

79

atmosphere would polymerize into complex molecules. These in turn might be collected on the fine dust particles, which may be expected to float through the atmosphere of any relatively small celestial body, and eventually settle on its surface. If the matrix of the Cold Bokevelt meteorite contains extremely small dust particles coated with hydrocarbons, this might possibly explain the difficulties experienced in its de-ashing.

The strikingly high oxygen percentage of the extracts, when compared with that of the average terrestrial hydrocarbons, might be connected with their non-biological origin. On the surface of the earth the green plants convert carbon dioxide into organic molecules of relatively lower percentage of oxygen and free oxygen through the process of photosynthesis. The plant and animal debris is further deoxygenated through micro-organisms and the physico-chemical processes of sedimentation. Without the activity of green plants and micro-organisms however, the initial deoxygenation does not take place, leaving the organic molecules richer in oxygen. According to general organic chemical experience, these acid substances are as a rule more stable, against loss of oxygen, than terrestrial plant debris that consists mainly of cellulose.

The presence of N, S, and Cl in the extracts implies, if our hypothesis proves to be the correct one, that the primitive atmosphere surrounding the original condensation contained these elements in relatively high proportions. The chlorine might have been present as HCl, when it could have entered the olefinic molecules only, or in elemental form, in which case it would easily displace H atoms from all types of hydrocarbons.

From the above considerations it seems that there may be a distinct similarity between the composition of the carbonaceous complexes on different celestial bodies, whether these are with or without life as we know it here on earth; and the role of organisms in the history of organic substances may not be so fundamental as is generally believed. Under a given set of physical conditions, reasonably similar substances may be formed either through biological or non-biological processes; the presence of bituminous substances of apparently magmatic origin in pegmatite veins seems to support the above suggestion. It would then seem that the most important factors in the formation of complicated organic molecules, in a system containing carbon, are the temperature and pressure of that system.

3. *The use of meteoritic carbonaceous complexes as "natural thermometers"*

Spectroscopic investigation of interstellar space indicates that the elements which readily enter carbon chains, that is O, H, N, S, etc., are present there in elemental form or as simple compounds with carbon. If we consider a condensing body within such a gaseous system of discrete atoms, simple radicals, or molecules, we see that different types of equilibria can develop according to the conditions prevailing in the condensing body. We are led to conclude that the carbonaceous complexes of both terrestrial rocks and meteorites possess considerable value as indicators of temperature and also possibly other physical and chemical conditions prevailing during their formation. This value as indicators is not seriously affected by the presence of living organisms.

8

Experience from a variety of terrestrial rocks clearly indicates that what may be symbolized as the "C—(H, O, N, S) equilibrium" in any given rock was generally reached when the rock was at the highest temperature in its history, and was as a rule little affected during subsequent epochs at lower temperatures. For example finely divided carbonaceous complexes that decompose in the range 100 to 500°C have been recorded so far only in sedimentary rocks. In the slightly metamorphosed slates, on the other hand, graphite is the only form of carbon that has been so far found. The metamorphism of carbonaceous sediments needs more careful study than it has yet received, but the incomplete data at present available suggest that carbonaceous substances yielding more than a trace of volatiles are absent from all the metamorphic rocks. Nor have finely divided hydrocarbons been recorded in the igneous rocks though, as previously mentioned, bituminous substances of magmatic origin are found in numerous pegmatite veins, which traverse igneous rocks. While studying the alteration of hydrocarbons by mineralizing solutions that has occurred in Derbyshire, I recently found that the bitumens are of relatively higher initial boiling points in those veins, which are believed to have been formed at a relatively higher temperature, on grounds of their high fluorite and low calcite percentage. Thus it seems, that the system C—(H, O, N, S) can be used as a "natural thermometer" with reasonable confidence, at least as a first approximation.

We can now consider the Cold Bokevelt meteorite, with its carbonaceous complex of low decomposition temperature, and with over 1% its total mass extractable by organic solvents. The implication is that this meteorite (and possibly others of similar properties) was formed at a low temperature, and never reached a high temperature after this first consolidation. From the table of decomposition of extracts it seems likely that the temperature did not exceed 200–350°C, providing the pressure was relatively low. Even under a very high pressure, however (which seems to be rather unlikely in view of most theories of the formation of meteorites), the temperature could not have risen as high as 600°C, for the complex molecules of high percentage of oxygen would then have decomposed, whatever the pressure.

Other chondrites and all the achondrites are likely, to judge from available records, to contain amorphous carbon, and on heating they give off gases, dominated as a rule by methane and carbon monoxide. Published data do not seem to rule out the presence of small quantities of carbonaceous substances of very low volatile proportion. This type of C—(H, O, N, S) equilibrium would put the maximal values of temperature throughout the history of the stones say between 500 and 1200°C.

Finally, the iron and stony iron meteorites rich in metal often contain graphite and yield mainly hydrogen on being heated. These are indications that they have had a high temperature epoch of over 1000°C during or after their condensation.

The above considerations suggest, then, that the carbonaceous chondrites are not igneous melts, but products of condensation at low temperature. The most likely supposition is that they originated as a mass of compacted fine dust, in which the chondrules and occasional phenocrysts were embedded. Neither the chondrules nor the phenocrysts appear to contain carbonaceous substances,

9

81

confirming the conclusions of independent mineralogical investigations, that both kinds of inclusion solidified at relatively high temperatures, presumably from melts. Thus the carbonaceous meteorites are fragmental aggregates.

By analogy it would seem that the "non-carbonaceous" chondrites and the achondrites condensed at a higher temperature, permitting the partial fusion of their dust particles, and hence their usually much firmer structure. It is interesting to note, in this connection, that G. P. Merrill (1921) postulated a tuff structure for the chondrites on petrological grounds.

By this study of carbonaceous complexes we seem to have found a clue to the approximate maximum temperature prevailing during and after the consolidation of a particular type of meteorite. This consideration should now be applied to all the general theories on the origin of meteorites, but a critical survey of hypotheses of meteorite formation in the light of the results of research on the carbonaceous complexes is outside the scope of the present paper.

I would like to thank Dr. W. Campbell Smith of the British Museum (Natural History), London, for his encouragement and kindness in placing at my disposal parts of the Cold Bokevelt Meteorite, and also Dr. M. H. Hey of the British Museum for providing facilities for the chemical work. Finally, I would like to convey my deep appreciation to Professors S. E. Hollingworth and E. P. Hughes of University College London, and Professor H. C. Urey, For. Mem. R.S., of University of Chicago of their constructive and highly encouraging criticism.

References

Cohen, E.	1894	Meteoritekunde. Stuttgart.
Hardinger, W.	1859	Sitzb. Math. Naturwiss. Akad. Wien **34**, 7
Merrill, G. P.	1921	Bull. Geol. Soc. Amer. **32**, 395
Spielmann, P. E.	1924	Nature, Lond. **114**, 276
Wöhler, F.	1859	Sitzb. Math. Naturwiss. Akad. Wien **33**, 205
Wöhler, F. and Hardinger, W.	1859	*ibid*, **34** 5

10

Reprinted from *Annals N.Y. Acad. Sci.*, **93**, 27–35 (June 5, 1961)

MASS SPECTROSCOPIC ANALYSIS OF THE ORGUEIL METEORITE: EVIDENCE FOR BIOGENIC HYDROCARBONS

Bartholomew Nagy*, Warren G. Meinschein†, and Douglas J. Hennessy*

*Department of Chemistry, Fordham University, New York, N. Y.

†Esso Research and Engineering Company, Linden, N. J.

Introduction

A carbonaceous chondrite fell at 8 P.M. on May 14, 1864, at Orgueil, Montauban, Tarn-et-Garonne in France (43° 53′ N, 1° 23′ E). After the appearance of a luminous meteor and detonations, both of which were observed by the local residents, approximately 20 stony fragments, the largest the size of a man's head but most of them as large as a fist, fell over an area of approximately 2 sq. mi. The fragments were collected shortly after the fall. This fact is ascertained by the observation that this meteorite disintegrates when contacted with water due to the dissolution of water soluble salts (see below); consequently, the fragments could not have been lying on the ground undetected and been subjected to rain or moisture. A "burnt" crust is clearly visible on all fragments, including the one analyzed in this study.

The Orgueil meteorite is well-documented (Daubrée, 1864, Laussedat, 1864, Kesselmeyer, 1864). There appeared 47 references in the literature between the years 1864 and 1894. Samples of the meteorite were distributed and are present in several collections. The approximate combined weight of all the fragments is 11,523 gm.; the largest sample, reported to weigh 9,266 gm., is in the Paris Museum.

The unusual chemical composition of this meteorite was noticed by Cloëz (1864*a*). TABLE 1 lists the chemical analysis.

Other analyses, such as the one by Pisani (1864) gave similar results, with the exception of the iron and magnesium oxides. Pisani listed 17 per cent magnesium oxide and 29.90 per cent combined iron oxides. This analyst also noted the presence of magnetite and a "serpentinelike" mineral. It is to be noted that iron-nickel metal has not been found in the meteorite. The organic components received early attention. Cloëz (1864*b*) and, later, Berthelot (1869) studied the organic constituents; the latter refers to the "coal-like" material in the Orgueil meteorite.

The composition of the meteorite, which includes hydrous silicates, iron oxide, elementary sulfur, water soluble salts, organic compounds, and a considerable amount of water of low-bonding energies is, of course, unlike the compositions of sedimentary or igneous rocks on earth. There

---

* This paper was presented at a joint meeting of the Section of Chemical Sciences and the Section of Geological Sciences on March 16, 1961.

TABLE 1

Component	Per cent composition	
	Not dried	dried at 110° C. temperature
SiO_2	24.475	26.0310
Fe_2O_3	13.324	14.2360
FeO	17.924	19.0630
MgO	8.163	8.6711
S	4.369	4.6466
SO_4	2.195	2.3345
NiO	2.450	2.6057
CaO	2.183	2.3220
Al_2O_3	1.175	1.2498
Na_2O	1.244	1.3230
K_2O	0.307	0.3265
MnO	1.815	1.9302
Cr_2O_3	0.225	0.2392
CoO	0.085	0.0904
Cl	0.073	0.0776
P	Trace	Trace
NH_3	0.098	0.1042
H_2O^-	5.975	--
H_2O^+	7.345	7.8120
Organic	6.027	6.4100
Total	96.442	99.4728

H_2O^+ designates water obtained above 110° C. temperature.

appears to be no olivine present in the stone (Mason, 1960, pointed out that olivine occurs in some carbonaceous chondrites). This suggests that the interior of the meteorite was not subjected to elevated temperatures. The chemical composition of the organic and mineral matter, as shown below, seems to preclude temperatures much in excess of 200° C.

The purpose of the present study was to identify some of the organic compounds in the meteorite through the application of modern methods of analysis. The results of the analysis have no meaning unless it can be shown, that (1) the sample analyzed was a meteorite and (2), the organic constituents are not the products of contaminations acquired after the fall and during storage. The first point needs no discussion because the authenticity of the meteorite sample has been fully established In order to be able to comment on the question of contaminations it is necessary to consider the organic compositions described below. Saturated hydrocarbons were found to be abundantly present. These compounds are estimated to fall in the parts per thousand range of solid meteorite, a concentration far higher than commonly found in soils on earth. Other volatile organic compounds are far less abundant. If the organic material analyzed had been caused by contaminations, such as bacteria growing in the sample, one would expect a higher concentration of nonhydrocarbon compounds because living things contain an abundance of proteins, carbohydrates, and fats, but only traces of hydrocarbons. The common, labile products of biogenesis have not been identified in the sample. The hydrocarbon composition preclude the possibility of contaminations from petroleum products or from mere dust accumulating on the sample during storage The fact that the hydrocarbon constituents can not be removed from the meteorite unless the sample is heated to above 100° C. temperature in vacuum indicates some type of an association between organic and mineral matter. Again, at least a part of the organic matter should be readily removable without the application of vigorous separation methods if it were only a loose accumulation of recent contaminants.

Analyses of the Meteorite Fragments and of the Solvent Extract Fractions

In the present study, a sample of the Orgueil meteorite obtained from the collection of the Museum of Natural History, New York, N. Y., has been subjected to solvent extraction, infrared, and ultraviolet analyses, and to mass spectrometric analyses of fragments and distillates as well as to thermal balance and X-ray diffraction studies.

In all experiments precautions were taken to exclude impurities. Glassware was acid-cleaned and baked out under vacuum. Mortars and pestles were ignited to red-hot temperature for at least a one-half hour period; no rubber or plastic implements or stopcock grease were used and only spectral grade or doubly distilled solvents were used.

Investigations with the thermal balance have shown that the ground meteorite kept losing weight up to 825° C. temperature, where the experiment was terminated. The sample lost altogether approximately 25 per cent weight during the experiment. X-ray diffraction analysis has shown a broad and weak diffraction effect between 7 to 14 Å. Two sharp peaks, one of 2.97 Å and another one at 2.53 Å were always present. Chlorites or serpentine have been reported to be present in some stony meteorites (Kvasha, 1948 and Mason, 1960). The X-ray diffraction study shows the presence of some magnetite. It also suggests, together with the chemical analysis, that layer lattice silicates are present. The 14 Å diffraction effect may be caused by chlorites. Chlorite minerals are, of course, known to occur in many sedimentary rocks.

Finely ground particles of the meteorite appeared as irregular shaped, angular, and greenish-brown grains under the petrographic microscope. A few single crystals of elongated to fibrous habit were noted. The irregular particles slowly disintegrated when a drop of water was added to the meteorite powder on the microscope slide. During the evaporation of the water, well-developed hexagonal or pseudohexagonal crystals, showing low birefringence, crystallized out from the solution. These crystals dissolved immediately upon the addition of a second drop of water to the slide. Mason (personal communication, 1961) observed that approximately 20 per cent of an Orgueil meteorite sample was dissolved in water at $100°$ C. temperature.

A carbon disulfide extract of the ground sample showed a weak infrared absorption at 3.46 μ at a 10.0 μ. programmed slit opening with a Perkin-Elmer Model 21 infrared spectrometer, using NaCl optics. There was no absorption to indicate the presence of hydroxyl or carbonyl-containing molecules in the extract. Carbon tetrachloride extracts of the ground sample, diluted with 2,2,4-trimethylpentane to a volume/volume ratio of at least one-fiftieth gave indications of a general gradient on the absorption curve but showed no sharp absorption effects.

Two samples of the meteorite were subjected to distillation in a closed system that was under an initial pressure of 10^{-3} mm. Hg. The distillation apparatus was constructed from 8 mm. o.d. glass tubing, and it had the shape of an H. One arm, containing the sample, was subjected to heating; the other arm acted as a trap and was immersed into liquid air and it was connected to Hyvac and mercury vapor pumps. The system was sealed off from the pumps after evacuation.

The first sample, which weighed 0.7 gm., consisted of approximately equal quantities of fine powder and coarse particles of 20 to 40 mesh size. A fraction that distilled during 1 hour between $100°$ and $210°$ C. and condensed at room temperature was collected and analyzed by mass spectroscopy. The second sample of the meteorite that was distilled weighed 1.8 gm. and consisted entirely of the coarser particles. Mass spectroscopy was conducted on fractions that distilled during one-half hour between $150°$ and $250°$ C. and during 1 hour between $250°$ and $400°$ C., both condensing at room temperatures. A further sample was collected that distilled between $400°$ and $510°$ C. This last fraction has not been analyzed. It was noted that the pressure in the still was slightly above atmospheric at the conclusion of the distillations. The liquid air trap following the air condenser removed practically all of the relatively large amounts of distilling water from the less volatile organic distillates.

Analyses of the Volatile Fractions of the Meteorite

Mass spectrometric analyses were run on solid particles and on distillates of the meteorite that were vaporized in the heated inlet system of the mass spectrometer. Twelve mass spectra of the various samples were obtained. Compositely, these spectra covered the 10- to 600-mass range, and fragment and molecular (parent) ions were recorded at all unit-saturated hydrocarbon and saturated hydrocarbon fragment masses from 12 to 428. Above mass 428, no peaks of significant size occurred in the mass spectra of the meteorite samples. The largest hydrocarbon, a C_{31} tetracyclic compound of mass 428, and *n*-paraffins as large as C_{26} appeared only in the mass spectra of the 250° to 400° C. distillate fraction (sample 7). Because evidence of the biogenic origin of the saturated hydrocarbons on earth (Meinschein, 1960) is best established in molecules containing 17 to 35 carbon atoms, subsequent discussions will deal with the mass spectra and molecular species in sample 7.

Significantly, in the mass spectrum showing the mass peaks of the highest molecular weight hydrocarbon, the peak heights of the compounds with low parent ion sensitivities (such as *n*-paraffins and monocyclic naphthenes) decrease rapidly above C_{24} and are of negligible size above C_{27}; while compounds with high parent ion sensitivities have mass peaks of significant size for molecules containing up to 31 carbons. This discrepancy in carbon number between the largest *n*-paraffin, and the largest tetracyclic naphthene indicates that the upper mass limit recorded in the spectrum of sample 7 was probably controlled primarily by the distillation procedure used to obtain the volatile organic fraction rather than by the absence of high molecular weight biogenic hydrocarbons in the meteorite. Apparently, only the hydrocarbons containing 25 or less carbon atoms were quantitatively distilled at temperatures up to 400° C. in the sealed distillation unit described above.

TABLE 2 presents the parent ion peak heights for the C_{15} to C_{25} *n*-paraffins from butter. Recent sediments from earth and the Orgueil meteorite. The hydrocarbons from butter and Recent sediments are presented as composites of hydrocarbons from a variety of biogenic sources.

TABLE 2

n-Paraffin carbon number	Peak heights of saturated hydrocarbons in		
	Butter	Recent sediments	Meteorite
15	31	123	180
16	43	158*	170
17	117	143	198
18	334*	163	221*
19	61	181	176
20	62	203	113
21	119*	232	65
22	58	238	55
23	101*	271*	205*
24	58	243	186

* Peaks larger than the peaks of their homologs which contain either one more or one less carbon atoms (that is, peaking).

The peaking at 18 carbons of the meteorite *n*-paraffins is typical of the *n*-paraffins found in some biogenic products on earth as shown by the butter data; and the C_{23} peaking in the meteorite hydrocarbon fraction is observed in all saturated hydrocarbon mixtures in Recent sediments (Evans *et al.*, 1957) as well as all biogenic hydrocarbons that contain C_{22} through C_{24} *n*-paraffins.

Peak heights of the monocyclic naphthenes from butter, Recent sediments, and the meteorite are shown in TABLE 3.

As shown in TABLE 3, monocyclic naphthenes from living things and thus in Recent sediments frequently show peaking in the C_{18}- to C_{20}- range. Whereas the monocyclics from the meteorite do not display peaking in TABLE 3, the decrease in peak height with increasing carbon number is less between C_{17} and C_{18} than between any two successive carbon numbers in the C_{15} to C_{22} range.

TABLE 3

Monocycloalkane carbon number	Peak heights of saturated hydrocarbons in		
	Butter	Recent sediments	Meteorite
15	591	507	667
16	554	465	563
17	442	465	498
18	452*	462	445
19	373	504*	356
20	371	488	281
21	274	428	185
22	258	395	155
23	192	276	135
24	167	199	100

* Peaking.

TABLE 4 presents the peak heights of the bicycloalkanes in the three samples which are considered.

In the C_{17} to C_{18} interval, the rate of decrease of peak heights with increasing carbon number reaches a minimum in the Recent sediment hydrocarbons, while the C_{18} bicycloalkane in the meteorite sample shows peaking. The C_{20} peak in the butter is probably a diterpane, a saturated relative of the diterpenes which are present in many plants.

Tetracycloalkanes, previously referred to as steranes (Meinschein, 1959, 1960) appear in all naturally-occurring hydrocarbon mixtures. These compounds have distinctive cracking patterns in the mass spectrometer (Meinschein and Kenny, 1957) that resemble those of parent sterol hydrocarbons. Furthermore, the steranes in naturally occuring hydrocarbon mixtures on earth and in the meteorite show peaking in the C_{27}-C_{29} range (the carbon number range of the most abundant plant and animal steroids).

TABLE 4

Bicycloalkane carbon number	Peak heights of saturated hydrocarbons in		
	Butter	Recent sediments	Meteorite
15	138	954	267
16	118	791	215
17	97	692	225
18	83	626	288*
19	84	515	149
20	375*	478	137
21	83	327	103
22	60	232	102
23	48	157	64
24	34	102	61

* Peaking.

TABLE 5 shows the C_{15} to C_{29} peak heights for the tetracyclic hydrocarbons in the three samples.

The peaking at C_{16} shown in TABLE 5 is typical of all hydrocarbon mixtures containing steranes (Meinschein and Kenny, 1957), and many hydrocarbon fractions from Recent sediments and crude oils show, as do the hydrocarbons from the meteorite, peaking at the C_{24} tetracyclic. It is noteworthy that further fractionation of the butter sample yields a tetracyclic concentrate that, likewise, has C_{24} peaking. The fact that the meteorite tetracycloalkanes peak at C_{27}, rather than at C_{29} as do most naturally occurring hydrocarbons on earth, again suggests that the hydrocarbons may not have been quantitatively distilled from the meteorite.

TABLE 5

Tetracycloalkanes carbon number	Peak heights of saturated hydrocarbons in		
	Butter	Recent sediments	Meteorite
15	218	243	231
16	245*	437*	578*
17	215	283	302
18	141	216	230
19	108	161	99
20	117*	117	70
21	65	86	46
22	50	87*	39
23	43	85	45
24	41	129*	56*
25	34	79	50
26	35	67	44
27	59	101*	53*
28	59	100	46
29	68*	135*	39

* Peaking.

In addition to the hydrocarbon analyses, a calculation of the approximate S^{32}/S^{34} ratios in the meteorite was made using the S_6, S_7, and S_8 molecular ions in a mass spectrum of sample 7. These calculations gave values for S^{32}/S^{34} that ranged from 22.1 to 22.6. More accurate determinations of S^{32}/S^{34} in meteorites give this ratio the value of 22.2 (Ault and Kulp, 1959).

Conclusions

Although the distillates analyzed probably did not contain all the high molecular weight hydrocarbons in the meteorite and the distillates were not subjected to further fractionation, the mass spectrometric analyses reveal that hydrocarbons in the Orgueil meteorite resemble in many important aspects the hydrocarbons in the products of living things and sediments on earth. Based on these preliminary studies, the composition of the hydrocarbons in the Orgueil meteorite provide evidence for biogenic activity.

Acknowledgments

We acknowledge our appreciation to Brian Mason of the Museum of Natural History for making a sample of the Orgueil meteorite available for study and for his comments and suggestions. We are also thankful to F. R. Taylor, R. J. Morrissey, and Josefina Lugay for their assistance in the research.

References

AULT, W. V. & J. L. KULP. 1959. Isotopic geochemistry of sulfur, Geochim. et Cosmochim. Acta. 16: 201.

BERTHELOT, P. 1869. La matiere charbonneuse de la météorite d'Orgueil, purifiee autant que possible par les dissolvants, s'est ensuite oxydée entierément, J. prakt. Chem. 106, 254.

CLOËZ, S. 1864a. Note sur la composition chimique de la pierre météorique d'Orgueil. Compt. rend. 58: 986.

CLOËZ, S. 1864b. Analyse chimique de la pierre météorique d'Orgueil. Compt. rend. 59: 37.

DAUBREE, G. A. 1864. Noveaux renseignements sur le bolide du mai 1864. Compt. rend. 58: 1065.

EVANS, E. D., G. S. KENNY, W. G. MEINSCHEIN & E. E. BRAY. 1957. Distribution of n-paraffins and separation of saturated hydrocarbons from recent marine sediments. Anal. Chem. 29: 1858.

KVASHA, L. G. 1948. An investigation of the stony meteorite Staroe-Boriskino. Meteoritika. 4: 83.

KESSELMEYER, P. A. 1864. Der Meteorsteinfall zu Orgueil und Nohic bei Montauban in Südfrankreich am 14. Mai 1864, Pogg. Ann. Phys. Chem. 122: 654.

LAUSSEDAT, A. 1864. Sur la méthode employée pour déterminer la trajectoire du bolide du 14 mai. Compt. rend. 58: 1100.

MASON, B. 1960. The origin of meteorites, J. Geophys. Research. **65**: 2965.

MEINSCHEIN, W. G. & G. S. KENNY. 1957. Analyses of a chromotographic fraction of organic extracts of soils. Anal. Chem. **29**: 1153.

MEINSCHEIN, W. G. 1959. Origin of petroleum. Bull. Am. Assoc. Pet. Geol. **43**: 925.

MEINSCHEIN, W. G. 1960. Living Things — Major Producers of Petroleum Hydrocarbons, Paper presented at the Annual Meeting of the Geochemical Society in Denver, Colo. Prepared for publication.

PISANI, F. 1864. Etude chimique et analyse de l'aerolithe d'Orgueil. Compt. rend. **59**: 132.

Printed in the United States of America

11

Reprinted from *Geochim. Cosmochim. Acta*, **31**, 1395–1440 (1967) with permission of Microforms International Marketing Corporation as exclusive copyright licensee of Pergamon Press journal back files.

Organic constituents of meteorites—a review*

J. M. Hayes†

The Enrico Fermi Institute for Nuclear Studies
The University of Chicago, Chicago, Illinois, 60637
and
Department of Chemistry
Massachusetts Institute of Technology
Cambridge, Massachusetts, 02139

(*Received* 12 *January* 1967; *accepted in revised form* 13 *March* 1967)

Abstract—All meteoritic organic chemical analyses published since 1900 are critically reviewed. Inactive during the first half of the century, this field has been revived by the modern realities of spaceflight and exploration, but work has been concentrated on searching the organic constituents of the carbonaceous chondrites for compounds possibly indicative of extraterrestrial life. It is one theme of this paper that this is a poorly focussed effort. A second theme is that many compound identifications reported in the literature are, by present standards, insufficiently supported.

The characteristics and thermal histories of the various classes of chondrites are very briefly reviewed. It is shown that certain classes other than carbonaceous chondrites might reasonably be expected to contain organic material. The distribution of carbon among these classes is discussed, and the forms which carbon takes in each case are noted. It is shown that significant amounts of extractable organic material may be expected not only in the carbonaceous chondrites but also in the unequilibrated ordinary chondrites and in the ureilites. Evidence indicating that volatile materials, possibly including carbon and its compounds, are heterogeneously distributed in chondrite specimens is considered. Amounts of organic material extracted or volatilized from various meteorite specimens are tabulated.

Various crude but informative studies undertaken using unfractionated extracts are discussed. Data indicating the elemental composition and general chemical nature of meteorite extracts are tabulated and the infrared and ultraviolet absorption spectra of meteorite extracts and crude fractions are described and discussed. Isotope ratio analyses of extracts, volatilized and combusted materials, and whole stones are discussed and the carbon and hydrogen isotope ratios found in carbonaceous chondrite analyses are tabulated. Analytical studies proving or claiming to prove the presence in meteorites of particular organic compounds are discussed critically and in detail. Evidence indicating that certain analytical studies of hydrocarbons and of amino acids have been crucially affected by the presence of contaminating materials is considered. It is concluded that optically active compounds have not been proven to be present in meteorites.

Possible origins of meteorite organic compounds and the relations of these origins to theories of meteorite origin are briefly discussed.

Contents

* This review is drawn in part from the author's Ph.D. thesis, Department of Chemistry, M.I.T., 1966. Experimental work from this thesis will be published elsewhere (Hayes and Biemann, 1967a).

† Present address: Exobiology Division National Aeronautics and Space Administration, Ames Research Center, Moffett Field, California, 94035.

INTRODUCTION

Scope and literature coverage

A REVIEW of organic constituents of meteorites is, by default, essentially a review or organic constituents of carbonaceous chondrites. Among the carbonaceous chondrites, work has been concentrated on a few popular specimens. These unfortunate limitations notwithstanding, this paper attempts to review all twentieth-century analyses of organic constituents of meteorites. This task is simplified by the startling fact that not a single paper appeared on this subject between 1899 and 1953. For a summary of early analyses, the reader is directed to the excellent review by COHEN (1894; a brief supplement appears in Volume II, 1903). BRIGGS and MAMIKUNIAN (1963) comment briefly on these early analyses, as does NAGY (1966a) in his review of chemical and microscopic investigations of the Orgueil carbonaceous chondrite.

The closing date for the literature survey in this review is approximately 1 December 1966. Every effort has been made to evaluate carefully and objectively all relevant papers. This examination has occasionally shown that compound identifications accepted as conclusive by the original investigators and even by recent uncritical reviewers do not meet the standards of modern organic analysis.

Detailed consideration of the organic analyses has required that tangential material be kept to a minimum. Thus, the reader will not find "organized elements" (possible microfossils found in some meteorites) or abiogenic synthesis experiments discussed in this review. Inorganic chemistry, mineralogy, classification, and origin of meteorites are discussed only in a very limited way, the coverage being designed to provide necessary background and introductory information. When an occasional paragraph stops short of including an interesting but incompletely supported generalization or interpretation, the reason is that the variety and general nature of meteorites make such speculations very perishable. At present, progress in this field is best aided by a detailed, critical, and unified exposition of the analytical facts.

General references

The reader seeking a general introduction to the field of meteoritics will be best served by the book by MASON (1962). WOOD (1963b) has reviewed the chemistry and physics of meteorites in much more condensed form. ANDERS (1964) has discussed the origin, age, and composition of meteorites; and PEPIN and SIGNER (1965) have published a general review on the subject of primordial rare gases in meteorites. Previous reviews dealing with organic constituents are those of COHEN (1894, 1903), BRIGGS and MAMIKUNIAN (1963), and UREY (1966a). Finally, theories of meteorite origin are changing so rapidly that it is desirable to cite a few of the most recent publications: UREY (1966b), LARIMER and ANDERS (1967), WOOD (1966), and DODD, VAN SCHMUS, and KOFFMAN (1967) are all relevant.

Classification and composition

Meteorites are classified both by method of discovery (observed *falls* vs. *finds*) and by chemical and mineralogical composition. It is worth noting here that the mere fact that a certain meteorite is classified as a fall cannot be taken as proof of unaltered condition. Two carbonaceous chondrites provide good examples: there are many specimens of Murray, some of which remained on the ground (and were thus subject to alteration) for some time before they were recovered; and the initial report of the fall of the Bells carbonaceous chondrite (MONNIG, 1963) reads as follows, "The first fragment hit the roof of a house and was picked up next morning in a perfect, fresh state. The remainder was found after the passage of Hurricane Carla and some five inches of rain . . . ."

The chemical and mineralogical classification systems are described by MASON (1962) and WOOD (1963b) and will not be discussed here except in the case of the chondrites, for which a new classification scheme has recently been proposed. VAN SCHMUS and WOOD (1967) have described a new system which retains the well established chemical groups as divisions along one axis of what they call a "two-dimensional" classification system. The divisions along the second axis (*types*) are mineralogical and textural and thus reflect the degree of metamorphism experienced. While it can be misinterpreted (as described below), this system will be followed in this review because it provides the least confusing classification of the interesting chondrites containing approximately 0·3% carbon—the type III carbonaceous chondrites and the unequilibrated chondrites.

The chemical groups are as follows:

E. The enstatite chondrites (MASON, 1966).

C. The carbonaceous chondrites (MASON, 1963).

H, L. The high-iron and low-iron chondrites, originally recognized by UREY and CRAIG (1953). KEIL and FREDRIKSSON (1964) have clarified the boundaries of these groups which contain, in addition to the bronzite (H) and hypersthene (L) chondrites (MASON, 1962), a number of the unequilibrated chondrites described by DODD, VAN SCHMUS and KOFFMAN (1967).

LL. A group firmly established by KEIL and FREDRIKSSON (1964) and containing the amphoteric chondrites (MASON and WIIK, 1964) as well as a few of the unequilibrated chondrites.

The classification of a chondrite according to petrologic type is determined by the characteristics summarized in Table 1 (taken from VAN SCHMUS and WOOD, 1967). These types are not intended to imply the existence of a single metamorphic sequence six steps long. The E, H, L, and LL groups have examples only in type 3–6; and while there is evidence for a metamorphic relationship among stones of different types within each ordinary chondrite group, this is as far as the idea may be carried. In fact, the classification system does not depend on the existence of any metamorphic links, since, as the authors state, "The fact that rock A is more metamorphosed than rock B does not require that A be derived from B." Outside the ordinary chondrites, which are distinguished by the absence and presence (see Table 1) of chondrules noting that the best example of this is the impossibility of any link between the C1 and C2 chondrites, which are distinguished by the absence and presence (see Table 1) of chondrites These first two petrologic types have been created especially to make room for the type I and type II carbonaceous chondrites as described by WIIK (1956). Thus, they imply no metamorphic relationship at all but are included only to provide an integrated classification of all chondrites.

Thermal history

A high-temperature history for a given meteorite is obviously incompatible with the preservation of any indigenous organic compounds of earlier origin. Thus, it is well worth reviewing what is known of the thermal histories of meteorite classes in which organic compounds have been reported or are expected to be found. This discussion may be divided into three parts: first, metamorphism and other "parent-body" and formation effects; second, the temperature of a meteorite in space; and third, the temperature of a meteorite during its fall to earth.

In the C1 and C2 chondrites evidence of any metamorphism is in general lacking. In fact, there are some "thermometers" which place upper limits on temperatures within the meteorite since their insertion. BOATO (1954) showed that water of non-terrestrial isotopic composition is lost above 180°C. Studying devitrification of strained glass fragments found in the Mighei C2 chondrite, DU FRESNE and ANDERS (1961) have shown that this meteorite could not have been subjected to a temperature of 180°C for longer than two weeks, of 250°C for more than a fraction of an hour, and never to a temperature of 300°C.

WOOD (1966) has studied the distribution of nickel in the metal grains of nine C-group chondrites and has shown that the patterns found are due to equilibration in the temperature range 1400–1535°C (that is, before accretion of the chondritic material in its present form) and that cooling from that temperature was so rapid that taenite (γ nickel–iron) and kamacite (α nickel–iron) phases, characteristic of equilibration at lower temperature, have not formed.

DODD, VAN SCHMUS and KOFFMAN (1967) have described a class of ordinary chondrites which also show very little evidence of metamorphism. These have become known as the unequilibrated chondrites and include all the H, L and LL group members of type 3 as well as a few of lower types (see filled-in symbols, Fig. 1). The authors note in particular that Ngawi and Semarkona (LL3); Krymka, Bishunpur and Hallingeberg (L3); and Sharps and Tieschitz (H3) closely approximate unmetamorphosed material. Krymka and Bishunpur were also examined by WOOD (1966), who noted that these two chondrites, along with Vigarano (C3), appeared to be unmetamorphosed. Stones showing more evidence of metamorphism (WOOD, 1966) included Felix (C3) and Lancé (C3), and calculated cooling rates for these stones (temperature range 500–400°C) are in the range 0·2–0·6 C°/10^6 yr. Since this indicates a period of

Table 1. Summary of petrologic types (Van Schmus and Wood, 1967)

	Petrologic types					
	1	2	3	4	5	6
(i) Homogeneity of olivine and pyroxene compositions	—	Greater than 5% mean deviations		Less than 5% mean deviations to uniform	Uniform	
(ii) Structural state of low-Ca pyroxene	—	Predominately monoclinic		Abundant monoclinic crystals	Orthorhombic	
(iii) Degree of development of secondary feldspar	—	Absent		Predominately as microcrystalline aggregates		Clear, interstitial grains
(iv) Igneous glass	—	Clear and isotropic primary glass; variable abundance		Turbid if present	Absent	
(v) Metallic minerals (maximum Ni content)	—	(<20%) Taenite absent or very minor	kamacite and taenite present (>20%)			
(vi) Sulfide minerals (average Ni content)	—	>0.5%	<0.5%			
(vii) Overall texture	No chondrules	Very sharply defined chondrules		Well-defined chondrules	Chondrules readily delineated	Poorly defined chondrules
(viii) Texture of matrix	All fine-grained, opaque	Much opaque matrix	Opaque matrix	Transparent microcrystalline matrix	Recrystallized matrix	
(ix) Bulk carbon content	~2.8%	0.6–2.8%	0.2–1.0%		<0.2%	
(x) Bulk water content	~20%	4–18%		<2%		

* The enstatite chondrites constitute an important exception to this and a few of the other criteria noted here.

about 3×10^8 yr at temperatures above 400°C, it is important to point out that these measured cooling rates may be in error due to a failure of this analytical technique when applied to stony or stony-iron meteorites instead of irons, SHORT and KEIL (1966) having observed that cooling times longer than the age of the solar system are derived when the technique is applied to some stony-irons.

Evidence of metamorphism does not absolutely preclude the presence in a meteorite of indigenous extraterrestrial organic compounds, since such compounds could be formed by reaction of trapped gases after the high-temperature phase had passed (WOOD, 1966). On this basis it is not unreasonable to note that even the organic compounds reported in graphite nodules obtained from iron meteorites (NOONER and ORÓ, 1967; VDOVYKIN, 1964a) might possibly be indigenous.

ANDERS (1964) has discussed the temperature of meteorites in space as determined by studies of the rare gas contents, noting that "chondrites and achondrites spent the major part of their history at temperatures below about 40°C," a conclusion based on the fact that above this temperature helium losses would become serious. BUTLER (1966) has calculated the equilibrium temperatures of various types of meteorites at various orbital distances, finding the equilibrium temperatures at 1 A.U., for example, to be about 300°K for a dark chondrite, 270°K for a light chondrite, and 380°K for an iron.

Phenomena of fall in general depend primarily on meteorite size (MASON, 1962). Meteorites with pre-atmospheric masses of less than one ton completely lose all cosmic velocity while passing through the atmosphere and reach the surface of the earth in free fall. The time required for deceleration and its accompanying heating is very short, and ablation functions effectively. HOUTERMANS (1958) has shown by means of thermoluminescence studies that the effects of atmospheric heating in Holbrook, for example, extend to a depth of only 10–15 mm below the surface crust. Energy of impact is a significant source of heat only for meteorites which have retained part of their cosmic velocity.

In stones where the presence of organic material is well established, its existence is occasionally used to set a temperature limit. It might be said for example that the presence of nonane limits the temperature to 150°C. This argument is not at all convincing. The method of entrapment of volatiles is secure enough to retain noble gases boiling very much lower than nonane, and on this basis absurdly low maximum temperatures could be stated. It is more to the point to note that the organic compounds are not pyrolysed, but that does not place too stringent a limit.

QUANTITIES OF ORGANIC MATERIAL

Carbon and its distribution among forms

Figure 1 graphically summarizes the distribution of carbon in chondrites. It is, of course, not enough simply to know how much carbon is present—a stone containing several weight per cent of carbon could easily contain no organic compounds, the carbon being instead in the forms of graphite, carbides, carbonates and even diamond. Table 2 summarizes available information on the distribution of carbon among its possible forms.

In the carbonaceous chondrites most of the carbon is in the form of organic material. In C1 and C2 chondrites about 30% of this organic material is extractable with water and organic solvents. The remaining 70%, unextractable, is presumably polymeric. This unextractable material is definitely not graphite—a statement based on X-ray diffraction studies (VINOGRADOV and VDOVYKIN, 1964; BRIGGS and KITTO, 1962) and on electron spin resonance studies of the isolated carbonaceous material (SCHULZ and ELOFSON, 1965). There is no evidence for the presence of amorphous carbon, and various references to its presence (SISLER, 1961; WOOD, 1963b) have probably been intended to refer to the organic polymer. A small portion, definitely less than 5%, of the carbon in C1 chondrites is present in the

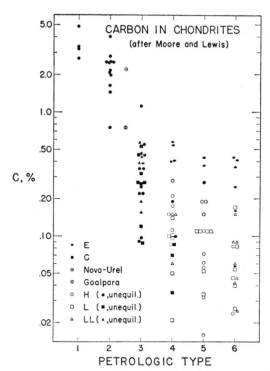

Fig. 1. Carbon in chondrites. Data for the carbonaceous chondrites are based on analyses by WIIK (MASON, 1963). Data for the enstatite chondrites are from MOORE and LEWIS (1966a), and data for the ordinary chondrites are from MOORE and LEWIS (1965, 1966b). The petrologic classifications are those proposed by VAN SCHMUS and WOOD (1967).

Table 2. Forms of carbon in various chondrite classes*

Carbon form	C	E	Ord.	Unequil.	Ureilites
Carbonate	—	—	—	—	—
	(+, C1)				
Extractable organic	+	(+, E4)	—	+	+
		(−, E6)			
Polymeric organic	+		(+)	(+)	(+)
Graphite	—	+	+	+	+
	(+, C3, 4)				
Diamond	—	—	—	—	+
Carbides	—	+	—	—	

* See text for references and details.
+ Present in unknown or variable amounts.
− Never found or definitely excluded.
Exceptions and doubtful notations are enclosed in parentheses.

form of carbonates (CLAYTON, 1963). The C3 and C4 chondrites are not as thoroughly studied, but individual members have been shown to contain graphite (Kainsaz, studied by VDOVYKIN, 1964b) and extractable organic compounds (for example, VDOVYKIN, 1964a, ORÓ et al., 1966; NOONER and ORÓ, 1967).

The principal form of carbon in the enstatite chondrites is graphite; cohenite, Fe_3C, having also been recorded (MASON, 1966). VDOVYKIN (1964a) was unable to extract organic material from Pillistfer (E6), and NOONER and ORÓ (1967) obtained only 1 $\mu g/g$ n-alkanes from Indarch (E4). There is not, however, any data available on total extractable organic material in an E3 or E4 chondrite.

As Fig. 1 shows, the amounts of carbon in ordinary chondrites are very small, and carbon forms are thus difficult to characterize. Carbonate has never been found in these metamorphosed stones, and extractable organic material is not expected. Of seven ordinary chondrites extracted by VDOVYKIN (1964a) only Ghubara (H5) yielded a measurable amount of material. Two L6 chondrites in plentiful supply, Holbrook and Bruderheim, have been used as controls in carbonaceous chondrite studies and have given acceptable blanks (MEINSCHEIN, NAGY and HENNESY, 1963; HODGSON and BAKER, 1964; NOONER and ORÓ, 1967). Polymeric material may be provisionally accepted as present on the basis of reports by HAYES and BIEMANN (1967, found ~50 ug/g polymer in Holbrook) and VDOVYKIN (1964b, "bitumens" observed microscopically in Ghubara and Kulp). VDOVYKIN (1964b) reports graphite in Ghubara, Farmington (L5), and Kulp (H) and RAMDOHR (1963) observed graphite in a number of ordinary chondrites. Searches for diamond in ordinary chondrites have produced only negative reports (UREY et al., 1957; SARMA and MAYEDA, 1961), and a recent report of diamonds in ordinary chondrites (VDOVYKIN, 1964b) is in error (LIPSCHUTZ, 1965).

The unequilibrated ordinary chondrites contain amounts of carbon ranging up to 1%. The only reported forms are extractable organic material and graphite. the presence of these forms makes it quite likely that some form of polymeric material is present, a speculation strengthened by VDOVYKIN's (1964b) observation of "bitumens" in Krymka (L3). VDOVYKIN (1964a) also reports the presence of extractable organic material in Krymka and, perhaps significantly, its absence in the more highly metamorphosed Mezö-Madaras (L3, see DODD, VAN SCHMUS, and KOFFMAN, 1967). NOONER and ORÓ (1967) extracted 2·7 $\mu g/g$ n-alkanes from a sample of Chainpur (LL3). Graphite is reported in Krymka by VDOVYKIN (1964b) and in Grady 1937 (H3) by RAMDOHR (1963), who made the interesting observation that the graphite resembled "that produced by thermal decomposition of hydrocarbons."

The ureilites are unique among the stone meteorites in that they contain diamond (UREY et al., 1957) accounting for about 50% of the carbon in the case of Novo-Urei (see COHEN, 1894, p. 141). VDOVYKIN (1964a) performed extractions of all three ureilites, obtaining an extract accounting for about 5% of the total carbon in each case. VDOVYKIN (1964b) reports the presence of graphite and "bitumens" in Dyalpur.

Some iron meteorites, particularly coarse octahedrites, contain substantial amounts of carbon. Typical analyses (e.g. BRENTNALL and AXON, 1962, Canyon Diablo 0·6%) indicate 0·2–0·6%, but in general amounts are not accurately known because large graphite nodules are frequently excluded from analytical samples. Cohenite (Fe_3C) is present mainly in irons containing 6–8% Ni (BRETT, 1966). COHEN (1894) has summarized reports indicating the presence of carbon in the form of dissolved carbon monoxide, and there are several reports of extractable hydrocarbons in graphite nodules (COHEN, 1894; VDOVYKIN, 1964a; NOONER and ORÓ, 1967).

Thus, at least two meteorite classes in addition to the carbonaceous chondrites—the unequilibrated ordinary chondrites and the ureilites—contain extractable organic material. Even ordinary chondrites seem to contain polymeric organic material, and it remains to be shown that the hydrocarbons in graphite nodules are definitely terrestrial contaminants. It is evident that more analytical effort should be devoted to the organic material in chondrites outside the C1 and C2 classes. From the standpoint of general meteoritics, investigations of these compounds may be particularly helpful, perhaps shedding light on relationships between chondrite classes and thus helping to elucidate the origin of meteorites.

Heterogeneity of volatile distribution

Evidence is rapidly accumulating that the distribution of volatile materials, possibly including carbon and its compounds, is, at least in some meteorites, very heterogeneous. REED and JOVANOVIC (1967) have obtained values ranging through two orders of magnitude for the mercury concentration in Orgueil (C1). ORÓ *et al.* (1966) find approximately the same range in Orgueil n-alkane concentrations. It is well established (PEPIN and SIGNER, 1965) that certain brecciated chondrites show great heterogeneity in rare gas distribution, and it now appears (MAZOR, HEYMANN, and ANDERS, manuscript in preparation) that similar heterogeneity is evident in at least eight carbonaceous chondrites including samples from each type. This heterogeneity may be attributed to different degrees of retention for different minerals or to shock-induced combination of fragments with different initial gas concentrations. Shock is the currently favored explanation, and it is therefore important to note that FREDRIKSSON (cited by MUELLER, 1966) has identified an inclusion of C1 chondrite material in Sharps (H3). MUELLER (1966) finds some C2 ("bunch of grapes") chondrules in Orgueil, and notes that some apparently C3 material appears in Mighei (C2), some C1 in Renazzo (C2), some C3 in Cold Bokkeveld (C2), and even some C3 in Orgueil. Thermogravimetric determination of the volatile content in eight grains of Mokoia (C2), a stone in which BRIGGS and KITTO (1962) observed carbon contents of from 0·5 to 2·1%, gave results ranging from 2·58–13·52% volatiles (MUELLER, 1966). The samples were from well below the fusion crust and the analyses are therefore apparently indicative of substantial heterogeneity. Five samples from Murray fell within the range 12·4–17·2% volatiles. While the only *direct* evidence for heterogeneous distribution of organic material is that obtained by ORÓ *et al.*, evidence for heterogeneity of any kind in these small bodies carries potentially broad significance.

Amounts of extractable organic material

Making the reasonable assumption that the organic material is roughly 70% C, calculation shows that organic material is present to the extent of about 50 mg/g in C1 chondrites and about 35 mg/g in C2 chondrites.

As has been mentioned above, by no means all of this organic material is extractable. All extractions shown in Table 3, with the exception of series 6, are of C2 chondrites. Table 3 shows that even a fairly complete series of solvents like those of series 3 succeeds in removing only 12·1 mg material per gram of meteorite. This is only about one third of the estimated total organic material, and there is no assurance that the weighed extracts are exclusively organic. The high values are given mainly by polar solvents (water, ethanol, methanol), and some dissolution of inorganic material can be expected. Therefore it is questionable that a value of

Table 3. Carbonaceous chondrites: amounts extractable by various solvents (mg/g). Numbers in parentheses indicate the orders in which the solvents were used

Solvent	1	2	3	4	5	6	7
Water			8·1(4)	9·8(4)		1·6(4)	
MeOH	5·1(1)	0·3(4)					
EtOH	2·0(2)	4·2(3)	1·0(5)	1·8(5)	0·17(4)	0·7(5)	0·14(4)
i-BuOH					0·08(5)		0·08(5)
Bz + EtOH(1:1)					0·68(3)		0·37(3)
Petr. Ether					0·17(1)		0·11(1)
CHCl$_3$	1·6(3)	0·2(2)	0·4(2)	0·8(2)	0·08(7)	0·4(2)	0·04(7)
CCl$_4$			2·2(1)	3·3(1)	0·04(6)	0·9(1)	0·03(6)
C$_6$H$_6$	2·4(4)	6·6(1)	0·4(3)	0·6(3)	0·34(2)	0·2(3)	0·19(2)
Totals	11·1	11·3	12·1	16·3	1·56	3·8	0·96

1. C2, Cold Bokkeveld, Soxhlet extractions, Mueller (1953).
2. Same, different solvent order.
3. C2, Murray, sample and solvent refluxed together, Calvin and Vaughn (1960).
4. C2, Haripura, sample and solvent refluxed together, Briggs (1963b).
5. C2, Mighei, sample and each solvent shaken 240 hr at 20°C, Vdovykin (1962).
6. C2, Mokoia, sample and solvent refluxed together, Briggs (1963b).
7. C3, Grosnaja, sample and each solvent shaken 240 hr at 20°C, Vdovykin (1962).

12·1 mg/g is representative, particularly since 9·1 mg were removed by water and ethanol. Clearly, these figures are to be taken with a grain of salt.

Table 4 summarizes available data on organic solvent extractions of various meteorites. All are chondrites with the exception of the three ureilites.

Amounts of volatilizable material

Using thermogravimetric analysis, Mueller et al. (1965) have studied total amounts of volatiles present in carbonaceous chondrites. The situation is complicated by water of hydration of silicates, oxidation of iron, and other inorganic effects. However, an interesting result is the finding that all but 2·5% of the carbon and 36·8% of the sulfur in Orgueil can be volatilized within the temperature range studied, 0–1000°C. In most C2 chondrites, 4% of the carbon and 55% of the sulfur remain after heating to 1000°C. Murray is anomalously low in this respect, only 2·6% of the carbon and 44·8% of the sulfur remaining.

Analyses of Crude Extracts

Elemental analyses

In the first paper published in the twentieth century on the analysis of organic constituents of carbonaceous chondrites, Mueller (1953) reported an elemental

Table 4. Total amounts of organic solvent extractable material in meteorites (extract weights before sulfur removal)

Meteorite	Amount, (mg/g)	Technique†	Reference
Orgueil, Cl[1]	5·2	9:1 Bz–MeOH 6 hr, Soxhlet	a
Orgueil, Cl[2]	16·4	CHCl₃, 16 hr THF, 20 hr ~22°C	b
Orgueil, Cl[3]	1·5	CHCl₃, 8 hr, ~22°C	b
Orgueil, Cl[4]	30·0	Benzene	g
Boriskino, C2	2·8	5	c
Cold	11·1	1	d
Bokkeveld, C2	11·3	2	d
	2·5	Benzene	g
	4·0	Benzene	g
	1·5	5	c
Haripura, C2	16·3*	4	e
Mighei, C2	1·6	5	c
Murray, C2	0·6	9:1 Bz–MeOH 6 hr, Soxhlet	a
	4·4	5	c
	12·1*	3	f
Mokoia, C2	3·8*	6	e
Dyalpur, Ureilite	0·2	5	c
Goalpara, Ureilite	0·4	5	c
Novo-Urei, Ureilite	0·7	5	c
Grosnaja, C3	1·0	5	c
Kainsaz, C3	0·1	5	c
Krymka, L3	0·1	5	c
Mezö-Madaras, L3	—	5	c
Ghubara, H5	0·4	5	c
Kulp, H	0·8	5	c

Also negative results (Technique 5, ref. c) on Farmington (L5), Gilgoin (H5), Kissij (H), Kunashak (L6), Lixna (H), Pillistfer (E6), and Sevrukovo (L).

Orgueil sample sources:

[1] American Museum of Natural History, New York
[2] Musée National d'Histoire Naturelle, Paris
[3] Musée d'Histoire Naturelle, Montauban, Tarn-et-Garonne
[4] British Museum (Natural History), London

References: a. MEINSCHEIN, NAGY and HENNESSY (1963).
          b. MURPHY and NAGY (1966).
          c. VDOVYKIN (1964a).
          d. MUELLER (1953).
          e. BRIGGS (1963b).
          f. CALVIN and VAUGHN (1960).
          g. COMMINS and HARINGTON (1966).

* Indicates an extract obtained with a solvent sequence which included water.
† Technique numbers refer to columns of Table 3.

analysis of material obtained by organic solvent extraction of two samples of Cold Bokkeveld C2 chondrite (see columns 1 and 2, Table 3). This analysis is shown as column 1 in Table 5. MUELLER removed some, but not all, of the elemental sulfur crystals in the extracts [BRIGGS and MAMIKUNIAN (1963), erroneously report that MUELLER's extract was free of elemental sulfur]. The original analysis showed 18·33% ash; the figures in Table 5 have been recalculated as ash-free. Similar elemental analyses made by BRIGGS (1963b) are shown in columns 4 and 6, Table 3. Sulfur was removed from these extracts by the use of a colloidal copper column (BLUMER, 1957). BRIGGS notes that the analyses showed 15–22% ash. VDOVYKIN

Table 5. Elemental compositions of meteorite extracts

	1	2	3	4
% C	24·26	46·43	48·50	38·21
% H	8·12	5·03	5·03	8·54
% N	4·00	2·13	4·87	
% S	8·78	3·99	2·75	53·25
% Halogen, assumed Cl	5·89	1·06	1·74	
% Remainder, O	48·95	39·36	37·32	

1. MUELLER (1953) extracts of Cold Bokkeveld, C2. Solvents used: methanol, ethanol, chloroform, benzene, carbon disulfide.
2. BRIGGS (1963b) extracts of Mokoia, C2. Solvents used: carbon tetrachloride, chloroform, benzene, water, ethanol.
3. BRIGGS (1963b) extracts of Haripura, C2. Solvents used same as for Column 2.
4. VDOVYKIN (1962) $CHCl_3$ extract of Grosnaja, C3.

(1962) has reported the partial elemental analysis of a chloroform extract of Grosnaja. The original data showed 7·89% ash.

The large H/C ratios in Table 5, columns 1 and 4, probably indicate the presence of substantial amounts of water. The assumed chlorine content (columns 1–3) has been considered an important fact by a number of writers who have failed to note that these analyses definitely do not constitute proof of indigenous organic chlorine. In fact, it is possible that the halogen is chlorine from chlorinated solvents used in extractions (as has been noted by MUELLER, 1963). This assertion is made more plausible by noting that more detailed methods of analysis have not turned up significant quantities of chlorine-containing compounds and that BRIGGS performed his extractions by refluxing solvents together with meteorite powders. MUELLER, however, used a Soxhlet extractor. Finally, MUELLER's analysis indicates about 655 ppm organic Cl in total meteorite, a value almost twice the 350 ppm Cl (all forms) in total meteorite obtained by REED and ALLEN (1966) for the C2 chondrite Mighei.

HAYES and BIEMANN (1967) searched high resolution mass spectra obtained from direct volatilization of Murray (C2) organic constituents in order to determine what heteroelements were present. Nitrogen, sulfur, and oxygen were all found in significant quantities. Chlorine was present in barely detectable amounts, certainly not 1–6%. No ions possibly containing silicon, phosphorus, bromine, or iodine could be found.

Raia (1966) has reported elemental analyses of non-extractable organic material. These will be covered later in the summary of work on meteoritic "polymers."

Extract fractionations

In his analyses of meteorite organic constituents, Meinschein has followed a standard method evolved during analyses of many terrestrial sediments, soils, and crude oils (Meinschein and Kenny, 1957). As is noted in Table 4, the sample is extracted with a single solvent mixture (benzene–methanol). A Soxhlet

Table 6. Percentage distributions of extracts among chromatographic fractions (silica gel adsorbent)*

	Orgueil, C1				Murray, C2	
Eluant	1	2	3	4	5	6
$n\text{-}C_7H_{16}$	5	9·5	9·8	11	5	6
CCl_4	3	6·0	5·3	4·8	10	1
C_6H_6	10	3·1	3·3	3·1	5	6
CH_3OH	70	61	63	65	70	60
On column	10	20	18	17	10	30

* Meinschein, Nagy and Hennessy (1963).
Extracts:

1. 6·0 mg obtained from 1·7 g sample (3·5 mg/g) from American Museum of Natural History. A portion of the extract solution was lost due to a solvent "bump". This is the sample used to obtain the mass spectra discussed in this review and summarized in Table 9.
2–4. Triplicate fractionation of 75·0 mg extract obtained from 14·5 g stone (5·2 mg/g).
5. 1·1 mg extract obtained from 1·9 g stone (0·6 mg/g).
6. 10·2 mg extract obtained from 10·2 g stone (0·7 mg/g).

apparatus is used. A stream of purified nitrogen is used to evaporate the solvent and to bring the extract to constant weight at 40°C. According to Meinschein, Nagy, and Hennessy (1963), this also removes any sample compounds with vapor pressures greater than that of n-tridecane. If, however, results obtained by Nooner and Oró (1967) are indicative, the problem can be more serious, with at least partial losses of n-alkanes up to octadecane. After solvent evaporation, the crude extract is chromatographically fractionated on a silica gel column. Four fractions, an n-heptane eluate, a carbon tetrachloride eluate, a benzene eluate, and a methanol eluate, are obtained. The n-heptane and carbon tetrachloride eluates usually contain free sulfur (most of the sulfur is in the latter fraction), and this is removed using a colloidal copper column (Blumer, 1957). The heptane fraction contains alkanes and cycloalkanes; the carbon tetrachloride fraction contains sulfur, some olefins, and simple aromatics; the benzene fraction contains aromatic hydrocarbons; and the methanol fraction contains virtually all non-hydrocarbons in the extract (Meinschein and Kenny, 1957; Meinschein, Nagy and Hennessy, 1963). Table 6 presents fractionation data for extracts of Orgueil and Murray. Meinschein (1963a, 1963b) has compared these distributions with those observed

4

in various terrestrial samples, such as sediments and alkane-containing biological waste products. He feels that the observed similarities constitute evidence for a biological origin of meteoritic compounds.

VDOVYKIN (1962, 1963, 1965) has presented data of a similar type but showing distinctly different distributions. The results summarized in Table 7 show much greater proportions of paraffinic hydrocarbons than found by MEINCHEIN, NAGY and HENNESSY (1963), and considerable variation from chondrite to chondrite. The reason for the disagreement almost undoubtedly lies in the grossly different techniques employed: actual fractionation in the case of MEINSCHEIN and "fractional luminescence analysis" in the case of VDOVYKIN. The two sets of data are not comparable. On the other hand, VDOVYKIN's results for various different

Table 7. "Fractional luminescence analyses" of chondrite extracts
(as % of extract)*

Sample	"Oils"	"Tars", or "Resins"	"Asphalt"
Boriskino, C2	74	17	9
Cold Bokkeveld, C2	68	23	9
Murray, C2	61	27	12
Mighei, C2	46	34	20
Grosnaja, C3	46	34	20

* VDOVYKIN (1965).

chondrites must be mutually comparable; and in that case the observed groupings of Boriskino, Cold Bokkeveld, and Murray vs. Mighei and Grosnaja are intriguing.

Spectral investigations of whole extracts or crude fractions

Infrared spectrophotometry. Infrared spectrophotometry is not a particularly informative method when applied to complex mixtures. However, bleak as the picture is from an analytical standpoint, some conclusions can be drawn—not the least of which is that a complex mixture is present. Some interesting comparisons can be made between the infrared absorption spectra of extracts of different meteorites and a few terrestrial mixtures.

In terms of what is present, it can only be said that reported absorption spectra of aqueous and organic solvent extracts alike show the presence of various aliphatic and aromatic carbon–hydrogen bonds, carbonyl bonds, and hydroxyl groups (BRIGGS, 1963a; CALVIN and VAUGHN, 1960; MEINSCHEIN, 1963a, 1963b; MEINSCHEIN, NAGY and HENNESSY, 1963). There is evidence for all types of branching within the carbon–hydrogen absorption regions. Considering the 2960 cm^{-1} (methyl) and 2930 cm^{-1} (methylene) absorption peaks in a carbon tetrachloride extract of Murray, CALVIN and VAUGHN (1960) arrived at a figure of between 10 and 15 for the methylene to methyl ratio. Absorption in the carbonyl region is generally so intense and complex that no estimates on the types of carbonyls present can be made.

Separate i.r. spectra of the previously described four fractions obtained from liquid–liquid chromatography of the benzene–methanol extract not only show the trend from alkanes to aromatics but also indicate that virtually all compounds possessing heteroatomic functional groups are restricted to the methanol eluate fraction (MEINSCHEIN, 1963a, 1963b; MEINSCHEIN, NAGY and HENNESSY, 1963).

Some interesting comparisons between terrestrial mixtures and meteorite extracts have been made by MEINSCHEIN (1963a, 1963b). Only gross features can be considered, but spectra obtained from crude oil and a commercial Fischer–Tropsch product are markedly different from that of the meteorite extract. Crude oil shows virtually no carbonyl absorption, and, since it was produced under conditions adjusted for production of aliphatics, the particular Fischer–Tropsch product studied does not show the aromatic substitution bands prominent in the meteorite extracts.

An indication of the complexity of a multicomponent mixture can be given by the presence or absence of well resolved absorption peaks in an infrared spectrum of that mixture, a featureless spectrum resulting from the presence of a jumble of related compounds. Spectra of sediment extracts show considerably more detail than that of the meteorite extract (MEINSCHEIN, 1963a, 1963b), and on this basis it seems that the benzene–methanol extract of Orgueil is a mixture more complex than the extract of a typical recent or ancient sediment. On this same basis it appears that the organic compounds present in Murray (C2) make up a more complex mixture than those present in Orgueil (MEINSCHEIN, NAGY and HENNESSY, 1963). This difference is particularly evident in the 700–850 cm^{-1} region of the spectrum where Orgueil shows well resolved so-called "oil bands" (which seem to be aromatic substitution bands). Presumably this indicates that in particular the aromatic fraction of Orgueil is simpler than that of Murray (this is borne out by ultraviolet and mass spectral data on the same fractions). It is surprising that infrared spectra published by KAPLAN, DEGENS and REUTER (1963) differ so widely from those published by MEINSCHEIN. The principal difference is the presence of intense bands at 1260–1290 cm^{-1} and 980–1120 cm^{-1} in extracts obtained by KAPLAN, DEGENS and REUTER. These bands can be attributed to α-β unsaturated carbonyl systems and to alcohols and ethers, respectively. Such compounds may be artifacts of the extraction procedure in which the benzene-acetone solvent mixture was refluxed over the powdered sample for 24 hours. Acetone is an unfortunate solvent choice, and the presence of the powdered sample in the boiling solvent mixture may have had catalytic effects. The authors note that their solvents were not redistilled.

Spectra of relatively lower quality than those already cited have been presented by SISLER (1961, Murray solid suspension), VDOVYKIN (1962, 1963; 1:1 benzene-ethanol extracts of Grosnaja and Mighei) and SZTROKAY *et al.* (1961, acid-insoluble residue from Kaba). VINOGRADOV *et al.* (1966) have shown that the infrared absorption spectrum of the chloroform extract of Novo-Urei indicates the presence only of hydrocarbons; carbonyl bands, usually intense in meteorite extracts, are conspicuously absent.

Ultraviolet spectrophotometry. U.v. spectra of meteorite extracts and fractions are about as useful as i.r. spectra: a few interesting general conclusions can be reached, but this is all.

Aqueous extracts show generally featureless end absorption, the cutoff occurring at about 220 mμ (CALVIN and VAUGHN, 1960; BRIGGS, 1961). BRIGGS (1961) has noted a slight pH sensitivity in the absorbance at 260 mμ of an aqueous extract of Mokoia (C2).

Unfractionated organic solvent extracts invariably contain much sulfur, its absorption obscuring any due to organic compounds (CALVIN and VAUGHN, 1960; HIMES, 1960; HAYATSU, 1965).

The u.v. absorption spectra of the heptane, carbon tetrachloride, benzene, and methanol fractions (see Table 6) of the Orgueil (MEINSCHEIN, 1963a, 1963b; MEINSCHEIN, NAGY and HENNESSY, 1963) and Murray (MEINSCHEIN, NAGY and HENNESSY, 1963) meteorites have been reported. It is clear in both cases that the heptane fractions contain no u.v. absorbing compounds. Since mass spectra of these fractions show intense peaks in the homologous series C_nH_{2n-2}, C_nH_{2n-4}, etc. this is important evidence for the presence of cycloalkanes rather than olefins in this fraction. After sulfur removal it is evident that the carbon tetrachloride fraction probably contains small amounts of olefins and simple aromatics. The benzene fraction of Orgueil gives a remarkably simple spectrum containing many well resolved absorption maxima. MEINSCHEIN (1963a) compares this spectrum to those of benzene fractions of sediments and crude oils, showing that although the total Orgueil mixture seems to be more complex than terrestrial mixtures, the aromatic fraction must be much simpler than its terrestrial counterpart. In particular, the Orgueil aromatics spectrum bears a remarkable resemblance to the spectrum of a fraction obtained by further separation of the benzene fraction of recent sediments. This terrestrial fraction consists of compounds with chromatographic properties similar to those of phenanthrenes, chrysenes, and pyrenes. The u.v. absorption spectrum of the benzene fraction of Murray (MEINSCHEIN, NAGY and HENNESSY, 1963) indicates that it is a mixture more complex than the benzene fraction of any terrestrial sediments. The methanol fraction of Orgueil shows distinct absorption maxima at 245 and 255 mμ. MEINSCHEIN (1963a) mentions that this spectrum resembles that of fluorenone, although this remark is obviously intended as a description of the spectrum, not as an identification of fluorenone. At an equal concentration level the methanol fraction of Murray absorbs two to three times more strongly than that of Orgueil, but shows no features at all.

Isotope ratio studies

Some forms of meteoritic carbon and hydrogen have isotope ratios which lie outside terrestrial ranges. This is, of course, good evidence for an extraterrestrial origin of these materials. Table 8 summarizes results of isotope ratio studies on carbon and hydrogen in Orgueil, Haripura, Murray and Mokoia.

The carbonate in Orgueil is greatly enriched in ^{13}C, the measured ratio lying outside the terrestrial range. CLAYTON (1963) discusses this result and the indicated difference of 70 per mil between the oxidized and reduced forms of carbon. He details possible abiogenic methods of achieving such a large fractionation. On earth such ^{13}C depletion in the organic phase would, rightly or wrongly, be regarded as evidence for biogenicity. Two further observations, however, call for caution in this interpretation: first, biological processes are not the only possible mechanism for isotopic enrichment or depletion; and second, the roles of the constituents are reversed—on earth carbonates are the major carbon reservoir while in the meteorite they are only minor constituents. BRIGGS (1963a) supplies no experimental details or supporting data allowing evaluation of his results, and the experiments performed by BOATO are by now rather old. Further study is certainly needed. Using a very small sample, CLAYTON (1963) obtained enough carbon dioxide for a single measurement of the isotope ratio for Ivuna (C1) carbonates, which he found to be isotopically indistinguishable from those of Orgueil. NAGY (1966a) has recently reported a measurement of $\delta^{13}C = -23\cdot8$ per mil for the carbon in the "saponifiable" fraction of the Orgueil benzene–methanol extract.

Table 8. Isotopic compositions of meteoritic carbon and hydrogen*

	Orgueil, C1	Haripura, C2	Murray, C2	Mokoia, C2	Reference
		$\delta^{13}C(\permil)$			
Extractable organics	$-10\cdot1$	$-3\cdot1$	$-4\cdot0$	$-17\cdot9$	BRIGGS (1963a)
				$-17\cdot4$	
Combustible C	$-11\cdot4$	$-3\cdot7$	$-3\cdot9$	$-18\cdot8$	BOATO (1954)
Carbonate C	$+60$				CLAYTON (1963)
		$\delta D(\%)$			
Extractable organics	$+27\cdot5$	$-1\cdot5$	$+8\cdot4$	$+21\cdot2$	BRIGGS (1963a)
				$+25\cdot9$	
Combined H_2O	$+29\cdot0$	$-3\cdot3$	$+9\cdot6$	$+23\cdot5$	BOATO (1954)

* Referred to the standards described by CRAIG (1957).

TROFIMOV (1950) has presented results showing that the carbon isotope ratios in carbonaceous chondrites are in the same range as those found in irons, stony-irons, and ordinary chondrites. However, the $\delta^{13}C$ measurements given in Table 8 for Haripura and Murray lie outside this range, so that the situation is, at the moment, unclear. It seems that the carbon in carbonaceous chondrites may be richer in ^{13}C than that in ordinary chondrites. This is consistent with the two ordinary chondrite measurements reported by BOATO (1954).

The tabulated values of hydrogen isotope ratios show that both forms (water H, extractable organic H) of hydrogen studied are enriched in deuterium to the point that the measured compositions are outside the terrestrial range. The difference is greatest in Orgueil water, amounting to about 25%. BOATO (1954) studied the hydrogen in water evolved during heating of powdered meteoritic material under an oxygen pressure of 50 torr. Since there is about four times as much water as carbon in these stones, BOATO's analyses are regarded as indicative of the deuterium content of meteoritic water. If BRIGGS (1963a) results may be relied upon, a simple comparison of these data strongly suggests that the organic and aqueous phases drew their hydrogen from the same extraterrestrial reservoir. EDWARDS (1955) has criticized BOATO's (1954) analyses on the basis that the quantitative results for water are in disagreement with those obtained by WIIK (1956). This discrepancy is due to a difference in procedure and has been explained by BOATO (1956).

Meteoritic sulfur and oxygen do not show such marked deviations from terrestrial ranges. Sulfur isotopes in organic material have been studied by BRIGGS (1963a). Sulfur isotopes in inorganic material have been studied as a function of valence state by MONSTER, ANDERS and THODE (1965) and KAPLAN and HULSTON (1966). NAGY, MEINSCHEIN and HENNESSY (1961) have estimated the $^{32}S/^{34}S$ ratio in meteoritic free sulfur. Oxygen isotopes in Orgueil carbonates have been studied by CLAYTON (1963), and those in silicate minerals of six carbonaceous chondrites by TAYLOR et al. (1965). The silicate phase is the richer (for Orgueil) in ^{18}O, and an interesting pattern is found, with the C3, C2 and C1 chondrites becoming progressively richer in ^{18}O.

IDENTIFICATIONS OR INDICATIONS OF SPECIFIC COMPOUNDS OR GROUPS OF COMPOUNDS

Aliphatic hydrocarbons

Amounts and types. The n-heptane eluate obtained from a silica gel column used for chromatographic fractionation of a meteorite extract contains virtually

Table 9. Saturated hydrocarbon type analysis for Orgueil and Murray*

Molecular type	Calculated from the mass spectrum of the n-heptane fraction of:	
	Orgueil (%)	Murray (%)
Paraffins	17·6	16·5
Cycloalkanes, 1 ring†	28·9	17·3
2 rings	17·2	18·1
3 rings	11·8	13·9
4 rings	11·3	13·2
5 rings	6·4	10·1
6 rings	6·8	11·0

* MEINSCHEIN, NAGY and HENNESSY (1963).
† Includes multi-ring species in which the rings are not fused together. For example, 1, 3-dicyclohexylhexane would appear as a 1-ring type, while decalin would appear as a 2-ring type.

all the saturated hydrocarbons, including branched and cyclic types. It is this fraction which has received closer attention than any other in meteorite extracts. According to Table 6, this fraction contains from 5 to 10% by weight of the total extract. Calculated concentrations of saturated hydrocarbons (in μg of organic material/g of meteorite) are thus approximately 510 $\mu g/g$ for Orgueil and 40 $\mu g/g$ for Murray. The quantitative distribution of compound types in this fraction may be estimated from mass spectral data. Table 9 shows results of an application of a type analysis calculation to mass spectra (resolution ≤ 1000) of heptane fractions. In this calculation, groups of peaks regarded as characteristic of the types listed are compared in intensity, and the resultant analysis is derived through simple matrix calculations. This form of analysis, described by LUMPKIN (1956), has been widely used in the petroleum industry for the analysis of such fractions. Such an analysis does not prove the presence of any one specific compound, although it does certainly indicate the presence of some compounds in each of the types listed.

Oró *et al.* (1966), have presented data which are in marked disagreement with those given above. They find greatly variable concentrations in Orgueil, with amounts of saturates ranging from 6 to 89 $\mu g/g$ (instead of 510 $\mu g/g$). Murray varies less, with two analyses showing 156 and 336 $\mu g/g$ (instead of 40 $\mu g/g$). Oró's estimates are based on gas chromatographic curve integration, and he notes that they "may be in error by as much as 50–100%." Such large errors, however, would still not account for the major disagreement: Oró finds that Murray contains at least twice as much as Orgueil, while Meinschein's data show that Murray contains one-tenth as much as Orgueil. Nooner and Oró (1967) have presented quantitative analyses of the n-alkanes and isoprenoid hydrocarbons in twenty-nine meteorites (some of which were non-carbonaceous controls). Figure 2 graphically summarizes the surprising findings of this very thorough study. The trend initially noted in the simple comparison of Orgueil and Murray applies to virtually all the samples studied: in general C2 chondrites contain more n-alkanes and isoprenoids than C1 chondrites. In fact, it seems that C3 chondrites have still more extractable hydrocarbons and that the concentration variation (C3 > C2 > C1) is the reverse of what might have been expected. Hayes and Biemann (1967) have obtained results indicating approximately 300 $\mu g/g$ aliphatic hydrocarbons in Murray. While this may be taken as confirmation of the analysis by Oró *et al.*, some caution is called for because of the very different procedure used. Instead of solvent extraction, Hayes and Biemann simply volatilized the organic material directly from the stone, examining the compounds by use of a directly coupled gas chromatograph and high resolution mass spectrometer (Hayes and Biemann, 1967). Orgueil was not studied in this way.

The distribution of compound types may also be estimated from gas chromatograms. The predominant features in most of these chromatograms are a series of major peaks due to well-resolved (separated) n-alkanes and a large base-line hump due to the presence of many subtly different, less abundant compounds in a mixture too complex to allow resolution into separate peaks. By comparing the area under the n-paraffin peaks to the area enclosed by the complete chromatogram (including the base-line hump), Oró *et al.* have estimated that n-paraffins account for from 1–28% of the total saturates in Orgueil. Considerable qualitative (1–28%) and quantitative (6–89 $\mu g/g$) heterogeneity is thus indicated. Using the same method of estimation with the Murray gas chromatograms, Oró *et al.* find that n-paraffins account for from 17 to 26% of the saturated hydrocarbons. These data are summarized in the third column of Table 10.

N-alkane distributions. The first reported quantitative analysis of the n-alkanes was that by Nagy, Meinschein and Hennessy (1961). Investigating the Orgueil meteorite, they reported a bimodal distribution of n-alkanes with maxima at C_{18} and C_{23}. A dry distillation was used rather than extraction, the pressure in the sealed distillation apparatus eventually exceeding 1 atm at 400°C. The mass spectrometrically analyzed distillate was not fractionated, and in this case the use of low resolution mass spectrometry coupled with the assumption that all peaks were due to hydrocarbons proved to be extremely dangerous. The analytical results have not been duplicated in any subsequent study, probably because of the contributions of non-hydrocarbons in the low resolution mass spectrum. This work has also been criticized by ANDERS (1962).

More recent studies have relied, for the most part, on solvent extraction techniques. While solvent extraction, along with later extract fractionation, has the

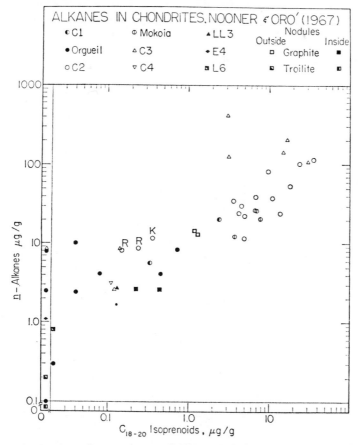

Fig. 2. N-alkanes and C_{18-20} isoprenoid hydrocarbons in chondrites. Note that the logarithmic scales are broken in order to include a zero point for blank analyses and analyses in which no isoprenoids were found. Two analyses of Orgueil in which isoprenoids could not be quantitated are not plotted, the n-alkane contents were $25 \cdot 0$ and $1 \cdot 0 \mu g/g$. All samples are plotted according to the classifications of VAN SCHMUS and WOOD (1967). The authors note that "the amounts and distribution of hydrocarbons given . . . do not necessarily represent all the aliphatic hydrocarbons in the original sample because of losses during solvent removal after (i) extraction and (ii) fractionation."

advantage of producing a well-defined sample which has had minimum opportunity for chemical alteration, it is a technique best suited to the study of compounds less volatile than (approximately) octadecane. More volatile compounds can be quantitatively retained if solvent removal is very carefully controlled, but a recent publication by NOONER and ORÓ (1967) suggests that this is very difficult and indicates that *some or all of the extracted n-alkane distributions discussed below may have been materially affected by the solvent removal process* and that the true distributional maximum (most abundant n-alkane) may lie several carbon numbers lower than indicated. For example, an initial investigation of the Mokoia (C2) n-alkanes placed the mode at C_{19}, but a later analysis in which care was taken to avoid complete solvent removal placed the mode at C_{15}. This single example might be attributed to sample heterogeneity, but the observation of similar mode shifts with other stones (Vigarano, C_{19}–C_{17}; Orgueil, C_{19}–C_{15}) makes this unlikely. Of course

Table 10. Distributions of meteorite n-alkanes

Reference	Method	Quantity $(\mu g/g)$	Mode(s)	Sample source
		Orgueil, C1		
NAGY, MEINSCHEIN and HENNESSY (1961)	Distillation —MS	Not measured	C_{18}, C_{22}	American Museum of Natural History
MEINSCHEIN, NAGY and HENNESSY (1963)	Extraction fractionation —MS	90*	C_{23}	American Museum of Natural History
MEINSCHEIN (1964)	Extraction fractionation GLC	Not measured	C_{19}	Montauban Museum
ORÓ et al. (1966)	Extraction fractionation GLC	Approx. <0·1–25	C_{19}	Montauban Museum
NOONER and ORÓ (1967)	Extraction fractionation GLC-MS	4·1	C_{15}†	Montauban Museum
		Murray, C2		
MEINSCHEIN, NAGY and HENNESSY (1963)	Extraction fractionation MS	6·6	C_{23}	Institute of Meteoritics Univ. of New Mexico
MEINSCHEIN (1964)	Extraction fractionation GLC	Not measured	C_{17}	U.S. National Museum
ORÓ et al. (1966)	Extraction fractionation GLC	Approx. 40–59	C_{21}	U.S. National Museum
HAYES and BIEMANN (1967)	Volatilization GLC HRMS	Very approx. 60	C_{15}	U.S. National Museum

* Quantity estimated from columns 2, 3, 4 of Table 6. Sample used in this analysis was "Orgueil 1" of Table 6.
† In this analysis an effort was made to retain the lighter alkanes during solvent evaporation.

extraction procedures provide little information on the amounts and types of hydrocarbons present below C_{15}.

MEINSCHEIN, NAGY and HENNESSY (1963) have presented data from a technique employing extraction and MEINSCHEIN's standard method of fractionation. The results given in Table 9 were obtained through type analysis of n-heptane fraction mass spectra, which the authors published in full. It appears that the maximum in the alkane distribution in both Orgueil and Murray is at C_{23}. In each case, n-alkanes in the range C_{14}–C_{20} are less than 25% as abundant as those in the range C_{22}–C_{24}. There is no evidence of a bimodal distribution in this second analysis.

Independent gas–liquid chromatographic analyses of the n-heptane fraction of Orgueil have been made by MEINSCHEIN (1964a) and ORÓ et al. (1966). These two sets of results are in remarkably good agreement but are in marked disagreement with those noted above. There is once again no evidence for a bimodal distribution, but in both gas chromatographic analyses the n-alkane distribution has its mode at C_{19}, not at C_{23}. Agreement between the two analyses extends even to small gas chromatographic peaks appearing between those due to n-alkanes. Nooner and Oró (1967) performed an analysis of Orgueil in which they minimized the loss of sample alkanes by not evaporating the solvent–sample mixtures to "dryness" at any point in their analysis. In this case a very different distribution was observed in which the order of n-alkane abundances was approximately $C_{17} > C_{18} > C_{19} > C_{20} > C_{15} \simeq C_{16} > C_{21} > C_{22} > C_{23}$.

Similar independent analyses of Murray have failed to produce any agreement. Meinschein (1964b) finds that the n-alkane distribution has its mode at C_{17}. Oró et al. find it at C_{21}. Both of these analyses are in disagreement with the earlier mass spectrometric one in which the mode was at C_{23}. The results of the direct volatilization-gas chromatographic experiment performed on Murray by Hayes and Biemann (1967) indicate a mode at C_{15} and provide the only information available on the low molecular weight hydrocarbons. The distribution observed was essentially symmetrical about C_{15}, C_{20} and C_{10} being the highest and lowest homologs observed.

In another mass spectrometric analysis of meteoritic alkanes, Studier Hayatsu and Anders (1965) report the presence of large amounts of methane in Orgueil, Murray and Cold Bokkeveld. They conclude that ethane and other light alkanes are present with an abundance at least three orders of magnitude lower.

The quantitative and qualitative distributions are summarized in Table 10. In the case of Orgueil, it seems that the first distribution may represent a combination of the second and third. Oró et al. have suggested that the distribution with its mode at C_{23} is due to contamination by paraffin wax hydrocarbons like those which they found in dust. If this is true, the first, dry distillation, analysis is indicative of both contamination and true meteorite alkane content; the second analysis succeeded in extracting only contaminants from the meteorite; and the later analyses have been made on uncontaminated samples. Table 10 shows that this rather confusing situation makes more sense if quantities and sample sources are taken into consideration. In the second analysis the amount of n-alkanes is so great that the C_{19} mode could be submerged in the contaminant distribution. Considering the source, one would expect the first analysis to show a similar pattern rather than two modes. There are at least two possible explanations: (1) the first mode is due to contributions from non-hydrocarbons or other artifacts of this crude analytical technique; (2) this first sample is less contaminated than the second, possibly because it was in the American Museum of Natural History collection for a shorter time than the second (Nagy, Meinschein and Hennessy, 1963). The agreement between the third and fourth analyses is only momentarily pleasing: the amounts column shows that there was at least one sample in which almost no n-alkanes were found ($<0\cdot1\ \mu g/g$), and the fifth analysis (the most reliable) emphasizes the possibility that solvent evaporation techniques have substantially affected the results of previous analyses. Does the meteorite contain any indigenous alkanes at all? This particularly bleak possibility seems unlikely because of the failure to find dust or human hand hydrocarbon distributions which account for the meteoritic distributions in the range C_{16}–C_{21} (Oró et al., 1966).

It seems that no amount of discussion can succeed in explaining the discrepancies encountered in the Murray analyses. For one thing, the C_{23} (contaminant?) mode occurs in the sample with the smallest amounts of n-alkanes. Possibly the quantitative distribution of hydrocarbons in Murray is as variable as that found in Orgueil. For Orgueil the evidence is much more clear cut, being embodied in a series of directly comparable experiments made by ORÓ et al., while for Murray comparison must be made using results from various laboratories and methods. This postulated heterogeneity might be related to other evidence that Murray has a variable distribution of rare gases (MAZOR, HEYMANN and ANDERS, manuscript in preparation). The low mode position found by HAYES and BIEMANN is very probably explained by the fact that their analysis could not have been affected by any solvent removal process.

Figure 2 summarizes the n-alkane and C_{18}–C_{20} isoprenoid concentrations reported by NOONER and ORÓ. In addition to these quantitative data, the authors present chromatograms showing the aliphatic hydrocarbon distribution patterns in each sample. It is perhaps surprising to observe that the highest alkane concentrations were found not in the C1 chondrites but in the C3 chondrites. (The highest C3 point represents an analysis of Grosnaja. The authors note that the hydrocarbon distribution in this sample suggests contamination—a C_{23} mode was observed. Relative to other C3 chondrites this sample shows an anomalously large n-alkane/isoprenoid ratio, and this may be further evidence of contamination.)

The C1 measurements, all but two made on Orgueil, scatter widely. There is no apparent regularity in the n-alkane/isoprenoid ratio or even in the presence or absence of isoprenoids. Four of the measurements bear a strong resemblance to the graphite nodule measurements used as controls.

The C2 measurements were made on a wide variety of samples,* Mokoia being the single chondrite with more than two replicates. A definite pattern is observed, with the average n-alkane/isoprenoid ratio being approximately 4, very different from that observed in the other types. The authors follow the classification of WIIK (1956) and tabulate Mokoia as a type III carbonaceous chondrite. VAN SCHMUS and WOOD (1966) list Mokoia as a C2 chondrite. While the question of classification cannot be decided on the basis of these measurements, it may at least be observed that Mokoia fits nicely into the group of C2 measurements shown here. A very different n-alkane/isoprenoid ratio of approximately 30 is observed for the three C2 chondrite measurements showing the lowest isoprenoid concentrations. Two of these measurements (marked "R") represent duplicate analyses of Renazzo (in fact, two sample sources) and the third (marked "K") represents the single measurement made on Kaba. It is extremely interesting to note that these two chondrites happen to contain more free metal than other meteorites in this class.

Even if the Grosnaja measurement is reduced by a factor of ten, there still seems to be some evidence for the presence of two groups in the type 3 chondrites. The upper group consists of (in order of decreasing n-alkane concentration)

---

* The Bells sample was a heavily weathered stone (O. E. MONNIG, B. MASON, private communications). The results of its analysis are not plotted in Fig. 2, and the observed hydrocarbon distribution should definitely not be taken as necessarily representing the indigenous constituents of a freshly fallen C2 chondrite.

Grosnaja, Vigarano, Lancé, Felix and Vigarano (duplicate). The lower group, found among the Orgueil measurements, consists of (same order) Ornans (no isoprenoids found), Warrenton, Chainpur (LL3) and Kainsaz. Karoonda, the only C4 chondrite measured, would also fit nicely in the second group.

Branched-alkanes. Gas chromatograms of meteoritic alkanes frequently show a large base line hump, presumably due to a complex mixture of cyclic and branched alkanes present in such profusion that they cannot be separated. There are, however, a few non-n-alkane peaks which appear regularly. Two of these have been identified as the saturated polyisoprenoids pristane (2,6,10,14-tetramethylpentadecane) and phytane (2,6,10,14-tetramethylhexadecane). Oró et al. (1966) based their identification on retention times, accepting a peak as identified if its retention time was correct on three different capillary columns: Carbowax 20M, Polysev, and Apiezon L. Using this technique they identified pristane and phytane in nineteen of the twenty meteorites they studied. These compounds were taken to be indigenous. They did not appear in dust blanks and they did not seem to be covariant with peaks in the C_{23} mode (possible contaminants). Meinschein (1964a,b) has also found evidence for pristane and phytane in Orgueil and Murray.

Additional evidence is provided by Nooner and Oró (1967) in the form of mass spectra of the compounds in the "pristane" and "phytane" peaks of the gas chromatograms. The agreement of these spectra with those of standard pristane and phytane is not perfect, and the slight discrepancy serves as a reminder that, when dealing with compounds like pristane or phytane, the identification is not on perfectly safe ground until the characteristics of subtly different isomers have been investigated and shown to be significantly different. For example, Calvin and McCarthy (1967) have shown that 2,6,10-trimethylhexadecane has gas chromatographic and mass spectrometric characteristics very much like those of 2,6,10,14-tetramethylpentadecane. Nooner and Oró have also obtained evidence for the presence of the lower homolog norpristane (2,6,10-trimethylpentadecane). The isoprenoid concentrations tabulated in Fig. 2 represent in each case the sum of phytane, pristane and norpristane. The authors speculate that an unidentified compound prominent in well-resolved chromatograms of heptane fractions of Orgueil, Mokoia and Vigarano, may also be an isoprenoid. If so, it would be the C_{16} homolog.

Cyclic alkanes. No specific cyclic alkanes have ever been identified, either by gas chromatography or mass spectrometry, but the type analysis gives substantial evidence for their presence, as do the many cyclic or olefinic type ions observed by Hayes and Biemann (1967) in the direct introduction and gas chromatographic high resolution mass spectra. Oró et al. (1966) suggest that cyclohexylalkanes are present in Orgueil. From the mass spectrum of a specimen of the n-heptane eluate (from a silica column) of the benzene–methanol extract of the Orgueil meteorite, Meinschein, Nagy and Hennessy (1963) suggest the possible presence of cholestane-like species among the four ring cycloalkanes. The only evidence supporting this is slightly larger peaks at m/e 378 and 363, the M^+ and $(M-15)^+$ ions of cholestane. Other peaks expected to be large, $(C_{16}H_{26})^+$ and $(C_{16}H_{25})^+$, are preceded in their homologous series by still larger peaks at $(C_{15}H_{24})^+$ and $(C_{15}H_{23})^+$, indicating that a substantially different ring system may be present. MEINSCHEIN (1963b) has presented a mass spectrum obtained from an alumina chromatographic fraction of the *n*-heptane fraction. This fraction is enriched in the larger cyclic alkanes, and its mass spectrum shows more detail than previous spectra of the unfractionated eluate. Even in this case, however, interpretations suggesting the presence of any specific compound are dangerous. Unfortunately, NAGY (1966a) has cited the crude data available from the dry distillation mass spectra (NAGY, MEINSCHEIN and HENNESSY, 1961) as evidence for the presence of "18,19-dimethyl-cyclopentanoperhydrophenanthrenes."

Aromatic hydrocarbons

MEINSCHEIN, NAGY and HENNESSY (1963) have reported analyses of the fraction eluted with benzene during chromatographic fractionations of Orgueil and Murray extracts. According to Table 6, the benzene fraction contains about 5% of the material extractable from Murray and Orgueil. For Orgueil this indicates an aromatic hydrocarbon concentration of about 260 μg/g and for Murray a concentration of about 30 μg/g. It should be remembered that the Orgueil sample involved is the same one presently thought to be contaminated (see the above discussion of *n*-alkanes).

Mass spectra were obtained at low ionizing potential (12 V), and the results were summarized in aromatic type analyses. The specific compound names used in the type analysis tabulation are based on experience with terrestrial samples. For example, an abundant ion of mass 178 if observed in the low ionizing potential mass spectrum of a terrestrial sediment aromatic hydrocarbon fraction may be safely attributed to phenanthrene, not anthracene. Thus, the type analysis computer program presents all $C_{14}H_{10}$ ions as phenanthrene. The u.v. absorption spectrum of the meteorite fraction does indicate that phenanthrene predominates over anthracene, but it is impossible to reinforce every name used in the type analysis with evidence from a single u.v. spectrum, particularly when the analysis shows 20–50% "other." What are tabulated as pyrenes may equally well be fluoranthenes.

Mass spectra of a similar Murray fraction were, unfortunately, obtained at an intermediate ionizing potential. The predominance of aromatic molecular ions is thus not as marked, and the spectra are more difficult to interpret.

It is, however, extremely important to point out that these spectra (the raw spectra, not the type analyses) may be of great confirmatory or comparative value in later investigations, and may indicate what compounds should be sought. For example, the mass spectrum of the Orgueil aromatic hydrocarbon fraction indicates the possible presence of dihydrobenzfluoranthenes, phenylnaphthalenes, picenes and tri- and tetra-phenylalkanes and alkenes *in addition to* most of the compounds listed by the authors.

Strikingly, COMMINS and HARRINGTON (1966) fail to find measurable amounts of aromatic hydrocarbons in Orgueil (MEINSCHEIN *et al.*, 1963: 260 μg/g). On the other hand, they report quantities of various aromatic hydrocarbons ranging from 7·2 μg/g (phenanthrene, South African Museum sample) to 0·008 μg/g (anthanthrene, British Museum sample) in two samples of the Cold Bokkeveld C2 chondrite.

The Orgueil sample was obtained from the British Museum (Natural History), and the authors note that the extract was "largely elemental sulfur." Individual compounds were isolated chromatographically and estimated spectrophotometrically. This work indicates that the concentration trend of C2 > C1 noted in the alkanes may carry over into the aromatic fraction, and to this extent these surprising results do fit into the present picture.

The Russian group (VINOGRADOV and VDVOYKIN, 1964; VDOVYKIN, 1964a) has reported the presence of anthracene, 3,4-benzopyrene, 1,12-benzoperylene and perylene in the C2 chondrites Mighei, Boroskino and Cold Bokkeveld. These identifications are based on low temperature fluorescence data obtained from benzene extracts. Some of the correlations with standard spectra are not good, and it seems that these reports should not be accepted as definite identifications but rather as indications of the possible presence of these specific compounds. FLOROVSKAYA et al. (1965) have compared the observed meteoritic distributions with those found in various terrestrial rocks.

STUDIER, HAYATSU and ANDERS (1965) have reported the presence of benzene in Murray, Cold Bokkeveld and Orgueil. Toluene was found in both Orgueil and Cold Bokkeveld. A variety of alkylbenzenes, naphthalene and anthracene or phenanthrene were also found in Orgueil. The identifications were, for the most part, made from time-of-flight mass spectra of mixtures volatilized directly from solid samples. A few of the less volatile compounds reported in Orgueil were found in crude extracts. While the misidentification of benzene, toluene, or naphthalene is very unlikely, the identifications of specific alkylbenzenes should probably be considered as tentative.

STUDIER, HAYATSU and ANDERS make the interesting generalization that "lighter" (C_6–C_{14}?) hydrocarbons seem to be all aromatics. They were unable to find n-alkanes in this range. This generalization has been attacked by BURLINGAME and SCHNOES (1966) on the basis that peaks with the molecular weight of the C_4–C_8 alkanes are listed by STUDIER as "unidentified," but HAYES and BIEMANN (1967) have since reported analyses based on high resolution mass spectra of gas chromatographically separated volatiles which, at least for the case of Murray, support the contention of STUDIER et al. The lowest alkane observed was decane, and it was definitely true that hydrocarbons observed in the C_6–C_{11} range were predominantly aromatic. Specifically, HAYES and BIEMANN obtained evidence for the presence in Murray of alkylbenzenes with alkyl substituents totalling up to seven carbon atoms, and naphthalene and alkylnaphthalenes with alkyl substituents up to C_3. Spectra of compounds volatilized from the stone in the temperature range 254–409°C gave evidence for the presence of biphenyl or acenaphthene and phenanthrene or anthracene. High resolution mass spectra obtained of the complex mixtures volatilized directly from the crushed stone gave evidence for the compounds noted above and, in addition, acenaphthalenes and series of indenes, and tetralins or indanes.

Oxygen containing compounds

Alcohols. STUDIER, HAYATSU and ANDERS (1965) have made the only report of alcohols. They investigated volatile constituents of Murray, Cold Bokkeveld and

Orgueil and reported alcohols as minor constituents in Murray. No indication of which alcohols are present is given, and this should certainly have been possible if the data permitted identification of mass spectral peaks characteristic of alcohols.

Phenolic compounds. Phenolic compounds (aromatic alcohols, acids, esters and similar compounds) have been reported to be present in aqueous extracts of Orgueil, Haripura (C2), and Mokoia (BRIGGS, 1961, 1963b), to be absent in the ether extract of Orgueil (RAIA, 1966); and to be plentiful in hydrolysates of meteoritic material (KAPLAN, DEGENS and REUTER, 1963).

BRIGGS (1961) inferred from paper chromatograms of aqueous extracts of Mokoia (C2) that phenolics might be present. He later presented results (BRIGGS, 1963b) of a more detailed study of the water extracts of Mokoia and Haripura. Thin layer chromatography was used to tentatively identify 2-, 3- and 4-hydroxybenzoic acid; 3,5-dihydroxybenzoic acid; 4-hydroxyphenylacetic acid; and 3-methoxy-4-hydroxybenzoic acid. Another report (BRIGGS, 1963c), apparently of the same work, lists in addition Orgueil as a sample and adds 3-methoxy-5-hydroxybenzoic acid; 3,4-dihydroxybenzoic acid; and 3-methoxybenzoic acid to the list of identified compounds, no indication being given as to which compounds were found in which meteorite.

HAYES and BIEMANN (1967) observed ions with elemental compositions corresponding to furans, alkylphenols and alkylbenzofurans in the direct introduction high resolution mass spectra obtained from samples of Murray (C2). These oxygenated species were observed primarily in the temperature range 255–285°C and thus may represent the degradation products of indigenous thermally labile compounds. The ions were of very low relative abundance.

The phenolics found in hydrolysates will be discussed in the section dealing with the polymeric material.

Fatty acids. MEINSCHEIN (1963b) suggested that fatty acids might be indicated in the refined Orgueil n-heptane fraction spectrum presented in his discussion of meteoritic cycloalkanes. A series of peaks corresponding to fatty acid molecular ions is present. This series shows pronounced even carbon preference and has its distribution maximum at C_{24}. (No explanation is given for the appearance of the relatively polar fatty acids in the alkane fraction.)

NAGY and BITZ (1963) analyzed two large Orgueil stones from the Montauban Museum. After fractionation, the extract of a 12·901 g stone yielded 8·17 mg of methyl esters of carboxylic acids (675 μg/g) of which 1·62 mg were methyl esters of straight chain fatty acids (125 μg/g). An 11·387 g stone yielded 6·6 mg of carboxylic acid methyl esters (580 μg/g). Gas-liquid chromatography of a urea adduct of the ester fraction indicated the presence of C_{14}, C_{16-20}, C_{22}, C_{24} and C_{26-28} fatty acids. There was no resemblance between the meteoritic fatty acid distribution and the fatty acid distributions observed in any terrestrial samples (NAGY and BITZ, 1963). The major difference was the presence of odd carbon number (17, 19, 27) fatty acids in the meteorite. If the slight even carbon number preference observed in the meteorite is attributed completely to contamination, no more than half the observed fatty acid concentration can be discounted as due to contamination.

There is no correspondence between the distribution in the homologous series mentioned by MEINSCHEIN (1963b) as possibly due to fatty acids and the fatty acid distribution found by NAGY and BITZ (1963). There is a good correlation between the homologous series in MEINSCHEIN's mass spectrum and the fatty acids found by NAGY and BITZ in a recent soil. This may be further evidence of contamination in the samples used by MEINSCHEIN, NAGY and HENNESSY (1963) in their studies of hydrocarbons.

While attempting to duplicate NAGY's results on optical activity, HAYATSU (1965) repeated the procedure given above and obtained i.r. spectra and quantitative results in very close agreement with those shown by NAGY and BITZ. Both HAYATSU (1965) and NAGY (1966a) regard this as a confirmation of NAGY's finding of fatty acids in spite of the fact that HAYATSU did not use urea adduction and gas chromatography. Thus, his results cannot be regarded as absolute confirmation.

Sugars. DEGENS and BAJOR (1962) reported 23 $\mu g/g$ free sugars (glucose and mannose) in an 80% ethanol extract* of Murray, and 8 $\mu g/g$ in Bruderheim, an ordinary chondrite. Acid hydrolysates of the same samples yielded 47 $\mu g/g$ and 12 $\mu g/g$ respectively. The combined sugars found in Murray included arabinose as well as glucose and mannose. KAPLAN, DEGENS and REUTER (1963) published a supplementary paper reporting 8 $\mu g/g$ of sugars in Orgueil, 24 $\mu g/g$ in Murray, and 26 $\mu g/g$ in Hvittis, an enstatite chondrite. One-dimensional paper chromatography was used to identify specific sugars in the non-ionic fraction of aqueous extracts. As was noted in the section on i.r. spectra, contamination may have been a major problem in these analyses.

Sulfur containing compounds

Considering the amounts of elemental sulfur present in meteorites, it seems strange that so little work has been directed toward the analysis of any organic sulfur compounds which might be present. Presumably there is a reluctance to work on this problem for fear that any results will be dismissed as artifacts due to the reaction of elemental sulfur with indigenous hydrocarbons. HAYES and BIEMANN (1967) contend that this possibility has been over-emphasized and discuss evidence, obtained in the high resolution direct introduction mass spectra of samples of the Murray C2 chondrite, for the presence of various sulfur-containing compounds. The authors argue that since all free sulfur has been volatilized from the sample at temperatures below 170°C and since no evolution of hydrogen sulfide is observed, the sulfur compounds found in the mass spectra must be indigenous constituents of the meteorite. MURPHY and NAGY (1966) have suggested a possible mechanism of accidental synthesis which HAYES and BIEMANN (1967) have discussed in detail, concluding that it probably has not affected their results.

Heating a sample of Murray under high vacuum (10^{-6} torr) in the temperature range 160–200°C, HAYES and BIEMANN (1967) observed ions with elemental compositions corresponding to thiophenes and benzothiophenes. The sulfur-containing compounds were less abundant than any of the unsubstituted aromatic hydrocarbons simultaneously observed. In the temperature range 255–286°C

---

* Experimental details are not given in the original paper but are mentioned in a later publication (KAPLAN, DEGENS, and REUTER, 1963) with specific reference to the earlier paper.

alkylthiophenes were observed and were about one-third as abundant as alkyl-benzenes. Ions with elemental compositions corresponding to alkylbenzothio-phenes, dibenzothiophenes, and possibly some thiophenols were also observed in this temperature range, as were $C_6H_4S_2$ and $C_7H_6S_2$, possibly the molecular ions of two compounds common as products of the pyrolysis of sulfur compounds. No aliphatic sulfur compounds were observed.*

In their mass spectrometric study of volatiles present in Orgueil, Murray and Cold Bokkeveld, STUDIER, HAYATSU and ANDERS (1965) found indications of the presence of carbon oxysulfide in Orgueil and Cold Bokkeveld, sulfur dioxide in Orgueil, and carbon disulfide in Murray and Orgueil. A study of the mass spectrum of a "solvent extract" of Orgueil suggested the presence of "sulfonic esters." Other than that they were indicated by fragment ions at m/e 207, 281 and 355, no specification of the type of sulfonic esters thought to be present was given. Because the ions mentioned are generally taken by mass spectroscopists as an indication of silicone grease contamination of the high-vacuum system, BURLINGAME and SCHNOES (1966) strongly criticized the interpretation of these ions in terms of sulfonic esters. It seems very likely that this is a case of mistaken identity and that the ions observed due to contaminating silicones.

MURPHY and NAGY (1966) report that thiols are not present in the Orgueil meteorite. Extracts were obtained by shaking the powdered meteorite samples with chloroform at room temperature. The fraction expected to contain thiols was separated from free sulfur and polar compounds by thin layer chromatography and then examined mass-spectrometrically. The conclusion that thiols were not present is based on the absence of long chain thiol molecular ions above mass 230 (the molecular weight of a C_{14} thiol; there was a peak at this mass). The "hydro-carbon fraction" contained an unaccounted for 1·6% sulfur, and the authors conclude without further evidence that it is present as sulfides.

Halogen containing compounds

The earliest basis for the presence of organic chlorine compounds in meteorites is MUELLER's (1953) elemental analysis of an extract of Cold Bokkeveld with various solvents including boiling chloroform. MUELLER assumed that the halogen which appeared in his samples was chlorine. BRIGGS (1963b) followed MUELLER's techniques and obtained extracts containing much lower amounts of chlorine (see Table 5) from Mokoia and Haripura. RAIA (1966) obtained a non-extractable fraction from Orgueil which contained 1·22% Cl and 1·25% F. This non-extractable residue had been treated three times with concentrated hydro-chloric acid, twice with 48% hydrofluoric acid, once with 6 N hydrochloric acid; digested once with 6 N hydrochloric acid; and refluxed once with 6 N hydrochloric acid. All these elemental analyses have been performed on materials which con-ceivably could have been halogenated by the isolation procedures used. It appears that the small amounts of chlorinated compounds identified (see below) do not

---

* The gas chromatography-mass spectrometry experiments described by HAYES and BIEMANN (1967b) were conducted with the heated sample at atmospheric pressure, with sulfur therefore distilling from the sample at a much higher temperature, greatly increasing the chance of accidental synthesis of sulfur compounds. The results of these experiments are not considered in the paragraph above.

5

by any means account for the large amounts of chlorine indicated as present by some elemental analyses.

STUDIER, HAYATSU and ANDERS (1965) report the mass spectrometric observation of chlorobenzene in the volatile fraction of Orgueil, dichlorobenzene in the volatile fraction of Cold Bokkeveld, and C_{12}–C_{18} alkyl chlorides in both meteorites. Supporting data are not published, but the characteristic isotopic multiplets indicating the presence of chlorine are distinctive mass spectral features which STUDIER, HAYATSU and ANDERS probably used in their identifications. The dichlorobenzene spectrum includes an intense molecular ion, but the alkylchloride spectra (particularly for long chain chlorides) do not. The presence of alkylchlorides is open to question.

HAYES and BIEMANN (1967) observed barely detectable amounts of dichlorobenzene in eight of ten Murray samples examined. There is no doubt as to the identity, two members of the molecular ion multiplet ($C_6H_4{}^{35}Cl_2$ and $C_6H_4{}^{35}Cl{}^{37}Cl$) were consistently observed and found to coincide with the expected mass within 0·003 mass units. The $C_6H_4Cl^+$ (loss of one chlorine atom) ion was also observed in four of the eight spectra. No evidence was obtained in support of STUDIER et al.'s report of alkylchlorides.

BURLINGAME and SCHNOES (1966) mention WADSWORTH's finding of a polychloronaphthalene in Murray. They consider it a contaminant because such compounds are common constituents of institutional floor waxes.

Nitrogen-containing compounds

Amino acids. It would be extremely interesting if amino acids could be shown to be indigenous constituents of meteorites. In early work, BRIGGS (1961, 1963c) failed to detect ninhydrin positive substances in Mokoia, and CALVIN and VAUGHN (1960) obtained a similar result with Murray. Using paper chromatographic separation of 80% ethanol extracts,* DEGENS and BAJOR (1962) found amino acids present in both Murray (16 μg/g) and Bruderheim (8 μg/g), an ordinary chondrite. Chromatographic separation of an acid hydrolysate yielded double these quantities in each case, but no change in qualitative distribution was observed. A supplementary publication by KAPLAN, DEGENS and REUTER (1963) reported similar results in an extended study of Orgueil, four C2 and three C3 stones, and also Karoonda (C4), Abee (E4), Hvittis (E6) and Norton County (an aubrite). The C group stones all showed total (free + combined) amino acid concentrations within a factor of two of 60 μg/g; the other meteorites contained less by a factor of about four. Lancé (C3) was a striking exception, containing 527·7 μg/g of amino acids. In comparing a sample of Orgueil which had been contaminated with collagen-based glue to one presumably containing ordinary amounts of amino acids, ANDERS et al. (1964) found that their analysis of the free amino acids in the ordinary sample was in good agreement with that of KAPLAN, DEGENS and REUTER. VALLENTYNE (1965) has presented an analysis of free and combined amino acids which qualitatively agrees with that of KAPLAN, DEGENS and REUTER but which shows a total amount smaller by a factor of five. On the other hand, RAIA (1966)

---

* Experimental technique was not given in the original paper but was described in a later paper (KAPLAN, DEGENS and REUTER, 1962) with specific reference to the earlier publication.

has searched for amino acids in Orgueil, Murray, Mokoia and Lancé. Only Murray contained amino acids, and this was the same sample which also contained viable terrestrial bacteria (ORÓ and TORNABENE, 1965).

After considering the qualitative distributions reported for amino acids in meteorites, P. B. HAMILTON was first to point out that "What appears to be the pitter-patter of heavenly feet is probably instead the print of an earthly thumb."

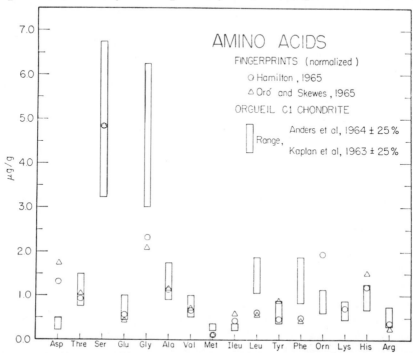

Fig. 3. Amino acids in the Orgueil C1 chondrite and in fingerprints. The finger-print analyses have been normalized to serine = 4·85 μg/g, the average of the meteorite analyses. This was done to facilitate comparison of the qualitative distributions and corresponds to approximately ten dry fingerprints per gram of sample. The meteorite analyses show that the stone contains in addition to the compounds tabulated here β-alanine, a compound not found in fingerprints. The large range is allowed on the meteorite analyses because of the imprecise analytical method: paper chromatography followed by estimation based on spot size.

Since then, both HAMILTON (1965) and ORÓ and SKEWES (1965) have published analyses of the amino acids deposited in a single human thumb-print under various conditions. The latter authors have also investigated amino acids in dust. The qualitative distribution in thumbprints closely resembles that reported in mete-orites, serine, glycine and alanine being the most abundant amino acids. Figure 3 presents the correspondence more clearly, comparing the two free amino acid analyses with the two fingerprint analyses. The ±25% error allowance is made because the meteorite amino acid determinations were based only on paper chromatographic estimations. The two sets of fingerprint data (each obtained using ion exchange chromatography and spectrophotometry) are normalized to serine = 4·85 μg/g, the average of the meteoritic values. According to HAMILTON, the amount of serine in a single dry thumbprint is approximately 0·5 μg. That is, to account for the observed meteorite concentrations, only ten fingerprints— two handfuls—on the sample itself or on any glassware that the reagents ever contacted are required per gram of meteorite analysed.

VALLENTYNE (1965), commenting on the analyses of combined amino acids, observed a "general similarity in the pattern of amino acids found in the two meteorite samples [Orgueil and Holbrook (L6)] as compared with that of 'hand-picked sand' [a blank]. All three contained about the same relative amounts of urea, cysteic acid, and different amino acids."

Serine, the most abundant amino acid in meteorites and fingerprints, is thermally the most unstable (VALLENTYNE, 1964). HAMILTON (1965) has noted that the presence of ornithine is particularly suspicious, and that its appearance in samples other than body fluids may indicate contamination. ORÓ and SKEWES (1965) have pointed out that the amino acid contents found by KAPLAN, DEGENS and REUTER are inversely proportional to sample size (Lancé, for example, was a relatively small sample), and that results are in better agreement if reported as quantities per sample. This could indicate contamination due to handling during analysis in addition to storage contamination.

It is not quite an open and shut case: Fig. 3 shows that the meteorite might contain indigenous glycine and phenylalanine (an excess of the latter is particularly noticeable in both combined amino acid analyses), and the meteorite also seems to contain β-alanine (KAPLAN, DEGENS and REUTER, 1963; ANDERS et al., 1964), a compound not found in fingerprints.

Purines pyrimidines, and similar compounds. CALVIN and VAUGHN (1960) first noted that the u.v. absorption spectrum of an aqueous extract (ion-exchange chromatographically fractionated) of Murray was pH sensitive. In the course of describing this behavior, the authors mentioned that it was similar to the spectrum of cytosine. In virtually the same sentence it was pointed out that such a statement was not meant to imply even a tentative indication of the presence of cytosine. Nonetheless, this report is occasionally cited as an "identification" of cytosine. Using ion-exchange materials like those used by CALVIN and VAUGHN, ORÓ (1963) has shown that very similar ultraviolet absorptions can be observed in the materials stripped from the ion-exchange beads by hot eluants. He has thus asserted that the cytosine-like absorption features observed by CALVIN and VAUGHN were really artifacts of the ion-exchange fractionations. The very close resemblance of the spectrum makes this certainly seem possible. However, a similar feature was observed by BRIGGS (1961) in a Mokoia aqueous extract which had not been exposed to ion-exchange materials. Since no biochemical names appear in BRIGGS' report, it has not received as much attention as CALVIN and VAUGHN's. BRIGGS (1961) also reported paper chromatographic separation of compounds which responded to normal (very non-specific) purine and imidazole spotting reactions.

NAGY et al. (1963b) found that the 260 mμ absorption of a mineralized organized element (possible microfossil) from Orgueil could be halved by treating the sample with ribonuclease. Presumably this treatment might have broken down indigenous polynucleotides and allowed some of the degraded material to escape. The authors are careful to avoid so speculative an interpretation, however.

The only published analysis of purines, pyrimidines, and similar compounds in meteorites is that by HAYATSU (1964). Using a 16 g sample of Orgueil which had previously been extracted with a series of nine organic solvents, HAYATSU obtained an organic nitrogen compound fraction by performing a 6-hr, 120°C, 3 N hydrochloric acid sealed tube hydrolysis. This hydrolysate was acetylated and fractionated on a solubility basis. I.r. and u.v. spectra of all fractions indicated the presence of nitrogen containing heterocycles and pH sensitive u.v. chromophores. The chloroform soluble fraction was further fractionated on an alumina column. The i.r. and u.v. spectra of the second of three fractions obtained suggested the presence of acetylated melamine-like compounds. Thin layer chromatography

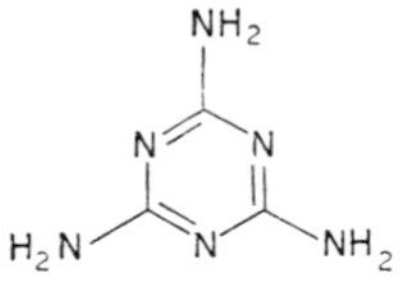

Melamine

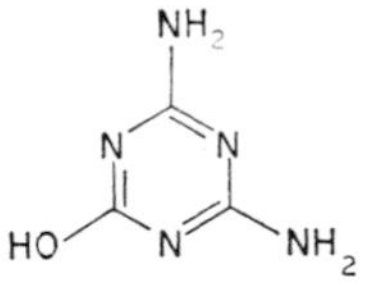

Ammeline

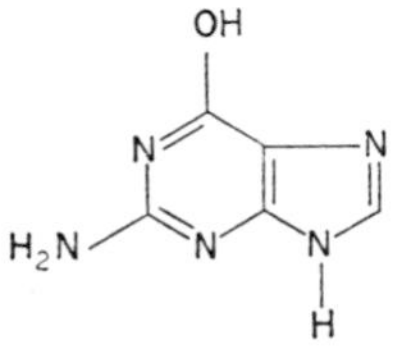

Guanine

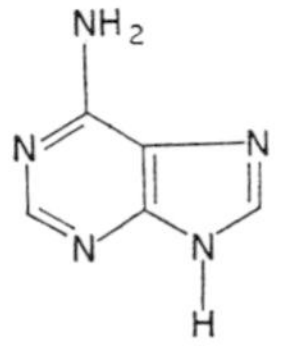

Adenine

of the free bases obtained from this fraction and comparison with thirty-eight standards led HAYATSU to report the presence of 35 $\mu g/g$ melamine, 28 $\mu g/g$ ammeline, 15 $\mu g/g$ adenine, and 20 $\mu g/g$ guanine. Unfortunately this basically preliminary report has neither been expanded on by the author nor confirmed by the work of another group. In view of the potentially great significance of these findings further work along these lines is most desirable.

In a report notable for its organization and completeness, HODGSON and BAKER (1964) summarize data indicating the possible presence of porphyrins in Orgueil at a level of 0·01 $\mu g/g$. A twice fractionated benzene–methanol extract showed a slight shoulder in its u.v. absorption spectrum at 410–412 mμ, the region expected for esterified vanadyl porphyrins. The spectral shifts in five solvents were as expected for an esterified vanadyl porphyrin, and the absorption was destroyed under the same conditions which decomposed authentic terrestrial samples. Consideration of some possible sources (fingerprints not included) of terrestrial contaminants showed that most would be expected to introduce chlorins at ten times the concentration of porphyrins. No chlorins were found in the meteorite extracts.

BAKER and HODGSON (1965) have reported an improved analytical method of higher sensitivity which they plan to apply to C2 chondrites, specifically, Cold Bokkeveld, Murray and Mokoia.

Other N-*containing compounds.* HAYES and BIEMANN (1967b) report the observation of an ion with the elemental composition of picoline in a high-resolution mass spectrum of a gas chromatographic fraction of compounds volatilized from the Murray C2 chondrite in the temperature range 25–146°C (atmospheric pressure). No nitrogen-containing ions were observed in the spectra obtained in the sample temperature range 146–254°C. A zinc dust distillation was carried out in the temperature range 254–409°C, and ions with elemental compositions corresponding to those of alkylpyrroles, alkylpyridines (or anilines), benzonitrile and indole (or methylbenzonitrile) were observed. Similar compounds were observed in the 255–285°C (10^{-6} torr pressure) sample temperature range when the powdered meteorite sample was introduced directly into the mass spectrometer ion source.

LIPMAN (1932b) has reported the presence of organic nitrogen compounds in a

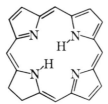

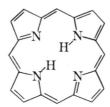

Chlorin ring system,
substituents omitted

Porphyrin ring system,
substituents omitted

number of ordinary chondrites. The report is based on the presence of N as detected in elemental analyses of meteorite fragments and the apparent absence of ammonia or its salts, nitrates, or nitrites. BUDDHUE (1942) has challenged these analyses, asserting that the amounts of inorganic nitrogen compounds he finds are sufficient to account for all the nitrogen.

Optically active compounds

To establish the presence of optical activity in meteorite organic compounds would be, by most standards, to establish the biogenicity of these compounds. Many papers have been written on this subject, and it is unfortunate that a discussion of them cannot fail to take on some of the aspects of a military history, since the groups in Chicago and LaJolla are in complete disagreement, and have been for three rounds of the battle. Because the initial investigation (NAGY et al., 1964) reported optical activity in a saponified sample, both sides have continued to use this technique, each trying to win its original point. Saponification is useful in these studies only as a method of fractionation (that is—it is impossible for a simple hydrolysis to produce optical activity*), and it is an unfortunate choice because it can severely degrade other compounds of interest which may be present. Other methods of fractionation may prove to be more useful, and would have the

---

* Barring the extremely remote possibility that the optical rotations of the acid and alcohol moieties of the esters happen to exactly cancel. A less remote possibility is that hydrolytic conditions would lead to racemization of some labile optically active centers, thus unbalancing a cancellation. Such destruction is by no means desirable—some other means of fractionation could be as effective and more informative.

advantage of not subjecting the samples to hydrolytic conditions. Because large samples are required for these studies, and because the measurement of optical activity is completely non-destructive, the investigation of optical activity of saponified fractions should be carried no farther at this time, another method of fractionation sought instead, and the studies coordinated with qualitative analysis of the compounds present.

Following the combined extraction and saponification technique previously used for isolation of fatty acids (NAGY and BITZ, 1963), NAGY et al. (1964) reported optical activity in the "saponified" fraction of the benzene–methanol (6:4) extract of the Orgueil chondrite. This fraction was found to be levorotatory, giving a maximum observed rotation of −23 ± 5 millidegrees (mdeg.) at a wavelength of 435 mμ. It is certain that this rotation was observed at a wavelength at which the absorbance of the sample was rather high: the authors note that previous failures to observe this optical activity may have been "related to the opacity of the solutions at the lower spectral range and to the apparent restriction of some of the easily detectable rotation to a narrow range very adjacent to this lower limit of transparency." Measurements in the range 500–650 mμ showed a constant rotation of −7 ± 5 mdeg. Absorbance in this range was presumably very low. The reliability of these measurements has been strongly challenged by HAYATSU (1965, 1966). By noting a large and rapidly growing collection of publications concerned with the occurrence of artifacts in polarimetry and by performing control experiments of his own, HAYATSU has concluded that it is likely that the rotation found near the transmission cutoff of the samples is spurious, a criticism which seems justified.

In addition to raising the question of artifacts, the first paper by HAYATSU (1965) reported an attempted confirmation of the results obtained by NAGY et al. A substantially different, but not illogical, procedure was employed. Separate extractions with benzene, chloroform and methanol were used, the extracts being combined. Sulfur was removed before saponification by precipitation at −9°C, and the nonsaponifiable material was treated with colloidal copper in order to remove the last traces of sulfur, a procedure which may have resulted in the loss of some of the polar constituents of this fraction. In addition to the saponifiable and nonsaponifiable fractions, the methyl esters of the saponifiable fraction, the organic material precipitated with the sulfur (after treatment with colloidal copper), and the acidic fraction of an extracted-stone hydrolysate were all examined polarimetrically. No rotation greater than −2 mdeg. was found within the 560–440 mμ range in which NAGY et al. found rotations as large as −20 mdeg. Some control samples, notably one which was intended to represent the probable blank from the LaJolla procedure, showed spurious rotations ranging up to −5 mdeg. at 440 mμ.

NAGY (1965) has asserted that this failure to find optical activity is not proof of error in his results because of the substantially different procedures employed. Since both procedures employed benzene and methanol as solvents, it seems unlikely that substantially different mixtures would have been extracted. It is necessary to suppose that the optically active compounds were somehow lost in HAYATSU's procedure. For the samples treated with colloidal copper, this is conceivable. But the saponifiable material, which should represent the bulk of the organic material in the fraction in which NAGY et al. found optical activity, was *not* treated with colloidal copper.

NAGY (1966) has reported a new set of measurements on Orgueil fractions obtained using the same procedures employed by NAGY et al. (1964). All measurements were made with sample absorbance less than 0·155. The saponifiable fraction showed rotations of from −4·5 to −1·8 mdeg. over the wavelength range 410–600 mμ. The nonsaponifiable fraction showed rotations of from −5·2 to −4·5 mdeg. over the wavelength range 500–605 mμ.

HAYATSU (1966) has also challenged these results. The observed rotations, if spurious, cannot be due to absorption artifacts, but HAYATSU has pointed out that colloidal or turbid solutions may produce spurious rotations by scattering light. He has run control experiments on spectropolarimeters of the type used by NAGY and has observed spurious rotations larger than the rotations observed by NAGY in the meteorite samples. However, the controls had absorbances of 0·27 (greater than NAGY's samples) and HAYATSU's attempt to prove the presence of a scattering component in NAGY's samples is not convincing.

The technique employed by the LaJolla group is clearly deficient;* and, until some confirming results are obtained using a different procedure, the presence of optically active compounds in Orgueil extracts cannot be accepted as proven.

Earlier reports of lack of optical activity are not relevant in the above discussion. MUELLER (1953) was working with a different meteorite and a much less sensitive instrument. BRIGGS (1963b) and KAPLAN, DEGENS and REUTER (1963) were working with aqueous extracts.

MEINSCHEIN et al. (1966) have reported a thorough investigation of the optical activity found in benzene–methanol extracts of ordinary chondrites. The initial extract of a 2·5 kg sample of the Homestead L5 chondrite was fractionated first by chromatography on silica gel. The methanol eluate of the silica gel column was further fractionated by gradient elution on an alumina column. A yellow band eluted with 4–8% methanol in benzene showed a complete Cotton effect centered on 340 mμ. The fraction weighed 19·7 mg., but the concentration of the optically active compound in the fraction was judged to be small. Similar activity was found in the same fraction of an extract of a 114 g sample of the Holbrook L6 chondrite. It also appeared that the compound responsible was present in the same fraction of hand-washings, although it was not the predominant optically active compound in this fraction. An extremely well executed experiment found the same material in the same fraction of an extract of a 2·5 kg fragment of the Holbrook chondrite and provided strong evidence that the material was on the surface of the stone. In fact, less than 200 μg of organic material were extracted from the interior of the stone. A final series of experiments showed that the same activity was present

---

* The final sample contains a poorly defined mixture of organic compounds, hydrocarbons as well as acids being present according to the thin layer chromatograms shown in NAGY et al. (1964). The final sample, in spite of having passed through several aqueous phase separations, contains "at least 50% sulfur," (NAGY, 1966) and there is evidence that this may be colloidal ("dyes and colloidal sulfur were added to some of the blanks in an attempt to reproduce the sample absorptions," NAGY et al., 1964).

in a sample used in geology classes for 65 yr, but absent in a series of six samples collected for organic geochemical investigations and protected from handling. Quantitative data concerning observed specific rotations and experimental conditions are carefully supplied.

Polymeric material

At most 30% of the carbon present in carbonaceous chondrites appears in solvent extracts. There is good evidence that the remaining carbon is tied up in an organic polymer. If 70% of the carbon present is in this polymer, and if the polymer is about 70% C, then the concentrations of polymeric material in C1 and C2 chondrites are about 35 and 25 mg/g, respectively. A complete elemental analysis of the polymeric material isolated from Orgueil by RAIA (1966) is as follows: 70·39% C, 4·43% H, 1·59% N (Dumas method), 9·80% O (Unterzaucher method), 6·91% S, 1·22% Cl, 1·25% F and 4·58% residue (ash). The presence of the halogens is discussed in the earlier section on halogen containing compounds. Excluding the halogens, the above elemental analysis corresponds to an empirical formula of $C_{10.0}H_{7.2}N_{0.2}O_{1.0}S_{0.4}$.

BRIGGS and KITTO (1962) and VDOVYKIN (1964a) have independently given very similar descriptions of the morphology of the unextractable carbonaceous material. Examining a thoroughly extracted sample of Mokoia, BRIGGS and KITTO found flat, translucent flakes "a few microns in size." These flakes charred on heating. VDOVYKIN found that the carbonaceous matter in Mighei was present as small, 2–3 μ inclusions mixed with silicate particles of the same size.

Various forms of spectral evidence indicate that the polymer is highly substituted, irregular and aromatic. BRIGGS and KITTO (1962) and VDOVYKIN (1964a) failed to observe any regular X-ray diffraction pattern, but VINOGRADOV et al. (1965) note some evidence for crystallinity in the carbonaceous material in Mighei, Cold Bokkeveld, and Boriskino (all C2). VINOGRADOV and VDOVYKIN (1964) obtained i.r. spectra indicating a carbonyl substituted highly condensed aromatic structure for the material in Mighei (C2). Electron spin resonance studies have been carried out by VINOGRADOV et al. (1964), VILLÉE et al. (1964), DUCHESNE et al. (1964a,b) and by SCHULZ and ELOFSON (1965). These studies are in substantial agreement and indicate that some sort of aromatic skeleton permitting extensive electron delocalization is present.

Before discussion the results of degradative studies of isolated polymeric material, it is worth briefly considering other experiments, for the most part described earlier in this review, in which some degradation of the polymeric material may have taken place. Certainly the volatilization studies of Murray by HAYES and BIEMANN (1967) are among these, and while there is no way of detecting that point in the spectra at which volatilization has declined and pyrolytic degradation predominates, the results obtained at high sample temperatures may be of use in future analyses of the polymeric material. Some extraction studies have involved acid hydrolysis steps which might have degraded the polymer to some extent. HAYATSU's (1964) results on organic nitrogen compounds in Orgueil are among these and have already been discussed. KAPLAN, DEGENS and REUTER (1963) found phenolic materials in a composite sample including portions of the

24 hr reflux, 5 N hydrochloric acid hydrolysates of Murray, Orgueil, Mokoia and Felix. The sample was desalted using ion-exchange columns and examined using two-dimensional paper chromatography. Seventeen spots (phenols, amines) were visible after treatment of the chromatogram with diazotized p-nitroaniline. Fifteen spots were visible under u.v. light. Three spots were observed both ways, so a total of twenty-nine compounds was resolved. Among the seventeen spots showing a color reaction, by far the most abundant had the R_f and color properties of 3-hydroxybenzoic acid. Other compounds tentatively identified included 4-hydroxybenzoic acid, 4-hydroxyphenylacetic acid and 4-hydroxy-3-methoxy-benzoic acid. These same compounds were among those found by BRIGGS (1963b) in extracts of Mokoia and Haripura.

Table 11. Fractionation of oxidation products of Orgueil C1 chondrite polymeric material* (mg/g).

Fraction	10·1 g Orgueil	63·3 g shale	12·4 g coal
A			3·2
B	0·35	0·02	3·3
C	24·3	12·3[1]	10·3
D			42·0
E	0·19	0·04	6·9
F	3·43	0·10	26·6
Total	28·27	12·46	92·3[2]

  * BITZ and NAGY (1966)
  [1] This fraction contained a large amount of elemental sulfur.
  [2] This number, recorded as 52·3 in the original paper, is correct as given here (NAGY, private communication).

BRIGGS (1963c) has noted that a good deal of the material considered unextractable is in fact soluble in "hot alkali." He speculates that this fraction is composed of high molecular weight acids. MUELLER (1963) has noted that after his first extraction of Cold Bokkeveld (MUELLER, 1953) he tried nitration as a method of degrading or otherwise making extractable the organic material remaining in the sample. Extraction of the nitrated stone yielded no additional material.

BITZ and NAGY (1966) have reported the only application of sophisticated degradation techniques to isolated meteoritic polymer. These authors treated the polymeric material with alkaline hydrogen peroxide and with ozone (followed with a second hydrogen peroxide treatment to insure complete conversion of the ozonolysis products to acids). The oxidation products were in each case divided into three fractions: compounds insoluble in the acidified aqueous oxidation mixture (fraction A from the H_2O_2 oxidation, D from the ozonolysis), compounds extracted from the acidic aqueous solution into ether (B, H_2O_2; E, ozonolysis), and compounds which remained in the aqueous phase through both the acidification and the ether extraction (C, H_2O_2; F, ozonolysis). Table 11 shows the fraction weights obtained from the three samples studied. Depending on the amount of elemental sulfur in fraction C from the shale, correspondence in terms of relative distribution of compounds may be close between the shale and the meteorite. The coal is very different.

The acid fractions were esterified with diazomethane, fractionated by gas chromatography, and isolated fractions examined using low resolution mass spectrometry. In this preliminary report the only meteorite results described were obtained during examination of fraction F. The authors present mass spectrometric evidence for the presence of benzene tetra- and pentacarboxylic acids. A spectrum attributed to the trimethyl ester of benzenetricarboxylic acid (mol. wt. 252) may be correctly identified, but the presence of an unidentified ion at m/e 270 raises questions. A spectrum attributed to dimethyl adipate does give some evidence for the presence of that compound, but is definitely the spectrum of a mixture containing substantial amounts of other, unidentified, compounds. The spectrum attributed to the methyl ester of a pyridinecarboxylic acid is probably misidentified, showing many ions at masses above the presumed "molecular ion," and containing several very intense ions at lower mass which cannot be derived from the methyl ester of any of the pyridinecarboxylic acids.

HAYES and BIEMANN (1967) have reported evidence for the presence of an aromatic polymeric material ($\sim$50 $\mu g/g$) in the Holbrook L6 chondrite. Vacuum pyrolysis of the untreated stone yielded (compounds identified only by their elemental compositions as determined by high resolution mass spectrometry) principally aromatic hydrocarbons (280–525°C). Oxygen-, nitrogen- and sulfur-containing ions were also observed, and probably indicate the presence of these compounds in the polymer, although at least in the case of sulfur this is an open question, since the temperatures required for pyrolysis make secondary reactions (with FeS?) possible.

Bacteria

Until some distinctly non-terrestrial bacterium is observed, viable bacteria observed in meteorites must be considered as definite terrestrial contaminants. In the most recent report, BITZ and NAGY (1966) describe bacteria cultured from samples of Orgueil. While the bacteria are unidentified, no exhaustive attempt at characterization was made. ORÓ and TORNABENE (1965) have identified contaminating terrestrial bacteria in samples of Murray and Mokoia but have found their sample of Orgueil to be sterile under aerobic conditions. SISLER (1961) has described bacteria apparently present in Murray. In earlier work, LIPMAN (1932a) reported the presence of bacteria in a number of ordinary chondrites. ROY (1935) has pointed out that the bacteria found by LIPMAN were all terrestrial contaminants.

ORIGIN OF METEORITE ORGANIC COMPOUNDS

This highly controversial subject is, at its simplest level, a question of whether the compounds observed are biogenic or abiogenic. Evidence for optical activity or the presence of recognizable microfossils* is at present inconclusive. A consideration

---

* "Microfossils" in meteorites—organized elements—have been dealt with in great detail in the literature. Pertinent references on the biogenic side of the question are NAGY, CLAUS and HENNESSY (1962); NAGY *et al.* (1963a); and CLAUS, NAGY and EUROPA (1963); on the abiogenic side FITCH and ANDERS (1963a,b). The subject has been reviewed by BRIGGS and MAMIKUNIAN (1963), by NAGY (1966a) and by UREY (1966a). UREY (1962) has reported on a discussion of some of the earlier findings.

of the extent to which the reported compounds may constitute evidence for bio-genicity is thus in order.

Amino acids are probably not present. The only analysis reporting good evidence for the presence of nucleic acid bases (HAYATSU, 1964) also reported the presence of several compounds of no known biological significance (melamine and ammeline). The amounts of porphyrins found (HODGSON and BAKER, 1964) are very, very small. On the other hand, the isoprenoid hydrocarbons pristane and phytane have been identified (MEINSCHEIN, 1964a), independently confirmed (ORÓ et al., 1966; NOONER and ORÓ, 1967), and a close correlation between mete-oritic and terrestrial sedimentary hydrocarbons observed.

One form of evidence for an abiogenic origin of meterite organic compounds is easily summarized: every compound found in meteorites, with the exception of pristane and phytane, has been synthesized in some type of abiogenic synthesis (NAGY, 1966a). The classic example in this field is MILLER's (1953) synthesis of amino acids from methane, ammonia and water. The n-alkanes may definitely be produced by Fischer–Tropsch syntheses (STUDIER et al., 1967), but at present the same cannot be said for the isoprenoid hydrocarbons. Indeed, abiogenic synthesis of hydrocarbons has been little studied relative to the compositions of terrestrial and meteoritic hydrocarbons. This disparity makes it impossible to make the necessary three-way comparison.

The wide distribution of carbon in meteorites (Fig. 1), and its quantitative correlation with trace element concentrations (LARIMER and ANDERS, 1967) both suggest that the distribution of carbon (and possibly, therefore, of organic material) was determined by some primary process in the formation of the solar system. It is of course possible that after this primary distribution biological processes modified the carbon compounds present in some meteorites and that some extra-terrestrial organic compounds are biogenic while others are not. Because of this possibility it is important to search for evidence of similarities or differences between the organic constituents of carbonaceous chondrites and of unequilibrated ordinary chondrites. Relationships between the unequilibrated ordinary chondrites and the ordinary chondrites are well established (DODD, VAN SCHMUS and KOFFMAN, 1967). The organic material of unequilibrated ordinary chondrites is presumably thus representative of the initial state of a tremendous amount of extraterrestrial organic material.

In general, theories proposing an asteroidal origin of meteorites (LARIMER and ANDERS, 1967; WOOD, 1966; RINGWOOD, 1966; ANDERS, 1964; WOOD, 1963a) include no hypotheses of extraterrestrial biospheres.* The general nature of the processes and conditions postulated makes the presence of biogenic material implausible. On the other hand, if it is true that meteorites are from the moon (UREY, 1966b; 1965), and if it is possible for organic material to have been trans-ferred from the earth to the moon at the time of its capture or ejection, then it is possible that the organic compounds observed in meteorites are remnants of some early terrestrial biosphere. ORÓ et al. (1966) have presented a striking comparison

---

* There is an *ad hoc* suggestion by ANDERS (1963) on a way in which life could temporarily have existed in an asteroid.

of meteoritic and terrestrial hydrocarbons which indicates that this could be the case. In general, however, evidence is sound that carbonaceous chondrites may not be created simply by the addition of water, organic compounds and trace elements (in cosmic proportions) to ordinary chondrites (ANDERS, 1963; FREDRIKSSON and KEIL, 1964; TAYLOR *et al.*, 1965; DODD, VAN SCHMUS and KOFFMAN, 1967). Any organic chemical evidence to the contrary must be very strong, and must in addition account for the differing distributions of organic material observed in at least the C1, C2 and C3 chondrites.

CONCLUSION

What was noted in the introduction—that work has been limited to a certain few carbonaceous chondrites—is by now obvious. It is equally clear that the compounds sought have been primarily those of terrestrial biochemical interest. While these experiments are important and interesting, pursuit of only this most obvious and spectacular possibility has at least partially obscured the fact that the organic constituents of meteorites, if examined on a broader scale, could be instrumental in helping to elucidate the origin of meteorites and the solar system.

The outstandingly complete study of hydrocarbons by NOONER and ORÓ (1967) raises some questions which demonstrate possibilities in this area:

Does the evidence indicating that C3 chondrites are richer in n-alkanes than are the C1 chondrites indicate that the n-alkanes are in some way associated with the "high temperature" fraction (LARIMER and ANDERS, 1967) which also becomes more abundant in the sequence C1, C2, C3; or does it simply indicate that the C3 chondrites are more retentive than the C1 chondrites?

Is the fact that, among the C2 chondrites, Renazzo and Kaba are lowest in n-alkane concentration and are distinctly higher in n-alkane/isoprenoid ratio related to the fact that these chondrites have greater amounts of metal present in their matrices?

Progress in general meteoritics makes it possible to ask some interesting questions about the general nature of organic constituents:

Can any relationship be observed between organic constituents and degree of metamorphism (WOOD, 1966; DODD, VAN SCHMUS and KOFFMAN, 1967) in the unequilibrated and related chondrites? If so, to what may the observed trend be attributed? To alteration or destruction of indigenous organic materials? To differing initial distributions and concentrations?

Is there any difference between the organic compound distributions in H and L unequilibrated chondrites?

Do the ordinary chondrites generally contain organic material or is the presence of polymeric material in Holbrook (HAYES and BIEMANN, 1967) due to the extremely fast cooling of that stone (50C°/10⁶ yr, WOOD, 1966)? Do the reheated chondrites (WOOD, 1966) contain some unique type or distribution or organic material or compounds?

The answers to these and better questions will, no doubt, be forthcoming. It is hoped that they may signal the change of the field of meteorite analysis from extraterrestrial life search to an energetic study of organic cosmochemistry.

Acknowledgement—I am indebted to Professors EDWARD ANDERS and KLAUS BIEMANN for encouragement, helpful discussions, and constructive criticisms. I am grateful to the many authors who promptly and helpfully answered questions concerning their papers, and to D. W. NOONER and J. ORÓ, who kindly made a manuscript copy of their 1967 publication available. This work was supported in part by NASA Grants NsG 211-62 and NsG-366.

REFERENCES

ANDERS E. (1962) Meteoritic hydrocarbons and extraterrestrial life. *Ann. N.Y. Acad. Sci.* **93**, 649–664.

ANDERS E. (1963) On the origin of carbonaceous chondrites. *Ann. N.Y. Acad. Sci.* **108**, 514–533.

ANDERS E. (1964) Origin, age, and composition of meteorites. *Space Sci. Rev.* **3**, 583–714.

ANDERS E. *et al.* (1964) Contaminated meteorite. *Science* **146**, 1157–1161.

BAKER B. L. and HODGSON G. W. (1965) Abstract of papers presented at Autumn Meeting of National Academy of Sciences, 1965. Improved porphyrin analysis for carbonaceous meteorites. *Science* **150**, 369.

BITZ M. C., and NAGY B. (1966) Ozonolysis of "Polymer-type" material in coal, kerogens, and in the Orgueil meteorite: A preliminary report. *Proc. Nat. Acad. Sci.* **56**, 1383–1390.

BLUMER M. (1957) Removal of elemental sulfur from hydrocarbon fractions. *Anal. Chem.* **29**, 1039–1041.

BOATO G. (1954) The isotopic composition of hydrogen and carbon in the carbonaceous chondrites. *Geochim. Cosmochim. Acta* **6**, 209–220.

BOATO G. (1956) Meaning of deuterium abundances in meteorites. *Nature* **177**, 424–425.

BRENTNALL W. D. and AXON H. J. (1962) The response of Canyon Diablo meteorite to heat treatment. *J. Iron and Steel Inst. (London)* **200**, 947–955.

BRETT R. (1966) Cohenite in meteorites: A proposed origin. *Science* **153**, 60–62.

BRIGGS M. H. (1961) Organic constituents of meteorites. *Nature* **191**, 1137–1140.

BRIGGS M. H. (1963a) Evidence of an extraterrestrial origin for some organic constituents of meteorites. *Nature* **197**, 1290.

BRIGGS M. H. (1963b) Organic extracts of some carbonaceous chondrites. *Life Sciences* **2**, 63–68.

BRIGGS M. H. (1963c) Unpublished results cited in BRIGGS and MAMIKUNIAN (1963).

BRIGGS M. H. and KITTO G. B. (1962) Complex organic microstructures in the Mokoia meteorite. *Nature* **193**, 1126–1129.

BRIGGS M. H. and MAMIKUNIAN G. (1963) Organic constituents of the carbonaceous chondrites *Space Sci. Rev.* **1**, 647–682.

BUDDHUE J. D. (1942) Nitrogen and its compounds in meteorites. *Pop. Astron.* **50**, 561–563.

BURLINGAME A. L. and SCHNOES H. K. (1966) Organic matter in carbonaceous chondrites. *Science* **152**, 104–106.

BUTLER C. P. (1966) Temperatures of meteoroids in space. *Meteoritics* **3**, 59–70.

CALVIN M. and McCARTHY (1967) The isolation and identification of the C_{17} isoprenoid hydrocarbon 2,6,10-trimethyl tetradecane: Squalane as a possible precursor. *Tetrahedron* (to be published).

CALVIN M. and VAUGHN S. K. (1960) Extraterrestrial life: some organic constituents of meteorites and their significance for possible extraterrestrial biological evolution. In *Space Research* (editor H. K. Bijl), Vol. I, pp. 1171–1191. North-Holland Publishing Co., Amsterdam.

CLAUS G., NAGY B. and EUROPA D. L. (1963) Further observations on the properties of the "organized elements" in carbonaceous chondrites. *Ann. N. Y. Acad. Sci.* **108**, 580–605.

CLAYTON R. N. (1963) Carbon isotope abundance in meteoritic carbonates. *Science* **140**, 192–193.

COHEN E. (1894, 1903, 1905) *Meteoritenkunde*, Vols. I–III. E. Schweizerbart, Stuttgart.

COMMINS B. T. and HARINGTON J. S. (1966) Polycyclic aromatic hydrocarbons in carbonaceous meteorites. *Nature* **212**, 273–274.

CRAIG H. (1957) Isotopic standards for carbon and oxygen and correction factors for mass-spectrometric analysis of carbon dioxide. *Geochim. Cosmochim. Acta* **12**, 133–149.

DEGENS E. T. and BAJOR M. (1962) Amino acids and sugars in the Brudeheim and Murray meteorites. *Naturwiss.* **49**, 605–606.

DODD R. T., VAN SCHMUS W. R. and KOFFMAN D. M. (1967) A survey of the unequilibrated ordinary chondrites. *Geochim. Cosmochim. Acta* **31** (to be published).

DUCHESNE J., DEPIREUX J. and LITT C. (1964a) Concerning the nature of free radicals in the Cold Bokkeveld meteorite. *Geochemistry International* **1**, 1022–1024.

DUCHESNE J., DEPIREUX J. and LITT C. (1964b) Organic free radicals in the meteorites Mighei and Nogoya. *Compt. Rend.* **259**, 4776–4780.

DuFRESNE E. R. and ANDERS E. (1961) The record in the meteorites. V. A thermometer mineral in the Mighei carbonaceous chondrite. *Geochim. Cosmochim. Acta* **23**, 200–208.

EDWARDS G. (1955) Isotopic composition of meteoritic hydrogen. *Nature* **176**, 109–111.

FITCH F. W. and ANDERS E. (1963a) Observations on the nature of the "organized elements" in carbonaceous chondrites. *Ann. N. Y. Acad. Sci.* **108**, 495–513.

FITCH F. W. and ANDERS E. (1963b) Organized element: possible identification in Orgueil meteorite. *Science* **140**, 1097–1100.

FLOROVSKAYA V. N. *et al.* (1965) Comparative characteristics of polycyclic aromatic hydrocarbons in carbonaceous chondrites, rocks, and minerals of endogenic origin. *Meteoritika* **26**, 169–176.

FREDRIKSSON K. and KEIL K. (1964) The iron, magnesium, calcium, and nickel distribution in the Murray carbonaceous chondrite. *Meteoritics* **2**, 201–217.

HAMILTON P. B. (1965) Amino acids on hands. *Nature* **205**, 284–285.

HAYATSU R. (1964) Orgueil meteorite: Organic nitrogen contents. *Science* **146**, 1291–1292.

HAYATSU R. (1965) Optical activity in the Orgueil meteorite. *Science* **149**, 443–447.

HAYATSU R. (1966) Artifacts in polarimetry and optical activity in meteorites. *Science* **153**, 859–861.

HAYES J. M. and BIEMANN K. (1967) High resolution mass spectrometric investigation of the organic constituents of the Murray and Holbrook chondrites. Submitted to *Geochim. Cosmochim. Acta.*

HIMES S. V. (1960) Further investigation of organic matter in meteorite Murray. University of California Radiation Lab. Report, UCRL 9208, 12–27.

HODGSON G. W. and BAKER B. L. (1964) Evidence for porphyrins in the Orgueil meteorite. *Nature* **202**, 125–131.

HOUTERMANS F. G. (1958) Natural thermoluminescence of silicate meteorites. In *Cosmological and Geological Implications of Isotope Ratio Variations*, pp. 31–36. NAS-NRC publication 572, Nuclear Science Series. report 23, Washington.

KAPLAN I. R., DEGENS E. T. and REUTER J. H. (1963) Organic compounds in stony meteorites. *Geochim. Cosmochim. Acta* **27**, 805–834.

KAPLAN I. R. and HULSTON J. R. (1966) The isotopic abundance and content of sulfur in meteorites. *Geochim. Cosmochim. Acta* **30**, 479–496.

KEIL K. and FREDRIKSSON K. (1964) The iron, magnesium, and calcium distribution in coexisting olivines and rhombic pyroxenes of chondrites. *J. Geophys. Res.* **69**, 3487–3515.

LARIMER J. and ANDERS E. (1967) Chemical fractionations in meteorites. II. Abundance patterns and their interpretation. Submitted to *Geochim. Cosmochim. Acta.*

LIPMAN C. B. (1932a) Are there living bacteria in stony meteorites? *Am. Museum Novitates*, 588.

LIPMAN C. B. (1932b) Discovery of combined nitrogen in stony meteorites (aerolites). *Am Museum Novitates*, 589.

LIPSCHUTZ M. E. (1965) Search for diamonds in the chondrite Ghubara, *Nature* **206**, 1145–1146.

LUMPKIN H. E. (1956) Determination of saturated hydrocarbons in heavy petroleum fractions by mass spectrometry. *Anal. Chem.* **28**, 1946–1948.

MASON B. (1962) *Meteorites.* Wiley.

MASON B. (1963) The carbonaceous chondrites. *Space Sci. Rev.* **1**, 621–646.

MASON B. (1966) The enstatite chondrites. *Geochim. Cosmochim. Acta* **30**, 23–39.

MASON B. and WIIK H. B. (1964) The amphoterites and meteorites of similar composition. *Geochim. Cosmochim. Acta* **28**, 533–538.

MEINSCHEIN W. G. (1963a) Benzene extracts of the Orgueil meteorite. *Nature* **197**, 833–836.

MEINSCHEIN W. G. (1963b) Hydrocarbons in terrestrial samples and the Orgueil meteorite. *Space Sci. Rev.* **2**, 653–679.

MEINSCHEIN W. G. (1964a) Development of hydrocarbon analyses as a means of detecting life in space. Quarterly Report, April 1964, NASA Contract NASW-508.

Meinschein W. G. (1964b) Development of hydrocarbon analyses as a means of detecting life in space. Quarterly Report, July 1964, NASA Contract NASW-508.

Meinschein W. G. and Kenny G. S. (1957) Analyses of a chromatographic fraction of organic extracts of soils. *Anal. Chem.* **29**, 1153–1161.

Meinschein W. G., Nagy B. and Hennessy D. J. (1963) Evidence in meteorites of former life. *Ann. N. Y. Acad. Sci.* **108**, 553–579.

Meinschein W. G. *et al.* (1966) Meteorites: Optical activity in organic matter. *Science* **154**, 377–380.

Miller S. L. (1953) A production of amino acids under possible primitive earth conditions. *Science* **117**, 528–529.

Monnig O. E. (1963) The Bells, Texas, meteorites. *Meteoritics* **2**, 67.

Monster J., Anders E. and Thode H. G. (1965) $^{34}S/^{32}S$ ratios for the different forms of sulfur in the Orgueil meteorite and their mode of formation. *Geochim. Cosmochim. Acta* **29**, 773–779.

Moore C. B. and Lewis C. (1965) Carbon abundances in chondritic meteorites. *Science* **149**, 317–318.

Moore C. B. and Lewis C. (1966a) The distribution of total carbon content in enstatite chondrites. Center for Meteorite Studies preprint No. 21.

Moore C. B. and Lewis C. (1966b) Total carbon content of ordinary chondrites. Submitted to *J. Geophys. Res.*

Mueller G. (1953) The properties and theory of genesis of the carbonaceous complex within the Cold Bokkeveld meteorite. *Geochim. Cosmochim. Acta* **4**, 1–10.

Mueller G. (1963) Organic cosmochemistry. In *Organic Geochemistry* (editor I. A. Breger), pp. 1–35. Macmillan.

Mueller G. (1966) Significance of inclusions in carbonaceous meteorites. *Nature* **210**, 151–155.

Mueller G., Shaw R. A. and Ogawa T. (1965) Interrelations between volatilization curves, elemental composition and total volatiles in carbonaceous chondrites. *Nature* **206**, 23–25.

Murphy M. T. J. and Nagy B. (1966) Analysis for sulfur compounds in lipid extracts from the Orgueil meteorite. *J. Am. Oil Chem. Soc.* **43**, 189–196.

Nagy B. (1965) Optical activity in the Orgueil meteorite. *Science* **150**, 1846.

Nagy B. (1966) A study of the optical rotation of lipids extracted from soils, sediments, and the Orgueil carbonaceous meteorite. *Proc. Nat. Acad. Sci.* **56**, 389–398.

Nagy B. (1966a) Investigations of the Orgueil carbonaceous meteorite. *Geol. for Förhandlingar (Stockholm)* **88**, 235–272.

Nagy B. and Bitz M. C. (1963) Long chain fatty acids from the Orgueil meteorite. *Arch. Biochem. Biophys.* **101**, 240–248.

Nagy B., Claus G. and Hennessy D. J. (1962) Organic particles embedded in minerals in the Orgueil and Ivuna carbonaceous chondrites. *Nature* **193**, 1129–1133.

Nagy B., Meinschein W. G. and Hennessy D. J. (1961) Mass spectroscopic analysis of the Orgueil meteorite: Evidence for biogenic hydrocarbons. *Ann. N. Y. Acad. Sci.* **93**, 27–35.

Nagy B., Meinschein W. G. and Hennessy D. J. (1963) Aqueous, low temperature environment of the Orgueil meteorite parent body. *Ann. N. Y. Acad. Sci.* **108**, 534–552.

Nagy B. *et al.* (1963a) Electron probe microanalysis of organized elements in the Orgueil meteorite. *Nature* **198**, 121–125.

Nagy B. *et al.* (1963b) Ultraviolet spectra of organized elements. *Nature* **200**, 565–566.

Nagy B. *et al.* (1964) Optical activity in saponified organic matter isolated from the interior of the Orgueil meteorite. *Nature* **202**, 228–233.

Nooner D. W. and Oró J. (1967) Aliphatic hydrocarbons in meteorites. Submitted to *Geochim. Cosmochim. Acta.*

Oró J. (1963) Ultraviolet-absorbing compound(s) reported present in the Murray meteorite. *Nature* **197**, 756–758.

Oró J. and Skewes H. B. (1965) Free amino-acids on human fingers: the question of contamination in microanalysis. *Nature* **207**, 1042–1045.

Oró J. and Tornabene T. (1965) Bacterial contamination of some carbonaceous meteorites. *Science* **150**, 1046–1048.

Oró J. *et al.* (1966) Paraffinic hydrocarbons in the Orgueil, Murray, Mokoia, and other meteorites. *Life Sciences and Space Res.* **4,** 63–100.

Pepin R. O. and Signer P. (1965) Primordial rare gases in meteorites. *Science* **149,** 253–265.

Raia J. (1966) M.S. Thesis, Dept. of Chemistry, University of Houston.

Ramdohr P. (1963) The opaque minerals on stony meteorites. *J. Geophys. Res.* **68,** 2011–2036.

Reed G. W. and Allen R. O. (1966) Halogens in chondrites. *Geochim. Cosmochim. Acta* **30,** 779–800.

Reed G. W. and Jovanovic S. (1967) Mercury in chondrites. *J. Geophys. Res.* (to be published).

Ringwood A. E. (1966) Genesis of chondritic meteorites. *Rev. Geophys.* **4,** 113–175.

Roy (1935) The question of living bacteria in stony meteorites. *Field Mus. Nat. Hist. Geol. Ser.*-6, **14,** 179–190.

Sarma D. V. N. and Mayeda T. (1961) Meteorite analysis: The search for diamond. *Geochim. Cosmochim. Acta* **22,** 169–175.

Schulz K. F. and Elofson R. M. (1965) Electron spin resonance studies of organic matter in the Orgueil meteorite. *Geochim. Cosmochim. Acta* **29,** 157–160.

Short J. M. and Keil K. (1966) Compositions and formation of metallic phases in 40 stony meteorites. Talk presented at 29th annual meeting of the Meteoritical Society, Washington, 3–5 November 1966.

Sisler F. D. (1961) Organic matter and life in meteorites. *Proc. Lunar and Planetary Exploration Colloquium* **2,** 4, 67.

Studier M. H., Hayatsu R. and Anders E. (1965) Organic compounds in carbonaceous chondrites. *Science* **149,** 1455–1459.

Studier H. M., Hayatsu R. and Anders E. (1967) Origin of organic matter in early solar system. I. Hydrocarbons. Enrico Fermi Institute for Nuclear Studies, Univ. of Chicago, Report EFINS 65–115. Submitted to *Science*.

Sztrokay K. I., Tolnay V. and Földvari-Vogi M. (1961) Mineralogical and chemical properties of the carbonaceous meteorite from Kaba, Hungary. *Acta Geologica* **7,** 57–103.

Taylor H. P., Jr. *et al.* (1965) Oxygen isotope studies of minerals in stony meteorites. *Geochim. Cosmochim. Acta* **29,** 489–512.

Trofimov A. V. (1950) Carbon isotope ratios in meteorites. *Doklady Akad. Nauk SSSR* **72,** 663–666.

Urey H. C. (1962) Lifelike forms in meteorites. *Science* **137,** 623–628. A more extensive transcript of this discussion has been reprinted in *Ann. N. Y. Acad. Sci.* **108,** 606–615.

Urey H. C. (1965) Meteorites and the moon. *Science* **147,** 1262–1265.

Urey H. C. (1966a) Biological material in meteorites: A review. *Science* **151,** 157–166.

Urey H. C. (1966b) Chemical evidence relative to the origin of the solar system. *Mon. Not. R. Astr. Soc.* **131,** 199–223.

Urey H. C. and Craig H. (1953) The composition of the stone meteorites and the origin of the meteorites. *Geochim. Cosmochim. Acta* **4,** 36–82.

Urey H. C., Mele A. and Mayeda T. (1957) Diamonds in stone meteorites. *Geochim. Cosmochim. Acta* **12,** 1–4.

Vallentyne J. R. (1964) Biogeochemistry of organic matter. II. Thermal reaction kinetics and transformation products of amino compounds. *Geochim. Cosmochim. Acta* **28,** 157–188.

Vallentyne J. R. (1965) Two aspects of the geochemistry of amino acids. In *The Origin of Prebiological Systems and of their Molecular Matrices* (editor S. W. Fox), pp. 105–125. Academic Press.

Van Schmus W. R. and Wood J. A. (1967) A chemical–petrologic classification for the chondritic meteorites. *Geochim. Cosmochim. Acta* **31,** 747–765.

Vdovykin G. P. (1962) Bitumens in the Grosnaja and Mighei carbonaceous chondrites. *Geochemistry* (English translation) 152.

Vdovykin G. P. (1963) Certain results from a comparison of the bitumens of carbonaceous chondrites with rock bitumens. *Meteoritika* (English Translation) **23,** 69–76.

Vdovykin G. P. (1964a) Carbonaceous matter in meteorites and its origin. *Geochemistry International* **1,** 693–697.

6

VDOVYKIN G. P. (1964b) Mineral composition of twelve meteorites containing carbon. *Meteoritika* **25**, 134–155.

VDOVYKIN G. P. (1965) On the origin of carbonaceous chondrites. *Meteoritika* **26**, 151–168.

VILLÉE F., DUCHESNE J. and DEPIREUX J. (1964) Free radicals in carbonaceous meteorites. *Compt. Rend.* **258**, 2376–2378.

VINOGRADOV A. P. and VDOVYKIN G. P. (1964) Multimolecular organic matter of carbonaceous chondrites. *Geochemistry International* **1**, 831–836.

VINOGRADOV A. P., VDOVYKIN G. P. and MAROV I. N. (1964) Free radicals in the Mighei meteorite. *Geochemistry International* **1**, 395–398.

VINOGRADOV A. P., VDOVYKIN G. P. and POPOV N. M. (1965) Investigation of carbonaceous meteorite matter by the method of high voltage electron diffraction. *Geokhimiya*, 387–389.

VINOGRADOV A. P. *et al.* (1966) Study of organic compounds and diamonds of the meteorite Novo Urei by the method of IR absorption spectroscopy. *Geokhimiya*, 1106–1109.

WIIK H. B. (1956) The chemical composition of some stony meteorites. *Geochim. Cosmochim. Acta* **9**, 279–289.

WOOD J. A. (1963a) On the origin of chondrules and chondrites. *Icarus* **2**, 152–180.

WOOD J. A. (1963b) Physics and chemistry of meteorites. In *The Solar System*, Vol. IV. *The Moon, Meteorites, and Comets* (editors G. P. Kuiper and B. M. Middlehurst), pp. 337–401. Univ. of Chicago Press.

WOOD J. A. (1966) Chondrites: Their metallic minerals, thermal histories, and parent planets. *Icarus* **6**, 1–49.

137

12

Reprinted from *Geochim. Cosmochim. Acta*, **34**, 981–994 (1970) with permission of Microforms International Marketing Corporation as exclusive copyright licensee of Pergamon Press journal back files.

Organic compounds in meteorites—IV. Gas chromatographic–mass spectrometric studies on the isoprenoids and other isomeric alkanes in carbonaceous chondrites

E. Gelpi* and J. Oró

Departments of Biophysical Sciences and Chemistry, University of Houston
Houston, Texas 77004

(Received 12 June 1969; accepted in revised form 4 May 1970)

Abstract—Nine homologous series of isomeric alkanes, in addition to the normal alkane series, have been identified by gas chromatographic and mass spectrometric techniques in the hydrocarbon fractions from the extracts of six carbonaceous chondrites (Essebi, Grosnaja, Mokoia, Murray, Orgueil and Vigarano). Two of these series show isoprenoid structures corresponding to 2,6,10-trimethyl and 2,6,10,14-tetramethyl alkanes ranging from C_{14} to C_{19} and C_{19} to C_{21}, respectively. The rest is made up of five series of monomethyl alkanes (2-, 3-, 4-, 5-, 6-methyl alkanes) and two of monocycloalkanes (cyclohexyl and cyclopentyl). The presence of members of the n-paraffin series between C_{11} and C_{26} inclusive was also confirmed by mass spectral data. The hydrocarbon content ranged from 6·7 to 82·8 ppm.

Isoprenoids are present in all samples analyzed. Usually they show a major maximum at C_{19} (pristane) and a secondary maximum at C_{16} (methyl farnesane) with a minimum at C_{17}.

INTRODUCTION

THE HYDROCARBONS in meteorites have been recently the subject of many investigations (NAGY *et al.*, 1961; NOONER and ORÓ, 1967; OLSON *et al.*, 1967; ORÓ and NOONER, 1967; ORÓ *et al.*, 1966, 1968, 1969; GELPI, 1968; see HAYES, 1967, for complete list of references) directed toward the elucidation of their nature and origin. As a consequence, we now have rather good knowledge of the identities and overall distributions of the normal hydrocarbons in these extraterrestrial samples; however, little is still known about most of their branched isomers (ORÓ *et al.*, 1968; ORÓ and GELPI, 1969; STUDIER *et al.*, 1968).

The possible presence of a great number of branched and cyclic alkanes in the n-heptane fraction of meteorite extracts had been anticipated from the base line hump usually observed in most gas chromatograms of meteoritic hydrocarbons (NOONER and ORÓ, 1967) and, also, in a more direct way, from the regular appearance and distribution pattern of a number of small peaks in these gas chromatograms. In spite of this, practically nothing was known about their nature, and at the time of the recent review of the subject by HAYES (1967) only three of the non-n-alkane components of the isomeric fraction (norpristane, pristane and phytane) had been tentatively identified (MEINSCHEIN, 1964; NOONER and ORÓ, 1967). Recently, STUDIER *et al.* (1968) presented data on the identification of small amounts of 2-methyl, 3-methyl, and other branched paraffins in the Orgueil and Murray carbonaceous chondrites. In general, isoprenoid hydrocarbons were reported absent in these two meteorites with the exception of a tentative identification of the C_{11}, C_{13} and C_{14} compounds in Murray.

---

* Present address: Space Sciences Laboratory, University of California, Berkeley, Calif. 94720.

Since an understanding of the nature and structural characteristics of all the alkanes in meteorites may provide a better insight into the much debatable question of their extraterrestrial origin, we have undertaken their mass spectrometric identification in several representative samples of carbonaceous chondrites (ORÓ *et al.*, 1968; ORÓ and GELPI, 1969).

A substantial amount of data obtained in this laboratory on the aliphatic and aromatic hydrocarbons in carbonaceous chondrites has already been presented (NOONER and ORÓ, 1967; OLSON *et al.*, 1967; ORÓ and NOONER, 1967; ORÓ *et al.*, 1966). The initial work had been directed mainly towards the gas chromatographic identification of these compounds, and in a few instances attempts were made with success to confirm part of the results by mass spectrometry. Mass spectra with the general characteristics of pristane and phytane were obtained for the Boriskino, Santa Cruz, Mighei, Mokoia, and Murray meteorites. Mass spectra corresponding to norpristane were also obtained for the Boriskino, Mokoia and Murray (NOONER and ORÓ, 1967).

The present work represents an extension of the earlier investigations by NOONER and ORÓ (1966), ORÓ and NOONER (1967), ORÓ *et al.* (1968) and ORÓ and GELPI (1969). All efforts have now been concentrated in the mass spectrometric identification of most of the paraffinic compounds previously detected by gas chromatography. Emphasis has been placed in particular on the characterization of isoprenoid structures.

EXPERIMENTAL METHODS

The analytical techniques used in connection with this work have already been reported in some detail in the first part of this series (NOONER and ORO, 1967). In essence, the method is based on extraction of the powdered samples with organic solvents, fractionation by silica gel chromatography, and analysis of the alkane eluate by gas chromatography and mass spectrometry. All necessary precautions were taken to avoid contamination in much the same manner as previously described (NOONER and ORÓ, 1967). Essentially all the control blanks showed either negative results or negligible amounts of alkanes. The only three modifications introduced in the overall procedure were: (1) the use of higher quality methanol ("Pesticide Grade" instead of "Spectrograde") and benzene ("Nanograde" instead of "Spectrograde"), both from Matheson, Coleman, and Bell; (2) the extracts were not taken to complete dryness at any time (with the exception of two cases) thus reducing possible losses of the more volatile hydrocarbons; and (3) the quantification of the results by means of the use of a small on-line electronic integrator instead of the usual manual computation by triangulation of the GC peaks.

All the gas chromatographic analyses were made on an F & M Model 810 gas chromatograph equipped with a flame ionization detector. Polysev [*m*-bis-*m*-(phenoxy-phenoxy)-phenoxy-benzene, Applied Science Laboratories, Inc.] was used as the GC stationary phase. The values of the areas under each one of the peaks were obtained directly from an electronic digital integrator (Infotronics CRS-11AB/H/41). The gas chromatographic–mass spectrometric analyses were performed on a LKB 9000 gas chromatograph–mass spectrometer.

RESULTS

Qualitative analysis

The results recently obtained in this laboratory on the gas chromatographic–mass spectrometric analysis of the alkane fraction from six carbonaceous chondrites (Essebi, Grosnaja, Mokoia, Murray, Orgueil, and Vigarano) have increased the number of hydrocarbons identified to a new total of 68 (Table 1). Typical gas

Table 1. Isoprenoids and aliphatic hydrocarbons in meteorites—GC–MS identification

	Mo.	Mu.	O.	V.
nC$_{11}$				1
nC$_{12}$	0·1	1		1
1 Decahydronaphthalene	1			1
2 2,6,10 TriMeC$_{11}$		1		0·1
3 2MeC$_{12}$		1		1
4 3MeC$_{12}$		1		1
5 4,8 DiMeC$_{12}$		1		
nC$_{13}$	1	1		1
6 2,6,10 TriMeC$_{12}$	1	1		1
7 2MeC$_{13}$	1	1		1
8 3MeC$_{13}$	1	1		1
9 4,8 DiMeC$_{13}$		1		1
nC$_{14}$	1	1		3
(10) 2,6,10 TriMeC$_{13}$	1	1		1
11 4MeC$_{14}$		1		
(12) 2MeC$_{14}$	1	1		3
(13) 3MeC$_{14}$	1	1		3
14 4,8 DiMeC$_{14}$		1		
nC$_{15}$	2	2	1	4
(16) 2,6,10 TriMeC$_{14}$	1	1		0·1
17 5MeC$_{15}$		1		1
18 4MeC$_{15}$	1	1		

	E.	Gr.	Mo.	Mu.	O.	V.
(19) 2MeC$_{15}$	1	1	1	1		1
(20) 3MeC$_{15}$	1	1	1			1
nC$_{16}$	1	1	4	6	1	4
(21) 2,6,10 TriMeC$_{15}$	1	1	2	2		2
(23) 6MeC$_{16}$			1	1		
24 5MeC$_{16}$			0·1	1		1
(25) 4MeC$_{16}$			1	1		
(26) 2MeC$_{16}$	1	1	1	1	1	1
(27) 2,6,10 TetraMeC$_{15}$	2	1	6	4	1	3
(28) 3MeC$_{15}$	1	1	1			1
nC$_{17}$	3	1	8	14	3	8
(30) 2,6,10 TriMeC$_{16}$			1			1
(31) 4,7,11 TriMeC$_{16}$			0·1	1		1
32 Decylcyclohexane			1	1	1	
(33) 6MeC$_{17}$				2	1	
(34) 5MeC$_{17}$			1	1		
(35) 4MeC$_{17}$	1		1	1		
(36) 2MeC$_{17}$	1	1	2	1	1	1
(37) 2,6,10 TetraMeC$_{16}$	3	1	4	3	1	2
(38) 3MeC$_{17}$	1	1	1	1	1	1
nC$_{18}$	6	1	7	9	3	5
(39) Dodecylcyclopentane				1		0·1
(40) Undecylcyclohexane			3	1	1	

	E.	Gr.	Mo.	Mu.	O.
41 5MeC$_{18}$			1		
(42) 2,6,10,14 TetraMeC$_{17}$	1	0·1	1	1	
43 4MeC$_{18}$			1	1	
(44) 2MeC$_{18}$	1		2	1	
(45) 3MeC$_{18}$	1		1	1	
nC$_{19}$	5	2	6	8	2
(46) Tridecylcyclopentane					
(47) Dodecylcyclohexane			1	1	
48 4MeC$_{19}$			1	1	
(49) 2MeC$_{19}$			1	1	
(50) 3MeC$_{19}$			1	1	
nC$_{20}$	1	2	3	5	1
(51) Tridecylcyclohexane			1	1	
52 2MeC$_{20}$				1	
(53) 3MeC$_{20}$				1	
nC$_{21}$	1	2	1	6	1
54 Tetradecylcyclopentane				1	
55 2MeC$_{21}$				1	
nC$_{22}$			1	4	1
nC$_{23}$				2	
nC$_{24}$				2	
nC$_{25}$					
nC$_{26}$					

E. (Essebi); Gr. (Grosnaja); Mo. (Mokoia); Mu. (Murray); O. (Orgueil); V. (Vigarano). The amounts given (in ppm) been obtained by rounding the actual values to the next higher number. Blanks in the columns of the table do not necess indicate the absence of certain alkanes, but rather that no useful mass spectra were obtained. Compounds other than n-alk have an identification number corresponding to that given in the gas chromatograms. When this number is placed in parenth it indicates that the compound has also been identified in graphite nodules from iron meteorites.

chromatograms are shown in Figs. 1, 2 and 3. The alkanes in Table 1 are given in order of their gas chromatographic elution times from capillary columns coated with Polysev.

Each assignment of structural identity for a compound has been confirmed by mass spectral data.

Ten well-defined homologous series of alkanes can be found in Table 1: The normal alkanes (in the C$_{11}$–C$_{26}$ range); the five monomethyl alkanes (2-, 3-, 4-, 5-, 6-methyl alkanes) with members in the C$_{13}$–C$_{22}$ range; the two isoprenoid homologies formed by the 2,6,10-trimethyl alkane or farnesane series (C$_{14}$–C$_{19}$) and the 2,6,10,14-tetramethyl alkane or phytane series (C$_{19}$–C$_{21}$); and finally the two monocyclo-alkane homologous series (cyclohexyl and cyclopentyl) with members in the C$_{16}$–C$_{20}$ range. The presence of a dimethyl alkane series (4,8-dimethyl alkanes from C$_{14}$ to C$_{16}$) would bring the total to eleven different series, but the mass spectrometric data should be considered tentative in this case.

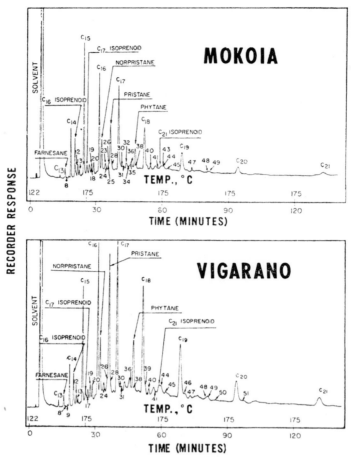

Fig. 1. Gas chromatographic separation of the alkanes from the Mokoia and Vigarano meteorites. Stainless steel capillary tubing, 115 m long by 0·076 cm i.d. coated with Polysev. Nitrogen carrier pressure: 1050 g/cm². F and M Model 810 gas chromatograph equipped with a flame ionization detector. Range 10^2, attenuation 1. Programming rate 2°C/min. Vigarano: 0·5 g extracted, $\frac{1}{5}$ injected. Mokoia: 0·5 g extracted, $\frac{1}{3}$ injected. (See Table 1 for peak number identification.)

The C_{19} member of the 2,6,10-trimethyl alkane series (2,6,10-trimethylhexadecane) and an apparently irregular C_{19} isoprenoid structure (4,7,11-trimethylhexadecane) have also been identified in some of the meteorites. The identification of these two compounds, especially the latter, is only tentative at this moment.

Quantitative analysis

Detailed quantitative evaluations of the hydrocarbons detected in four of the six meteorite samples are given in Table 2A. The members of the ten homologous series indicated above have been grouped into four major classes of related structures, namely the cycloalkanes (A), methyl alkanes (B) isoprenoids (C), and n-alkanes (D). The total values given in Table 2A correspond to the sum A + B + C + D. The

141

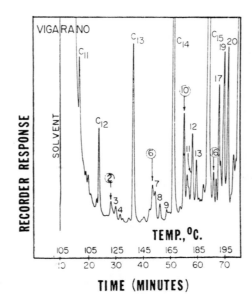

Fig. 2. Partial high resolution gas chromatogram of the hydrocarbons from Vigarano. Same conditions as in Fig. 1 with the following exceptions: Column length, 195 m; range, 10; attenuation, 2: Temperature held isothermally at 105°C for 20 minutes and then programmed to 195°C at 2°C/min. (See Table 1 for peak number identifications.)

"E" column is made up by the contribution of the components present in the gas chromatograms but not identified by mass spectrometry. In the case of the Murray and Grosnaja carbonaceous chondrites, where most of the compounds in the alkane fraction have been identified, "E" has a very small value. This is not so in the Mokoia and Vigarano chondrites, which apparently contain considerable amounts of un-identified high molecular weight materials.

In Table 2B a comparison is made between the total value of isoprenoids plus n-alkanes (C + D) given in Table 2A and those previously reported for the same meteorites by NOONER and ORÓ (1967). Taking into consideration the different methods employed for the quantitation of the results, peak areas from an electronic digital integrator vs. manual computation, there appears to be a close correlation between the values obtained in both investigations for the samples having a common source, such as the Murray and Mokoia specimens (Table 2B; a, d). The values obtained with the digital integrator appear to be higher than the ones obtained by NOONER and ORÓ (1967) by a factor of about 1·5. In any case, the deviation observed by any of the two methods is not great as demonstrated by the results of the replicate runs (Table 2B; d). However, great differences in the amounts of alkanes are obtained when the samples are from different donors probably indicating somewhat different histories (Vigarano and Grosnaja in Table 2B). The differences observed between the two samples of the Grosnaja meteorite (Table 2B; b and c) are particularly striking.*

* For gas chromatographic pattern order NAPS Document 01010 from ASIS National Auxiliary Publication Service, c/o CCM Information Sciences, Inc., 909 Third Avenue, New York, N.Y. 10022, remitting $2.00 for microfiche or $5.30 for photocopies.

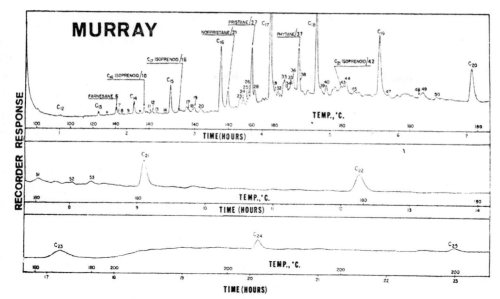

Fig. 3. High resolution gas chromatogram of the hydrocarbons from the Murray meteorite. Same conditions as in Fig. 1 with the following exceptions: Column length, 310 m; Range, 10; attenuation, 2. Temperature held isothermally at 100°C for about 70 min, then programmed to 140°C at 1°C/min and held at this temperature for 100 min after which it was programmed at 1°C/min to 180°C. After about 17 hr, the temperature was raised to 200°C to elute the last components. A total of 4·1473 g of the meteorite were extracted. One half of this extract was taken for this analysis and about 2/3 of the n-pentane eluate was injected. See Table 1 for peak number identifications. See Figs. 4 and 5 for the corresponding mass spectra.

The hydrocarbons in the three specimens of Vigarano (Table 2B; e, f, g), although differing widely in total content, all have gas chromatographic patterns identical to that shown as a representative example in Fig. 1B. Likewise, the hydrocarbon distribution in the four Mokoia specimens is similar to that of Vigarano (Fig. 1A).

From the data shown in Table 2A it becomes apparent that the cycloalkanes are always present in low amounts; the methyl alkanes are in general lower than the isoprenoids, and the ratio of n-alkanes to isoprenoids goes from 6·4 for the Grosnaja to 2·9 for Mokoia.

The relative isoprenoid distribution is given in Table 3. It is worthy of notice that (1) in all cases pristane is the major isoprenoid, (2) the concentration of the C_{17} is always the lowest among all the isoprenoids, and (3) there is a second maximum at the C_{15} orC_{16} isoprenoids.

High resolution gas chromatography–mass spectrometry

The significance of the mass spectral data obtained from very complex mixtures of alkanes containing many isomeric forms is limited by the resolving power of the gas chromatographic columns used.

Table 2. A. Gas chromatographic estimation of hydrocarbons in meteorites

Meteorite	Type	Cycloalkanes (A) (ppm)	(range)	Me-Alkanes (B) (ppm)	(range)	Isoprenoids (C) (ppm)	(range)	n-Alkanes (D) (ppm)	(range)	Maximum	Total (ppm) (A + B + C + D)	E
Murray	II	1·2	$(C_{16}-C_{20})$	10·5	$(C_{13}-C_{21})$	8·9	$(C_{14}-C_{21})$	55·1	$(C_{12}-C_{24})$	n-C_{17}	75·7	1·6
Grosnaja	III	0·4	(C_{16})	0·8	$(C_{16}-C_{18})$	0·9	$(C_{18}-C_{20})$	5·8	$(C_{16}-C_{21})$	n-C_{21}	7·9	1·2
Mokoia	III	—	—	—	—	10·4	$(C_{16}-C_{21})$	35·6	$(C_{12}-C_{21})$	n-C_{19}	46·0	39·1
		1·8	$(C_{17}-C_{18})$	6·6	$(C_{14}-C_{20})$	11·4	$(C_{15}-C_{21})$	33·4	$(C_{12}-C_{22})$	n-C_{17}	53·2	29·6
Vigarano	III	0·5	$(C_{17}-C_{19})$	3·8	$(C_{13}-C_{20})$	5·5	$(C_{14}-C_{21})$	29·3	$(C_{11}-C_{26})$	n-C_{17}	39·1	6·5

Paraffinic hydrocarbons

E—Represents the total value of unidentified compounds.

Table 2. B. Quantitative variations of n-alkanes plus isoprenoids depending on the source of the samples

Meteorite	Type	Total (C + D),* ppm	Total (C + D),† ppm	
Murray	II	64·0[a]	35·3[a]	
Grosnaja	III	6·7[b]	415·0[c]	
Mokoia	III	44·8[d]	29·3[d]	Duplicate
		46·0[d]	33·3[d]	analyses
Vigarano	III	34·8[e]	233·0[f]	
			108·3[g]	

* Values obtained in this investigation
† Values reported by NOONER and ORÓ (1967)
Samples obtained from (a) Henderson; (b) Vdovykin; (c) Mueller; (d) Moore; (e) Hey; (f) Olsen; (g) Nagy. See Acknowledgments for more details.
Duplicate analyses:
    Two identical samples of Mokoia were analyzed according to the same procedure to check for reproducibility of results.

Table 3. Relative isoprenoid distribution in carbonaceous chondrites

	Murray	Grosnaja	Mokoia	Vigarano
2,6,10-TriMeC$_{11}$	0·01	—	—	0·06
2,6,10-TriMeC$_{12}$	0·51	—	0·06	0·31
2,6,10-TriMeC$_{13}$	0·30	—	0·14	0·42
2,6,10-TriMeC$_{14}$	0·12	—	0·01	0·02
2,6,10-TriMeC$_{15}$	1·62	2·12†	1·65	2·47
2,6,10,14-TetraMeC$_{15}$‡	4·92	4·50	6·12	4·93
2,6,10-TriMeC$_{16}$	—	—	0·24	—
2,6,10,14-TetraMeC$_{16}$§	3·68	4·25	4·41	2·77
2,6,10,14-TetraMeC$_{17}$	0·25	0·37	1·02	0·27

Relative %*

* Of total isoprenoid content as given in Table 2A.
† This value represents an upper limit due to contributions from other unresolved components.
‡ Pristane.
§ Phytane.

In an effort to obtain a more representative mass spectrum for each individual component present in the alkane fraction of the meteorite extracts, part of the work previously carried out on low resolution gas chromatographic columns (90 m and 195 m long × 0·076 cm i.d., ORÓ et al., 1968) was repeated using higher efficiency columns (310 m long × 0·076 cm i.d.) with about 225,000 theoretical plates. Some of the results are shown in Fig. 2. The partial gas chromatographic pattern of Fig. 2 can be compared directly to that shown in Fig. 1B since both represent different analyses of the same meteorite. The effects of the higher efficiency column on this pattern can be seen clearly in Fig. 2, where the 2,6,10-trimethyl C_{12} or farnesane peak (No. 6) has now been partially resolved into three peaks. The pattern shown in Fig. 3 illustrates how difficult it is to separate certain isomeric alkanes even with very high resolution columns. However, in these cases, the combination of a gas chromatograph with a mass spectrometer affords the possibility of detecting and eventually identifying as many as three components within a poorly resolved gas chromatographic peak (GELPI and ORÓ, 1969). The mass spectra taken at some points along the upward slope of the peak, at its top, and at some points along its downward slope can usually tell each component apart from the others. This technique has been used in several occasions in this study whenever identification problems of this sort were encountered.

Mass spectrometric identifications

(a) *Isoprenoids.* Typical mass spectra for pristane and phytane from five carbonaceous chondrites analyzed in this laboratory have already been reported (ORÓ *et al.*, 1968).

To complement and support our data, the isoprenoids from the Murray meteorite, from C_{14} to the C_{21}, are displayed in Figs. 4 and 5. This series is representative of the mass spectra of the isoprenoids detected in the other meteorites.

The characteristic contribution of the chromatographically unresolved anteiso-C_{18} to the mass spectrum of phytane can be seen in Fig. 5, No. 37. The fragmentation pattern of the anteiso C_{18} hydrocarbon will exhibit as the most prominent feature a high M-29 peak at m/e 225. This would account for the high relative intensity of this ion in the spectra of phytane (Fig. 5). The almost perfect coincidence of their respective elution times has been demonstrated experimentally with the proper standards. In the case of pristane and the next lower isoalkane homolog (iso C_{17}), the contributions of the latter to the spectrum of pristane are less evident because pristane is usually present in larger amounts (Table 3) and so the total per cent contribution of the iso C_{17} to the isoprenoid peak is smaller. In general there is a very close correlation between the spectra of the standards and those of the identified compounds.

Two other isoprenoid-like structures not yet reported in any terrestrial sample have also been tentatively identified in three of the carbonaceous chondrites. These are the 2,6,10-trimethylhexadecane, which has been discussed in the literature (McCARTHY, 1967) in relation to the problem of structural mass spectrometric identification of pristane, and the irregular isoprenoid-like structure, 4,7,11-trimethylhexadecane.

The mass spectra obtained for both compounds seem to fit the breakdown pattern predicted for these two structures rather well, but both identifications cannot be final until the proper standards are synthesized and the mass spectral results compared. A mass spectral pattern consistent with the 2,6,10-trimethylhexadecane structure was observed for compound 30 (Table 1). The mass spectrum shows the characteristically intense C_8 and C_{13} fragments typical of the regular isoprenoid structure. Its chromatographic retention time also checks with the predicted value. As has been demonstrated by McCARTHY (1967) and by EGLINTON and CALVIN (1967)

MURRAY ISOPRENOIDS

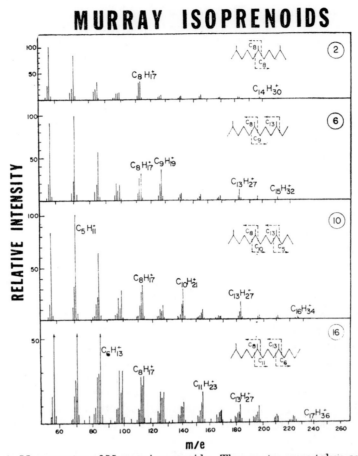

Fig. 4. Mass spectra of Murray isoprenoids. The spectra were taken as each individual component was eluted from a stainless steel capillary column (310 m long by 0·076 cm i.d.) coated with Polysev, and ionized by electron impact at 20 eV in an LKB-9000 gas chromatograph–mass spectrometer combination. Circled numbers to the right hand corner correspond to identification numbers in Table 1.

with the corresponding synthetic compound, those are practically the only features exhibited by this spectrum. In fact, it is so similar to the mass spectrum of pristane that these authors have noted the possibility of confusing one with the other.

The mass spectrum of 4,7,11-trimethylhexadecane would be expected to exhibit prominent C_5, C_7, C_9, C_{12}, C_{14}, and C_{16} peaks, which is in full agreement with the mass spectra obtained from compound 31 in Fig. 3 (see footnote, page 985, for supplementary material on isoprenoids).

(b) *Monomethyl alkanes.* The monomethyl branched alkanes reported in Table 1 were easily characterized by their mass spectrometric fragmentation patterns, which were found to be dominated by fragments corresponding to cleavage of the carbon bonds at either side of the methyl substituted carbon atom. The mass spectrometric characteristics of the methyl alkanes are well documented in the literature (BIEMANN,

MURRAY ISOPRENOIDS

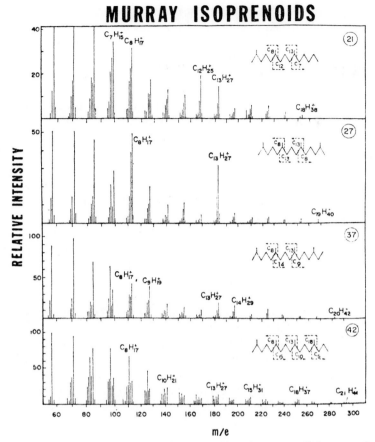

Fig. 5. Mass spectra of Murray isoprenoids. Same conditions as described in previous figure. Circled numbers to the right hand corner correspond to identification number in Table 1.

1962; McCarthy *et al.*, 1968; Gelpi, 1968) and will not be discussed here in any detail. Mass spectra of 2-methyl and 3-methyl alkanes in Vigarano have already been shown (Oró *et al.*, 1968).

The presence of the 4-methyl alkanes can be easily overlooked in routine GC analysis since they are only partially resolved from their isomeric 2-methyl alkanes even in high efficiency columns. This requires special attention since the overall similarities existent between their mass spectral fragmentation patterns (major ion at M-43 and high M-15 ion) makes it the more difficult to distinguish one from the other. However, a careful study of both MS patterns reveals enough minor differences to allow a rather reliable identification, namely, the higher relative intensity of the M-29 ion in the mass spectra of 4-methyl alkanes and the presence of a small M-71 fragment corresponding to the loss of the branched C_5 fragment. Another differantiating characteristic is the intensity of the olefinic M-44 ion which is greater, relative to the M-43 ion, in the spectrum of the 4-methyl alkanes. On this basis we have been able to differentiate between the two isomeric structures.

(c) *Cycloalkanes.* The mass spectra of the cycloalkanes identified in the Murray, Mokoia and Vigarano meteorites were found to be consistent with the fragmentation pattern characteristic of these structures. (For mass spectral data on meteorite cycloalkanes, see footnote, page 985.) The ions at m/e 82 and m/e 83, which dominate these spectra, arise from cleavage of the alkyl chain in cyclohexanes.

It appears that an unidentified homologous series, whose peaks are indicated by asterisks in Fig. 3, is made up by the lower members of the cyclohexyl series reported in Table 1. This is also supported by the relative intensities of their m/e 82 and 83 fragments. This class of compounds would be expected to be eluted after their corresponding n-alkanes with one more carbon atom (BESTOUGEFF, 1967). Thus, the cyclohexyl series for the Murray meteorite may be tentatively extended down to heptylcyclohexane. A similar fragmentation process takes place with the cyclopentyl alkanes resulting in abundant ions at m/e 69 and m/e 68, which were observed in the spectra of compounds 39 and 46 in Table 1.

(d) *Alkenes.* Although not reported in Table 1, some mass spectrometric evidence was obtained relative to their presence in these meteorites. Some olefin parent ions were detected, but in general all the olefins are present in very small amounts when compared to the other saturated compounds.

(e) *Dimethyl substituted alkanes.* The search for this type of compounds has not been particularly successful since only one possible homologous series, the 4,8-dimethyl alkane series, has been tentatively identified. (For partial mass spectrometric data on the dimethyl alkanes, see footnote, page 985.) There are also some indications of the presence of compounds, such as the 3,5-dimethyl alkanes among others, but in general the identification of the dimethyl alkanes has proved to be very difficult, in part because they are present in very low amounts and in part because of the practical limitations on the resolution of the gas chromatographic column for complex mixtures of alkanes.

DISCUSSION

Most of the detectable aliphatic hydrocarbons in meteorites have been fully characterized. That isoprenoid structures can be found in meteorite extracts seems certain now that the already existing tentative mass spectrometric evidence of the presence of the C_{18}, C_{19}, and C_{20} isoprenoids in the Mokoia and Murray meteorites (NOONER and ORÓ, 1967; ORÓ and NOONER, 1967; ORÓ *et al.*, 1966) has been confirmed by the latest data obtained for the meteorites Essebi, Orgueil, Vigarano (ORÓ and GELPI, 1968) and Grosnaja. This, together with the previous report from this laboratory (NOONER and ORÓ, 1967) on the identification of isoprenoids in Boriskino, Santa Cruz and Mighei, extends the count of carbonaceous chondrites containing pristane and phytane, as verified by the gas chromatographic and mass spectrometric analyses, to nine. Furthermore, gas chromatographic and mass spectrometric evidence has also been obtained for other isoprenoids such as the C_{15} (farnesane), C_{16}, C_{17}, C_{18}, C_{19} (trimethyl substituted) and C_{21} compounds (Figs. 4, 5). Concerning the irregular C_{19} isoprenoid structure (4,7,11-trimethylhexadecane), it is difficult to visualize how it is formed or its parent compound. Not much significance should be attached to it until further proof of its structure is obtained.

In general, these results agree with the reported identification of normal paraffins

4

above C_{10} as well as 2-methyl, and 3-methyl paraffins by Studier *et al.* (1968) on the Orgueil and Murray. However, it is surprising that while we find significant amounts of isoprenoids in the C_{14}–C_{21} range in most of the carbonaceous chondrites analyzed, only a tentative identification of the C_{11}, C_{13} and C_{14} isoprenoids in Murray was made by Studier *et al.* (1968).

In view of the general gas chromatographic similarities observed between the hydrocarbons in carbonaceous chondrites, in sediments, in petroleum crudes, and in other biological products (Nagy *et al.*, 1961), it is not too surprising that the mass spectrometric data have shown that the bulk of the isomeric alkane fraction is made up of isoprenoids (C_{14}–C_{21}), monomethyl substituted paraffins and cycloalkanes. It is well known that most of these compounds are constituents of living organisms (Mold *et al.*, 1963; Eglinton and Hamilton, 1963), petroleum (Bendoraitis *et al.*, 1962; Dean and Whitehead, 1961; Rossini, 1960; Bestougeff, 1967; Hills *et al.*, 1968), oil shales (Lawlor and Robinson, 1966; Gohring *et al.*, 1967) and sediments (Oró *et al.*, 1967; Johns *et al.*, 1966; Lawlor and Robinson, 1966).

Thus, the contribution of the biogenic products present in the terrestrial environment to the hydrocarbon content of various carbonaceous chondrites may be as significant as to mask any traces of indigenous extractable material, if any. This assumption is supported by the quantitative heterogeneities of samples of a particular meteorite from different sources and with different histories. For instance, the very high value (415 ppm) obtained in the first analysis of the Grosnaja meteorite (Nooner and Oró, 1967), together with the small amount of pristane and phytane, was taken as a possible indication that the sample had undergone a process of contamination by commercial waxes at some stage in its history. On the other hand the extremely low values (6·7 ppm) obtained in the more recent analysis of a sample of the same meteorite which was taken from a well preserved inside piece seem to strengthen the earlier assumption about the source of the hydrocarbon pattern found in this meteorite. (Vdovykin, G. P., personal communication. The sample, an inside piece, had been recently cut and it was analyzed shortly after its arrival at the laboratory.)

In line with these observations the recent analyses of the extractable organic matter of a freshly recovered meteorite, the Pueblito de Allende, show negligible amounts of aliphatic hydrocarbons (Han *et al.*, 1969; Simmonds *et al.*, 1970).

Conclusions

1. All the carbonaceous chondrites were found to contain n-alkanes (C_{11}–C_{26}), iso, anteiso and other methyl branched alkanes (C_{13}–C_{22}), plus two isoprenoids homologies, the 2,6,10-trimethyl alkane or farnesane series (C_{14}–C_{19}) and the 2,6,10,14-tetramethyl alkane or phytane series (C_{19}–C_{21}), as well as cyclohexyl (C_{16}–C_{20}) and cyclopentyl alkanes (C_{17}–C_{18}).

2. The n-alkane distribution modes show maxima centred around C_{17}.

3. The methyl alkanes are present in very low amounts, their individual concentrations being not higher than 1 ppm. The iso and anteiso alkanes appear to be present in the same relative amounts.

4. The high molecular weight cycloalkanes predominate over the shorter alkyl chain cyclohexanes.

5. The isoprenoids show a consistent pattern in all samples investigated, with a major maximum at C_{19}, a second maximum at C_{15} and a very pronounced minimum at C_{17}.

6. The contributions of our terrestrial environment to the aliphatic hydrocarbon patterns of carbonaceous chondrites may be significant enough as to mask the presence of any indigenous component.

Acknowledgments—This work was supported in part by research grants NGR-44-005-020, NGR-44-005-002, and contract NAS 9-8012 from the National Aeronautics and Space Administration. The authors thank the donors of the following meteorite samples: *Essebi*: J. JEDWAB, Univ. Libre de Bruxelles and M. G. DEMBE, Service Geologique Kinshasa, Rep. Dem. du Congo; *Vigarano*: M. H. HEY, British Museum Natural History; *Mokoia*: C. B. MOORE, Arizona State University, Tempe, Arizona, and G. MUELLER, Department of Crystallography, Birkbeck College, University of London, England; *Grosnaja*: G. P. VDOVYKIN, V. I. Vernadskii Inst. of Geochemistry and Analytical Chemistry, USSR Acad. of Sciences, Moscow; *Murray*: E. P. HENDERSON, Smithsonian Institution, U.S. National Museum, Washington, D.C.; *Orgueil*: A. CAVAILLE, Museum of Natural History, Montauban, France.

Extracted in part from the Ph.D. dissertation of E. GELPI.

REFERENCES

BENDORAITIS J. G., BROWN B. L. and HEPNER L. S. (1962) Isoprenoid hydrocarbons in petroleum. Isolation of 2,6,10,14-tetramethylpentadecane by high-temperature gas–liquid chromatography. *Anal. Chem.* **34**, 49–53.

BESTOUGEFF M. A. (1967) Petroleum hydrocarbons. In *Fundamental Aspects of Petroleum Geochemistry*, (editors B. Nagy and V. Colombo), Chap. 3, pp. 77–108. Elsevier.

BIEMANN K. (1962) The nature of mass spectra and their interpretation. In *Mass Spectrometry, Organic Chemical Applications*, p. 80. McGraw-Hill.

DEAN R. A. and WHITEHEAD E. V. (1961) Occurrence of phytane in petroleum. *Tetrahedron Letters* **21**, 768–770.

EGLINTON G. and CALVIN M. (1967) Chemical fossils. *Scient. Amer.* **216**, 32–43.

EGLINTON G. and HAMILTON R. J. (1963) The distribution of alkanes. In *Chemical Plant Taxonomy*, (editor T. Swain), Chap. 8, pp. 187–217. Academic Press.

GELPI E. (1968) Application of combined gas chromatography-mass spectrometry to the analysis of organic products of bio-geochemical significance. Ph.D. Dissertation, University of Houston, Houston, Texas.

GELPI E. and ORÓ J. (1969) Comparative mass spectrometric studies on the isoprenoids and other isomeric alkanes in terrestrial and extraterrestrial samples. *ASTM 17th Annual Conference—Mass Spectrometry*, Dallas, Texas. May 18–23, pp. 350–361.

GOHRING K. E. H., SCHENCK P. A. and ENGELHARDT E. D. (1967) A new series of isoprenoid isoalkanes in crude oils and cretaceous bituminous shales. *Nature* **215**, 503–505.

HAN J., SIMONEIT B. R., BURLINGAME A. L. and CALVIN M. (1969) Organic analysis of the Pueblito de Allende meteorite. *Nature* **222**, 364–365.

HAYES J. M. (1967) Organic constituents of meteorites. *Geochim. Cosmochim. Acta* **31**, 1395–1440.

HILLS I. R., SMITH G. W. and WHITEHEAD E. V. (1968) Hydrocarbons from fossil fuels and their relationship with living organisms. *Symp. Hydrocarbons from Living Organisms and Recent Sediments. Division of Petroleum Chemistry.* Amer. Chem. Soc. Atlantic City Meeting, September 8–13.

JOHNS R. B., BELSKY T., McCARTHY E. D., BURLINGAME A. L., HAUG P., SCHNOES H. K., RICHTER W. and CALVIN M. (1966) The organic geochemistry of ancient sediments—II. *Geochim. Cosmochim. Acta* **30**, 1191–1222.

LAWLOR D. L. and ROBINSON W. E. (1966) Alkanes and fatty acids in Green River formation oil shale. Presented at AOCS meeting, Los Angeles, California.

McCARTHY E. D. (1967) A treatise in organic geochemistry. Ph.D. Thesis, University of California, Berkeley, California.

McCARTHY E. D., HAN J. and CALVIN M. (1968) Organic geochemical studies—III. Hydrogen atom transfer in the mass spectrometric fragmentation patterns of saturated aliphatic hydrocarbons. *Anal. Chem.* **40,** 1475–1480.

MEINSCHEIN W. G. (1964) Development of hydrocarbon analyses as a means of detecting life in space. Esso Res. and Engr. Co. No. NASw-508, Quarterly Rep., July.

MOLD J. D., STEVENS R. K., MEANS R. E. and RUTH J. M. (1963) The paraffin hydrocarbons of tobacco; normal, iso, and anteiso homologs. *Biochemistry* **2,** 605–610.

NAGY B., MEINSCHEIN W. G. and HENNESSY D. J. (1961) Mass spectrometric analysis of the Orgueil meteorite: Evidence for biogenic hydrocarbons. *Ann. N.Y. Acad. Sci.* **93,** 27–35.

NOONER D. W. and ORÓ J. (1967) Organic compounds in meteorites—I. Aliphatic hydrocarbons. *Geochim. Cosmochim. Acta* **31,** 1359–1394.

OLSON R. J., ORÓ J. and ZLATKIS A. (1967) Organic compounds in meteorites—II. Aromatic hydrocarbons. *Geochim. Cosmochim. Acta* **31,** 1935–1948.

ORÓ J. and GELPI E. (1969) Gas chromatographic-mass spectrometric studies on the isoprenoids and other isomeric alkanes in meteorites. In *Meteorite Research*, (editor P. M. Millman), pp. 518–523. D. Reidel.

ORÓ J., GELPI E. and NOONER D. W. (1968) Hydrocarbons in extra-terrestrial samples. *J. Brit. Interplanet. Soc.* **21,** 83–98.

ORÓ J. and NOONER D. W. (1967) Aliphatic hydrocarbons in meteorites. *Nature* **213,** 1085–1087.

ORÓ J., NOONER D. W., ZLATKIS A. and WIKSTROM S. A. (1966) Paraffinic hydrocarbons in the Orgueil, Murray, Mokoia, and other meteorites. In *Life Sciences and Space Research IV*, (editors A. H. Brown and M. Florkin), pp. 63–100. Spartan Books.

ROSSINI F. D. (1960) Hydrocarbons in petroleum. *J. Chem. Educ.* **37,** 554–561.

SIMMONDS P. G., BAUMAN A. J., BOLLIN E. M., GELPI E. and ORÓ J. (1970) The unextractable organic fraction of the Pueblito de Allende meteorite: Evidence for its indigenous nature. *Proc. Nat. Acad. Sci. U.S.* **64,** 1027–1034.

STUDIER M. H., HAYATSU R. and ANDERS E. (1968) Origin of organic matter in early solar system—I. Hydrocarbons. *Geochim. Cosmochim. Acta* **32,** 151–173.

151

13

Copyright © 1970 by the American Association for the Advancement of Science

Reprinted from *Science*, **167**, 1367–1370 (Mar. 6, 1970)

Endogenous Carbon in Carbonaceous Meteorites

J. W. SMITH and I. R. KAPLAN

Abstract. *Seven carbonaceous chondrites have been analyzed for soluble organic compounds, carbonate, and residual carbon. Carbon-13/carbon-12 isotopic measurements on these fractions gave the following values relative to a marine carbonate standard: carbonate, +40 to +70 per mil; residual carbon, −15 to −17 per mil; soluble organic material, −17 to −27 per mil, with one value of −5.5 per mil. These values are interpreted to indicate that carbonate, residual carbon, and part of the extractable organic material are endogenous to these meteorites.*

Over a century ago, European scientists first isolated carbon compounds from the carbonaceous chondrites Kaba and Orgueil (*1*). Little attention was given to this accomplishment until the publication of Mueller's paper in 1953 (*2*). During the last decade a great effort has been exerted to separate and identify the organic molecules in meteorites (*3*). Both soluble and insoluble organic carbon have been identified, the former consisting partly of saturated hydrocarbons (*4*), and the latter, probably of highly condensed aromatic structures (*5*). Quantitative data were not well documented, and the origins of the organic carbon were not satisfactorily established (*6*).

It is generally accepted that the carbonaceous material remaining after extraction with organic solvents and demineralization with hydrochloric and hydrofluoric acids is representative, in chemical composition, of the meteoritic carbon prior to the entry of the meteorite into the earth's atmosphere. Which of the extractable organic components, if any, are endogenous is still undecided because of the ubiquitous distribution of the same compounds on this planet.

Boato (*7*) measured $^{13}C/^{12}C$ ratios in intact meteorites, and the results appeared to overlap those typical of terrestrial carbon (*8*). Clayton (*9*) showed that carbonate minerals from the Orgueil meteorite are highly enriched in ^{13}C relative to terrestrial carbonate. More recent studies (*10, 11*) have given $\delta^{13}C$ (*12*) values of −19 to −28 per mil for extractable organic components and −18 to −29 per mil for the carbonate-free fractions remaining after treatment with acid. Since these values are essentially the same as those for reduced carbon compounds on earth (*13*),

it has not been possible to use carbon isotope ratios as a criterion for differentiating between organic compounds of extraterrestrial origin and those of terrestrial origin. We believe that the results presented below provide the first evidence for the endogenous character of extractable organic compounds in carbonaceous chondrites.

Seven carbonaceous chondrites (two of type I, four of type II, and one of type III) were analyzed (*14*). An attempt was made to separate several groups of carbon compounds in order (i) to obtain a mass balance for carbon, (ii) to identify critical compounds that may be indicative of terrestrial contamination, and (iii) to measure the $^{13}C/^{12}C$ ratios of the separated compounds.

A small portion of each meteorite was analyzed to obtain the $\delta^{13}C$ value of the intact sample. This analysis was carried out by combusting the meteorite in an atmosphere of oxygen at 1050°C, capturing and purifying the resultant carbon dioxide, and subsequently measuring the $^{13}C/^{12}C$ ratio in a dual-collecting mass spectrometer (*8*). A separate sample was extracted with a benzene-methanol (80 : 20, by volume) solvent mixture under reflux. The extract was treated with freshly prepared copper turnings to remove sulfur and then with 2 percent (by weight) potassium hydroxide solution to extract the fatty acids as the water-soluble potassium salts. The acid-free extract was concentrated by evaporation and a portion was reserved for combustion and isotopic measurements. The remainder of the acid-free extract was chromatographed on a column packed with 60- to 100-mesh Florisil (Matheson, Coleman & Bell), and the fraction eluted with hexane was col-

lected and analyzed. The procedures used in the subsequent estimation and identification of normal alkanes, pristane, phytane, and saturated and unsaturated fatty acids have been described by Smith *et al.* (*15*). The insoluble residue from the solvent extraction was dried and then allowed to react with 85 percent phosphoric acid solution in evacuated flasks for several days at temperatures not exceeding 50°C. The liberated carbon dioxide was captured and purified by distillation over silver nitrate and passage through Dry Ice traps; the volume of carbon dioxide was measured manometrically. To determine whether organic matter was solubilized during this treatment, the acid was recovered by washing, concentrated by evaporation, and oxidized with chromic acid in evacuated flasks. The carbon dioxide evolved was collected, and the volume was measured as before. After removal of a portion for combustion, the residual meteorite was exhaustively extracted with concentrated hydrochloric acid on a boiling water bath, evaporated to dryness twice with 50 percent hydrofluoric acid under the same condition, and then was treated with hydrochloric acid and evaporated again. This procedure dissolved most of the minerals, leaving an organic carbon-rich residue. This residue, that remaining from the phosphoric acid treatment of the meteorite, and the organic fraction soluble in the benzene-methanol mixture were combusted (*8*); the volume of carbon dioxide formed was measured, and $^{13}C/^{12}C$ ratios were determined by mass spectroscopy.

The results (Table 1) indicate that the major portion of the identified carbon remains in the acid-insoluble carbonaceous residues. In the case of the type III carbonaceous chondrite, this amounts to 63 percent (by weight) of the original carbon, whereas in chondrites of types I and II less than 50 percent of the total identified carbon is generally found in this fraction. An extraction with the benzene-methanol mixture of these insoluble organic residues from the Mighei and Orgueil meteorites indicated that only traces of hydrocarbons, insignificant in the car-

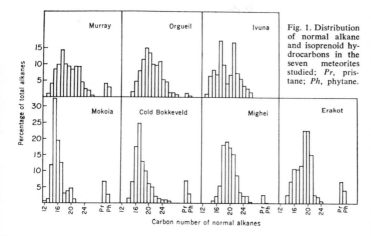

Fig. 1. Distribution of normal alkane and isoprenoid hydrocarbons in the seven meteorites studied; Pr, pristane; Ph, phytane.

bon balance, could be extracted from such residues. An x-ray diffraction pattern showed no evidence for graphite in the Orgueil residue.

The abundances of carbonate and soluble organic fractions were about the same (Table 1), together generally comprising less than 10 percent of the total meteorite carbon. The fraction soluble in phosphoric acid amounted to only about 2 percent of the total carbon. It thus appears that losses of unidentified volatile compounds have occurred during the analytical procedure which prevent one from determining a complete carbon balance at this point.

The yields and distributions of fatty acids extracted in potassium hydroxide are given in Table 2. In every case, the normal C_{16} and normal C_{18} acids were the two components present in largest amounts and were accompanied by significant quantities of the corresponding unsaturated acids. This distribution

probably results from bacterial contamination (16). Saturated fatty acids in the range C_{14} to C_{28}, with a predominance of an even number of carbon atoms, have been isolated from the Orgueil meteorites (17).

The distribution of the normal alkanes and the isoprenoids, pristane and phytane, in the solvent extracts from the meteorites is shown in Fig. 1. Both the yields and the distribution of the alkanes are in very reasonable agreement with published data (18). The normal alkanes from the two type I meteorites and from the Murray meteorite show a tendency toward a bimodal distribution; all other specimens exhibited a smooth distribution of normal alkanes with a pronounced maximum between C_{17} and C_{19}. There was a slight predominance of compounds with an odd number of carbon atoms relative to normal alkanes with an even number of carbon atoms only in the extract

from the Orgueil sample in the range of compounds above C_{21}. Pristane and phytane were present in all the samples examined; pristane was always more abundant (19). The identified normal alkanes and isoprenoids (Fig. 1) represent 2 to 12 percent of the total residue extracted from the benzene-methanol mixture.

Data for $^{13}C/^{12}C$ ratios are given in Table 3. Measurements on the intact meteorites give $\delta^{13}C$ values in the range -5.6 to -18.3 per mil. These values have a bimodal distribution of -5 to -12 per mil for type I and type II carbonaceous chondrites and -17 to -20 per mil for type III carbonaceous chondrites (7, 20). The particular distribution reflects the presence or absence of carbonate. The oxidized carbon is highly enriched in ^{13}C relative to terrestrial carbonate, as Clayton observed for the Orgueil meteorite (9). Our data show, however, that this oxidized fraction may be divided into two subgroups: (i) type I carbonaceous chondrites with $\delta^{13}C = +60$ to $+70$ per mil, and (ii) type II carbonaceous chondrites with $\delta^{13}C = +40$ to $+50$ per mil. The insoluble organic residues have $\delta^{13}C$ values falling in a narrow range of -14.8 to -17.1 per mil. The soluble organic material showed a wider range, varying from -5 to -27 per mil.

Before one can use data on meteoritic carbon compounds for interpretive purposes, it must first be established that these compounds are endogenous to the meteorite. Until now, the evidence for this has been inconclusive (3, 18). We believe that the data presented here suggest that the most abundant forms of carbon recognized in the carbonaceous chondrites are of extraterrestrial origin.

Our study also shows that some forms of carbon are most probably terrestrial contaminants. For example, the presence of the unstable unsaturated fatty acids in quantities amounting to almost 30 percent of the total recognizable alkanoic acids and the predominance of C_{16} and C_{18} acids indicate that these acids are recent biological contaminants introduced either from microorganisms present within the meteorites (16) or from handling.

Data for hydrocarbons are more difficult to assess, because (i) they are only a minor component of most living systems, (ii) they are present in a great variety of structures, and (iii) they can be formed both by biological and abio-

Table 1. Distribution of carbon in meteorite fractions; ppm, parts per million.

Meteorite	Carbonaceous chondrite type	Total (% by wt.)	Soluble organic* (% by wt.)	Carbonate (% by wt.)	Residue after H_3PO_4 treatment (% by wt.)	Residue after HCl + HF treatment (% by wt.)	Alkanes† (ppm)	Fatty acids‡ (ppm)
Ivuna	I	4.03	0.07	0.20	2.54	1.57	93	49
Orgueil	I	3.75	.11	.13		2.15	24	48
Mighei	II	2.85	.09	.21		0.72	71	91
Cold Bokkeveld	II	2.35	.11	.07	1.71	1.27	50	11
Erakot	II	2.30	.13	.05	1.54	0.89	61	23
Murray	II	2.24	.11	.13	1.72	1.06	46	3
Mokoia	III	0.74	.06	.00	0.61	0.47	66	11

* Includes alkanes and excludes organic acids. † Alkane content $\times 0.857$ (C/CH₂). ‡ Mean values calculated on the basis of the carbon content of the normal C_{17} acid.

logical processes. There is no doubt that some meteorites show definite contamination in the form of extractable normal alkanes (*18*), which accounts for a major portion of the carbon in the extract. In general, the abundance distribution of the alkanes is so similar to that present in geologically old sedimentary terrestrial rocks that uncertainty as to the origin of the normal alkanes has led to controversy regarding their formation by biological processes in other planetary bodies of the solar system (*4*). Unless a deliberate effort has been made to introduce hydrocarbon contamination, it may be assumed as a first-order approximation that, if terrestrial alkanes are of recent biological origin, they should be less abundant than fatty acids. Since the alkanes amount to ≤ 12 percent of the total soluble organic component, it may be reasoned that 80 to 90 percent of this latter component is endogenous to the carbonaceous chondrites.

This assumption can now be tested by evaluating the $\delta^{13}C$ values shown in Table 3. The insoluble organic residue has a median $\delta^{13}C$ value of about −16 per mil. All values for the solvent-extractable carbon compounds are isotopically lighter, with the exception of those from the Murray meteorite.

Carbon-hydrogen compounds on earth generally have $\delta^{13}C$ values of −15 to −60 per mil (*13*). Molecules containing carboxyl groups are isotopically enriched in ^{13}C relative to the "lipid" components, and appear at the heavier end of the range (*21*). Carbon from petroleum, and kerogen from Precambrian sedimentary rocks, have $\delta^{13}C$ values between −25 and −35 per mil (*22*). By comparison then, all the meteoritic, nonextractable carbon has measured $\delta^{13}C$ values outside the terrestrial range for equivalent complex reduced molecules and therefore appears to be endogenous to the meteorite. The similarity between the $^{13}C/^{12}C$ data for this component and for the compounds soluble in the benzene-methanol mixture, in the case of the Orgueil, Mighei, and Cold Bokkeveld meteorites, suggests there is a generic relationship between these fractions. These isotopic data also suggest that the extractable components from the Erakot, Ivuna, and Mokoia meteorites may be comprised of about 20 percent, 50 percent, and 75 percent, respectively, of a terrestrial contaminant with a $\delta^{13}C$ value of −30 per mil. The effect of contamination is gen-

erally greatest in the meteorites containing the lowest carbon content. Noncarbonaceous meteorites with carbon contents of ≤ 0.1 percent have $\delta^{13}C$ values (−25 to −30 per mil) in the range of common terrestrial contaminants (*20*). The contamination in the Ivuna meteorite may be partly related to the small size (230 mg) of the sample analyzed.

Three distinct $^{13}C/^{12}C$ groupings appear to be present in the carbon compounds of meteorites: (i) values of $\delta^{13}C$ greater than +40 per mil, representative of the carbonate fraction, (ii) $\delta^{13}C$ rang between −4 and −8 per mil, representing complex polynuclear insoluble carbon compounds and possibly extractable material, and (iii) graphite and iron carbide (cohenite) with a $\delta^{13}C$ range between − 4 and − 8 per mil (*11, 20, 23*). These compounds may have formed independently of each other in a synthetic reaction mixture in which they reached equilibrium or they may be the result of a series of sequential processes.

Fractionation factors between 1.06 and 1.09 (as measured between the carbonate and reduced organic phases of the meteorite) have been observed in nature for coexisting carbon dioxide gas and methane formed at low temperatures (10° to 30°C) during micro-

biological fermentation (*24*). They can also be accounted for theoretically by equilibration of carbon dioxide and methane at temperatures below 50°C (*25*). Therefore, an equilibration between gases containing oxidized and reduced forms of carbon may have occurred at an early stage in planetary formation. Terrestrial calcium carbonate is not known to equilibrate with coexisting complex reduced organic molecules (hydrocarbons or kerogen) after deposition in sediments. Since carbonates are seen only as minor components of the total carbon in meteorites, the isotopic value of the primordial carbon must have been closer to that of the presently identifiable reduced carbon compounds. The narrow range (2.3 per mil) of $\delta^{13}C$ values of the residual insoluble carbon suggests that this component may have formed rapidly and possibly under highly energetic conditions (high temperature or plasma irradiation) from reduced carbon (for example, methane) where there was minimum opportunity for equilibration among different molecular phases. The $\delta^{13}C$ values for graphite and iron carbide in meteorites either could represent carbon from a different source or could be the result of high-temperature equilibration processes that have resulted in the loss of isotopically lighter

Table 2. Yields and distribution of methylated alkanoic acids. Values for individual acids are expressed as percentages of the total.

Meteorite	Number of carbon atoms in chain acids										
	12	13	14	15	16*	16	17	18*	18	20	20†
Ivuna			8	10	16	32	3	13	11		
Orgueil	2		9	5	12	37	1	21	11		
Murray	8	2	9	4	9	32	4	16	10		
Erakot	9		12	4	10	25	2	12	11	2	2
Cold Bokkeveld					8	50		16	26		
Mighei	14		4	2	4	20	2	4	20		29
Mokoia	2		9	6	20	21	4	11	7		9

* Unsaturated. † Unidentified.

Table 3. Isotopic abundances of carbon-13 (as $\delta^{13}C$) for meteorite fractions relative to the Pee Dee belemnite standard. Values given are the averages of at least two measurements differing by <1 per mil.

Meteorite	Total (per mil)	Carbonate (per mil)	Soluble organic* (per mil)	Insoluble organic (per mil)
Ivuna	− 7.5	+65.8	−24.1	−17.1
Orgueil	−11.6	+70.2	−18.0	−16.9
Murray	− 5.6	+42.3	− 5.3	−14.8
Cold Bokkeveld	− 7.2	+50.7	−17.8	−16.4
Erakot	− 7.6	+44.4	−19.1	−15.1
Mighei	−10.3	+41.6	−17.8	−16.8
Mokoia	−18.3	None	−27.2	−15.8

* Excludes organic acids.

methane (26). The isotopically heavy extractable organic matter in the Murray meteorite ($\delta^{13}C$, -5.3 per mil) may have resulted from some such high-temperature equilibration.

We believe that the $^{13}C/^{12}C$ data presented here are the first indication of a nonrandom isotopic distribution of carbon in meteorites. This property can now be included among others in descriptions of the generic relationships of the carbonaceous chondrites. The data strongly suggest that both oxidized and reduced forms of carbon are indigenous to the meteorites. The data also show which fractions are the least contaminated by terrestrial carbon.

Institute of Geophysics and Planetary Physics, University of California, Los Angeles 90024

References and Notes

1. J. J. Berzelius, *Ann. Phys. Chem.* **33**, 113 (1834); F. Wohler, *Sitz. Akad. Wiss. Wien Math. Naturwiss. Kl.* **41**, 7 (1859).
2. G. Mueller, *Geochim. Cosmochim. Acta* **4**, 1 (1953).
3. It is not possible to reference all the published work here; two recent reviews are: H. C. Urey, *Science* **151**, 157 (1966); J. M. Hayes, *Geochim. Cosmochim. Acta* **31**, 1395 (1967).
4. W. G. Meinschein, B. Nagy, D. J. Hennessy, *Ann. N.Y. Acad. Sci.* **108**, 553 (1963).
5. M. C. Bitz and B. Nagy, *Proc. Nat. Acad. Sci. U.S.* **56**, 1383 (1966).
6. E. Anders, *Ann. N.Y. Acad. Sci.* **93**, 651 (1962); P. B. Hamilton, *Nature* **205**, 284 (1964); J. Oro and M. B. Skewes, *ibid.* **207**, 1042 (1965).
7. G. Boato, *Geochim. Cosmochim. Acta* **6**, 209 (1954).
8. H. Craig, *ibid.* **3**, 53 (1953).
9. R. N. Clayton, *Science* **140**, 192 (1963).
10. B. Nagy, *Proc. Nat. Acad. Sci. U.S.* **56**, 389 (1966); H. R. Krouse and V. E. Modzeleski, *Geochim. Cosmochim. Acta,* in press.
11. A. P. Vinogradov, O. I. Kropotova, G. P. Vdovykin, V. A. Grinenko, *Geokhimiya* **3**, 267 (1967).
12. The isotopic abundance may be defined as
$$\delta^{13}C \text{ mil} = 1000 \times \frac{^{13}C/^{12}C \text{ sample} - ^{13}C/^{12}C \text{ std. (PDB)}}{^{13}C/^{12}C \text{ std. (PDB)}}$$
where PDB is the Pee Dee belemnite standard.
13. E. T. Degens, *Geochemistry of Sediments* (Prentice-Hall, Englewood Cliffs, N.J., 1965).
14. Haripura meteorite was also analyzed, and the results are very similar to those for the Mokoia meteorite in all respects. Since it is generally considered that the Haripura meteorite is a type II carbonaceous chondrite, we may have analyzed a mislabeled specimen and therefore do not report the results here.
15. J. W. Smith, J. W. Schopf, I. R. Kaplan, *Geochim. Cosmochim. Acta,* in press.
16. T. G. Tornabene, thesis, University of Houston (1967).
17. B. Nagy and M. C. Bitz, *Arch. Biochem. Biophys.* **101**, 240 (1963).
18. D. W. Nooner and J. Oro, *Geochim. Cosmochim. Acta* **31**, 1359 (1967).
19. Because of the loss of sample during adduction, pristane and phytane were not determined in the Ivuna sample. The sharp maximum at C_{21} in the alkane distribution from this meteorite may be due to a branched or cyclic saturated hydrocarbon also seen in the crude extract from the Erakot meteorite.
20. T. Belsky and I. R. Kaplan, *Geochim. Cosmochim. Acta* **34**, 257 (1970).
21. P. H. Abelson and T. C. Hoering, *Proc. Nat. Acad. Sci. U.S.* **47**, 623 (1961); E. T. Degens, M. Behrendt, B. Gotthardt, E. Reppmann, *Deep-Sea Res.* **15**, 11 (1968).
22. S. R. Silverman, in *Isotopic and Cosmic Chemistry,* H. Craig, S. L. Miller, G. J. Wasserburg, Eds. (North-Holland, Amsterdam, 1964); T. C. Hoering, in *Researches in Geochemistry,* P. H. Abelson, Ed. (Wiley, New York, 1967), vol. 2.
23. D. A. Flory and J. Oro, personal communication.
24. W. D. Rosenfeld and S. R. Silverman, *Science* **130**, 1658 (1959).
25. Y. Bottinga, *Geochim. Cosmochim. Acta* **33**, 49 (1969).
26. M. O. Dayhoff, E. R. Lippincott, R. V. Eck, *Science* **146**, 1461 (1964); R. V. Eck, E. R. Lippincott, M. O. Dayhoff, Y. T. Pratt, *ibid.* **153**, 628 (1966).
27. We thank E. Ruth for valuable assistance. This study was performed under NASA grants NGR 05-007-221 and NGL 05-007-215. Contribution No. 823, Institute of Geophysics and Planetary Physics, UCLA.
* Permanent address: Division of Mineral Chemistry, Commonwealth Scientific and Industrial Research Organization, P.O. Box 175, Chatswood, New South Wales, Australia.

30 January 1970

Reprinted from Nature, **228**(5275), 923–926 (1970)

Evidence for Extraterrestrial Amino-acids and Hydrocarbons in the Murchison Meteorite

by

KEITH KVENVOLDEN
JAMES LAWLESS
KATHERINE PERING
ETTA PETERSON
JOSE FLORES
CYRIL PONNAMPERUMA

Exobiology Division,
Ames Research Center, NASA,
Moffett Field,
California 94035

I. R. KAPLAN

Department of Geology,
University of California,
Los Angeles,
California 90024

CARLETON MOORE

Center for Meteorite Studies,
Arizona State University,
Tempe, Arizona 85281

Organic molecules found in meteorites seem to have been formed before the meteorites reached Earth.

THE Oparin–Haldane[1,2] hypothesis of chemical evolution, postulating the formation of organic compounds before the appearance of life on Earth, has been substantiated by experiments in the laboratory[3-5]. Many of the molecules important to living processes can be synthesized in conditions which may have prevailed on a primitive planet[6,7]. The fundamental concept of chemical evolution would be further substantiated by the finding of compounds of biological significance in extraterrestrial conditions.

Recent studies of intergalactic space, using the techniques of radioastronomy, have shown the presence of water[8,9], ammonia[10], hydrogen cyanide[11], formaldehyde[12], carbon monoxide[13] and cyanoacetylene[14], which are considered to be precursors in the formation of many biochemicals. Some evidence has also been put forward for the possible presence of molecules such as porphyrins[15] and polyaromatic compounds[16,17] in the interstellar medium. The lunar samples from the Apollo 11 and Apollo 12 missions provided a further opportunity for the search for carbon compounds in extraterrestrial materials, but the analysis of the fines from Mare Tranquillitatis[18] and Oceanus Procellarum[19] revealed only minute traces of such compounds. No conclusive evidence is available for the presence of any biomolecules indigenous to the lunar surface.

On the other hand, meteorites have been analysed for organic compounds for over a century[20-22]. Berzelius[23] examined the Alais, Wöhler[24] the Kaba and Berthelot[25] the Orgueil, and reported the presence of substances of organic origin. These investigations have continued, and there is general agreement about the presence of polymeric organic matter in carbonaceous chondrites[20]. The inherent potentiality for contamination resulting from the ubiquitous distribution of biomolecules on Earth[26,27] leads us to believe that many of the results reporting the presence of organic compounds in meteorites are inconclusive, but the results of our present investigation seem to resolve some of these ambiguities and provide evidence for amino-acids and hydrocarbons of possible extraterrestrial origin.

The Murchison meteorite fell at about 11.00 a.m., local time, September 28, 1969, near Murchison, Victoria, Australia (lat. 36° 36′, long. 145° 12′)[28]. The parent object broke up during flight and scattered many fragments over an area of about 5 square miles. Most fractured surfaces on the individual pieces have fusion crusts. Notable and distinctive features of many of the individual stones are networks of deep cracks extending into their interiors. Several stones were picked up soon after the fall and many have been collected since that time particularly during February and March, 1970. Many of the stones broke on impact with the ground.

For this study we selected those stones which had the fewest cracks, the least exterior contamination, and which generally appeared to have a massive character. The samples we examined contained 2·0 weight per cent carbon and 0·16 weight per cent nitrogen. Chemically, the Murchison meteorite is a type II carbonaceous chondrite (C-2).

Amino-acids

An interior piece of the meteorite (10 g) was pulverized and refluxed with 50 ml. of triply distilled water for 20 h at about 110° C. The extract was recovered by centrifugation and the residue rinsed twice with 50 ml. of water. The extract and rinses were combined, evaporated to dryness, and hydrolysed with 50 ml. of 6 M HCl for 20 h at about 100° C. This hydrolysate was evaporated to dryness and desalted by ion exchange using Dowex 50 (H+) followed by Dowex 2 (OH-). Portions of the recovered product were analysed by conventional ion exchange chromatography (Beckman–Spinco 120), by capillary gas chromatography of amino-acid derivatives (Perkin Elmer 881), and by gas chromatography combined with mass spectrometry (CEC 21-491). The meteorite residue after water extraction was hydrolysed with 6 M HCl for 20 h at about 110° C. This hydrolysate was treated in the same manner as the water extract.

Conventional ion exchange chromatography of the hydrolysed water extract from the 10 g sample revealed a number of peaks having retention times similar to common amino-acids. Five abundant amino-acids suggested by the chromatograms were glycine (6 μg/g), alanine (3 μg/g), glutamic acid (3 μg/g), valine (2 μg/g) and proline (1 μg/g).

The presence of the amino-acids, suggested by ion exchange chromatography, and their enantiomeric distribution were established by gas chromatography of the N-trifluoroacetyl-D-2-butyl esters of the amino-acids[29-31]. In this procedure, amino-acids with an asymmetric centre were transformed to diastereomers and the enantiomeric distribution of these compounds was determined on two different capillary columns—'UCON 75-H-90,000' and 'XE 60'. Fig. 1 shows a representative chromatogram resulting from an analysis of amino-acid derivatives. Peaks of the common amino-acids identified on the chromatogram confirm the identification made by ion exchange chromatography. Moreover, both enantiomers of amino-acids with asymmetric centres are present. The percentage of D-amino-acid enantiomers present was calculated from the gas chromatograms to be about 50 per cent for alanine, 45 per cent for glutamic acid, 40–47 per cent for valine and 40–43 per cent for proline. The presence of 2-methylalanine and sarcosine, amino-acids not commonly found in biological systems, was suggested by the gas chromatogram.

To obtain an unambiguous identification, the compounds eluted from the gas chromatograph were introduced through a membrane separator into the mass spectrometer. The mass spectra obtained were compared with the spectra of known standards. In this manner the identities of the amino-acids, glycine, alanine, valine, proline and glutamic acid, were confirmed. Furthermore, the identity of each enantiomeric derivative of alanine, valine and proline was established. The D enantiomeric derivative of glutamic acid was confirmed but because of interfering compounds the spectra of the L enantiomeric derivative was equivocal.

The identities of 2-methylalanine and sarcosine were also established by mass spectrometry. Fig. 2 shows the spectra obtained from a standard N-TFA–D-2-butyl ester of 2-methylalanine compared with the derivatized compound from the Murchison meteorite sample. Similarly, Fig. 3 illustrates the spectra of a derivatized standard sample of sarcosine and a derivatized compound from the meteorite sample. As in the mass spectra of the N-TFA-*n*-butyl esters of amino-acids[33] the key peaks in the spectra of the N-TFA-D-2-butyl esters of 2-methylalanine and sarcosine are the $(M - C_4H_9COO)^+$ peak and the $(M - C_4H_9O)^+$ peak.

Amino-acids have been found previously in other meteorites[20,34] but because the distributional patterns of these amino-acids are similar to amino-acids in fingerprints, the occurrence of amino-acids in meteorites has commonly been attributed to contamination[20]. The total concentration of glycine, alanine, valine, proline and glutamic acid obtained from Murchison meteorite was about 15 µg/g, which is about twice the total concentration of these same compounds found free in Murray, another type II carbonaceous chondrite[34]. In previous work serine has usually been reported as among the most abundant amino-acid in both meteorites and fingerprints[20]. In this study serine was tentatively identified only by ion exchange chromatography and its concentration was an order of magnitude less than the concentration of glycine. The unique distribution of amino-acids in the Murchison meteorite cannot be explained satisfactorily on the basis of contamination by human hands.

The distribution of the enantiomers of alanine, valine, proline and glutamic acid also supports the idea that the amino-acids from the Murchison meteorite are not recent biological contaminants. Amino-acids in living systems are of the L configuration except in rare instances. In recent sediments the L configuration dominates[32].

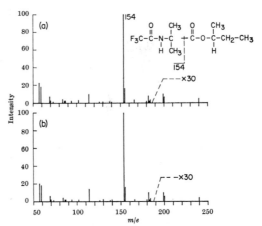

Fig. 2. Mass spectra of N-trifluoroacetyl-D-2-butyl ester of: *a*, 2-methylalanine standard; *b*, compound from the Murchison meteorite.

Studies of amino-acid configurations in other carbonaceous chondrites show that the distribution of enantiomers resembles those of recent sediments and soils[35]. The D and L enantiomers of amino-acids in the Murchison meteorite are almost equally abundant. Slight biological contamination might account for the small dominance of the L isomers of valine, proline and glutamic acid. The distribution of amino-acid enantiomers in Murchison can be accounted for either as a racemic modification of biologically produced amino-acids or as a racemic mixture of abiotically produced amino-acids. The discovery of 2-methylalanine and sarcosine, which are not commonly found in proteins, supports the hypothesis that the whole collection of amino-acids was produced abiotically. In chemical evolution experiments simulating the primitive atmosphere, both sarcosine and 2-methylalanine have been noted among the products[36].

Hydrocarbons

Ten g of the pulverized meteorite were extracted by sonication in a centrifuge tube three times successively, with about 20 ml. benzene–methanol (9 : 1). After centrifugation the extract was evaporated to dryness in a rotary evaporator. This residue was redissolved in 1–2 ml. hexane, which was applied to a 10 g activated silica gel column. The column was eluted with 15 ml. portions of hexane, benzene and benzene–methanol. Activated copper strips were placed in the hexane and benzene fractions and allowed to stand overnight to remove any free sulphur. After the removal of the copper the three fractions were evaporated to about 100 µl. About onetenth of the hexane fraction was analysed by gas chromatography using a 50 foot 'OV-1' support coated open tubular column in a Perkin–Elmer '880' gas chromatograph. The remainder of the sample was analysed by gas chromatography combined mass spectrometry ('CEC 21-491'). Mass spectra were taken every 3 min during the chromatography.

A second sample (15 g) was similarly extracted. After the removal of sulphur the hexane fraction was analysed by thin layer chromatography using silica gel plates impregnated with silver nitrate to separate saturated alkanes from the unsaturated[37]. Saturated alkanes were eluted from the silica gel with hexane, and one-twenty-

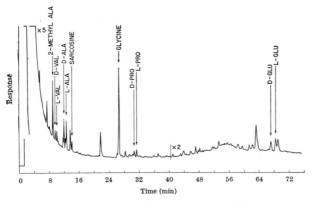

Fig. 1. Gas chromatogram of N-trifluoroacetyl-D-2-butyl esters of amino-acids in acid hydrolysed water extract of a 10 g sample of the Murchison meteorite. Gas chromatography: 0·02 inch × 150 feet capillary column coated with 'UCON 75-H-90,000'; temperature programmed from 100° to 150° C at 1° C/min.

fifth of the eluate was analysed by gas chromatography as before. A small fraction (approximately one-third) was volatilized directly into the source of a 'CEC 21-110' mass spectrometer for a series of high resolution photoplate spectra, and one-third was vaporized directly into the source of the 'CEC 21-110' mass spectrometer to obtain a series of low resolution spectra.

The chromatograms in Fig. 4 indicate that a very complex mixture of alkanes has been isolated from the two meteorite samples. A rough estimate of their abundance can be made by assuming that the arch shaped trace is an equilateral triangle composed of a series of peaks spaced at very close intervals. This estimate suggests that about 350 μg of alkanes were isolated from the 10 g sample and 175 μg from the 15 g sample.

Contamination seems to be limited to a small group of compounds eluted at 210°–249° C. These have been routinely observed in our laboratory and seem to be airborne contaminants.

The distribution pattern of the alkanes isolated from the 10 g sample is a symmetrical trace (Fig. 4) with a peak at 180° C, which corresponds to the elution temperature of n-C_{18}, with a shoulder in the 100°–140° C region. On this shoulder (130° C) is a peak with the retention time corresponding to a saturated n-C_{13} hydrocarbon, and represents about 0·1 μg of hydrocarbons. Other—much smaller—peaks on the chromatogram correspond, in retention time, to n-C_{18} and n-C_{22}-C_{23}. The mass spectra of the various GLC fractions taken at 3 min intervals were very similar and showed a predominance of fragment ions corresponding to the homologous series:

$$(C_nH_{2n-3})^+ > (C_nH_{2n-1})^+ > (C_nH_{2n-5})^+ > (C_nH_{2n+1})^+$$

Mass spectra of fractions eluting from the column at higher temperatures show an increasing predominance of the $(C_nH_{2n}-5)^+$ ion over the $(C_nH_{2n}-1)^+$ ion. The peak emerging at 130° C has not been completely characterized. The mass spectra of the species coinciding with the appearance of this peak on the gas chromatogram shows intense ions at m/e 127 and 182.

The gas chromatographic distribution pattern of the saturated alkanes isolated from the 15 g sample by thin

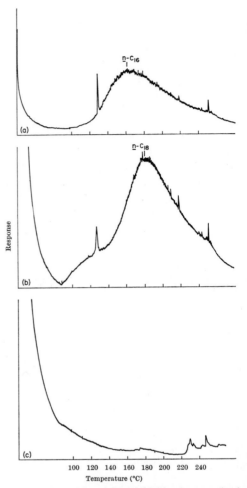

Fig. 4. Gas chromatograms of alkanes isolated from benzene–methanol extracts of two samples of the Murchison meteorite: *a*, 1/10 of alkane fraction isolated from 10 g sample (attenuation ×50); *b*, 1/25 of alkane fraction isolated from 15 g sample (attenuation ×20); *c*, chromatographic blank (attenuation ×20) gas chromatography: 50 feet support coated open tubular column with 'OV-1' liquid phase; temperature programmed from 100° to 240° C at 2° C/min.

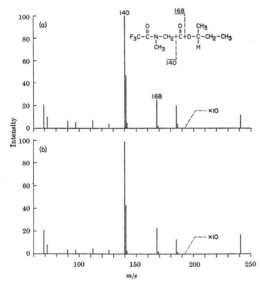

Fig. 3. Mass spectra of N-trifluoroacetyl-D-2-butyl ester of: *a*, standard of sarcosine; *b*, compound from the Murchison meteorite.

layer chromatography is also a symmetrical trace (Fig. 4), but it differs from the first sample, in that it reaches a maximum at 160° C (n-C_{16} retention time) and does not have a shoulder in the 100°–140° C region. This may result from removal of small amounts of unsaturated isomers by $AgNO_3$ chromatography. The low resolution mass spectra of this alkane mixture resembled those from the first sample. Several predominant molecular ions are present (Fig. 5), suggesting that the hydrocarbons in this particular sample are relatively stable to fragmentation on electron bombardment. These molecular ions (m/e 278, 292, 306) correspond to the bicyclic, homologous series C_nH_{2n-4}. A series of relatively intense peaks at m/e 221, 235 and 263 suggests that the loss of hydrocarbon fragments, ranging from $-CH_3$ to $-C_4H_9$, is a favourable process. This evidence, and the relatively large intensity of the m/e 151 peak relative to m/e 137 and m/e 165, suggests that these hydrocarbons closely

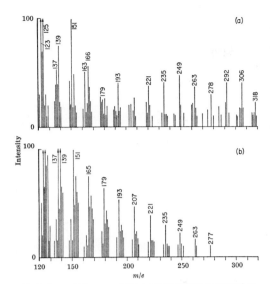

Fig. 5. Low resolution mass spectrum of alkane fractions obtained on volatilization at 125° C from the direct probe of a 'CEC 21–110' mass spectrometer. Ion intensities normalized to m/e 151. *a*, Fraction from 15 g sample of the Murchison meteorite; *b*, fraction from spark discharge in methane.

resemble decahydronaphthalenes containing short chain hydrocarbon substituents.

There is a marked resemblance between the gas chromatographic traces of the aliphatic hydrocarbons synthesized by the action of electrical discharges on methane[38] and those from the meteorite. The analysis of these samples by thin layer chromatography on silica gel plates impregnated with AgNO₃ (ref. 37) indicates that they consist largely of saturated alkanes. The mass spectral data reveal the same dominant homologous series in both samples (Fig. 5). The similarity between the aliphatic hydrocarbons from the meteorite and the spark discharge material, on these three counts, is suggestive of a possible abiogenic origin for the hydrocarbons in the Murchison meteorite.

Further investigations are in progress to determine the possible presence of aromatic and unsaturated hydrocarbons, carboxylic acids and porphyrins.

Carbon Isotope Distribution

Various carbon compounds were analysed for ¹³C/¹²C following the technique of Smith and Kaplan[39]. The results (expressed in the conventional manner as δ¹³C, relative to Peedee Belemnite) are given in Table 1. These results, together with the carbon content (2·5 per cent), place the meteorite as a type II carbonaceous chondrite[39]. The values obtained for the organic solvent extractable material are significantly more enriched in ¹³C than any previous measurements on such compounds in other carbonaceous chondrites. Extracts of terrestrial sediments and petroleum fall in the range δ¹³C = − 25 to − 35 parts per thousand[40,41]. Results obtained from the meteorite confirm the indigenous nature of the carbon and indicate that the amount of terrestrial contamination is insignificantly low.

Table 1

Form of carbon analysed	σ¹³C PDB (parts per thousand)
(i) Total carbon	−7·3, −7·1
(ii) Carbonate	+45·4
(iii) Benzene–methanol soluble compounds	+4·4, +4·8, +5·9
(iv) Insoluble carbon (after solvent extraction and demineralization of silicates with HF)	−10·6

The results of our investigation suggest the indigenous nature of amino-acids and hydrocarbons in the Murchison meteorite. The δ¹³C values of + 4·4 to + 5·9 for the extractable organic material of the meteorite fall into a range widely different from terrestrial organic matter, and the presence of the amino-acids glycine, alanine, valine, proline, glutamic acid, 2-methylalanine and sarcosine was unequivocally established. The presence of almost equal amounts of the D and L enantiomers of valine, proline, alanine and glutamic acid minimizes the possibility of terrestrial contamination and suggests a possible extraterrestrial origin. The presence of the two amino-acids, 2-methylalanine and sarcosine, which are not generally found in biological systems, is indicative of a possible abiogenic synthesis. The gas chromatographic pattern of distribution of hydrocarbons and their mass spectrometry fragmentation pattern similar to that obtained by an abiotic synthesis in the laboratory support the contention that the organic molecules identified here are abiotic and possibly extraterrestrial in origin.

We thank Mr Michael Romiez and Miss Cheryl Boynton of the Ames Research Center and Mrs Chari Petrowski of the Department of Geology, University of California, Los Angeles, for their assistance.

Received September 29; revised October 19, 1970.

¹ Oparin, A. I., *Proischogdenie Zhizni*. (Moscovsky Robotchii, Moscow, 1924).
² Haldane, J. B. S., *Rationalist Ann.*, 148 (1928).
³ *The Origin of Life on Earth* (edit. by Oparin, A. I.) (Pergamon, Oxford, 1959).
⁴ *The Origins of Prebiological Systems and of Their Molecular Matrices* (edit. by Fox, S. W.) (Academic Press, 1965).
⁵ *Proc. Third Intern. Conf. on the Origin of Life, Pont-a-Mousson, France* (1970) (edit. by Buvet, R., and Ponnamperuma, C.) (North-Holland, Amsterdam, in the press).
⁶ Ponnamperuma, C., and Gabel, N. W., *Space Life Sci.*, 1, 64 (1968).
⁷ Lemmon, R. M., *Chem. Rev.*, 70, 95 (1970).
⁸ Cheung, A. C., Rank, D. M., Townes, C. H., Thornton, D. D., and Welch, W. J., *Nature*, 221, 626 (1969).
⁹ Snyder, L. E., and Buhl, D., *Astrophys. J.*, 155, L65 (1969).
¹⁰ Cheung, A. C., Rank, D. M., Townes, C. H., Thornton, D. D., and Welch, W. J., *Phys. Rev. Lett.*, 21, 1701 (1968).
¹¹ Snyder, L. E., and Buhl, D., *IAU Circular*, No. 2251 (1970).
¹² Snyder, L. E., Buhl, D., Zuckerman, B., and Palmer, P., *Phys. Rev. Lett.*, 22, 679 (1969).
¹³ Wilson, R. W., Jefferts, K. B., and Penzias, A. A., *Astrophys. J.*, 161, L43 (1970).
¹⁴ Turner, B. E., *IAU Circular*, 2268 (1970).
¹⁵ Johnson, F. M., *Proc. Regional Symp. on Planetary Geology and Geophysics* (Amer. Astron. Soc. and New England Academic Community, 1967).
¹⁶ Donn, B., *Astrophys. J.*, 152, L129 (1968).
¹⁷ Donn, B., in *Exobiology* (edit. by Ponnamperuma, C.) (North-Holland, Amsterdam, 1970).
¹⁸ Lunar Science Conference, *Science*, 167, 751 (1970).
¹⁹ Chang, S., Kvenvolden, K., Lawless, J., Ponnamperuma, C., and Kaplan, I. R. (in the press).
²⁰ Hayes, J. M., *Geochim. Cosmochim. Acta*, 31, 1395 (1967).
²¹ Nagy, B., *Geol. for Förhandlingar* (Stockholm), 88, 235 (1966).
²² Studier, M. H., Hayatsu, R., and Anders, E., *Science*, 149, 1455 (1965).
²³ Berzelius, J. J., *Ann. Phys. Chem.*, 33, 113 (1834).
²⁴ Wöhler, M. F., and Hörnes, M., *Sitzber. Akad. Wiss. Wien, Math.-naturw. Kl.*, 34, 7 (1859).
²⁵ Berthelot, M., *CR Acad. Sci.*, 67, 849 (1868).
²⁶ Hamilton, P. B., *Nature*, 205, 284 (1965).
²⁷ Oro, J., and Skewes, H. B., *Nature*, 207, 1042 (1965).
²⁸ *Rep. Center for Shortlived Phenomena*, No. 779 (Smithsonian Inst., Cambridge, Mass., 1969).
²⁹ Gil-Av, E., Charles, R., and Fischer, G., *J. Chromatog.*, 17, 408 (1965).
³⁰ Pollock, G. E., Oyama, V. I., and Johnson, R. O., *J. Gas Chromatog.*, 3, 174 (1965).
³¹ Pollock, G. E., and Oyama, V. I., *J. Gas Chromatog.*, 4, 126 (1966).
³² Kvenvolden, K. A., Peterson, E., and Brown, F. S., *Science*, 169, 1079 (1970).
³³ Gelpi, E., Koenig, W. A., Gibert, J., and Oro, J., *J. Chromatog. Sci.*, 7, 604 (1969).
³⁴ Kaplan, I. R., Degens, E. T., and Reuter, J. H., *Geochim. Cosmochim. Acta*, 27, 805 (1963).
³⁵ Nakaparksin, S., *Diss. Abst.*, 70–4495, 4016B (1970).
³⁶ Miller, S. L., and Urey, H. C., *Science*, 130, 245 (1959).
³⁷ Murphy, M. T. J., in *Organic Geochemistry, Methods and Results* (edit. by Eglinton and Murphy), 75 (Springer-Verlag, 1969).
³⁸ Ponnamperuma, C., Woeller, F., Flores, J., Romiez, M., and Allen, W. V., *ACS Adv. Chem.*, 80, 280 (1969).
³⁹ Smith, J. W., and Kaplan, I. R., *Science*, 167, 1367 (1970).
⁴⁰ Degens, E. T., *Geochemistry of Sediments* (Prentice-Hall, Englewood Cliffs, 1965).
⁴¹ Silverman, S. R., in *Isotopic and Cosmic Chemistry* (edit. by Craig, H., Miller, S. L., and Warserburg, G. J.) (North-Holland, Amsterdam, 1964).

Erratum: The isotopic value −10.6 shown in Table 1 should read −13.8 (I. R. Kaplan, personal communication).

III
Primitive Environments— Hydrosphere and Atmosphere

Editor's Comments on Papers 15 Through 26

The geologic record of earth begins about 3.7 billion years ago. This is the age of the oldest radiometrically dated rocks; they have been found in the Godthaab district of western Greenland (Black et al., 1971). Between this time and the origin of the earth 4.6 billion years ago, the geologic record is obscured and as yet undecipherable. During the first billion years of earth history the crust must have been unstable. Reorganization of the earth from a homogenous mass of accreted planetary material to a structured body composed of core, mantle, and crust undoubtedly involved thermal processes which permitted a permanent crust to develop only near the end of the reorganization period. Geologic information about chemical evolution is generally lacking because many of these processes probably took place during a time for which there remains no rock record. Nevertheless, geochemists have been able to develop models of the primitive earth based on the geologic evidence gathered from the last 3.7 billion years of geologic time. The models of interest here concern the development of the hydrospheric and atmospheric environment in which life on earth began and evolved.

The first four papers of this section are concerned mainly with the development and composition of the primitive ocean. By considering the excess volatiles in the present bisphere, atmosphere, hydrosphere, and lithosphere, Rubey (Paper 15) concluded that their source must have been dominantly from plutonic gases escaping from volcanoes, fumeroles, hot springs, and igenous rocks. This outgassing resulted in a gradual accumulation of the ocean and of excess volatiles in the hydrosphere and atmosphere. By this reasoning the conclusion was reached that the composition of the ocean and atmosphere varied over geologic time only within relatively narrow ranges. The paper by Rubey is a classic example of geologic reasoning applied to a complex problem. Because of length constraints, only the abstract of this paper is given; fortunately the abstract is comprehensive. The basic conclusion that the atmosphere and hydrosphere of the primitive earth resulted from outgassing remains today, but ideas about composition of the primitive atmosphere have changed. In a more recent paper on the geologic history of seawater, Holland (1972) agreed that the chemistry of seawater has been highly conservative during the past 3.5 billion years; however, he concluded that Precambrian seawater had a higher silica content, and the early atmosphere had a lower oxygen pressure, than today.

It is commonly believed now that the primitive atmosphere was reducing in contrast to the present atmosphere, which is oxidizing. Results described in some of the papers in Part I were obtained from simulations utilizing reducing primitive atmospheres composed of methane, ammonia, and water. If ammonia was indeed a primitive atmospheric constituent, it would have an important effect on the composition of the primitive ocean because of its great solubility. Most of the ammonia would have dissolved in the ocean to form a mixture of ammonium ion and ammonia. Bada and Miller (Paper 16) have considered the possible ammonium ion concentration in the primitive ocean. They estimated that the range of ammonium ion concentration would have been between $10_{.2}$ and $10_{.3} M$ and the pH about 8. Their paper demonstrates the application of principles of inorganic and organic chemistry to attempt an understanding of primitive earth conditions.

The ocean apparently accumulated from the outgassing of the earth, and the volume of the oceans likely increased significantly during Precambrian times until, by the beginning of the Cambrian time, the oceanic volume approached its present value. Weyl (Paper 17) postulated that this accumulating ocean became statified. When the ocean became deeper than 300 m, a thermocline developed that played an important role in the origin and evolution of Precambrian life. His paper traces the development of life from the abiotic generation of organic molecules, which collected in layers or at the bottom of the primitive ocean, to the development of metazoans, which first occupied the floors of the continental shelves at the beginning of the Cambrian period.

The rocks that were deposited in the primitive oceans should reflect some of the conditions of that ocean. At least that is the premise followed by Perry (Paper 18), who studied the oxygen isotope chemistry of Precambrian cherts. He assumed that the cherts examined were of marine origin and that they were formed in approximate isotopic equilibrium with seawater. The oxygen isotopic compositions obtained showed a progressive change through time. This observation is consistent with progressive outgassing of the earth to form the ocean. His paper shows the application of isotopic studies to interpretation of the Precambrian record. The success of this kind of approach, however, can only be judged as new data are obtained and evaluated.

The next three papers deal with geological aspects of Precambrian rocks that provide evidence for the composition of the Precambrian atmosphere. Rankama (Paper 19) attempted to interpret the chemical composition of the Precambrian atmosphere by measuring the FeO and Fe_2O_3 content of 2-billion-year-old breccias and conglomerate schists from Finland. A relationships of FeO and Fe_2O_3 was interpreted to indicate that the weathering of some of the rock constituents took place in an oxygen-depleted atmosphere. Although such an interpretation of these data has been challenged (Rutten, 1971), this paper presents an interesting approach in which geochemistry is used to attempt to interpret paleoenvironmental conditions. Besides the abstract, only that part of Rankama's paper that deals specifically with the primitive atmosphere is included.

Studies by Ramdohr (Paper 20) on the pyrite and uraninite in early and middle Precambrian deposits of South Africa, Brazil, and Canada have had a major impact on geologic thinking with regard to the composition of the primitive atmosphere. Ramdohr considered a number of alternative explanations for the occurrences of pyrite and uraninite in these ancient rocks. The most interesting hypothesis from the point of view of the origin of life is one in which the minerals were thought to be detrital, having been deposited under an atmosphere inferred to be deficient in oxygen. Ramdohr cautiously hypothesized that the formation of the uraninite- and pyrite-bearing conglomerates "might have taken place when fundamentally different weathering conditions still existed—such as, for example, a methane or carbon dioxide atmosphere." This idea was strongly challenged by Davidson, who claimed that such a hypothesis departed from uniformitarianism, a concept fundamental to much geologic reasoning. Criticisms by Davidson of Ramdohr's work are given in the original paper (20), but because of length constraints these arguments have not been reprinted. Schidlowski (1966) has extended some of the observations of Ramdohr and has concluded that the evidence strongly supports concepts of a detrital origin for the uraninite. This mode of origin requires the existence of a reducing (oxygen-deficient) atmosphere during the time of deposition.

Detrital pyrite and uraninite grains have been found in sediments ranging in age from about 3 to 1.8 billion years, but are absent in younger sediments. It is now generally agreed that sediments older than 1.8 billion years were deposited under an anoxic atmosphere where detrital pyrite and uraninite could survive the weathering process. Only summary sections of Ramdohr's lengthy paper are included here. A detailed discussion of Ramdohr's work, including photomicrographs, can be found in Rutten (1971).

Besides the evidence obtained from considerations of detrital pyrite and uraninite, the worldwide distribution of iron formations has provided some persuasive arguments concerning the early atmosphere. In their important paper, Lepp and Goldich (Paper 21) noted distinct lithologic differences in iron formations deposited during the time period from 3.0 to 1.7 billion years and those younger that 1.2 billion years. The iron deposits of the early and middle Precambrian occur as banded cherty iron formations in which the principal iron mineral is magnetite. In the late Precambrian and younger record, iron formations are present as red beds in which the principal iron-containing mineral is hematite. In the older iron formations iron and silica were transported and deposited together as banded ironstone. The formation of the younger iron deposits involved chemical differentiation in which iron and silica were deposited separately. The origin of the banded iron formations took place during a time when there was an absence or deficiency of free oxygen in the atmosphere. As free oxygen became available, the development of banded iron formations stopped, and red beds became the lithologic expression of iron formations. Only the abstract and a portion of the original paper concerned with the Precambrian atmosphere are included here. A more extensive treatment can be found in Lepp (1974).

Holland (Paper 22) developed a three-stage model for the evolution of the earth's atmosphere. This truly classical presentation amalgamates the ideas of Rubey (Paper 15) and of Urey (Paper 1), along with considerations of chemical equilibria, to produce a geologically and geochemically satisfying reconstruction of the probable chemical composition of the atmosphere during its evolution. The first stage involved outgassing of a highly reducing atmosphere containing mainly hydrogen with some methane, nitrogen, ammonia, and water; oxygen was absent. During the second stage the volcanic gases became less reduced, and the composition of the atmosphere was made up mainly of nitrogen, water, and carbon dioxide; free oxygen was still absent. The third-stage atmosphere developed as free oxygen accumulated and a nitrogen–oxygen atmosphere evolved. The geologic evidence presented by Ramdohr (Paper 20) and Lepp and Goldich (Paper 21) suggests that the second-stage atmosphere ended about 1.8 to 1.2 billion years ago, at which time free oxygen began to accumulate in the earth's atmosphere.

Of interest to the question of the origin of life is the length of time that the highly reducing atmosphere was available in which the synthesis of organic compounds could have taken place; in an oxidizing environment organic compounds would be destroyed. In order to set some limit to the length of the first-stage atmosphere of Holland (Paper 22), Rasool and McGovern (Paper 23) calculated the exospheric temperature of the primitive atmosphere, which was considered to be in conducive equilibrium. They determined that, depending on the amounts of methane and hydrogen, the exospheric temperature would lie in the range 500–900°K. At a temperature of 600°K, a reducing atmosphere of methane, ammonia, nitrogen, and hydrogen could be stable against gravitational escape for as long as 1 billion years, thus providing a long period of time for the early processes in chemical evolution to take place on earth. In actual fact, the length of time needed to produce the prebiotic milieu of organic compounds could have been much shorter than 1 billion years.

The next two papers discuss directly the important question of the changing oxygen content of the atmosphere during Precambrian time. Although photodissociation of water produced oxygen during the early stages of atmospheric evolution, it is now generally believed that most of the oxygen has been generated after life began, through the process of photosynthesis. Berkner and Marshall, in a series of papers, the abstract of one of which is included here (Paper 24), focus on the rise of oxygen from primitive levels by means of photosynthetic activity. They developed a chronology wherein the oxygen level during Precambrian times rose from about 0.001 present atmosphere level (P.A.L.) to about 0.01 P.A.L. (the Pasteur point) at the beginning of the Cambrian Period and passing 0.1 P.A.L. at the end of the Silurian Period. Although the basic idea that most of the oxygen in the atmosphee has resulted from photosynthetic activity remains today, the time scale of the evolution of oxygen has been questioned.

Rutten (Paper 25) essentially agreed with the ideas of Berkner and Marshall (Paper 24) but disagreed with the time scale and with the importance of the oxygen level at the Pasteur point. According to Rutten, the Pasteur point (0.01 P.A.L.), at which the metabolism of microorganisms change from fermentation to respiration, was an important self-regulating factor in the history of atmospheric oxygen. He believed that 0.01 P.A.L. of oxygen was reached during early and middle Precambrian times. This self-regulation was overrun about 1.8 billion years ago and the oxygen level began to rise, reaching 0.1 P.A.L. during the Ordovician Period.

Another approach to the evolution of atmospheric oxygen has been made by Broecker (1970), who considered the carbon isotopic evidence in marine carbonates. Although his study only concerned rocks of the Phanerozoic, he felt that the work had important implications with regard to the Precambrian. He observed a constancy in the carbon isotopic composition of marine carbonates and concluded that during the Phanerozoic there was no great change in the available atmospheric oxygen; therefore, the level of oxygen in the early Paleozoic must have been the same as the oxygen level today. This conclusion is directly opposed to that of Berkner and Marshall (Paper 24), who suggested that the absence of land plants and animals in pre-Silurian time was due to atmospheric oxygen levels lower than exist today. Now Becker and Clayton (1972) have challenged Broecker's (1970) conclusions. They point out that although the carbon isotopic compositions of marine carbonates have remained fairly constant over a period of 3 billion years, it does not necessarily follow that the atmospheric oxygen levels were also constant. Perry and Tan (1972) have drawn similar conclusions.

Finally, Cloud (Paper 26) has developed a working model of the primitive earth in which he tries to put together the geological and biological record of the Precambrian (pre-Paleozoic). As he notes, this model is subject to refinements and change as more is learned. He divides geologic history into four parts. During each of these parts events occurred that dramatically changed conditions on the surface of the earth. That part of geologic history for which the record is obscure, from the origin of the earth to the oldest rocks, he calls Hadean time. During Archaen time, from about 3.6 to 2.6 billion years ago, chemical evolution led to the origin of life. The interval from 2.6 to 1.9 billion years is called Proterophytic time; procaryotic organisms only were present, and between 2.2 and 1.9 billion years ago free oxygen became available in the atmosphere. Before 2.2 billion years biologically produced oxygen was used up mainly in the oxidation of iron. During Proterozoic time, from 1.9 to about 0.68 billion years, the oxygen content of the atmosphere increased and eucaryotic organisms evolved. Metazoans entered the record about 0.68 billion years ago, signaling the end of Precambrian time. Cloud's paper helps put in perspective the processes and consequences of these processes as detailed by the papers included in this section.

Much of this book to this point has been concerned with chemistry. In contrast, this part has directed much attention to the geological aspects of the origin of life. Two individuals who have made and continue to make significant contributions to this subject are Heinrich D. Holland and Preston E. Cloud.

Holland was born in 1927 in Mannheim, Germany, and came to the United States in 1939. He studied geochemistry at Princeton University and Columbia University.

His research on the chemistry of ore-forming fluids is widely recognized, but of most interest for this book has been his work in planetary geology and the chemical evolution of the atmosphere and oceans. For 21 years he was on the faculty of Princeton University; he joined the Department of Geology at Harvard in 1972.

Cloud is a geologist and paleontologist who has done much to provide an increased understanding of the record of life on earth. He was born in 1912 in Massachusetts and studied at George Washington University and Yale. He was associated with the U.S. Geological Survey for 19 years and has held academic appointments at the University of Minnesota and the University of California at Los Angeles; presently he is a member of the faculty of the University of California at Santa Barbara and director of the Biogeology Clean Laboratory.

References

Becker, R. H., and R. N. Clayton (1972) Carbon isotopic evidence for the origin of a banded iron formation in Western Australia: *Geochim. Cosmochim. Acta*, **36**, 577–595.

Black, L. P., N. P. Gale, S. Moorbath, R. J. Pankhurst, and V. McGregor (1971) Isotopic dating of very early Precambrian amphibolite facies gneisses from the Godthaab District, West Greenland: *Earth Planet. Sci. Lett.*, **12**, 245–259.

Broecker, W. S. (1970) A boundary condition on the evolution of atmospheric oxygen: *Jour. Geophy. Res.*, **75**, 3553–3557.

Holland, H. D. (1972) The geologic history of sea water—an attempt to solve the problem: *Geochim. Cosmochim. Acta*, **36**, 637–651.

Lepp, H. (1974) *Geochemistry of Iron* (Benchmarks in Geology Series): Dowden, Hutchinson & Ross, Stroudsburg, Pa. (in press).

Perry, E. C., Jr., and F. C. Tan (1972) Significance of oxygen and carbon isotope variations in early Precambrian cherts and carbonate rocks of southern Africa: *Geol. Soc. Am. Bull.*, **83**, 647–664.

Rutten, M. G. (1971) *The Origin of Life by Natural Causes:* Elsevier, Amsterdam, 420p.

Schidlowski, M. (1966) Beiträge zur Kenntis der radioaktiven Bestandteile der Witwatersrand-Konglomerate I. Uranpecherz in den Konglomeraten des Oranje-Freistaat-Goldfeldes: *Neues Jahrb. Mineral. Abhandl.*, **105**, 183–202.

Additional Suggested Readings

Brancazio, P. J., and A. G. W. Cameron, eds. (1964) *The Origin and Evolution of Atmospheres and Oceans*: Wiley, New York, 314p.

Brinkmann, R. T. (1969) Dissociation of water vapor and evolution of oxygen in the terrestrial atmosphere: *Jour. Geophy. Res.* **74**, 5355–5368.

Cloud, P. E., Jr. (1968) Atmospheric and hydrospheric evolution on the primitive earth: *Science*, **160**, 729–736.

Cloud, P. (1973) Paleoecological significances of the banded iron-formation: *Econ. Geol.*, **68**, 1135–1143.

Eugster, H. P., and B. F. Jones (1968) Gels of sodium-aluminum silicate, Lake Magadi, Kenya: *Science*, **161**, 160–163.

Fanale, F. P. (1971) A case for catastrophic early degassing of the earth: *Chem. Geol.* **8**, 79–105.

Garrels, R. M., E. A. Perry, Jr., and F. T. Mackenzie (1973) Geneses of Precambrian iron-formations and the development of atmospheric oxygen: *Econ. Geol.*, **68**, 1173–1179.

Meadows, A. J. (1972) The atmosphere of the earth and the terrestrial planets: their origin and evolution: *Physics Reports* (Sec. C of *Phy. Lett.*) **5**, 197–236, North-Holland, Amsterdam.

Perry, E. C., J. Monster, and T. Reimer (1971) Sulfur isotopes in Swaziland System barites and the evolution of the earth's atmosphere: *Science*, **171**, 1015–1016.

Rubey, W. W. (1955) Development of hydrosphere and atmosphere, with special reference to probable composition of the early atmosphere: in *Crust of the Earth*, edited by A. Poldervaart: Geological Society of America, Washington, D.C., pp. 631–650.

Schidlowski, M. (1971) Probleme der atmosphärischen Evolution im Präkambrium: *Geol. Rundschau*, **60**, 1351–1384.

Van Trump, J. E., and S. L. Miller (1973) Carbon monoxide on the primitive earth: *Earth Planet. Sci. Lett.*, **20**, 145–150.

Viljoen, M. J., and R. P. Viljoen (1970) Archaen vulcanicity and continental evolution in the Barberton region, Transvaal: in *African Magmatism and Tectonics*, edited by T. N. Clifford and I. G. Gass: Hafner Press, New York, pp. 27–49.

15

Reprinted from *Geol. Soc. Am. Bull.*, **62**, 1111–1112 (Sept. 1951)

GEOLOGIC HISTORY OF SEA WATER
An Attempt to State the Problem

(Address of Retiring President of The Geological Society of America)

By William W. Rubey

Abstract

Paleontology and biochemistry together may yield fairly definite information, eventually, about the paleochemistry of sea water and atmosphere. Several less conclusive lines of evidence now available suggest that the composition of both sea water and atmosphere may have varied somewhat during the past; but the geologic record indicates that these variations have probably been within relatively narrow limits. A primary problem is how conditions could have remained so nearly constant for so long.

It is clear, even from inadequate data on the quantities and compositions of ancient sediments, that the more volatile materials—H_2O, CO_2, Cl, N, and S— are much too abundant in the present atmosphere, hydrosphere, and biosphere and in ancient sediments to be explained, like the commoner rock-forming oxides, as the products of rock weathering alone. If the earth were once entirely gaseous or molten, these "excess" volatiles may be residual from a primitive atmosphere. But if so, certain corollaries should follow about the quantity of water dissolved in the molten earth and the expected chemical effects of a highly acid, primitive ocean. These corollaries appear to be contradicted by the geologic record, and doubt is therefore cast on this hypothesis of a dense primitive atmosphere. It seems more probable that only a small fraction of the total "excess" volatiles was ever present at one time in the early atmosphere and ocean.

Carbon plays a significant part in the chemistry of sea water and in the realm of living matter. The amount now buried as carbonates and organic carbon in sedimentary rocks is about 600 times as great as that in today's atmosphere, hydrosphere, and biosphere. If only 1/100 of this buried carbon were suddenly added to the present atmosphere and ocean, many species of marine organisms would probably be exterminated. Furthermore, unless CO_2 is being added continuously to the atmosphere-ocean system from some source other than rock weathering, the present rate of its subtraction by sedimentation would, in only a few million years, cause brucite to take the place of calcite as a common marine sediment. Apparently, the geologic record shows no evidence of such simultaneous extinctions of many species nor such deposits of brucite. Evidently the amount of CO_2 in the atmosphere and ocean has remained relatively constant throughout much of the geologic past. This calls for some source of gradual and continuous supply, over and above that from rock weathering and from the metamorphism of older sedimentary rocks.

A clue to this source is afforded by the relative amounts of the different "excess" volatiles. These are similar to the relative amounts of the same materials in gases escaping from volcanoes, fumaroles, and hot springs and in gases occluded in igneous rocks. Conceivably, therefore, the hydrosphere and atmosphere may have come almost entirely from such plutonic gases. During the crystallization of magmas, volatiles such as H_2O and CO_2 accumulate in the remaining melt and are largely expelled as part of the final fractions. Volcanic eruptions and lava flows have brought volatiles to the earth's surface throughout the geologic past; but intrusive rocks are probably a much more adequate source of the constituents of the atmosphere and hydrosphere. Judged by the thermal springs of the United States, hot springs (carrying only 1 per cent or less of juvenile matter) may be the principal channels by which the "excess" volatiles have escaped from cooling magmas below.

This mechanism fails to account for a continuous supply of volatiles unless it also provides for a continuous generation of new, volatile-rich magmas. Possibly such local magmas form by a continuous process of selective fusion of subcrustal rocks, to a depth of several hundred kilometers below the more mobile areas of the crust. This would imply that the volume of the ocean has grown with time. On this point, geologic evidence permits differences of interpretation; the record admittedly does not prove, but it seems consistent with, an increasing growth of the continental masses and a progressive sinking of oceanic basins. Perhaps something like the following mechanism could account for a continuous escape of volatiles to

1111

the earth's surface and a relatively uniform composition of sea water through much of geologic time: (1) selective fusion of lower-melting fractions from deep-seated, nearly anhydrous rocks beneath the unstable continental margins and geosynclines; (2) rise of these selected fractions (as granitic and hydrous magmas) and their slow crystallization nearer the surface; (3) essentially continuous isostatic readjustment between the differentiating continental masses and adjacent ocean basins; and (4) renewed erosion and sedimentation, with resulting instability of continental margins and mountainous areas and a new round of selective fusion below.

Reprinted from *Science*, **159**, 423–425 (Jan. 26, 1968)

Ammonium Ion Concentration in the Primitive Ocean

JEFFERY L. BADA and STANLEY L. MILLER

Abstract. *If ion exchange on clay minerals regulated the cations in the primitive ocean as it does in the present ocean, the pH would have been 8 and the K+ concentration 0.01M. Since NH_4^+ and K+ are similar in their clay-mineral equilibria, the maximum NH_4^+ concentration in the primitive ocean would also have been 0.01M. An estimate of the minimum NH_4^+ concentration is $1 \times 10^{-3}M$, based on the reversible deamination of aspartic acid and the assumption that aspartic acid is necessary for the origin of life. The rate of this nonenzymic deamination is rapid on the geological time scale.*

Although the presence of NH_3 is considered important in the synthesis of organic compounds on the primitive Earth (1), there have been no quantitative estimates of the concentrations based on the organic chemistry or on detailed atmospheric models (2). The stable species of nitrogen would have been ammonia under the generally accepted reducing conditions, but most of this NH_3 would have dissolved in the ocean to form a mixture of NH_4^+ and NH_3, the ratio depending on the pH.

The available nitrogen places one upper limit on the NH_4^+ concentration. If all the nitrogen in the atmosphere (755 g/cm²) were placed in the present ocean (282 liter/cm²) as NH_4^+, the concentration would be 0.19M. If the primitive ocean were smaller or the nitrogen in the rocks is included, the concentration would be correspondingly higher.

This upper limit would be reduced by the clay minerals. The pH and the cations in the present ocean are regulated largely by the clay minerals, as was first suggested by Sillén (3) and discussed by others (4, 5). This regulation is based on the fact that the ion-exchange capacity of the oceanic sediments is large compared to the H+ and the buffer capacity of the ocean. The ion-exchange capacity also appears to be sufficient to regulate the concentrations of Na+, K+, Ca++, and Mg++. The ion-exchange equilibrium between H+ and K+, as well as the other cations, controls the pH of the oceans at 8.1.

We can expect that a similar regulation of cations took place in the primitive ocean as long as there was a sufficient amount of weathering and sedimentation. These considerations suggest that the pH of the primitive ocean also was about 8. At this pH the ammonia in the ocean would have been mainly NH_4^+, and NH_4^+ would have entered into the clay-mineral ion-exchange equilibria.

The exchange of NH_4^+ and K^+ on the clay minerals

$$NH_4^+ + clay \cdot K^+ = K^+ + clay \cdot NH_4^+$$

should have an equilibrium constant of about 1.0 since NH_4^+ and K^+ are very similar in clay-mineral and other silicate-mineral reactions (6). In particular, the equilibrium constant for this reaction on montmorillonite (7) is 1.40, while a similar equilibrium for Na^+ and K^+ is 0.25. These equilibrium constants are for 25°C, but they are almost independent of temperature ($\Delta H = 1$ to 3 kcal). The exchange with montmorillonite should be representative of most layered silicates.

The dynamic process for the regulation of potassium in the ocean involves uptake of the excess potassium entering from rivers (5). The molar ratio of sodium to potassium entering the ocean from rivers averages 4.6, but the ratio in seawater is 46. Most of this excess potassium from rivers is removed in formation of the framework of potassium-rich minerals, with the final adjustments of the potassium concentration taking place on the exchangeable sites of the clay. Although all the minerals involved in this process are not known, the net result of this ion exchange leads to a K^+ concentration in the present ocean of $0.01M$. A similar regulation of NH_4^+ would be expected on the primitive Earth unless the cations, alumina, and silica entering the oceans from the rivers were very different. If the NH_4^+ in the rivers was in greater concentration than the K^+, the NH_4^+ would have been taken up preferentially by the clay minerals. This uptake would continue, if the sediments were in excess, until the NH_4^+ and K^+ were approximately equal in concentration and close to the present $0.01M$ K^+. This excess NH_4^+ in the rivers could have come from the weathering of rocks containing NH_4^+, or from the NH_3 in the atmosphere.

If the NH_4^+ in the rivers was in lower concentration than the K^+, the NH_4^+ in the oceans also would have been lower than the K^+, with a corresponding decrease in the NH_4^+ content of the clay minerals. This is the

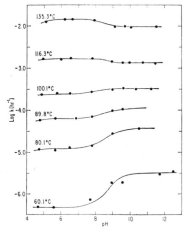

Fig. 1. Rate of deamination of aspartic acid as a function of pH and temperature.

situation today, where the concentration of NH_4^+ in the rivers is very low and variable. Therefore this ion-exchange process on the clay minerals establishes an upper limit on the concentration of NH_4^+ in the ocean, but it does not regulate NH_4^+ at low concentrations.

An estimate of the lower limit of the NH_4^+ concentration can be obtained from the prebiological organic chemistry since a number of these syntheses require NH_3. The kinetics and equilibria can fix the optimum concentrations for efficient synthesis of the compounds presumably needed for life to arise. A relevant equilibrium of this type is the decomposition of aspartic acid to ammonia and fumaric acid:

$$^-OOC\text{-}CH_2\text{-}CH\text{-}COO^- = \\ \underset{NH_3^+}{|} \\ NH_4^+ + {}^-OOC\text{-}CH = CH\text{-}COO^- \quad (1)$$

In decomposition aspartic acid differs from most amino acids. Abelson and Vallentyne (8) have shown that alanine, serine, phenylalanine, and glutamic acid

decompose by a slow ($t_{\frac{1}{2}} = 10^{11}$ yea at 25°C for alanine) irreversible d carboxylation.

Aspartic acid decomposes by a d amination, which is reversible and rap on the geological time scale. Of t amino acids that occur in proteins, or aspartic acid and asparagine are know to undergo this type of chemical d composition, although there are e zymes that can deaminate histidi phenylalanine, and tyrosine (9).

The equilibrium constant for Eq. 1

$$(NH_4^+) = K[(\text{DL-aspartate})/(\text{fumarate})]$$

has been measured enzymically (1 11) and nonenzymically (10) betwe 5° and 135°C and is given by the equ tion

$$\log K = +8.22 - 2276/T - 0.0106T$$

The equilibrium constant is essentia independent of pH between 5 and the aspartic acid is less stable outsi this pH range. Malic acid can formed from the fumaric acid (1 but we shall omit this complicatic

The likely syntheses of aspartic ac on the primitive Earth are from Strecker synthesis with NC-CH₂CHO its equivalent (13), from cyanoacetyle (14), from ammonium cyanide (1 or from hydrogen cyanide alone (1 Although the last synthesis does n require ammonia, any aspartic ac made by this process would ha decomposed until the equilibrium Eq. 1 was reached. If there was : NH_4^+ in the primitive ocean, the aspa tic acid would have entirely deco posed to fumarate.

The heterotrophic hypothesis of t origin of life assumes that the bas constituents of the first living organis were available in large quantities in t primitive ocean; we shall assume th aspartic acid was one of these const uents. Although the concentration aspartic acid in the primitive oce cannot be estimated, we shall assur

Table 1. Equilibrium concentrations of ammonia and hydrogen.

Item	Temperature (°C)		
	0	25	50
Upper limit from clay mineral equilibrium			
NH_4^+ (M)	0.01	0.01	0.01
NH_3 (M)	6.0×10^{-5}	4.2×10^{-4}	2.1×1
pNH_3 (atm)	2.9×10^{-7}	7.3×10^{-6}	1.0×1
pH_2 (atm)	1.6×10^{-7}	4.3×10^{-6}	6.3×1
Lower limit from aspartic acid equilibrium			
NH_4^+ (M)	1.0×10^{-3}	2.7×10^{-3}	5.8×1
NH_3 (M)	6.0×10^{-6}	1.1×10^{-4}	1.2×1
pNH_3 (atm)	2.9×10^{-8}	1.9×10^{-6}	5.9×1
pH_2 (atm)	3.5×10^{-8}	1.8×10^{-6}	4.5×1

that the ratio of aspartic to fumaric acid did not fall substantially below 1.0. This criterion may be too restrictive, and it is possible that life could have arisen if much less aspartic acid had been present on the primitive Earth. On the basis of these assumptions we can say that the minimum concentration of NH_4^+ in the primitive ocean would be given by the K from Eq. 1, 1.0 $\times 10^{-3}M$ at 0°C and 2.7 $\times 10^{-3}M$ at 25°C.

It remains to be shown that this equilibrium would have been attained in the time available, which was less than 10^9 years but probably several hundred million years, since the Earth was formed 4.5 $\times 10^9$ years ago and the earliest evidence of life is in rocks 3.5 $\times 10^9$ years old (*17*). We have measured the rate of deamination of aspartic acid as a function of pH (Fig. 1) between 60° and 135°C (*18*). Only deamination of aspartic acid was observed; there was less than 0.2 percent decarboxylation to α- or β-alanine. The equation for the deamination half-life between pH 5 and 8 is

$$\log t_{\frac{1}{2}} \text{ (years)} = -22.01 + 8048/T$$

giving 28 $\times 10^6$ years at 0°C and 96,000 years at 25°C. The half-life for the deamination for pH values greater than 10 is given by

$$\log t_{\frac{1}{2}} \text{ (years)} = -17.35 + 6249/T$$

The half-lives are 330,000 years at 0°C and 4100 years at 25°C.

There is a similar deamination of asparagine to fumaramic acid at neutral pH (*19*), the rate of which is 190 times faster than the aspartic acid deamination rate at pH 8 and 60°C (*18*). The rate of deamination of the nitriles of aspartic acid should be even faster. Since the probable prebiological precursor of aspartic acid is the dinitrile, the rate of deamination during the four hydrolytic steps to aspartic acid would be very rapid in the absence of ammonia.

β-Methyl aspartic acid, which does not occur in proteins, also deaminates (*20*):

DL-β-methyl aspartate = mesaconate + NH_4^+, $K_{25°C} = 0.12$

β-Methyl aspartic acid may be absent from proteins either because it is not functional or because it was not available in the primitive ocean. Since the maximum concentration of NH_4^+ allowed by the clay minerals is 0.01M, the maximum ratio of β-methyl aspartic to mesaconic acid is 0.08. β-Methyl

aspartic acid has not yet been synthesized in any prebiological experiment, but even if it was synthesized on the primitive Earth most of it would have deaminated to mesaconic acid. Preliminary experiments indicate that this deamination is comparable in rate to deamination of aspartic acid.

A single example of a NH_4^+-dependent equilibrium or prebiological synthetic pathway cannot by itself demonstrate that the concentration of NH_4^+ was greater than 1.0 $\times 10^{-3}M$. However, a number of such reactions would form a more convincing argument. Another example is the synthesis of amino acids and hydroxy acids by the Strecker and cyanohydrin syntheses from an aldehyde, HCN, and NH_3. Only hydroxy acids are obtained in the absence of NH_3, and only amino acids at high concentrations of NH_3. Preliminary data indicate that the NH_4^+ concentration must be greater than $10^{-3}M$ for production of equal amounts of amino acids and the corresponding hydroxy acids at pH 8.

From the value of NH_4^+ obtained from the clay mineral or the aspartic acid–fumaric acid equilibrium, the pNH_3 and pH_2 in the atmosphere can be calculated provided the pH and temperature are known. The equilibrium values of NH_4^+, NH_3, pNH_3, and pH_2, with the atmosphere and the ocean at the same temperature, are given in Table 1 for pH 8 and $pN_2 = 1$ atm. These values are calculated from the ionization of NH_4^+ (*21*), the volatility of NH_3 (*22*), and the equilibrium constant for the reaction (*23*)

$$\tfrac{1}{2} N_2 + \tfrac{3}{2} H_2 = NH_3$$

Equilibrium would be attained for the ionization of NH_4^+ and the exchange of NH_3 with the atmosphere and ocean. The extent to which N_2, H_2, and NH_3 approached equilibrium is a much more complicated problem, since this is a pressure-dependent equilibrium, and the ultraviolet flux would tend to decompose the NH_3, giving a steady-state concentration less than the equilibrium value. To maintain the calculated pNH_3 and in turn the NH_3 and NH_4^+ in the ocean, the pH_2 would have to be substantially greater than the equilibrium value, possibly by several orders of magnitude.

References and Notes

1. S. L. Miller and H. C. Urey, *Science* **130**, 245 (1959).
2. Rasool and McGovern [*Nature* **212**, 1225 (1966)] have calculated the stability and escape rate of hydrogen in an atmosphere containing CH_4 and small amounts of H_2 and NH_3.
3. L. G. Sillén, in *Oceanography*, M. Sears, Ed. (AAAS, Washington, D.C., 1961), p. 549; *Arkiv Kemi* **24**, 431 (1965); *Science* **156**, 1189 (1967).
4. R. M. Garrels, *Science* **148**, 69 (1965); R. T. Mackenzie and R. M. Garrels, *ibid.* **150**, 57 (1965); H. D. Holland, *Proc. Nat. Acad. Sci. U.S.* **53**, 1173 (1965).
5. F. T. Mackenzie and R. M. Garrels, *Amer. J. Sci.* **264**, 507 (1966).
6. H. P. Eugster and J. Munoz, *Science* **151**, 683 (1966); J. J. Hanway, A. D. Scott, G. Stanford, *Soil Sci.* **21**, 29 (1957); R. C. Erd, D. E. White, J. J. Fahey, D. E. Lee, *Amer. Mineralogist* **49**, 831 (1964); W. Vedder, *Geochim. Cosmochim. Acta* **29**, 221 (1965); D. S. Barker, *Amer. Mineralogist* **49**, 851 (1964); E. Gruner, *ibid.* **24**, 428 (1939); R. M. Barrer and P. J. Denny, *J. Chem. Soc.* **1961**, 971 (1961); R. M. Barrer and L. W. R. Dicks, *ibid.* **1966A**, 1379 (1966).
7. H. Martin and H. Landelout, *J. Chim. Phys.* **60**, 1086 (1963).
8. P. H. Abelson, *Mem. Geol. Soc. Amer.* **67**, 87 (1957); *Ann. N.Y. Acad. Sci.* **69**, 276 (1957); *Progr. Chem. Org. Nat. Prod.* **17**, 379 (1959); in *Researches in Geochemistry*, P. H. Abelson, Ed. (Wiley, New York, 1959), p. 79; J. R. Vallentyne, *Geochim. Cosmochim. Acta* **28**, 157 (1964); D. Povoledo and J. R. Vallentyne, *ibid.*, p. 731; D. Conway and W. F. Libby, *J. Amer. Chem. Soc.* **80**, 1077 (1958).
9. A. Meister, *Biochemistry of the Amino Acids* (Academic Press, New York, 1965), pp. 826, 921, 922.
10. J. L. Bada and S. L. Miller, in preparation.
11. K. P. Jacobsohn and J. Tapadinhas, *Biochem. Z.* **282**, 374 (1935); J. S. Wilkinson and V. R. Williams, *Arch. Biochem. Biophys.* **93**, 80 (1961).
12. L. T. Rozelle and R. A. Alberty, *J. Phys. Chem.* **61**, 1637 (1957); L. E. Erickson and R. A. Alberty, *ibid.* **63**, 705 (1959); M. L. Bender and K. A. Connors, *J. Amer. Chem. Soc.* **84**, 1980 (1962).
13. S. L. Miller, *Biochim. Biophys. Acta* **23**, 480 (1957).
14. R. A. Sanchez, J. P. Ferris, L. E. Orgel, *Science* **154**, 784 (1966).
15. J. Oró and S. S. Kamat, *Nature* **190**, 442 (1961); C. U. Lowe, M. W. Rees, R. Markham, *ibid.* **199**, 219 (1963); C. N. Matthews and R. E. Moser, *Proc. Nat. Ac. '. Sci. U.S.* **56**, 1087 (1966).
16. P. H. Abelson, *Proc. Nat. Acad. Sci. U.S.* **55**, 1365 (1966).
17. E. S. Barghoorn and J. W. Schopf, *Science* **152**, 758 (1966).
18. The rates were determined by study of the rate of appearance of ammonia and fumaric acid from buffered solutions of DL-aspartic acid. The initial aspartic solutions were deoxygenated and sealed under partial vacuum in glass ampules. The rates and mechanism of this reaction will be discussed in detail elsewhere.
19. E. A. Talley, T. J. Fitzpatrick, W. L. Porter, *J. Amer. Chem. Soc.* **81**, 174 (1959).
20. H. A. Barker *et al.*, *J. Biol. Chem.* **234**, 320 (1959); the equilibrium constant is for the threo isomer.
21. R. G. Bates and G. D. Pinching, *J. Res. Nat. Bur. Std.* **42**, 419 (1949); at an ionic strength of 0.1, $pKa = -0.492 + 2835.76/T + 1.225 \times 10^{-3} T$.
22. T. K. Sherwood, *Ind. Eng. Chem.* **17**, 745 (1925); M. E. Jones, *J. Phys. Chem.* **67**, 1113 (1963); $\log (P_{atm}/M_{NH3}) = +5.071 - 1903.5/T - 1.517 \times 10^{-3} T$.
23. W. M. Latimer, *Oxidation Potentials* (Prentice-Hall, New York, ed. 2, 1952), p. 91; $\log K = -5.177 + 2412.6/T$.
24. Supported by grant GB 2687 from the National Science Foundation.

11 December 1967

Reprinted from *Science*, **161**, 158–160 (July 12, 1968)

Precambrian Marine Environment and the Development of Life

PETER K. WEYL

Abstract. *The tropical thermocline must have existed since the ocean's depth exceeded 300 meters. The density gradient in this layer concentrated organic aggregates formed abiologically near the surface of the sea, and the low rates of diffusion across this layer permitted the accumulation of oxygen once the layer was populated by blue-green algae; thus the evolution of eukaryotes became possible within the layer. Because of rapid mixing over the shelves, the eukaryotes were restricted initially to the thermocline over deep water. The shelves could not be permanently inhabited by organisms requiring respiration until the oxygen level of the atmosphere was adequate. At this stage, the swimming Metazoa of the thermocline could adapt to a benthic environment on the shelves by developing exoskeletons.*

In order to investigate the origin and early evolution of life on Earth one must consider the Precambrian environment. Our knowledge of Precambrian paleooceanography is extremely limited, and so the Precambrian ocean has usually been characterized by a single value for its parameters as if it had been a well-mixed system. A more realistic reconstruction of the Precambrian ocean may lead to new insights into the early development of life on Earth.

The distribution of noble gases on Earth and in stars indicates that Earth initially lost its fluid envelope, and that the present ocean and atmosphere must have accumulated from outgassing of Earth's interior (1). Thus the volume of the oceans has increased with time during the last 4×10^9 years. If the ratio of outgassing of chlorine and water remained relatively constant, the salinity of sea water did not vary greatly; thus its density has probably always been a function of salinity as well as temperature.

The volume of the oceans increased significantly during Precambrian times and approached its present value at the beginning of the Cambrian. This increase in volume does not necessarily imply an increase in the fractional area of Earth covered by the oceans. If, as is probable, the evolution of continental crust paralleled the outgassing of the mantle, the ocean became deeper with time, while covering approximately the same area.

As a first approximation we can consider the ocean to consist of three superimposed layers: a seasonally variable mixed surface layer, a layer in which the density increases rapidly with depth, and a more uniform deep layer. In low latitudes the density gradient in the intermediate layer results primarily from a decrease in temperature (thermocline), while in high latitudes density gradient results from increase in salinity with depth (halocline). The two upper layers extend to a depth of only about 200 m, while the third layer comprises most of the ocean.

The density-gradient layer results from the latitudinal variation of the insolation at the top of the atmosphere, which in turn results from the spherical shape of Earth and the inclination of its axis to the plane of its orbit. The latitudinal variation of the heat received results in a poleward transport of heat by winds, as latent heat of vaporization, and by ocean surface; thus the temperature difference across the thermocline has varied significantly over geologic time. When the temperature gradient was great, the salinity gradient also was great; during times of low temperature gradient the more rapid circulation resulted in a lesser salinity gradient. Since heat and salt affect the density oppositely, the density gradient across the thermocline has varied relatively little.

During the Precambrian the gross density structure of the two upper layers probably differed little from the present configuration. As soon as the ocean became deeper than 300 m, the vertical layering must have been similar to that of today's ocean. The Precambrian density-gradient layer, particularly the thermocline between 30°N and 30°S, may have played an important role in the origin and Precambrian evolution of life.

The initial surface environment being devoid of oxygen (2), ultraviolet light from Sun penetrated the upper 10 m of the ocean (3). The ultraviolet irradiation of the reducing atmosphere and ocean led to abiotic photosynthesis of organic molecules (4). Bernal (5) has suggested that this organic matter became concentrated by absorption on mineral grains on the seashore where it was polymerized into coacervate drops. An alternative mechanism for concentration is provided within the ocean. In today's ocean, surface-active dissolved organic matter is swept to the surface by rising bubbles and compressed into lines of convergence by the Langmuir circulation, where it is polymerized into particulate organic matter (6).

In the present ocean, the dissolved organic matter is derived from organisms, and the particulate organic matter produced sinks in the zones of convergence, where it is consumed by zooplankton. In the early ocean, the organic molecules would have been produced by abiotic processes, and the aggregates would not have been consumed. Depending on their bulk densities, the aggregates would have been concentrated in the density-gradient layer and on the ocean bottom. The steepest density gradients, and hence the largest concentrations, would have accumulated in low latitudes on both sides of the equator, where the present density increases from 1.023 to 1.026 between 50 and 100 m. Once removed from the sea surface, the organic aggregates would have been shielded from ultraviolet radiation. Prokaryotic heterotrophs could have evolved either at the ocean bottom or within the density-gradient layer where there was a concentration of organic aggregates and where they were shielded from ultraviolet radiation.

Conditions for the evolution of photosynthetic prokaryotic autotrophs would have been optimum in the tropical thermocline which was shielded from ultraviolet radiation while visible sunlight penetrated this layer. Evolution would have proceeded at constant density, since cells that became too heavy would have sunk below the illuminated region, while a reduction in density would have carried the cells into the mixed surface layer where they would have become exposed to ultraviolet radiation. The thermocline is a more extensive and more stable environment than the bottom of the shallow seas.

Vertical convection across the density-gradient layer is very slow, so that significant oxygen concentrations would have accumulated in the thermocline once oxygen was evolved by the blue-green algae. By use of data from the low-productivity regions of the present tropical oceans, an oxygen production rate of 1 mole m^{-2} year^{-1} would have led to an oxygen partial pressure in the thermocline equivalent to about 3 percent of the present atmospheric level.

Oxygen that diffused from the thermocline would have been mixed rapidly into the atmosphere and the underlying sediment—where it would have been used for oxidation of reduced minerals. If the algae were primarily restricted to the density-gradient layer, that layer would have had a significant concentration of oxygen while the concentration in the rest of the ocean and the atmosphere would have been very low until the products of weathering were oxidized. Thus the thermocline probably was an extensive, stable, oxygenated environment for the evolution of eukaryotic cells. This environment was more extensive in time and space than the microenvironment postulated by Fischer (8).

Where the thermocline intersected the sea floor, mixing would have been enhanced by the breaking of internal waves and by upwelling. As a result of rapid diffusion, the oxygen content of the water over the continental shelf would not have been significantly greater than that of the atmosphere. Therefore the early animals probably were planktonic and restricted to the density-gradient layer over the deeper parts of the oceans. The evolution of a skeleton would have been strongly inhibited since such organisms would have sunk out of the oxygen-containing layer unless their bulk densities were maintained constant by the simultaneous evolution of flotation mechanisms.

Blue-green algae were able to carpet the bottoms of the continental shelves, where development of a shallow seasonal thermocline would have provided temporary concentrations of oxygen. Thus organisms that evolved active swimming mechanisms could have seasonally exploited the food resources on the continental shelves. As winter mixing reduced the oxygen concentration on the shelves, these organisms had to return to the deep-water thermocline region to survive. Only after the atmospheric concentration of oxygen became sufficiently high could the continental shelves have been permanently inhabited by animals. At that time the organisms could have adapted to a benthic habit and increased their bulk density by skeletogenesis. This transition supposedly occurred at the beginning of the Cambrian.

If the blue-green algae in the open ocean were concentrated in the tropical thermocline as I have suggested, an environment containing sufficient oxygen for the evolution of Metazoa existed earlier than 10^9 years before the Cambrian. The early animals could not permanently populate the continental shelves and could not readily evolve a skeleton; thus the probability of their preservation in the fossil record would have been small. Without a Precambrian thermocline, one is forced to assume a very rapid rate of evolution for the Metazoa once the concentration of oxygen in the atmosphere was adequate for respiration (9). The thermocline, however, provided an extensive offshore environment between about 50 and 150 m that probably was oxygenated and within which floating and (later) swimming animals were able to evolve. The existence of a seasonally oxygenated continental-shelf environment, carpeted by algae, would have offered an adaptive advantage to organisms that evolved swimming mechanisms. Once the atmospheric concentration of oxygen became sufficient, these swimming organisms could have adapted to a benthic habit on the shelves. At this stage, density would no longer have been a problem, and the organisms could have evolved skeletons. In the early stages these exoskeletons would have provided an adaptive advantage as ultraviolet shields and would have permitted the organisms to seal themselves from the environment to survive temporary anoxia. According to my hypothesis the beginning of the Cambrian marks the first time Metazoa could permanently occupy the floors of the continental shelves.

References

1. W. W. Rubey, *Bull. Geol. Soc. Amer.* **62**, 1111 (1951).
2. H. D. Holland, *Proc. Nat. Acad. Sci. U.S.* **53**(6), 1173 (1965).
3. L. V. Berkner and L. C. Marshall, *J. Atmos. Sci.* **22**, 225 (1965); **23**(2), 133 (1966).
4. S. L. Miller, "Formation of organic compounds on the primitive earth," in *The Origin of Life on the Earth*, F. Clark and R. L. M. Synge, Eds. (Pergamon, New York, 1959), vol. 1.
5. J. D. Bernal, in *Oceanography*, M. Sears, Ed. (AAAS, Washington, D.C., 1961), pp. 95–118.
6. W. H. Sutcliffe, Jr., E. R. Baylor, D. W. Menzel, *Deep-Sea Res.* **10**, 233 (1963).
7. W. J. Schopf, "Antiquity and evolution of Precambrian life," in *McGraw-Hill Yearbook of Science and Technology* (McGraw-Hill, New York, 1967).
8. A. G. Fischer, "Fossils, early life, and atmospheric history," in *Proc. Nat. Acad. Sci. U.S.* **53** (15 June 1965).
9. P. E. Cloud, "Premetazoan evolution and the origins of the Metazoa," in *Evolution and Ecology*, E. Drake, Ed. (Yale Univ. Press, New Haven, Conn., 1968).

29 April 1968

Copyright © 1967 by North-Holland Publishing Company

Reprinted from *Earth Planet. Sci. Lett.*, **3**, 62–66 (1967)

THE OXYGEN ISOTOPE CHEMISTRY OF ANCIENT CHERTS

Eugene C. PERRY Jr.
*Department of Geological Sciences, University of Minnesota,
Minneapolis, Minnesota, USA*

Received 26 July 1967

This is a report of a continuing study of oxygen isotope variations in ancient cherts and of the possible significance of these variations in determining the oxygen isotopic composition and perhaps the manner of development of the Precambrian ocean. Variation in oxygen isotopic composition of the ocean through time is important in paleothermometry, and several evaluations of this composition have been attempted beginning with a discussion by Silverman [1]. Silicates and carbonates formed in equilibrium with water at low temperatures are enriched in ^{18}O. Urey, Lowenstam, Epstein and McKinney [2] concluded that the abundance of ^{18}O in the ancient ocean should be high relative to the present ocean but surmised that essentially a steady state has existed since the Cambrian period. Both Degens and Epstein [3] and Weber [4] noted a reversed trend, that is, a decrease in the abundance of ^{18}O in ancient marine carbonates [3, 4] and in cherts associated with ancient marine carbonates [3] relative to modern examples. Weber ascribed this to a change in the ^{18}O abundance in the ocean, but Degens and Epstein attributed the result to exchange between the rocks and ground water, citing as evidence the fact that Compston [5] observed an ^{18}O abundance in unrecrystallized Permian fossils similar to that found in modern specimens. More recently Longinelli [6] studied ^{18}O abundances in phosphate and carbonate from Paleozoic and younger fossils with somewhat puzzling results.

In the present study oxygen was extracted from samples by the BrF_5 technique of Clayton and Mayeda [7]. Yields, after samples were thoroughly dried at 125°C, were 100 ± 1%. Isotopic analyses were performed on CO_2 gas with a 15 cm radius of curvative 60° gas mass spectrometer with a double collector

[8]. Seasonal high humidity was an annoying source of contamination during the period in which experimental work was performed.

Oxygen isotope ratios * of several cherts and the approximate ages of these cherts are given in table 1 and plotted in fig. 1 (line A). Data from [3] are plotted in fig. 1 for comparison (line B). All of the samples which I have analyzed are from thick chert beds or from formations in which chert is a major constituent. Samples from the Bitter Springs Formation, the Gunflint Iron-formation, and the Fig Tree Formation are from localities and stratigraphic horizons in which microfossils have been reported [10, 11]. Preservation of these delicate organic remains is taken as evidence that the enclosing chert has undergone little or no modification after crystallization.

It seems improbable that the cherts of table 1 have attained their present isotopic composition as the result of exchange with ground water of modern isotopic composition for the following reasons:

1. Chert samples from the correlative Gunflint and Biwabik Iron-formations (over a sampled distance of 450 km) give similar isotopic values. The scatter in values for samples from the Biwabik Iron-formation probably reflects the fact that these are from cherty iron-formation having variable but appreciable amounts of iron silicates and magnetite and that diagenesis has altered the $\delta(^{18}O)SiO_2$ in a way that depends on the bulk composition of the sample.

*Ratios are expressed as

$$\delta(^{18}O) = \left[\frac{(^{18}O/^{16}O) \text{ sample}}{(^{18}O/^{16}O) \text{ standard}} - 1 \right] \times 100$$

The standard for this study is standard mean ocean water [9].

Table 1
Oxygen isotope compositions and approximate ages of bedded cherts.

Sample No.	Collector and reference	Approx. age $(\times 10^{-6})$	^{18}O * (%)	Remarks
	[3]	75	+3.19	Upper Cretaceous samples analyzed by Degens and Epstein [2]. $\delta(^{18}O)$ range is from 30.1 to 33.8.
C 123	E. F. McBride [19]	~380	+2.84	Caballos Formation, Brewster Co. Texas. 2 m above base of 60 m chert section.
C 126	E. F. McBride [19]	~380	+2.99 ± 0.016 (2)	Same locality as C 123. 32 m above base of formation
16-67	E. S. Barghoorn [10]	700-1000	+2.66	Bedded chert from Bitter Springs Formation, 80 km E of Alice Springs, Northern Territory, Australia.
9-67	E. C. Perry [11, 20]	1600-2000	+2.07	Basal chert of Gunflint Iron-formation. From 6 km W of Schreiber, Ontario.
3-67	E. C. Perry [20]	1600-2000	+2.27 ± 0.001 (2)	Basal chert of Gunflint Iron-formation. From 2.5 km W of Nolulu, Ontario.
37-66	E. C. Perry [12, 20]	1600-2000	+2.21 ± 0.025 (3)	Microcrystalline quarts from Biwabik Iron-formation. Near Keewatin, Minn.
1-66	E. C. Perry [12, 20]	1600-2000	+1.96	Microcrystalline quartz from Biwabik Iron-formation. 4.5 km W of Mesabi, Minn.
11-67	E. S. Barghoorn [10, 21]	~3000	+1.41 ± 0.03 (3)	Bedded chert from Fig Tree Formation, Barberton Mountain Land, E Transvaal, South Africa.

* With mean deviation if applicable. Number of determinations in parentheses.

2. Contact metamorphism of part of the Biwabik Iron-formation occurred 10^9 years ago. A continuous transition in $\delta(^{18}O)SiO_2$ and in quartz-magnetite isotope fractionation exists between metamorphosed and essentially unmetamorphosed parts of the iron-formation [12]. This implies that oxygen isotope exchange has not occurred in these rocks for 10^9 years.

3. The oldest chert reported in this paper, from the Fig Tree Formation, has an exceptionally low $\delta(^{18}O)$ value typical of sandstones and metamorphic rocks, yet the fact that it has retained the impressions of delicate microfossils is evidence that it has been almost undisturbed since its formation. In the present hydrologic cycle it would have to exchange with water isotopically similar to glacial meltwater to attain the observed $\delta(^{18}O)$ value at earth-surface temperatures.

4. With the help of fig. 1 a comparison may be made between oxygen isotope ratios in ancient bedded cherts (line A) and in Phanerozoic cherts associated with carbonate rocks in which Degens and Epstein [3] have found evidence for exchange of both chert and carbonate (line B). Considering possible diagenetic variations, the slopes of the two time versus $\delta(^{18}O)$ curves are well defined, and they are distinctly different. An unusual coincidence would be required for the same phenomenon to produce these separate trends.

Assuming that the cherts listed in table 1 are of marine origin (cf. ref. [13]) and that they were formed in approximate isotopic equilibrium with sea water, I interpret the $\delta(^{18}O)$ variation in the cherts to represent progressive change in the oxygen isotopic composition of the ocean through time from $\delta(^{18}O)$ $\cong 1.8\% \ 3 \times 10^9$ years ago to $\delta(^{18}O) = 0\%$ today. Some variation of $\delta(^{18}O)$ of cherts of like age may result primarily from dehydration proceeding irreversibly at various temperatures and pressures as well as

177

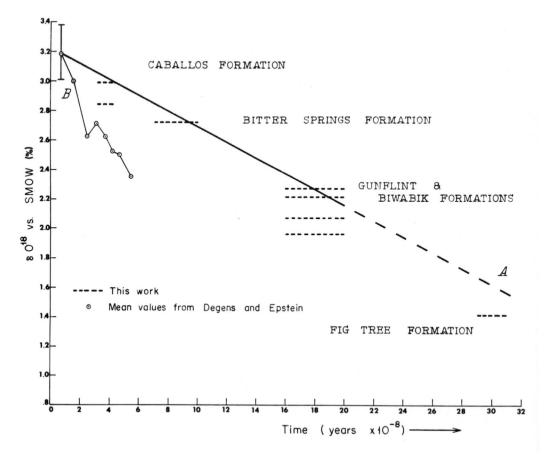

Fig. 1. Variation in $\delta(^{18}O)$ of cherts with age. Curve A is for thick bedded cherts; curve B (plotted from mean values in ref. [3])
is for cherts with carbonate rocks.

from diagenetic or low grade metamorphic exchange among minerals *. The dark envelope in fig. 1 may represent the approximate range of such variation in bedded cherts which have not recrystallized.

Line A in fig. 1 has been drawn through the two best established sets of points, young cherts (taken

* In addition to textural criteria that indicate little or no metamorphism in the samples of table 1, tentative measurement of quartz-magnetite fractionations in samples 1-66 and 37-66 indicate fractionations (1000 $\ln\alpha_{QM}$) of greater than 24 [12]. This probably corresponds to temperatures of less than 100°C.

from ref. [3]) and the maximum $\delta(^{18}O)$ for cherts from the iron-formations of Lake Superior. (The maximum observed value is used because almost any modification of the chert would tend to lower its $\delta(^{18}O)$.) The graph implies that $\delta(^{18}O)$ of Permian chert is about 0.1% lower than that of modern chert and, therefore, that the value for the Permian ocean was about 0.1% lower than the value for the present ocean. The change is equivalent to a change of less than 5°C on the $CaCO_3$-water paleotemperature scale [14]; it would be difficult to verify such a change by studying Paleozoic and younger sediments.

If the data in fig. 1 actually reflect a change in the oxygen isotope composition of the ocean through time, one may ask how this change came about and what it signifies. Although it has not been measured directly, the isotopic composition of water escaping from the earth's mantle is controlled by the bulk composition of the mantle and by temperature; a $\delta(^{18}O)$ of about 0.7% is a reasonable estimate [15]. Water having this isotopic composition could be modified at or near the earth's surface by low temperature isotopic exchange with carbonates and silicates. Virtually all such reactions result in a relative decrease in $\delta(^{18}O)$ of water [1, 16].

Thus a steady increase in $\delta(^{18}O)$ of about 0.18% in the last 3×10^9 years can be the result of addition of mantle-derived water through time, a decrease in the amount of crustal material reacting with a unit volume of water * and, perhaps, an increase in $\delta(^{18}O)$ of the reactive crust. The detailed process of isotopic mixing and exchange is undoubtedly a complicated one. To produce the modern ocean from a mixture of "juvenile" water (+0.7%) and water from the 3×10^9 year old ocean (−0.18%) would require a juvenile contribution of about 70% of the total without making any correction for the negative shift in $\delta(^{18}O)$ of ocean water produced by the precipitation of silicates and carbonates. A more difficult but more meaningful calculation would also consider high temperature reactions between isotopically heavy crustal rocks and "juvenile" water and between crustal rocks, particularly carbonates, and isotopically light meteoric water [17]. Both of these reactions would tend to increase $\delta(^{18}O)$ in the hydrosphere.

Progressive outgassing of the mantle is qualitatively suggested by fig. 1 and is entirely consistent with current hypotheses [18]. Closer study of the isotopic exchange of water with silicates through time may provide some information about the process of continental accretion. The regularity of the trend in fig. 1 suggests that the present ocean has evolved in a simple, direct way from the primitive ocean and that this evolutionary trend began more than 3×10^9 years ago.

---

* This probably corresponds to an increase in the amount of water per unit area of silicate.

ACKNOWLEDGEMENT

Support from NSF grant GA 912, from the Minnesota Geological Survey, and from the Graduate School of the University of Minnesota. Samples were provided by those people listed in table 1. C. G. Fisher, W. L. Griffin, J. W. Gruner, C. R. Kustra, G. B. Morey, W. C. Phinney, J. W. Morse and P. W. Weiblen assisted in this study. R. N. Clayton, G. D. Garlick, P. W. Gast and P. K. Sims have criticized one or more manuscript drafts of this paper. This acknowledgement does not necessarily imply that they agree with the conclusions presented.

REFERENCES

[1] S. R. Silverman, Geochim. Cosmochim. Acta 2 (1951) 26.

[2] H. C. Urey, H. A. Lowenstam, S. Epstein and C. R. McKinney, Geol. Soc. Am. Bull. 62 (1951) 399.

[3] E. T. Degens and S.Epstein, Am. Assoc. Petrol. Geol. Bull. 46 (1962) 534.

[4] J. N. Weber, Abstr. in Geol. Soc. Am. Special Paper 82 (1965).

[5] W. Compston, Geochim. Cosmochim. Acta 18 (1960) 1.

[6] A. Longinelli, Nature 211 (1966) 923.

[7] R. N. Clayton and T. K. Mayeda, Geochim. Cosmochim. Acta 27 (1963) 43.

[8] C. R. McKinney, J. A. McCrea, S. Epstein, H.A. Allen, H. C. Urey, Rev. Sci. Instr. 21 (1950) 724.

[9] H. Craig, Science 133 (1961) 1833.

[10] E. S. Barghoorn and J. W. Schopf, Science 152 (1966) 758;
J. W. Schopf and E. S. Barghoorn, Abstr. in Geol. Soc. Am. Annual Meeting Program (1966) p. 193;
J. W. Schopf and E. S. Barghoorn, Science 156 (1967) 508.

[11] E. S. Barghoorn and S. A. Tyler, Science 147 (1965) 563;
P. E. Cloud, Science 148 (1965) 27.

[12] E. C. Perry Jr. and B. Bonnichsen, Science 153 (1966) 525;
E. C. Perry Jr. and J. W. Morse, Abstr. in 13th Annual Institute on Lake Superior Geology, Michigan State University, East Lansing, Michigan, 1967.

[13] H. L. James, Econ. Geol. 49 (1954) 235;
G. J. S. Govett, Geol. Soc. Am. Bull. 77 (1966) 1191.

[14] S. Epstein, R. Buchsbaum, H. A. Lowenstam and H. C. Urey, Bull. Geol. Soc. Am. 64 (1953) 1315.

[15] H. Craig, in: Nuclear Geology and Geothermal Areas, ed. E. Tongiorgi (Consiglio Nazionale delle Richerche, Rome, 1963).

[16] J. R. O'Neil and R. N. Clayton, in: Isotopic and Cosmic Chemistry (North-Holland, Amsterdam, 1964); S. M. Savin and S. Epstein, Abstr. in Geol. Soc. Am. Annual Meeting Program (1966) p. 190.

[17] H. Craig, G. Boato and D. E. White, Natl. Acad. Sci., Nucl. Ser., Rept. 19 (1956) 29; R. N. Clayton, I. Friedman, D. L. Graf, T. K. Mayeda, W. F. Meents and N. F. Shrimp, J. Geophys. Res. 71 (1966) 3869.

[18] H. D. Holland, in: Petrologic Studies, Buddington Volume, eds. A. E. J. Engel, H. L. James and B. F. Leonard (Geol. Soc. Am., New York, 1962).

[19] A. Thomson, in: The Filling of the Marathon Geosyncline, Publication 64-9. Soc. of Econ. Paleontologists and Mineralogists, 1964;

P. B. King, U.S. Geol. Survey Prof. Paper 187 (1937).

[20] P. M. Hurley, H. W. Fairbairn, W. H. Pinson Jr. and J. Hower, J. Geol. 70 (1962) 489; S. S. Goldich, A. O Nier, H. Baadsgaard, J. H. Hoffman and H. W. Krueger, Minn. Geol. Survey Bull. No. 41 (1961); Z. E. Peterman, Geol. Soc. Am. Bull. 77 (1966) 1031. J. Kovach and G. Faure, Abstr. in Geol. Soc. Am. North-Central Section Meeting Program 1967.

[21] H. L. Allsopp, T. J. Ulrych and L. O. Nicolaysen, Abstr. in Geochronology of Stratified Rocks, eds. R. A. Burwash and R. D. Morton, University of Alberta, Edmonton, Canada (1967).

Errata:

The oxygen isotope chemistry of ancient cherts, Eugene C. Perry, Jr., *Earth and Planetary Science Letters*, **3** (1967), 62.

1. As printed, per mil and per cent (%) are both used to express $\delta(^{18}O)$. The ordinate of fig. 1 and the definition of $\delta(^{18}O)$ (footnote, p. 62) are expressed in per cent. Per mil is used in the remarks column of table 1, top row. All other $\delta(^{18}O)$ entries in table 1 and in the text are expressed in per cent.

2. p. 62, address should read:
Minnesota Geological Survey,
University of Minnesota,
Minneapolis, Minnesota 55455.

3. p. 63, table 1, sample 37-66, second word in remarks column is *quartz*.

4. p. 65, paragraph 2, the number on line 1 should be 1.8%; the number on line 10 should be −1.8%.

5. p. 66, ref. [17], fourth line should read "W. F. Meents and N. F. Shimp, *J. Geophys. Res.*, **71**...."

The Publisher regrets that these printing errors occur in Professor Perry's paper and offers his apologies for any inconveniences caused to author and readers.

19

Reprinted from *Geol. Soc. Am. Spec. Paper 62*, 651, 659–664 (1955)

Geologic Evidence of Chemical Composition of the Precambrian Atmosphere

KALERVO RANKAMA

Institute of Geology, University of Helsinki, Helsinki, Finland

ABSTRACT

A Precambrian breccia and conglomerate schists from Suodenniemi, Finland, with an age approaching 2 billion years, consist of fragments and pebbles of pre-Bothnian diorite embedded in a fine-grained granoblastic cement of varying degree of schistosity and consisting mainly of quartz, biotite, hornblende, and oligoclase. This breccia is interpreted as a genuine weathering breccia *in situ* (Sederholm), even though its cement may contain some material transported from the neighborhood (Eskola). Chemical analyses reveal the predominance of FeO over Fe_2O_3 in both unaltered diorite from a pebble in the conglomerate schist and the fine-grained schistose cement surrounding the pebbles in the conglomerate schist and the fragments in the breccia. Furthermore, there is no increase in the Fe_2O_3 content of the cement, as compared with the Fe_2O_3 content of the diorite. These results are interpreted to indicate that weathering of the diorite and the formation of the breccia and the conglomerates took place in an atmosphere devoid of oxygen. They appear to lend support to Urey's hypothesis of a reducing early Precambrian atmosphere. Absence of atmospheric oxygen also explains the occurrence of reduced native carbon as a finely disseminated pigment in the surrounding Bothnian phyllites and, by analogy, the common occurrence of carbonaceous slates and graphitic schists in early Precambrian terranes.

* * * * * * *

Editor's Note: The following sections have been omitted from this reprint for the sake of brevity: "Introduction," "Acknowledgments," "Description of Material," "Chemical Data," and "Discussion of Factors Affecting the Oxidation State of Iron."

CHEMICAL COMPOSITION OF THE PRIMORDIAL ATMOSPHERE

Various authors have suggested that the primitive atmosphere of the earth was reducing rather than oxidizing. This thesis is not new. Clarke (1924, p. 59–61) summarized early arguments and beliefs favoring a reducing atmosphere. Some geologic evidence in favor of the thesis has been presented also. Lane (1917, p. 44, 45) found that Precambrian rocks of the Great Lakes region in North America indicated the absence of an atmosphere consisting largely of oxygen in early Precambrian times. Red beds, which indicate oxidation of iron-bearing minerals, are absent in the early Precambrian (Pettijohn, 1949, p. 170). The earliest extensive red beds are in the Lake Superior region, where there is an abrupt change in metamorphic character between the red Keweenawan series and the underlying Huronian series, which lacks the reddish colors of the Keweenawan (Leith, Lund, and Leith, 1935, p. 8). Macgregor (1927, p. 158, 162, 171) considered formation beneath an atmosphere relatively or completely deficient in oxygen a ready explanation for the origin of certain carbonate-rich schists among the early Precambrian basement rocks of Southern Rhodesia. He suggested that the early atmosphere was rich in carbon dioxide. Furthermore, he (p. 163, 165, 171) found that the early Precambrian sediments in Rhodesia are characteristically rich in FeO and low in Fe_2O_3. He concluded that these rocks give no evidence of the contemporaneous oxidation which later sediments have undergone. Finally, he (p. 167, 171) suggested that Precambrian banded ironstones were formed when there was no oxygen but much carbon dioxide in the atmosphere.

Poole (1941, p. 350, 351, 363, 364) suggested that free oxygen developed slowly in the primordial atmosphere, first through photodissociation of water vapor and later through photosynthesis. He thought that methane, carbon dioxide, nitrogen, and water vapor were the main constituents of the primitive atmosphere. Rubey (1951, p. 1123) too felt that free oxygen was absent in the early atmosphere. But contrary views were presented as well. Chamberlin (1949, p. 251, 255), on the basis of the existence of Precambrian hematite ores and red beds, concluded that a supply of free oxygen is required throughout geologic time.

The conclusion that the earth's present atmosphere is almost entirely of secondary origin (Brown, 1949, p. 266; Suess, 1949, p. 600) proved very important for the discussion of problems of the early atmosphere. Urey (1951, p. 209, 218, 220, 245, 246; 1952a, p. 149–157; 1952b; 1953, p. 291) reopened the case and presented, largely on the basis of thermodynamic considerations and supported by data from many branches of science, a rather coherent and logical hypothesis of the formation of the earth or, at least, of its outer parts at low temperatures. A link in Urey's hypothesis consists of the thesis that the early atmosphere of the earth was reducing rather than oxidizing, devoid of free oxygen, and consisted chiefly of water vapor, hydrogen, methane, ammonia, and some hydrogen sulfide, as Bernal (1949, p. 545) had already suggested. Hydrogen was continuously lost from the gravitational field of the earth, water vapor was photochemically decomposed into hydrogen and oxygen, and the oxygen formed was consumed in the oxidation of ammonia into nitrogen and water and of methane into carbon dioxide and water. One must suppose, however, that juvenile carbon dioxide was released into the atmosphere very early in its history. Finally, an excess of free oxygen became available because it was not totally consumed in oxidation reactions. The appearance of free oxygen marks the turning point from a reducing to an oxidizing atmosphere. The reducing period in the history of the atmosphere also coincides with prebiologic time in the history of life, although possibly reducing conditions still existed after the first living matter had evolved.

Some experiments have been made to investigate whether highly complicated chemical reactions could take place in reducing surroundings before biologic processes existed. The results of such experiments are encouraging and indicate that complex chemical compounds may be produced by thermochemical, photochemical, and electrochemical processes (MacNevin, 1953a; 1953b). A number of photochemical reactions of probable importance in prebiologic chemistry were listed by Calvert (1953, p. 295, 296). Furthermore, organic compounds containing nitrogen—namely, amino acids and, possibly, porphyrins—form readily at low temperatures (20°–100°C) from methane, ammonia, water, and hydrogen under an electrical discharge, that is, in a gas mixture similar to the alleged primitive atmosphere (Caley, 1953; Miller, 1953). These results support the idea suggested by Oparin (1938, p. 62, 64, 101, 108, 109, 126, 127, 131, 133–135, 146, 159, 202, 248) and seconded by Bernal (1949, p. 545, 546; 1951, p. 32) and Urey (1952a, p. 149–157; 1952b, p. 356, 357) that numerous prebiologic organic compounds formed in a primitive reducing atmosphere, or before the advent of living organisms in general (Blum, 1951, p. 161, 163, 164; *see also* Strughold, 1953, p. 395).

Thode, Macnamara, and Fleming (1953, p. 235, 238, 240–242) concluded that fractionation of stable sulfur isotopes between sulfide sulfur and sulfate sulfur did not start until approximately 0.7 or 0.8 billion years ago. They felt that, even though biogenic matter and living organisms certainly existed at that time, such autotrophic organisms that obtain their free energy from the oxidation of hydrogen sulfide and sulfur had not yet made their appearance. Furthermore, they believed that their date probably preceded the beginning of large-scale photosynthesis. Urey (1952a, p. 153, 154; 1952b, p. 360) thought that this date might mark the transition from reducing conditions to oxidizing conditions in the atmosphere, and that it is just barely possible that reducing conditions were maintained until approximately 0.8 billion years ago. He (Urey, 1952b, p. 362) believed that the time of transition should be recorded in rocks.

TESTIMONY OF BOTHNIAN SCHISTS

Fe_2O_3/FeO ratios computed from Fe_2O_3 and FeO percentages presented in Table 1 indicate the predominance of ferrous iron in both the diorite and its weathering products. Conditions prevailing during subaerial weathering of the diorite and accumulation of weathering products evidently were such as to cause no oxidation of iron. From this result one might conclude that there was no free oxygen in the atmosphere during the time of weathering. But it would be too bold to base such a conclusion solely on the Fe_2O_3/FeO ratios, inasmuch as clay analyses quoted in Table 3 indicate that under special conditions ferrous iron may predominate among recent weathering products definitely formed beneath an oxidizing atmosphere. Furthermore, one should consider the possible role of metamorphism in affecting the oxidation state of iron.

As already discussed, the original Fe_2O_3/FeO ratios may be retained in rocks of sedimentary origin even during high-grade regional metamorphism. This result is further confirmed by analyses nos. 1, 5, and 11 in Table 1. The diorite pebbles have retained their original FeO preponderance indicated by several analyses of quartz diorites and granodiorites of the area. Moreover, the conclusion follows from analyses nos. 1 and 11 (Table 1) that the Fe_2O_3/FeO ratio of the matrix of the conglomerate schist has not changed during metamorphism, because the ratio has not changed in the rather small diorite pebble.

Analyses nos. 1, 10, 11, 12, 13, and 14 may now be used to support the thesis that free oxygen was absent when weathering of the diorites and accumulation of the weathering products took place. There is no increase in Fe_2O_3 content in the matrix of the weathering breccia, the matrix of the conglomerate schist, and the meta-arkose formed from the partly decomposed diorite gravel, as compared with the Fe_2O_3 content of fresh unaltered diorite. In this respect, analyses nos. 1 and 11 make a particularly illustrative pair, because the conglomerate schist consists of transported material that must have been even more affected by weathering than the material of the breccia lying *in situ*. Furthermore, metamorphism must have affected the small pebble and the surrounding matrix in a similar manner.

The writer believes the conclusion is justified that the breccia and the conglomerate schist are evidence of weathering beneath an atmosphere devoid of oxygen. The material on which this conclusion is based is admittedly very scarce and certainly will not satisfy present-day demands for a multitude of data that can be treated statistically. It is well to remember, however, that the early Precambrian rocks that may be used in evidence of physical and chemical conditions on the earth during those very remote times are choice morsels, no standard items, in the bill of fare of the research geologist. A maximum amount of information should be digested from such material.

The absence of atmospheric oxygen until late in the Precambrian would also explain the manner of formation of finely disseminated carbon pigment in slates and schists. Biogenic origin has been attributed to this carbon in the Precambrian formations of Finland (Eskola, 1932, p. 27, 67; Rankama, 1948), in the southern part of the Canadian Shield (Pettijohn, 1943, p. 948), and in the Superior Province in Manitoba (Rankama, 1954b, 1954c). Finely disseminated carbon in slates and schists of argillaceous origin seems to be much more common in Precambrian terranes than was previously believed. Such carbon may be assumed to represent the remains of primitive organisms capable of living under reducing conditions (Rankama, 1954c, p. 16). In the absence of atmospheric oxygen, carbon in the remains did not become oxidized to carbon dioxide but was reduced to native carbon.

It follows that the presence of a reducing atmosphere in early Precambrian time appears thoroughly possible geologically and is not contradicted by such evidence as can be collected from rocks. In fact, all evidence available strongly supports Urey's hypothesis. The writer, for one, is fully convinced that it will be possible to coax the ancient rocks to talk much more about the chemical composition of the early terrestrial atmosphere.

REFERENCES CITED

Bernal, J. D. (1949), The physical basis of life: Phys. Soc. London, Proc., v. A 62, p. 537; v. B 62, p. 597.

————— (1951), The Physical Basis of Life: Routledge and Kegan Paul, London, 80 p.

Blum, H. F. (1951), Time's Arrow and Evolution: Princeton Univ. Press, Princeton, N. J., 222 p.

Brown, H. (1949), Rare gases and the formation of the Earth's atmosphere: in The Atmospheres of the Earth and Planets, G. P. Kuiper, edit., Univ. Chicago Press, Chicago, p. 260–268. Also Revised Edition, 1952, p. 258–266.

Caley, E. R. (1953), Reactions induced by lightning in prebiological chemistry: in A Symposium on Prebiological Chemistry, Ohio State Univ., April 25, 1953. Mimeographed Rept., p. 22–34.

Calvert, J. G. (1953), The utilization of the solar energy through photochemical reactions: Ohio Jour. Sci., v. 53, p. 293–299.

Chamberlin, R. T. (1949), Geologic evidence on the evolution of the Earth's atmosphere: in The Atmospheres of the Earth and Planets, G. P. Kuiper, edit., Univ. Chicago Press, Chicago, Ill., p. 250–259. Also Revised Edition, 1952, p. 248–257.

Clarke, F. W. (1924), The data of geochemistry: U.S. Geol. Surv., Bull. 770. 5th ed., 841 p.

Collins, C. B., Russell, R. D., and Farquhar, R. M. (1953), The maximum age of the elements and the age of the Earth's crust: Canadian Jour. Physics., v. 31, p. 402–418.

Eskola, P. (1932), Conditions during the earliest geological times as indicated by the Archaean rocks: Acad. Sci. Fennicae Ann., A, v. 34, no. 4, 74 p.

————— (1936), Förhållandena på jordytan under de äldsta geologiska tiderna: Nordiska (19. skandinaviska) naturforskarmötet i Helsingfors, 1936, p. 145–153.

————— (1941), Erkki Mikkola und der heutige Stand der präkambrischen Geologie in Finnland: Suomi, Finnlandheft, Geol. Rundschau, v. 32, p. 452–482 (pub. 1942).

Goldich, S. S. (1938), A study in rock-weathering: Jour. Geol., v. 46, p. 17–58.

Goldschmidt, V. M. (1943), Oksydasjon og reduksjon i geokjemien: Geol. Fören. Förh. Stockholm, v. 65, p. 84–85.

————— (1954), Geochemistry: A. Muir, edit., Clarendon Press, Oxford, 730 p.

Huhma, A., Salli, I., and Matisto, A. (1952), Suomen geologinen kartta—Geological map of Finland, 1:100,000, Lehti—Sheet 2122, Ikaalinen, Kallioperäkartan selitys—Explanation to the map of rocks: Geologinen tutkimuslaitos, Helsinki.

Lane, A. C. (1917), Lawson's correlation of the pre-Cambrian era: Am. Jour. Sci., 4th ser., v. 43, p. 42–48.

Leith, C. K., Lund, R. J., and Leith, A. (1935), Pre-Cambrian rocks of the Lake Superior region. A review of newly discovered geologic features with a revised geologic map: U. S. Geol. Surv., Prof. Paper 184, p. 34.

Macgregor, A. M. (1927), The problem of the Precambrian atmosphere: South African Jour. Sci., v. 24, p. 155–172.

MacNevin, W. M. (1953a), Ohio State Labs Recreate Prebiological Era: Chem. Eng. News, v. 31, p. 1546.

————— (1953b), Evidences for pre-biological chemistry: in A Symposium on Prebiological Chemistry, Ohio State Univ., April 25, 1953. Mimeographed Rept., p. 3–9.

Mäkinen, E. (1914), Ytterligare om kontakten vid Naarajärvi i Lavia: Geol. Fören. Förh. Stockholm, v. 36, p. 185–203.

————— (1915), Ein archäisches Konglomeratvorkommen bei Lavia in Finnland: Geol. Fören. Förh. Stockholm, v. 37, p. 385–421.

Mason, B. (1949), Oxidation and reduction in geochemistry: Jour. Geol., v. 57, p. 62–72.

Miller, S. L. (1953), A production of amino acids under possible primitive Earth conditions: Science, v. 117, p. 528–529.

Nanz, R. H. Jr. (1953), Chemical composition of pre-Cambrian slates with notes on the geochemical evolution of lutites: Jour. Geol., v. 61, p. 51–64.

Oparin, A. I. (1938), The origin of life: 2nd ed., 1953, Dover Publications, Inc., New York, 270 p.

Patterson, C., Tilton, G., and Inghram, M. (1954), The age of the Earth: Manuscript.

Pettijohn, F. J. (1943), Archean sedimentation: Geol. Soc. America Bull., v. 54, p. 925–972.

————— (1949), Sedimentary rocks: Harper & Brothers, New York, 526 p.

Poole, J. H. J. (1941), The evolution of the atmosphere: Roy. Dublin Soc., Sci. Proc., v. 22, no. 36, p. 345–365.

Rankama, K. (1948), New evidence of the origin of pre-Cambrian carbon: Geol. Soc. America Bull., v. 59, p. 389–416.

————— (1954a), A calculation of the amount of weathered igneous rock: Geochim. Cosmochim. Acta, v. 5, p. 81–84.

————— (1954b), Early pre-Cambrian carbon of biogenic origin from the Canadian Shield: Science, v. 119, p. 506–507.

————— (1954c), Origin of carbon in some early pre-Cambrian carbonaceous slates from southeastern Manitoba, Canada: Soc. géol. Finlande, C. R., v. 27; Comm. géol. Finlande, Bull., v. 166, p. 5–20.

—————, and Sahama, Th. G. (1950), Geochemistry: Univ. Chicago Press, Chicago, Ill., 912 p.

Rowledge, H. P. (1934), A new method for the determination of ferrous iron in refractory silicates: Jour. Roy. Soc. Western Australia, v. 20, p. 165–199, 1933–1934 (published 1934).

Rubey, W. W. (1951), Geologic history of sea water. An attempt to state the problem: Geol. Soc. America, Bull., v. 62, p. 1111–1147.

Salminen, A. (1933), On the chemical composition of clays in a vertical profile through layers of different ages: Soc. géol. Finlande, C. R., v. 6; Comm. géol. Finlande, Bull., v. 101, p. 79–81.

————— (1935), On the weathering of rocks and the composition of clays: Acad. Sci. Fennicae Ann., A, v. 44, no. 6, 149 p.

Sederholm, J. J. (1897), Über eine archäische Sedimentformation im südwestlichen Finland und ihre Bedeutung für die Erklärung der Entstehungsweise des Grundgebirges: Comm. géol. Finlande, Bull., v. 6, 254 p.

————— (1913a), Kontakten mellan de bottniska sedimenten och deras underlag vid Naarajärvi i Lavia: Geol. Fören. Förh. Stockholm, v. 35, p. 163–195.

————— (1913b), Suomen geologinen yleiskartta. Lehti B 2, Tampere: Vuorilajikartan selitys. Geologinen toimisto, Helsinki, 122 p.

————— (1915), De bottniska skiffrarnas undre kontakter: Geol. Fören. Förh. Stockholm, v. 37, p. 52–118.

————— (1931), On the Sub-Bothnian unconformity and on Archaean rocks formed by secular weathering: Comm. géol. Finlande, Bull., v. 95, 85 p.

————— (1932), On the geology of Fennoscandia with special reference to the pre-Cambrian. Explanatory notes to accompany a general geological map of Fennoscandia: Comm. géol. Finlande, Bull., v. 98, 30 p.

Seitsaari, J. (1951), The schist belt northeast of Tampere in Finland: Comm. géol. Finlande, Bull., v. 153, 120 p.

Sharp, R. P. (1940), Ep-Archean and Ep-Algonkian erosion surfaces, Grand Canyon, Arizona: Geol. Soc. America Bull., v. 51, p. 1235–1269.

Simonen, A. (1948), On the petrology of the Aulanko area in southwestern Finland: Comm. géol. Finlande, Bull., v. 143, 66 p.

————— (1949), Suomen geologinen kartta—Geological map of Finland, 1:100,000, Lehti—Sheet

2131, Hämeenlinna, Kallioperäkartan selitys—Explanation to the map of rocks: Geologinen tutkimuslaitos, Helsinki, 45 p.

—————— (1952), Suomen geologinen kartta—Geological map of Finland, 1:100,000, Lehti—Sheet 2124, Viljakkala-Teisko, Kallioperäkartan selitys—Explanation to the map of rocks: Geologinen tutkimuslaitos, Helsinki.

—————— (1953), Stratigraphy and sedimentation of the Svecofennidic, early Archean supracrustal rocks in southwestern Finland: Comm. géol. Finlande, Bull., v. 160, 64 p.

——————, and Kouvo, O. (1951), Archean varved schists north of Tampere in Finland: Soc. géol. Finlande, C. R., v. 24; Comm. géol. Finlande, Bull., v. 154, p. 93–114.

——————, and Neuvonen, K. J. (1947), On the metamorphism of the schists in the Ylöjärvi area: Soc. géol. Finlande, C. R., v. 20; Comm. géol. Finlande, Bull., v. 140, p. 247–260.

Strughold, H. (1953), Comparative ecological study of the chemistry of the planetary atmospheres: Jour. Aviation Med., v. 24, p. 393–399, 464.

Suess, H. E. (1949), Die Häufigkeit der Edelgase auf der Erde und im Kosmos: Jour. Geol., v. 57, p. 600–607.

Thode, H. G., Macnamara, J., and Fleming, W. H. (1953), Sulphur isotope fractionation in Nature and geological and biological time scales: Geochim. Cosmochim. Acta, v. 3, p. 235–243.

Urey, H. C. (1951), The origin and development of the Earth and other terrestrial planets: Geochim. Cosmochim. Acta, v. 1, p. 209–277.

—————— (1952a), The Planets. Their origin and development: New Haven, Yale Univ. Press, 245 p.

—————— (1952b), On the early chemical history of the Earth and the origin of life: Nat. Acad. Sci., Proc., v. 38, p. 351–363.

—————— (1953), On the concentration of certain elements at the Earth's surface: Roy. Soc. London, Proc., v. A 219, p. 281–292.

Väyrynen, H. (1933), Discussion. in: Compte Rendu de la Réunion Internationale pour l'Étude du Précambrien et des Vieilles Chaînes de Montagnes; Comm. géol. Finlande, Bull., v. 102, p. 20–21.

Wickman, F. E. (1954), Preliminära åldrar av några svenska pegmatiter: Geol. Fören. Förh. Stockholm, v. 76, p. 336 (abstract).

Wilson, M. E. (1931), Life in the pre-Cambrian of the Canadian Shield: Roy. Soc. Canada, Trans., v. 25, 4, p. 119–126.

20

Reprinted from *Abhandl. Deut. Akad. Wiss. Berlin, Kl. Chem. Geol. Biol.*, **3**, 1, 6, 27–29, 33–35 (1958)

PAUL RAMDOHR

Die Uran- und Goldlagerstätten Witwatersrand— Blind River District—Dominion Reef—Serra de Jacobina: erzmikroskopische Untersuchungen und ein geologischer Vergleich

* * * * * * *

Editor's Note: The following sections have been omitted from this reprint for the sake of brevity: "Einleitung": "Erganzende Beobachtungen zum Witwatersrands," "Blind River District, Onterio," "Dominion Reef," "Serra de Jacobina, Brasilien": "Kritik an einer Arbeit von C. F. Davidson," and "Abbildungsanhang."

GEOLOGISCHE GEGEBENHEITEN

Allgemeines. Witwatersrand, Blind River Distrikt, Dominion Reef System und Serra de Jacobina stehen heute als in der Produktion von Gold und Uran sehr wesentliche, in den nachgewiesenen Reserven führende Lagerstättengebiete im Mittelpunkt des Interesses.

Gemeinsam ist allen, daß die bauwürdige Metallführung überall geknüpft ist an eine Mehrzahl, im Witwatersrand große Vielzahl von Konglomerathorizonten, bevorzugt aufgebaut aus Quarzgeröllen. Der Urangehalt liegt in verschiedener Form vor, aber in den unverwitterten Teilen allemal in U^{+4}-Bindung; das Gold kann reichlich (Witwatersrand) bis sehr spärlich sein (Blind River) und ist fast stets fein verteilt. In allen Fällen ist als Begleitmineral viel Pyrit vorhanden, oft so, daß zwischen der Menge von Pyrit und den bauwürdigen Mineralien eine klare Parallelität herrscht, anderswo besteht die Parallelität aber nicht. Wenngleich größere Gold- oder Uranmengen wohl immer von Pyrit begleitet sind, so braucht Pyrit seinerseits nicht immer nennenswerte Gehalte zu bedeuten.

Alle Vorkommen gehören dem Präkambrium an. Die altersmäßige Einstufung macht, wie wir später leicht verstehen werden, große Schwierigkeiten. Heute weiß man, daß erhebliche Altersunterschiede bestehen, wahrscheinlich — hier muß ich mich ganz auf das Schrifttum und auf eine Unterhaltung mit Kollegen L. H. AHRENS verlassen — ist das Dominion Reef mit roh 3000 Millionen Jahren weitaus am ältesten, das Witwatersrandsystem etwa 1800, die Blind River Konglomerate 1200 Millionen Jahre alt — alle Angaben natürlich nur als *ganz* roher und erster Hinweis (weiteres später!).

Jedem, der auch nur oberflächlich Proben angesehen hat, fällt die große Ähnlichkeit in allen Fällen auf, eine Ähnlichkeit, die bei gründlicherer Kenntnis sich nur immer wieder bestätigt und die von den großen Ähnlichkeiten im geologischen Bild bis in feine erzmikroskopische Einzelheiten geht[1]. Wenn im folgenden besonders Unterschiede hervorgehoben werden, so darf das keinen falschen Eindruck erwecken — das Übereinstimmende sagt man nur einmal, Verschiedenes muß mehrfach gebracht und genauer ausgeführt werden! Insgesamt sind die Übereinstimmungen so, daß man mit allergrößter Wahrscheinlichkeit die Bildung in allen Fällen als sehr ähnlich wird annehmen müssen. Das ist denn auch, ganz gleichgültig welche Schulmeinung die betr. Autoren vertraten, von wohl allen, die bisher im Schrifttum vergleichende Betrachtungen anstellten, betont worden.

Da die einzelnen Vorkommen, um die Vollständigkeit einigermaßen zu gewährleisten — ganz ist das wegen vertikaler wie horizontaler Verschiedenheiten auch in den Einzelgebieten nie möglich —, mit ihrem gesamten Erzmineralbestand und Strukturen beschrieben sind, sind Wiederholungen nicht vermeidbar.

Im folgenden werden Witwatersrand *W. W. R.*, Blind River Gebiet *Bl. R.*, Dominion Reef *D. R.* und Serra de Jacobina *S. J.* abgekürzt. — Die Reefs der Black Reef Serie werden mit dem W. W. R. zusammen behandelt, da sie räumlich meist dahin gehören und abgesehen von gewissen „jugendlichen" Merkmalen ihm auch in der Vererzung ähneln. Da das Black Reef erheblich jünger als das W. W. R.-System ist, ist das natürlich eine an sich nicht korrekte Vereinfachung.

---

[1] Ich selbst würde, trotz reichlicher Beschäftigung mit der Materie, mich keineswegs im Stande fühlen, etwa unetikettierte oder verwechselte Proben fundortmäßig ohne mikroskopische Untersuchung in allen Fällen zu bestimmen.

*   *   *   *   *   *   *

GEOLOGISCHE ÜBERLEGUNGEN

* * * * * * *

Wenn wir versuchen, den geologischen wie mikroskopischen Befund, wie er sich als *gesichertes Beobachtungsmaterial* in allen zur Besprechung stehenden Lagerstätten darstellt, zu einer genetischen Deutung auswerten zu wollen, so ist es am bequemsten zu sagen: ,,Viele Tatsachen können nicht ohne weiteres erklärt werden, manche weiteren erscheinen sogar widerspruchsvoll, schließlich ist auch einiges mit mehreren Hypothesen tragbar zu deuten — also stellen wir fest, daß wir es nicht wissen!"

Damit wird sich aber ein Wissenschaftler nur im äußersten Notfall abfinden, nie aber zufriedengeben!

Gesichert ist jedenfalls, daß die Hauptmenge von *Pyrit, Uranpecherz, Arsenkies* u. a. genau wie der *Quarz, Zirkon, Chromit, Diamant, Iridosmium* und (im Dominion Reef) *Spinell, Granat, Monazit* u. a. Geröllnatur besitzen.

Sicher ist *ebenso*, daß *Gold, Kupferkies, Zinkblende, Bleiglanz, Magnetkies* im Konglomerat unter *hydrothermalen Bedingungen* umgelagert sind. Das letztere trifft auch zu auf weitaus die überwiegende Menge von *Brannerit*. Weiterhin ist die sehr frühe Anwesenheit von ,,kohliger Substanz" im W. W. R. erwiesen, da sie bereits während der Zeit der Bildung mindestens jüngerer Konglomerate im W. W. R. Uranpecherz ,,korrodiert" oder ,,verdaut" hatte — umgelagerte Gerölle solcher ,,kohliger Substanz" finden sich ja schon in einigen der höheren Konglomerate.

Schließlich ist *erwiesen*, daß im W. W. R. der Vulkanismus der *Ventersdorplaven* bereits fertige Konglomerate, die Pyritgerölle, z. T. typische buckshots führten, vorfand. Durch diese Laven sind ja solche Buckshots (bis zu > 0.5 cm Gr. mindestens) hier unter Erhaltung der Strukturen in Magnetkies verwandelt.

Ebenso war das Dominion Reef in etwa der jetzigen Form schon bei der Entstehung des Witwatersrandsystems vorhanden, da für das D. R. charakteristische Gerölle sich umgelagert in den ältesten Reefs des W. W. R. wiederfinden.

Jeder Deutungsversuch muß mit diesen Tatsachen sich auseinandersetzen. Eine Hypothese, die an ihnen vorbeigeht, ist nicht nur eine leere Spekulation, sondern von vornherein falsch, so falsch wie etwa eine noch so geistreiche Überlegung, die zur Folgerung hätte, daß die Pferde fünf Beine haben.

Wie in meiner letzten Arbeit möchte ich mich ganz vorsichtig äußern: Hydrothermale Zufuhr aus magmatischen Quellen ist ausgeschlossen. Der Befund für die oben zuerst genannten Mineralien ist unbedingt der einer Seife. Die ,,umgelagerten" Sulfide sind in ihrer Herkunft zunächst schleierhaft. Da aber für die Pyrit- und Arsenkiesgerölle eine Herkunft zum Teil aus Gängen, zum Teil Sulfidlagern, z. T. aus Bildungen in Tonen der ,,Euxinischen Fazies" angenommen werden muß — was beides aus Dutzenden mikroskopischer Kriterien gesichert ist — in den ersteren aber Kupferkies, Zinkblende, Bleiglanz reichlich sind, könnten unter den gleich näher zu besprechenden Voraussetzungen auch diese Mineralien zunächst als Gerölle, bzw. feiner Grus in die Konglomerate gelangt sein. Hier sind gerade sie — und das entspricht allen Beobachtungen z. B. in Zerrklüften tektonisch beanspruchter Lagerstätten — besonders zu Umlagerungen ,,pseudohydrothermaler" Natur oder unter ,,hydrothermalen" Bedingungen oder in ,,leeren Thermen", oder wie man denselben Vorgang noch anders nennen mag, geeignet. Schließlich entstammt ja wohl die Hälfte aller schönen Kristalle dieser Mineralien in unseren Sammlungen solchen Vorgängen! Für Gold gilt mit geringer Abwandlung dasselbe.

Daß Pyrit wenig, Uranpecherz (abgesehen von der Branneritbildung) kaum, Diamant, Zirkon, Chromit, Iridosmium gar nicht sich bei solchen Bedingungen umlagern, ist keinesfalls überraschend — falls sie zuerst einmal intakt in das Konglomerat gelangt sind.

Hier setzt die einzige ernstzunehmende Kritik der Hydrothermalisten ein, derzuliebe sie allen noch so klaren Beobachtungsbefund glauben übersehen zu dürfen: Sie behaupten, daß heutzutage Pyrit und erst recht Uranpecherz so schnell bei Wassertransport verwittern, daß eine Ablagerung in irgendwie gearteten Seifen (und Sedimenten überhaupt) unmöglich ist. Der Augenschein in rezenten Sanden und Konglomeraten gibt ihnen mit seltenen Ausnahmen bei Uranpecherz, mit häufigeren bei Pyrit, recht. Da nun aber der Beobachtungsbefund sowohl Pyrit wie Uranpecherz als Gerölle ausweist, müssen wir diskutieren, welche Möglichkeiten trotzdem bestünden, diese Tatsache zu erklären. *Alle* diese Möglichkeiten zu erfassen wird uns vielleicht nie gelingen!

1.) Es könnte junges Uranpecherz, das noch wenig durch radioaktiven Zerfall bedingte Gitterstörungen hat, viel widerstandsfähiger sein als gealtertes. Das trifft eindeutig zu z. B. in den rezenten Seifen der Indusquellflüsse, die aus ganz jugendlichen Graniten genährt werden (Abb. 75). Es ist aber das für uns kaum *allein* der entscheidende Punkt, der ja *vier* verschiedene Lagerstätten, *eine* davon (W. W. R.) mit einer wahrscheinlich sehr langen Bildungsdauer, ganz ähnliche Verhältnisse zeigen und es natürlich recht unwahrscheinlich ist, daß in *allen* Fällen gerade junges Uranpecherz zur Verfügung stand.

2.) Die Umgebung war so, daß die Verwitterung auf ein Minimum zurückgedrängt war. Das wäre denkbar (aber unwahrscheinlich) bei einer Sedimentation in einem etwa durch Faulschlammbildung reduzierenden Gewässer. Eher könnte man denken an Ablagerungen vor einer abschmelzenden Eiskappe in einem Schelfmeer während einer Eiszeit oder besser mehreren aufeinanderfolgenden. Gewisse Anzeichen deuten am W. W. R. auf solche Verhältnisse, sind aber vorläufig für die ganze, sicher sehr langdauernde W. W. R.-Zeit schwer vorstellbar. Außerdem müßten am Bl. R., S. J., D. R. zu ganz anderen Zeiten und in anderen Gegenden ähnliche Bedingungen geherrscht haben — was naturgemäß sehr unwahrscheinlich, wenn auch nicht ausgeschlossen ist.

3.) Die Bildung der Konglomerate: 3000 Mill. Jahre Dom. Reef, 1800 Mill. Jahre W. W. R., 1100 Mill. Jahre Bl. R. wäre erfolgt, als noch prinzipiell andere Verwitterungsverhältnisse — also z. B. eine Methan- oder Kohlensäureatmosphäre — bestanden. Damit würden wir — so meint z. B. DAVIDSON — den Aktualismus im Sinne von LYELL verlassen. Verlassen wir ihn wirklich?! Die moderne Astronomie weiß schon recht lange, daß z. B. der Planet Venus eine Kohlensäureatmosphäre besitzt; es ist also nur eine Art von „Aktualismus höherer Ordnung", wenn wir diese Annahme machen. Die Geologen exemplifizieren und schließen schon längst aus den Meteoriten auf das Erdinnere — tun sie damit nicht genau dasselbe?[1]

Weitere Möglichkeiten, auch Kombinationen der Genannten untereinander oder mit eventuell neuen gibt es sicher, wir kennen sie aber nicht. Daß es sie aber geben kann, sollte uns abhalten uns zu energisch auf Fall 1, 2 oder 3 festzulegen. Wir können nur hoffen, daß neue Erkenntnisse — nicht by considerations, sondern by observations! — uns später weiterhelfen!

---

[1] Nicht unerwähnt bleiben soll in diesem Zusammenhang eine höchst bemerkenswerte Arbeit von P. ESKOLA [3], deren in einer Tabelle dargestellten Schlüsse geradezu verblüffend auf unsere Lagerstätten passen: D. R. würde mit 3000 Mill. nach ESKOLA gerade in die Zeit allererster Ausscheidung von Kohlenwasserstoffen durch anaerobische Organismen fallen, W. W. R. mit 1800 Mill. in den allerersten Beginn der Photosynthese, Bl. R., 1100 Mill. in den frühen Bereich freien Sauerstoffes. — Ich halte diese Arbeit, geschrieben aus der Erfahrung eines langen Lebens von einem genialen Gelehrten, der gleichzeitig einer der besten Kenner des Präkambriums ist, für überaus lesenswert — ohne mich aber imstande zu fühlen, sie in positivem oder negativem Sinn zu kritisieren.

Die Frage der „pseudohydrothermalen"[1]) authigenen Umlagerung bzw. Bildung von Gold, Bleiglanz, Magnetkies, Zinkblende, Pentlandit, Millerit, Skutterudit, den Fortwachsungen von Pyrit, macht in einem Medium, wo auch der Quarz des Bindemittels sich umlagerte, Sericit und Chloritoid sich bildeten, und Ilmenit wie in alpinen Klüften sich in Rutilnetze und -Aggregate umbildete, keinerlei Schwierigkeit. Sind doch einfach durch die Versenkung, aber auch — in Hinblick auf mehrere „Generationen" — durch Aufheizung durch Eruptivgänge Temperaturen von ~ 300° (lokal neben Dykes sicher auch mehr!) und sehr hohe Drucke selbstverständlich. Durch lokale Abröstung von Schwefel (Pyrit → Magnetkies) wird der Schwefelhaushalt (Umbildung z. B. des Eisens des Black Sand zu Pyrit) verständlich. Ein Teil des S mag auch der Tätigkeit von S-Bakterien entstammen, die in dunkleren tonigen Schichten existierten.

* * * * * * *

ZUSAMMENFASSUNG UND VERGLEICH DER BEOBACHTUNGEN IN DEN VIER LAGERSTÄTTEN

Schon die Besprechung der Einzelvorkommen wie auch der Abschnitt „Geologische Überlegungen" mit der Auseinandersetzung mit DAVIDSON brachte so oft Gelegenheit und Notwendigkeit sich mit Vergleichen zu befassen, daß ich mich hier kurz fassen kann und muß. Dabei ist des besseren Überblickes wegen weitgehend generalisiert:

Geologisch sind die Vorkommen ähnlich; D. R., Bl. R., wahrscheinlich S. J. haben weitgehend den Charakter unmittelbarer und demgemäß nach Stoff, Korngröße und Geröllform relativ wenig klassierter Transgressionskonglomerate. Der Witwatersrand gibt in seinen älteren Teilen noch z. T. dasselbe Bild, während er in den jüngeren Reefs mehr und mehr eine immer strenger werdende Auswahl aus dem Altmaterial darstellt, die durch lokal wechselnde, oft durch Teiltransgressionen oder Erosionen („Channel"bildungen z. B.) beeinflußte Neuzufuhren von außen, wie aus den zwischenliegenden mehr sandig-tonigen Schichten etwas modifiziert, lokal verwischt ist. Alle Einzelbeobachtungen lassen sich zwanglos in dieses Schema eingliedern.

Das *geologische Alter* ist verschieden. Wenn ich in dieser Arbeit ein Alter für D. R. 3000, W. W. R. 1800, Bl. R. 1100 Millionen Jahre angebe, so sind das Richtwerte, die zwar 1.) die Altersfolge relativ zueinander, 2.) die ungefähre Größenordnung sicher korrekt darstellen, aber durch die Tatsache, daß die Altersbestimmungen sich auf radioaktive Mineralien gründen, die dem Untergrund entstammen, bei genauerer Untersuchung sich ändern werden (nach meiner Überzeugung zu etwas niedrigeren Werten für die Konglomeratbildung selbst). — Auch die Beobachtungen, aus denen implicite auf geologisches Alter, absolutes wie relatives, geschlossen werden kann, also z. B. der hohe PbS-Gehalt im Uranpecherz des Dominion Reefs, wie die Spärlichkeit, beinahe schon das Fehlen kohliger Substanz und die weitaus stärkere Regionalmetamorphose in diesem stimmen ausgezeichnet damit. Altersbestimmungen der S. J.-Konglomerate sind mir nicht bekannt. Nach Art und Zustand der Gerölle möchte ich hier ein relativ geringes Alter, wie etwa Bl. R. annehmen. —

In allen vier Gebieten liegen *mehrere Konglomerathorizonte* vor. In drei Fällen, D. R., Bl. R., S. J. folgen diese aber so schnell, im vertikalen Abstand und sicher auch zeitlich aufeinander, daß man gesetzmäßige Unterschiede nicht wird erwarten können. Wenn, wie z. B. in dem erheblich höheren Pyritgehalt eines der Horizonte an der S. J., einmal Unterschiede vorliegen, sind sie sicher zufälliger Natur wie sie eben bei allen Transgressionsfolgen vorkommen. Im Witwatersrand ist das anders: Hier verschwinden zunehmend die empfindlicheren Komponenten — natürlich nur in großer Sicht gesehen — und die Gerölle, sowohl die großen aus Quarz, wie die kleinen aus Chromit, Uranpecherz, Pyrit werden durchschnittlich runder. Daß dabei immer etwas feinster Abrieb in die mehr sandigen Partien geraten wird, entspricht der Erwartung durchaus.

Der *Mineralbestand* ist für die Hauptkomponenten: Quarz, Pyrit, zu Rutil pseudomorpho-sierter Ilmenit, Uranpecherz sehr ähnlich. Die Nebenkomponenten geben ein Bild wechselnder Zuzugsgebiete und wechselnder Bedingungen. Wenn also am Dominion Reef für sie eine große Vielgestaltigkeit der Paragenese offensichtlich ist, so bedeutet das, daß unter Berück-sichtigung der relativ geringen Aufarbeitung — eben das Abtragungsgebiet wechselvolle Ge-steine und Lagerstätten enthalten haben muß. In dem außerordentlich großen Areal, das die W. W. R.-Sedimente bedecken, und der vielfachen Umlagerung im Großen wie im Kleinen ist eine vielgestaltige Paragenese natürlich nicht überraschend. Am Bl. R. war die Umlage-rung gering, der Mineralbestand ist demgemäß noch offenkundiger abhängig vom Untergrund, bzw. von dem — wie bei jedem schlecht klassierten Konglomerat — nicht weit entfernten Zuzugsgebiet als z. B. am D. R. Der starke lokale Wechsel von U:Th ist eine Folge davon, auch der recht eintönige Charakter in den Typen des Pyrits. Am eintönigsten ist bisher S. J., doch kann das an der ungünstigen Auswahl der untersuchten Proben liegen.

In allen Fällen ist der Grad der Metamorphose gering, am stärksten noch am Dominion Reef, wo ausgesprochene Kristallisationsschieferung vorkommt[1]. In allen Fällen haben natür-lich kleine Teilbewegungen stattgefunden, die sich in Abscherungen besonders an den Grenzen dickerer Konglomerathorizonte gegenüber den nachgiebigeren Liegendschichten äußern. Sie sind im allgemeinen aber sehr gering, oft nur durch Sericitfilme dargestellt.

Das *Uranpecherz* ist am Witwatersrand durchschnittlich besser gerundet als im D. R. oder Bl. R. Daß das aber nur eine Folge stärkerer und wiederholter Umlagerung ist, zeigt die Tat-sache, daß in den älteren Reefs die Rundung durchschnittlich geringer ist. Im Konglomerat neugebildetes Uranpecherz ist in allen Fällen sehr selten oder fehlt ganz. Wo es einmal beob-achtet ist, stellt es oft Fortwachsungen von Geröllen dar, also ganz analog den gewöhnlichen Fortwachsungen des Pyrits (übrigens auch des Quarzes), oder ist mit größter Wahrscheinlich-keit aus „Brannerit" zurückgebildet. Das Gerölluranpecherz ist, wo sichere Kennzeichen vorliegen, stets „Uraninit". Da in gewissen Primärlagerstätten kolloidal gefällte Pechblenden vorkommen, bei denen die kennzeichnenden Merkmale wenig deutlich sind, möchte ich das Vorkommen solcher unter den Geröllen immerhin nicht ausschließen — wahrscheinlich er-scheint es mir nicht!

Brannerit[2]), gebildet aus Rutil + Uranpecherz im Konglomerat selbst, ist im Bl. R., D. R., W. W. R. in ganz ähnlicher Weise nachgewiesen und tatsächlich überall einer der Haupt-träger des Urans. — Auf den Gegensatz des detritogenen Urans und des durch Umlagerung im Brannerit gebundenen sei nachdrücklich hingewiesen. Alle solche Lagerstätten werden bei einer radioaktiven Altersbestimmung besondere und bisher nie berücksichtigte Schwierig-keiten machen.

Alle Beobachtungen passen vorzüglich zu einer Deutung aller unserer Lagerstätten als fossile Seifen mit einer Umlagerung der Mineralien, die auch sonst unter „hydrothermalen Bedingungen" bekanntermaßen leicht wandern. Die Einwände von DAVIDSON gehen an den Tatsachen völlig vorbei.

Die Gedanken von BATEMAN, die meinen das für hydrothermale Zufuhr aus Gängen magma-tischer Abfolge charakteristische Isotopenverhältnis des Schwefels zu finden, berücksichtigen die Tatsache nicht, daß der Pyrit Geröllnatur hat und mehr oder weniger, manchmal sicher überwiegend, aus aufgearbeiteten Hydrothermallagerstätten stammt. — Sie stehen also *nicht* im Gegensatz zu den von mir in der vorausgehenden wie vorliegenden Arbeit mitgeteilten

---

[1]) Auch hier könnte das, im Hinblick auf die geringe Ausdehnung des aufgeschlossenen Areals, nur eine lokale Erscheinung sein.

[2]) Im Sinne der konventionellen Definition! Daß sich „Brannerit" vielleicht einmal als ein Mischungsglied der Columbitfamilie im weitesten Sinn erweisen wird, ähnlich wie Euxenit u. v. a., stehe hier nicht zur Dis-kussion.

Beobachtungen. Sie haben aber damit in den beiden zur Diskussion stehenden Ansichten keine Beweiskraft.

Die Sedimentation von Uranpecherz in schweren Sanden ist nicht beispiellos. So findet es sich rezent reichlich im Quellgebiet des Indus. Hier hat auch Pyrit in großem Unfang den Transport bei wechselnd guter Abrollung überstanden, ebenso z. B. ged. Wismut und Arsenkies. Trotzdem scheinen dem Verf. für die Erklärung von gleich vier so großen Lagerstätten von verschiedenem Alter und in verschiedenen Weltgegenden besondere Verhältnisse nötig zu sein. Sie in einer Kohlensäureatmosphäre oder wenigstens einer kohlensäurereichen Atmosphäre zu suchen, scheint plausibel im Hinblick auf die Erkenntnisse der Astronomen. Daß das nur eine Arbeitshypothese ist, ist mehrfach nachdrücklichst betont.

LITERATURVERZEICHNIS

Witwatersrand, Blind River, Dominion Reef, Serra de Jacobina

[1] BATEMAN, A. M., Recent uranium developments in Ontario. Econ. Geol. 50, 361—372, 1955.

[2] DAVIDSON, CH. F., On the occurence of uranium in ancient conglomerates. Econ.Geol. 52, 668—693, 1957.

[3] ESKOLA, P., On the geological eras and the factors controlling organic evolution. Verhandl. Koninklijk Nederlandsch Geol. Mijnbouwkund. Genotschap. Deel XVI, 85, 1956.

[4] FRONDEL, CL., Mineralogy of Thorium. U. S. Geol. Survey, Prof. Pap. 300, 567—579, 1956.

[5] GEIJER, P., Pre-Cambrian atmosphere: evidence from the Pre-Cambrian of Sweden. Geochimica and Cosmochimica Acta 10, 304—310, 1956.

[6] HOLMES, S. W., Geology and mineralogy of the Pronto uranium deposit, District of Algoma, Ontario, Canada. Econ. Geol. 51, 116, 1956 (nur kurzes Referat).

[7] DE KOCK, W. P., The Carbon Leader of the Far West Rand with special reference to the proportion of the gold content of this horizon recovered in drilling. Transact. Geol. Soc. S. Afr. 51, 213—238, 1948.

[8] LIEBENBERG, W. R., The occurrence and origin of gold and radioactive minerals in the Witwatersrand System, the Dominion Reef, the Ventersdorp Contact Reef and the Black Reef. Trans. Geol. Soc. S. Africa, 1956.

[9] McDOWELL, J. P., Sedimentary petrology of the Mississagi Quartzite in the Blind River Area, Ontario. Progr. Amer. Meeting Geol. Soc. Amer. 94, 1957.

[10] NUFFIELD, E. W., Brannerite from Ontario, Canada. Amer. Min. 39, 520—522, 1954.

[11] PABST, A., Brannerite from California. Amer. Min. 39, 109—117, 1954.

[12a] RAMDOHR, P., Die Erzmineralien und ihre Verwachsungen. Akademieverlag Berlin. 1955.

[12b] —, Neue Beobachtungen an Erzen des Witwatersrandes in Südafrika und ihre genetische Bedeutung. Abhandl. Akad. Berlin, Kl. Chemie No. 5, 1955.

[12c] —, Neue Beobachtungen über radioaktive Höfe und radioaktive Sprengungen. Abhandl. Akad. Berlin. Kl. Chemie No. 2, 17 S., 1957.

[12d] —, Die Pronto-Reaktion,. Neues Jahrbuch, Mineral. Monatsh. 1957, 217—222.

[12e] —, Diskussionsbemerkungen zu DAVIDSON. Econ. Geol. 53, 1958. (Im Druck!)

[12f] —, Weitere Untersuchungen über radioaktive Höfe und andere radioaktive Einwirkungen auf natürliche Mineralien. Abhandl. Akad. Berlin, Kl. Chemie 1958.

[13a] ROSCOE, S. M., The Blind River, Ontario, Uranium Area. Institute Lake Superior Geology 40—48, Houghton 1957.

[13b] —, Geology and uranium deposits, Quirke Lake-Elliot Lake-Blind River Area, Ontario. Geol. Survey of Canada Paper 56/7, 1—2, 1957.

[14] STIEFF, L. R., STERN, T. W., C. M. CIALELLA u. J. J. WARR, Preliminary age determinations of some uranium ores from the Blind River area, Algoma District, Ontario, Canada. Bull. Geol. Soc. Amer. 67, 1736—1737, 1956.

[15] TRAILL, R. J., Preliminary account of the mineralogy of radioactive conglomerates in the Blind River region, Ontario, Can. Min. Journ. 75, 63—68, 1954.

[16] WHITE, M. G., Uranium in the Serra de Jacobina, State of Bahia, Brazil. Peaceful Uses of Atomic Energy (U. N. O.) Vol. 6 p. 140—142, 1956.

[17] ANONYM, Blind River, Algoma District, Ontario. Canad. Geol. Survey Map (1925 und 1955), Publ. No.1970.

[18] —, („Prepared by the Commitee") Mining, Metallurgy and Geology in the Algoma uranium area. 6. Commonwealth Mining & Metallurgical Congress Toronto 1957.

3*

21

Reprinted from *Econ. Geol.*, **59**, 1026, 1047–1060 (1964)

Origin of Precambrian Iron Formations

HENRY LEPP and SAMUEL S. GOLDICH

ABSTRACT

A statistical study of the chemical composition of the Precambrian iron formations of the Canadian Shield affords a new approach to the origin of these unusual formations. The average total iron content of 2,200 samples from the literature and from unpublished mining company analyses is 26.7 percent Fe. The average Fe content for 16 iron formations in the United States and Canada ranges from 24.5 to 34.1 percent. Low contents of Al_2O_3, TiO_2, P_2O_5, and CaO characterize the Precambrian iron formations compared to the relatively large amounts of these constituents in the post-Precambrian iron-bearing sediments. The chemical data emphasize that whereas iron, manganese, and silica were transported and deposited together in the cherty iron formations of the Precambrian, these same elements were chemically differentiated in younger geological time in large but separate deposits of iron and silica.

Isotopic age determinations indicate that cherty iron formations were deposited during a long interval of geologic time from approximately 1,700 to 3,000 million years ago. A model is proposed to explain the origin of the iron formations of the Lake Superior type based on the absence or marked deficiency of free oxygen in the atmosphere prior to the Late Precambrian. Lateritic weathering under these conditions permitted the transport of iron and manganese together with silica. The weathered mantle effectively retained aluminum, titanium, phosphorus, and colloidal clay.

Graphitic material of biogenic origin is closely associated with the Precambrian iron formations. Although it is uncertain whether iron was precipitated directly through biologic processes, the removal of CO_2 and the liberation of oxygen to the sea water through photosynthesis of primitive plants undoubtedly influenced the energy relationships among the iron minerals. As a result of the variable conditions the iron formations commonly are characterized by nonequilibrium mineral assemblages.

In Late Precambrian time a critical level of free oxygen in the atmosphere was attained permitting a marked acceleration in plant growth and in accretion of oxygen. This stage in the development of an oxygenated atmosphere was reached at least 1,200 million years ago and effectively curtailed the development of cherty iron formations of the Lake Superior type.

* * * * * * *

Editor's Note: The following sections have been omitted from this reprint for the sake of brevity: "Introduction," "Chemistry of Iron Formations," "Ages of Precambrian Iron Formations," and "Acknowledgments."

ORIGIN OF PRECAMBRIAN IRON FORMATIONS

General Statement

The statistical data presented emphasize chemical properties and differences between Precambrian iron formations of the Canadian Shield and post-Precambrian iron formations that must be considered in any hypothesis of origin. Although similar detailed chemical data for other areas are not available, descriptions and analyses given in the literature indicate that the iron formations of the various Precambrian shields closely resemble those of North America. Some notable features of the chemical composition of iron formations may be summarized.

(1) The Precambrian iron formations of the Canadian Shield are characterized by a surprisingly uniform tenor of iron and small contents of Al_2O_3, TiO_2, and P_2O_5.

(2) Within a district the iron contents of the different mineralogic and lithologic varieties or facies are also surprisingly uniform. Apparent differences in the relative weight percentages of iron can be accounted for in large part by the formula differences of the iron minerals. In terms of the major chemical constituents there are no great chemical differences between carbonate, silicate, and oxide facies, or between thin- and thick-bedded slaty or cherty members.

(3) Silica and iron are intimately and genetically related in the Lake Superior type of iron formation. In later geologic time silica and iron were geochemically separated in sediments.

(4) Isotopic age determinations indicate that banded cherty iron formations were deposited in many parts of the world over a long period in Early and Middle Precambrian time.

Precambrian Atmosphere

Early investigators suggested that the origin of the Precambrian iron formations was related to the unusual composition of the Precambrian atmosphere. Some held that carbon dioxide was more abundant than in the present atmosphere and was a factor in accelerating weathering processes; others suggested that oxygen was deficient or lacking. Macgregor (59) stated that the stumbling block in the theories advanced to explain the origin of the Precambrian iron formation was in the assumption that the atmosphere of the time was oxygenated.

It is now generally held that the primitive atmosphere probably was strongly reducing. The probable course of the evolution of this primitive atmosphere to the present oxidizing atmosphere fortunately is one of interest to a number of sciences, and reviews by Kuiper (56), Rubey (84), Urey (97), and Holland (38) contain references to the extensive literature.

Holland (38) considers three stages in the evolution of the present-day atmosphere. In the earliest or pre-core stage metallic iron probably was present in the upper part of the mantle. Volcanic gases evolved during this period contained large amounts of hydrogen, and the atmosphere, as a result, was in a reduced state with carbon in the form of methane and nitrogen in ammonia. Holland suggests that the first stage was of relatively short duration, probably 500 million years. This period of atmospheric development is well beyond the presently known geologic record.

The second stage of the atmosphere is estimated by Holland to have lasted until approximately 1,800 million years ago, an interval corresponding to the Early and Middle Precambrian eras as used in this paper. Basaltic and granitic rocks with isotopic ages between 3,500 and 2,500 m.y. closely resemble modern igneous rocks chemically. Volcanic gases in equilibrium with Early Precambrian magmas can be assumed to have been similar to the gases evolved from modern lavas. As Holland (38, p. 466) points out, carbonates and hematite in the ancient sedimentary rocks also indicate a relatively low p_{H_2} during this second stage. Methane, ammonia, and hydrogen, important constituents of the primitive atmosphere, were largely eliminated, and the atmosphere was composed principally of N_2 with minor amounts of CO_2 and H_2O. During the second stage the production of oxygen through processes such as photodissociation of H_2O probably was augmented through photosynthesis as will be considered in a later section. We agree with Holland, however, that during Early and Middle Precambrian time oxygen was present in the atmosphere only in trace and nonequilibrium amounts.

The third and present stage of atmospheric development was initiated in Late Precambrian time. The sedimentary mantle formed since then is oxidized. Secondary ferric oxides and sulfates are widely distributed in rocks ranging in age from Recent to Keweenawan. Red beds are common in Late Precambrian sequences in many parts of the world, such as the Keweenawan of the Lake Superior region, the Jotnian of the Baltic region, and the Torridonian of Scotland (9). The red-colored sandstone and shale of the Late Precambrian such as the Fond du Lac Formation (24, p. 93) of the Lake Superior region, indicate an oxygenated atmosphere. More ancient are the red-colored sandstone and shale intercalated with the flows of the North Shore Volcanic Group (24, p. 81). The flows were intruded by the gabbro at Duluth and by diabase at Beaver Bay, Minnesota which have been dated (24) at approximately 1,100 m.y.

Red sericitic argillite (pipestone) occurs as layers in the Sioux Quartzite of southwestern Minnesota and adjacent parts of South Dakota and Iowa. A chemical analysis of the pipestone quoted by Berg (6) gave 3.06 percent Fe_2O_3 and no FeO. Diaspore is common in the quartzite as well as in the argillite of the Sioux Quartzite, and Berg (5) attributed the diaspore to differential leaching. It is unlikely, however, that leaching could produce diaspore after the formation was exposed by erosion. A more reasonable explanation is that the diaspore was formed during folding and was derived from original gibbsite or boehmite. A K-Ar age of 1,200 m.y. (21) probably is the approximate time of folding of the Sioux Quartzite. The time of deposition of the oxidized materials cannot be limited closer than 1,700 to 1,200 m.y., but the blood-red argillite and red-colored quartzite indicate that an oxidizing atmosphere had been established before 1,200 million years ago.

Lateritic Weathering

The lateritic weathering model (Fig. 5) is designed to explain the derivation and accumulation of the Precambrian iron formations under an atmosphere lacking free oxygen. Under these conditions the Eh-sensitive elements iron and manganese accompanied silica.

Chemical Differentiation.—Aluminum and titanium were largely retained in the weathered mantle as oxides and hydrous oxides. Phosphorus was fixed in insoluble compounds with ferric iron and aluminum, and lateritic soils commonly contain relatively large amounts of P_2O_5 (22, p. 70; 23, p. 75; 40, p. 64). Ferric iron and appreciable amounts of silica were also retained, the silica undoubtedly combined in clay minerals which would be expected to form in the less well-drained parts of the profile and below the water table.

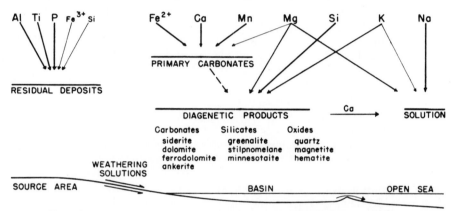

Fig. 5. Lateritic weathering model showing chemical differentiation to produce Precambrian iron formations.

Elements liberated through the lateritic weathering process include Fe^2, Mn^2, Mg, Ca, Na, K, and large amounts of Si. These were transported in true solution; Si as H_4SiO_4, and the others mainly as bicarbonates. Our model thus explains the chemical features of iron formations because under these conditions Fe^2 and Mn^2 behave very much like calcium and magnesium. Although differential adsorption, particularly of potassium, and precipitation of the elements liberated in the source region may have occurred locally, the bulk of the soluble cations probably reached the sea.

Stability of Source Region.—A second important feature of the lateritic weathering model is that it provides for stability in the source region to minimize mechanical erosion. Lateritic soils are characterized by their remarkable porosity and permeability (23, p. 74). Rainfall is quickly absorbed and drained off through the porous laterite, and as a result sheet wash and erosion are limited. We prefer the assumption that the region of weathering was one of moderate rather than of low relief as has been favored by previous writers. Relief is required to provide drainage and to permit rapid removal of silica and soluble elements from the weathered source rocks. Low relief and poor drainage favor kaolinization rather than lateritization. Not only is a large amount of silica thus tied up in silicates, but the resulting clay minerals from an impervious surface, and heavy vegetation would be required to prevent erosion.

Gruner (31) suggested the presence of abundant land vegetation of a low form which aided in the rapid decay of the rocks and provided the organic matter to stabilize the ferric oxide hydrosols and permit transportation of iron. Anaerobic bacteria may have lived in the weathering mantle, but a surface vegetation adequate to prevent erosion under heavy rainfall is most improbable. The problem of the selective transport of the ferric oxide hydrosols has been generally avoided in earlier discussions. Under the conditions of lateritic weathering postulated here, colloidal hydrosols were of minor importance. The permeability of the weathering profile inhibited erosion, and colloidal oxides and clay minerals were filtered out of the groundwaters moving through the residual mantle.

Deposition and Mineralogic Development

Basins.—Woolnough (107, 108) suggested that chemical sediments, including the Precambrian iron formations, were deposited in barred basins, and James (46) also indicated deposition in restricted basins. Although we have followed Woolnough and James in including a barrier in Figure 5, we cannot fully defend this position. White (103) and Cullen (13) whose studies of iron formations are primarily stratigraphic did not feel the need for a circulation barrier.

Facies Approach.—James (46, p. 240) argued that the iron minerals are "equilibrium products" and that differences in primary minerals reflect different depositional environments. He related mineralogical facies primarily to depth of water and to degree of oxygenation on the assumption that the Precambrian atmosphere did not differ appreciably from that of today. James based his explanation in part on the work of Krumbein and Garrels (55) who calculated the stability fields of some of the sedimentary iron minerals in terms of pH and Eh. Huber (42) developed pH-Eh stability diagrams for hematite, siderite, pyrite, and iron sulfide. Garrels (18) has presented a series of new diagrams. These studies are valuable, and the facies concept has greatly advanced our understanding of the iron formations. Our studies, however, suggest that the facies approach to date has fostered an oversimplification of the problem of iron formation origin.

Chemical Approach.—The chemical data afford an opportunity to examine the mode of deposition of iron formations and the development of mineralogical facies that is not bound by assumptions of primary equilibrium mineral assemblages. The statistical study shows that the Precambrian iron formations are remarkably uniform in iron content considering that they are products of natural processes. This we explain as being the result of a fundamental control in the sources of materials:

Iron and calcium are present in the lithosphere in approximately mole proportions of 1:1. These elements in the weathering cycle (Fig. 5), which we postulate was controlled by the absence or marked deficiency of free oxygen in the atmosphere, would be transported in solution to the depositional basins in approximately their proportions in the source rocks. We suggest that the principal minerals deposited were siderite and calcite. Possibly some primary dolomite, ferrodolomite, and ankerite also may have been formed.

At the time of deposition, or following closely, a number of reactions between the primary carbonates and the elements in solution were possible. Dolomite, ferrodolomite, and ankerite were formed by replacement as was also chert. The silicates, greenalite, stilpnomelane, and possibly also minnesotaite, were formed by diagenetic reactions. Small amounts of clay particles or volcanic glass that found their way into the deposition basins probably acted as nuclei for crystallization of the iron silicates. Magnetite and hematite also were formed, for the most part, during diagenesis. As a result of these various interactions carbon dioxide was liberated, and calcium was returned to solution. Calcium and sodium, together with magnesium and potassium, not fixed in a mineral phase, were carried to the open sea by tides and currents (Fig. 5).

Chert.—One of the main arguments employed by James (46) and later by Huber (43) for a primary origin of chert is that the amount of chert varies with the facies. James (46, p. 273) states ". . . for example, the amount of chert in. carbonate iron-formation is 30 to 35 percent, whereas that in oxide iron-formation is about 40 percent." We have shown that whereas the tenor of iron is essentially constant, the SiO_2 and CO_2 vary inversely. If we start with an original mixture of 1 $CaCO_3$:1 $FeCO_3$ and replace $CaCO_3$ by SiO_2 on a volume basis, the resulting combination of chert and siderite would contain approximately 45 percent SiO_2. Averages for carbonate iron formation from the Marquette district and the Gogebic district are 42.4 and 39.3 percent SiO_2, respectively (Table 8). If siderite were altered to hematite with no other change except for the replacement of $CaCO_3$ by SiO_2, the resulting rock would contain about 55 percent SiO_2. Only the composite analysis representing the Main Mesabi district (Table 2, No. 2) with 51.1 percent SiO_2 approaches the calculated value of 55 percent. The computed values for SiO_2 are high because in nature the replacement is not so simple; unreplaced carbonate and other mineral phases normally present in the formation were not considered. Of significance, however, is that the difference in SiO_2 contents for carbonate and hematite facies, calculated on the assumption that the chert formed by replacement, is of the order of magnitude exhibited by the rocks.

Ferric-Ferrous Iron Relationships.—In the simplified model the original mud of 1 $CaCO_3$:1 $FeCO_3$ would contain 26 percent Fe. If the $CaCO_3$ were replaced by SiO_2 on a volume basis, the iron content would not be appreciably changed. The mean iron content for carbonate iron formations is 26 percent (Table 4), and the total Fe in composites for four districts ranges from 24.7 to 25.3 percent (Table 8). If the siderite were altered to hematite with no change other than the replacement of $CaCO_3$ by SiO_2, the resulting mixture of hematite and chert would contain approximately 31 percent Fe. The calculation for magnetite in place of hematite gives 32 percent Fe. Our data

(Tables 4, 6) show that the average iron content of magnetite iron formations actually is less than that for hematite iron formations; however, an average of approximately 31 is a good value for the oxide iron formations. The excellent agreement between observed iron contents of carbonate and oxide iron formations with values for these facies predicted from our model is in part fortuitous; nevertheless, the agreement supports our contention that there are no fundamental differences in iron content of the principal iron facies. The oxide iron formations contain no more iron than the carbonate facies when the formulae differences have been considered. If the environment controlled the deposition of the iron-bearing minerals, the broad range of Eh suggested by James (46) should have resulted in great variations in iron content of the facies.

Mineral Relationships.—The ferric-ferrous relationships, and, hence the minerals of iron formations, are obviously related to Eh and pH, and Garrels' (18) diagrams are valuable for an understanding of these relationships. Garrels has pointed out many of the difficulties in applying equilibrium principles to natural occurrences in which reaction rates may be exceedingly low at surface temperatures. Metastable phases commonly are developed, and some may persist for long periods of geologic time.

Huber (42) has made a significant contribution in emphasizing the lack of equilibrium among iron-bearing minerals in natural occurrences. Assemblages indicating disequilibrium have been noted by many writers. Gruner (33, p. 201) has called attention to the commonplace occurrence of mixtures of silicates, oxides, and carbonates in iron formations. Tyler and Barghoorn (94, p. 428) in describing chlorite appendages on magnetite grains in granular ferruginous chert from the Ironwood Iron Formation of Michigan, as incidental information, state that "the rock is composed of chert, siderite, magnetite, hematite, and chlorite." The lack of equilibrium among the minerals of iron formations is commonly so pronounced that it suggests a number of contributing causes. In fact, there is a marked lack of agreement among students of iron formations as to which minerals are primary, diagenetic, or metamorphic. The assemblages, therefore, could represent the sum total of conditions that change appreciably between the time of initial deposition and of final lithification.

Siderite is characteristically deposited in a reducing environment. Under certain conditions an increase in pH would result in magnetite becoming the stable mineral phase. Similarly hematite can be formed in the absence of free oxygen. These reactions can be represented in simple chemical equations for which Garrels (18) has calculated the energy relationships. A number of considerations suggest that in addition to changes in pH changes in Eh effected through biological processes were important in the formation of the oxides, magnetite and hematite. The most logical explanation for the development of the oxides seems to us to demand the presence of free oxygen in the environment. Sources such as the formation of H_2O_2 by ultraviolet irradiation of water can be considered, but a ready source is oxygen formed through photosynthesis.

The red jasper and hematite of the iron formations, however, are not satisfactory evidence for the presence of free oxygen in the atmosphere, because the aqueous environment in which the ferric oxide minerals were formed was not in equilibrium with the atmosphere. The strongly reducing euxinic environment of certain present-day lacustrine and marine basins serves to illustrate this point. Tyler and Barghoorn (94, p. 430) observe that the pyrite and organic matter in the basal Gunflint Iron Formation of Ontario and in the Biwabik Iron Formation of Minnesota indicate "that a reducing or euxinic environment existed on a regional scale at the time these sediments were deposited. The red algal jaspers, which have a restricted occurrence, are presumably the product of local oxidizing environments that existed as microenvironments within the euxinic environment."

Tyler and Barghoorn (94, p. 426) also note the presence of "siderite and ankerite grains and masses as well as siliceous oolites." Some of the oolites have jasper cores. Pettijohn (79, p. 685) states that ferric oolites indicate deposition in turbulent aerated waters and could not have been developed in Precambrian iron formations if the atmosphere was reducing. Pettijohn thus finds it more difficult to conceive of local oxidizing environments under an atmosphere deficient in oxygen than of reducing environments under an oxygenated environment. In fairness to Tyler and Barghoorn it should be said that they do not specifically advocate an oxygen-deficient atmosphere at the time of deposition of the Gunflint and Biwabik Iron Formations.

Biologic Factors.—Graphitic material in argillites and slates, together with structures interpreted as being of biological origin, have long been considered by geologists as evidences of life in the Precambrian. Numerous occurrences of graphitic argillites closely associated with iron formations have been described. Anthracite coal has been reported from the Michigamme Slate of northern Michigan (95)...Most remarkable has been the discovery of well-preserved organic forms in chert of the Gunflint Iron Formation near Port Arthur, Ontario, reported by Tyler and Baghoorn (93). Additional fossiliferous material from this area has been described by Moorhouse and Beales (69). Older rocks associated with Early Precambrian iron formations also contain structures and carbon of possible biologic origin, although the interpretations of these occurrences have been questioned from time to time. Dark-colored graphitic slates are abundant in the Knife Lake Group of Minnesota and in the correlative Seine Series of Ontario, and massive graphite was found in the iron mine at Soudan, Minnesota. These rocks were involved in folding 2,500 million or more years ago. Similarly of ancient age, more than 2,500 m.y., are the structures of the Bulawayan Limestone in Southern Rhodesia described as algal by Macgregor (60).

Hoering (36) isolated organic compounds from a number of Precambrian rocks including the coal from the Michigamme Formation and the graphite from the Soudan mine. Preliminary results indicate that the Precambrian carbon was originally kerogen and studies of the isotopic composition of the carbon in coexisting carbonate and carbon show a relative enrichment of C^{13} in carbonates and of C^{12} in reduced carbons. From these results Hoering (37, p. 191) concludes that photosynthesis and biological activity are indicated in the oldest rocks of the Precambrian.

Oparin (74), a long time researcher on the origin of life, postulated that life originated under a strongly reducing atmosphere. This concept in the light of available geologic evidence leads to the conclusion that life must have originated on earth long before the oldest geologic record and well before 3,500 million years ago. A complex history involving an abiogenic origin of organic matter followed by the development of anaerobic heterotrophic bacteria prior to the appearance of photosynthesis is outlined by Oparin (74). It is generally suggested that early photosynthesis utilized H_2 or H_2S rather than H_2O. Hoering's work does not indicate the probable path of the Precambrian photosynthesis.

Harder (35) first emphasized the probable role of bacteria in the development of the Precambrian iron formations. Gruner (31) and also others have suggested that "iron bacteria" played a significant part. A somewhat extreme view of the biologic factor in the origin of the iron formations has been developed by Moorhouse and Beales (69). They regard the "Gunflint rocks as an immense organically controlled accumulation." In their hypothesis iron-secreting organisms reached a major development in the Precambrian, and this unique development is said to have been the controlling factor in the origin of the Precambrian iron formations. Moorhouse and Beales (69, p. 109) rule out "any drastic differences in composition of atmosphere and ocean of Animikie time and of the present." The hypothesis of the accumulation of the Precambrian iron formation being controlled by the activity of organisms, however, fails in a number of respects. The most serious of the shortcomings is the assumed "abundant supply of iron and silica" (69, p. 108). Kaplan, Emery, and Rittenberg (48) have shown that sea water is an inadequate source to explain the development of authigenic pyrite in recent marine sediments of southern California. They conclude that the iron must be derived from detrital sediments.

It seems reasonable to assume that low forms of life such as bacteria and algae may have been important in the precipitation of iron. Whether or not the organisms made direct use of the iron in their life cycle is not demonstrable with the presently available information. The abstraction of CO_2 from sea water and the control of pH and Eh by the primitive organisms certainly would have been effective. One could speculate at considerable length on probable reactions and mechanisms. It may be noted that the distribution of oxygen generated through photosynthesis in sea water might be subject to many of the controls of present-day marine environments in terms of depth and circulation of water. A few of the factors that might be involved are depth of sunlight penetration, rates of sedimentation, and availability of nutrients for plant life. It is useful to consider the biologic factors as part of the total physical environment in which the iron formations were developed. Biologic controls of pH and Eh can be used to explain the mineralogic facies of the iron formations, but the primary mineral equilibrium assemblages are not necessarily indicated. As Harder (35) pointed out many years ago, the formation of iron silicates cannot be explained through the agency of iron-precipitating bacteria, and neither can the silicates be considered a direct precipitation from sea water.

Accepting the time of origin of life as being well back in the Early Precambrian, the rate of evolution of life was exceedingly slow during an interval of two billion or more years. This slow development can be correlated with a slow rate of accumulation of free oxygen in the atmosphere. During this long interval, organisms that developed the faculty of utilizing H_2O as a source of hydrogen, liberating oxygen, had a precarious existence. The production of oxygen through photosynthesis was gradually augmented to a point where a positive balance was established in which the rate of production exceeded the rate of destruction of oxygen in the atmosphere. This condition was well established at least 1,200 m.y. ago as is indicated by the pipestone of the Sioux Quartzite and other clastic sediments of the Late Precambrian. At that time a level of oxygen production had been achieved that permitted a rapid development of plant life. An exceedingly rapid development of invertebrate and vertebrate fauna followed.

CONCLUDING REMARKS

The complexity of the natural laboratory requires no special emphasis, but it is well to call attention to the variable conditions that undoubtedly existed in the different areas and throughout the period of deposition of the Precambrian iron formations. This variability is indicated in the ranges for chemical constituents found in the statistical study. It is important to note that for a period of over a billion years conditions during the Precambrian were favorable for deposition of the chemical iron formations. We may dismiss hypotheses of origin that postulate unusual or freakish geologic conditions that prevailed on the earth once only or for a limited time.

We are of course, advocating a secular change in sedimentation effected by a change in the oxygen content of the atmosphere. Pettijohn (79, p. 689) has given an excellent critique of apparent secular changes in the nature and character of the sedimentary rocks. His analysis emphasizes the need for valid sampling, for a valid chronology, and for the distinction between postdepositional alteration and real differences in the environment of deposition. Many of the Middle Precambrian iron formations have undergone only low-grade metamorphism; hence, their chemical composition probably has not been altered appreciably except for the postdepositional replacements and alterations that are commonly attributed to diagenesis. The higher grade of metamorphism and replacement of the iron formation by crystalline specularite exhibited by Early Precambrian iron formations, however, suggests possible oxidation of iron by reaction with H_2O at high temperature.

Many Precambrian iron formations are inadequately dated and sampled. Geologic relationships indicate an Early Precambrian age for the iron formation in the Atlantic district of Wyoming. Four analyses (89, p. 18) give 43, 42, 39, and 27 percent of Fe. The disparity between these values points up the difficult problem of sampling. A recent analysis given by Bayley (4, p. C-10) quotes 56 percent of SiO_2 and 29 percent of Fe, values that fit well within the range for iron formations of the Canadian Shield. We feel that the data presented in our statistical study are representative of Precambrian iron formations. If there are important differences between the

Precambrian iron formations of the different continents, their nature and extent must await further studies.

We do not argue the point that some Late Precambrian and post-Precambrian iron formations resemble the Lake Superior type. Possibly the cherty iron formations such as the Morro do Urucum of Brazil represent a transitional type. The reverse relationship for post-Precambrian minette-type of iron formation should also be mentioned. Precambrian iron formations that obviously contain detrital materials and resemble chemically the post-Precambrian iron formations should be expected. Such deposits occur in South Africa and have been described by Wagner (102). These deposits have been excluded from the present discussion.

The origin of the iron formations is a very old problem, and the informed reader is aware that the ideas incorporated in this paper have been anticipated in some form or other by previous writers. It is probably inevitable that our emphasis and references have slighted some valuable contributions to the geology of the iron formations. Our treatment of the evolution of the atmosphere and of the origin of life is of necessity brief. Rutten (85) gives a more elaborate presentation of "the geological aspects of the origin of life on earth." His division of exogenic geologic processes into actualistic and pre-actualistic suggests a rather abrupt transition from an anoxygenic to the present oxygenic atmosphere. The data considered as a whole are quite against an abrupt change from a primitive reducing atmosphere to the present-day oxidizing atmosphere. Biologic factors in the deposition and development of iron formations have received little attention since Harder's contribution some 40 years ago. It is encouraging to note increasing interest in the field that may be called geomicrobiology.

It may be noted that our model for the origin of the Precambrian cherty iron formations is not opposed to the concept of uniformitarianism. The products of lateritic weathering are found the world over in rocks ranging from Recent to Late Precambrian, if we include the Sioux Quartzite. Conditions of source rocks, climate, topography, and related factors, however, must be favorable. Similarly iron formations of the Lake Superior type were developed the world over under locally favorable conditions. These deposits range from Late Precambrian (~ 1.7 b.y.) to the oldest known sedimentary sequences (~ 3 b.y.) in age.

Some aspects of the origin of the Precambrian iron formations present unusually difficult problems because of the antiquity of the deposits and the complicated geologic history subsequent to deposition. The major problem of origin, however, can be seen logically only in proper perspective against a background of knowledge of the Precambrian. The problems of the evolution of the continents and of the ocean basins, of the origin of the atmosphere and the hydrosphere, and of life on the earth, are now attracting wider attention. Considerable progress toward a better understanding of the Precambrian is a reasonable expectation.

*   *   *   *   *   *   *

with many iron company geologists. We are indebted to these gentlemen and to the companies who have made available to us their files of chemical and mineralogic data. J. V. N. Dorr II, R. M. Grogan, and R. W. Marsden read and criticized the manuscript at several stages in its preparation. Their friendly criticism and encouragement are appreciated. The writers, however, are solely responsible for shortcomings of the data and of the presentation. W. J. Croke, J. R. Haigh, and L. E. Warren assisted in the compilation under a National Science Foundation program for undergraduate research participation at the University of Minnesota, Duluth. The financial support of the Graduate School of the University of Minnesota and of the National Science Foundation is gratefully acknowledged.

UNIVERSITY OF MINNESOTA, DULUTH,
  DULUTH, MINNESOTA,
PRESENT ADDRESS : MACALESTER COLLEGE,
  SAINT PAUL, MINNESOTA.
  AND
U. S. GEOLOGICAL SURVEY,
  WASHINGTON, D. C.
  March 27, 1964

REFERENCES

1. Aldrich, H. R., 1929, The geology of the Gogebic iron range of Wisconsin : Wisconsin Geol. Nat. Hist. Survey, Econ. Ser., Bull. 71, 279 p.
2. Alevandrov, E. A., 1955, Contributions to studies of origin of Precambrian banded iron ores : ECON. GEOL., v. 50, p. 459–468.
3. Alling, H. L., 1947, Diagenesis of the Clinton hematite ores of New York : Geol. Soc. America Bull., v. 58, p. 991–1017.
4. Bayley, R. W., 1963, A preliminary report on the Precambrian iron deposits near Atlantic City, Wyoming : U. S. Geol. Survey Bull. 1142-C, 23 p.
5. Berg, E. L., 1937, An occurrence of diaspore in quartzite : Am. Mineralogist, v. 22, p. 997–999.
6. Berg, E. L., 1938, Notes on catlinite and the Sioux quartzite : Am. Mineralogist, v. 23, p. 258–268.
7. Bien, G. S., Contois, D. E., and Thomas, W. H., 1958, The removal of soluble silica from fresh water entering the sea : Geochim et Cosmochim. Acta, v. 14, p. 35–54.
8. Blondel, F., 1955, Iron deposits of Europe, Africa, and the Union of Soviet Socialistic Republics, *in* Survey of World Iron Ore Resources : United Nations, Dept. Economic and Social Affairs, New York, p. 224–264.
9. Brinkmann, R., 1960, Geologic evolution of Europe : Hafner Publ. Co., New York, 161 p.
10. Burwash, R. A., Baadsgaard, H., and Peterman, Z. E., 1962, Precambrian K-Ar· dates from the Western Canada sedimentary basin : Jour. Geophys. Research, v. 67, p. 1617–1625.
11. Cahen, L., 1961, Review of geochronological knowledge in middle and northern Africa : New York Acad. Science Annals, v. 91, p. 535–566.
12. Craig, J. J., 1961, The use of iron ore pellets : Skillings' Mining Review, v. 50, p. 1, 4, 5.
13. Cullen, D. J., 1963, Tectonic implications of banded ironstone formations : Jour. Sed. Petrology, v. 33, p. 387–392.
14. Dixon, W. J., and Massey, F. J., Jr., 1951, Introduction to Statistical Analysis : McGraw-Hill Co., New York, 370 p.
15. Dorr, J. V. N., 2d., 1945, Manganese and iron deposits of Morro do Urucum, Mato Grosso, Brazil : U. S. Geol. Survey Bull. 946-A, 47 p.
16. Durum, W. H., and Haffty, J., 1963, Implications of the minor element content of some major streams of the world : Geochim. et Cosmochim. Acta, v. 27, p. 1–11.
17. Dutton, C. E., 1955, Iron ore deposits of North America and the West Indies, *in* Survey of World Iron Ore Resources : United Nations, Dept. Economic and Social Affairs, New York, p. 179–208.
18. Garrels, R. M., 1960, Mineral equilibria : Harper and Brothers, New York, 254 p.
19. Geijer, P., and Magnusson, N. H., 1952, The iron ores of Sweden, *in* Symposium sur les gisements de fer du Monde : 19th International Geol. Congr., Algiers, v. 2, p. 477–499.

20. Gill, J. E., 1927, Origin of the Gunflint iron-bearing formation: Econ. Geol., v. 22, p. 687–728.
21. Goldich, S. S., Baadsgaard, H., Edwards, G., and Weaver, C. E., 1959, Investigations in radioactivity-dating of sediments: Am. Assoc. Petroleum Geologists Bull., v. 43, p. 654–662.
22. Goldich, S. S., and Bergquist, H. R., 1947, Aluminous lateritic soil of the Sierra de Bahoruco area, Dominican Republic, W. I.: U. S. Geol. Survey Bull. 953-C, p. 53–84.
23. Goldich, S. S., and Bergquist, H. R., 1948, Aluminous lateritic soil of the Republic of Haiti, W. I.: U. S. Geol. Survey Bull., 954-C, p. 63–109.
24. Goldich, S. S., Nier, A. O., Baadsgaard, H., Hoffman, J. H., and Krueger, H. W., 1961, The Precambrian geology and geochronology of Minnesota: Minnesota Geol. Survey Bull. 41, 193 p.
25. Goodwin, A. M., 1956, Facies relations in the Gunflint iron formation: Econ. Geol., v. 51, p. 565–595.
26. Goodwin, A. M., 1960, Gunflint iron formation of the Whitefish Lake area: Ontario Dept. Mines, v. 59, p. 41–63.
27. Graves, R. W., Jr., 1954, Geology of Hood Spring quadrangle, Brewster County, Texas: Texas Univ. Bur. Econ. Geology Rept. Inv., no. 21, 51 p.
28. Graves, R. W., Jr., Letter, March 1963.
29. Gross, G. A., 1959, Metallogenic map for iron in Canada: Geol. Survey Canada Map 1045 A-M4.
30. Grout, F. F., Gruner, J. W., Schwartz, G. M., and Thiel, G. A., 1951, Precambrian stratigraphy of Minnesota: Geol. Soc. America Bull. 62, p. 1017–1078.
31. Gruner, J. W., 1922, The origin of sedimentary iron formations: the Biwabik formation of the Mesabi Range: Econ. Geol., v. 17, p. 407–460.
32. Gruner, J. W., 1946, The mineralogy and geology of the taconites and iron ores of the Mesabi Range Minnesota: Office of the Commissioner of the Iron Range Resources and Rehabilitation, St. Paul, Minn., 127 p.
33. Gruner, J. W., 1956, The Mesabi Range, in Precambrian of northeastern Minnesota: Geol. Soc. America Guidebook ser., Minneapolis Meeting, p. 182–215.
34. Gunderson, J. N., 1960, Lithologic classification of taconite from the type locality: Econ. Geol., v. 55, p. 563–573.
35. Harder, E. C., 1919, Iron-depositing bacteria and their geologic relations: U. S. Geol. Survey Prof. Paper 113, 89 p.
36. Hoering, T. C., 1962, The isolation of organic compounds from Precambrian rocks: Carnegie Inst. Washington Year Book 61, p. 184–187.
37. Hoering, T. C., 1962, The stable isotopes of carbon in the carbonate and reduced carbon of Precambrian sediments: Carnegie Inst. Washington Year Book 61, p. 190–192.
38. Holland, H. D., 1962, Model for the evolution of the earth's atmosphere, in Petrologic Studies: Geol. Soc. America Buddington Volume, p. 447–477.
39. Holliday, R. W., and Lewis, H. E., 1962, Iron ore, in 1961 Minerals Yearbook, U. S. Bur. Mines, v. 1, p. 649–680.
40. Hose, H. R., 1963, Jamaica type bauxites developed on limestone: Econ. Geol., v. 58, p. 62–69.
41. Hough, J. L., 1958, Fresh-water environment of deposition of Precambrian banded iron formations: Jour. Sed. Petrology, v. 28, p. 414–430.
42. Huber, N. K., 1958, The environmental control of sedimentary iron minerals: Econ. Geol., v. 58, p. 123–140.
43. Huber, N. K., 1959, Some aspects of the origin of the Ironwood iron foramtion of Michigan and Wisconsin: Econ. Geol., v. 54, p. 82–118.
44. Irving, R. D., and Van Hise, C. R., 1892, The Penokee iron-bearing series of Michigan and Wisconsin: U. S. Geol. Survey Mon. 19, 534 p.
45. James, H. L., 1951, Iron formation and associated rocks in the Iron River district, Michigan: Geol. Soc. America Bull., v. 62, p. 251–266.
46. James, H. L., 1954, Sedimentary facies of iron formations: Econ. Geol., v. 49, p. 235–293.
47. James, H. L., 1958, Stratigraphy of pre-Keweenawan rocks in parts of northern Michigan: U. S. Geol. Survey Prof. Paper 314-C, p. 27–44.
48. Kaplan, I. R., Emery, K. O., and Rittenberg, S. C., 1963, The distribution and isotopic abundance of sulphur in recent marine sediments off southern California: Geochem. et Cosmochim. Acta, v. 27, p. 297–331.
49. Kidd, D. J., 1949, Iron occurrence in the Peace River region, Alberta: Research Council Alberta, Prelim. Rept. 59-3, 38 p.
50. King, P. B., 1937, Geology of the Marathon region, Texas: U. S. Geol. Survey Prof. Paper 187, 148 p.

51. Klinger, F. L., 1956, Geology of the Soudan Mine and vicinity, *in* Precambrian of northeastern Minnesota: Geol. Soc. America Guidebook ser., Minneapolis Meeting, p. 120–134.
52. Krauskopf, K. B., 1956, Dissolution and precipitation of silica at low temperatures: Geochim. et Cosmochim. Acta, v. 10, p. 1–26.
53. Krishnan, M. S., 1955, Iron ore deposits of the Middle East and of Asia and the Far East, *in* Survey of World Iron Ore Resources: United Nations, Dept. Economic and Social Affairs, New York, p. 265–334.
54. Krishnan, M. S., 1960, Pre-Cambrian stratigraphy of India, *in* Pre-Cambrian stratigraphy and correlations: 21st Internat. Geol. Congr. (Norden), Copenhagen, pt. IX, p. 95–107.
55. Krumbein, W. C., and Garrels, R. M., 1952, Origin and classification of chemical sediments in terms of pH and oxidation-reduction potentials: Jour. Geology, v. 60, p. 1–33.
56. Kuiper, G. P., 1952, Planetary atmospheres and their origin, *in* The Atmospheres of the Earth and Planets: Univ. Chicago Press, 434 p.
57. Lepp, H., 1963, The relation of iron and manganese in sedimentary iron formations: ECON. GEOL., v. 58, p. 515–526.
58. Lepp, H., Goldich, S. S., and Kistler, R. W., 1963, A Grenville cross section from Port Cartier to Mount Reed, Quebec, Canada: Am. Jour. Sci., v. 261, p. 693–712.
59. Macgregor, A. M., 1927, The problem of the pre-Cambrian atmosphere: South African Jour. Sci., v. 24, p. 155–172.
60. Macgregor, A. M., 1940, A pre-Cambrian algal limestone in Southern Rhodesia: Geol. Soc. South Africa, Trans., v. 43, p. 9–16.
61. Macgregor, A. M., 1951, Some milestones in the Precambrian of Southern Rhodesia: Geol. Soc. South Africa, Trans., v. 54, p. 27–71.
62. Mann, V. I., 1961, Iron formations in the southeastern United States: ECON. GEOL., v. 56, p. 997–1000.
63. Marmo, V., 1956, Banded ironstones of the Kangari Hills, Sierra Leone: ECON. GEOL., v. 51, p. 798–810.
64. Miles, K. R., 1942, The blue asbestos-bearing banded iron formations of the Hamersley Ranges, Western Australia, Part 1: Geol. Survey Western Australia Bull. 100, p. 5–37.
65. Miles, K. R., 1946, Metamorphism of the jasper bars of Western Australia: Geol. Soc. London Quart. Jour., v. 102, p. 115–154.
66. Moore, E. S., 1918, The iron-formation on Belcher Islands, Hudson Bay, with special references to its origin and its associated algal limestones: Jour. Geology, v. 26, p. 412–438.
67. Moore, E. S., and Armstrong, H. S., 1946, Iron deposits in the district of Algoma: Ontario Dept. Mines 55th Ann. Rpt., v. 55, 118 p.
68. Moore, E. S., and Maynard, J. E., 1929, Solution, transportation, and precipitation of iron and silica: ECON. GEOL., v. 24, p. 272–303; 365–402; 506–527.
69. Moorhouse, W. W., and Beales, F. W., 1962, Fossils from the Animikie, Port Arthur, Ontario: Royal Soc. Canada, Trans., v. 56, p. 97–110.
70. Nalivkin, D. V., 1960, The Geology of the U. S. S. R., *in* Internat. Ser. Monographs on Earth Science (translation): Pergamon Press, New York, 170 p.
71. Nicolaysen, L. O., 1962, Stratigraphic interpretation of age measurements in southern Africa, *in* Petrologic Studies: Geol. Soc. America Buddington Volume, p. 569–598.
72. Nicolaysen, L. O., de Villiers, J. W. L., Burger, A. J., and Strelow, F. W. E., 1958, New measurements relating to the absolute age of the Transvaal System and of the Bushveld igneous complex: Geol. Soc. South Africa Trans., v. 61, p. 137–163.
73. Oftedahl, C., 1958, A theory of exhalative-sedimentary ores: Geol. Fören. Stockholm Förh., v. 80, p. 1–19.
74. Oparin, A. I., 1961, Life, Its Nature, Origin, and Development (translated): Academic Press, Inc., New York, 207 p.
75. O'Rourke, J. E., 1961, Paleozoic banded iron-formation: ECON. GEOL., v. 56, p. 331–361.
76. Parsons, A. L., 1921, Economic deposits in the Thunder Bay district: Ontario Dept. Mines Rept., v. 30, p. 27–38.
77. Patterson, C. C., 1956, Age of meteorites and the earth: Geochim. et Cosmochim. Acta, v. 10, p. 230–237.
78. Percival, F. G., 1955, Nature and occurrence of iron ore deposits, *in* Survey of World Iron Ore Resources: United Nations, Dept. Economic and Social Affairs, New York, p. 45–76.
79. Pettijohn, F. J., 1957, Sedimentary Rocks, 2nd ed., Harper and Brothers, New York, 718 p.
80. Polkanov, A. A., and Gerling, E. K., 1960, The Pre-Cambrian geochronology of the Baltic Shield, *in* Pre-Cambrian stratigraphy and correlations: 21st Internat. Geol. Congr. (Norden), Copenhagen, Pt. IX, p. 183–191.

81. Quirke, T. T., Jr., 1961, Geology of the Temiscamie iron-formation, Lake Albanel iron range, Mistassini Territory, Quebec, Canada: Econ. Geol., v. 56, p. 299–320.
82. Quirke, T. T., Jr., Goldich, S. S., and Krueger, H. W., 1960, Composition and age of the Temiscamie iron-formation, Mistassini Territory, Quebec, Canada: Econ. Geol., v. 55, p. 311–326.
83. Rhoden, H. N., 1961, Paleozoic banded iron-formations: Econ. Geol., v. 56, p. 1473.
84. Rubey, W. W., 1955, Development of the hydrosphere and atmosphere, with special reference to probable composition of the early atmosphere, in Crust of the earth, Geol. Soc. America Special Paper 62, p. 631–650.
85. Rutten, M. G., 1962, The Geological Aspects of the Origin of Life on Earth: Elsevier Publishing Co., Amsterdam, 146 p.
86. Sakamoto, T., 1950, The origin of the pre-Cambrian banded iron ores: Am. Jour. Sci., v. 248, p. 449–474.
87. Sarkar, S. N., and Saha, A. K., 1963, On the occurrence of two intersecting pre-Cambrian orogenic belts in Singhbhum and adjacent areas: India: Geol. Mag., v. 100, p. 69–92.
88. Semenenko, N. P., Rodionov, S. P., Usenko, I. S., Lichak, I. L., and Tsarovsky, I. D., 1960, Stratigraphy of the Pre-Cambrian of the Ukrainian Shield, in Pre-Cambrian stratigraphy and correlations: 21st Internat. Geol. Congr. (Norden), Copenhagen, Pt. IX, p. 108–115.
89. Spencer, A. C., 1916, The Atlantic gold district and the north Laramie Mountains, Fremont, Converse, and Albany Counties, Wyoming: U. S. Geol. Survey Bull. 626, 85 p.
90. Stockwell, C. H., 1963, Second report on structural provinces, orogenies, and time-classification of rocks of the Canadian Precambrian Shield: Geol. Survey Canada, Paper 62-17, p. 123–133.
91. Swenson, W. T., 1960, Geology of the Nakina iron property, Ontario: Am. Inst. Min. Eng. Trans., v. 217, p. 451–457.
92. Taylor, J. H., Davies, W., and Dixie, R. J. M., 1952, The petrology of the British Mesozoic ironstones and its bearing on problems of beneficiation, in Symposium sur les gisements de fer du Monde, 19th Internat. Geol. Congr., Algiers, v. 2, p. 453–471.
93. Tyler, S. A., and Barghoorn, E. S., 1954, Occurrence of structurally preserved plants in pre-Cambrian rocks of the Canadian Shield: Science, v. 119, p. 606–608.
94. Tyler, S. A., and Barghoorn, E. S., 1963, Ambient pyrite grains in Precambrian cherts: Am. Jour. Sci., v. 261, p. 424–432.
95. Tyler, S. A., Barghoorn, E. E., and Barrett, L. P., 1957, Anthracite coal from Precambrian Upper Huronian black shale of the Iron River district, northern Michigan: Geol. Soc. America Bull., v. 68, p. 1293–1304.
96. Tyler, S. A., and Twenhofel, 1952, Sedimentation and stratigraphy of the Huronian of Upper Michigan: Am. Jour. Sci., v. 250, p. 1–27; 118–151.
97. Urey, H. C., 1959, The atmospheres of the planets, in Handbuch der Physik, Springer-Verlag, Berlin, p. 363–418.
98. Van Hise, C. R., and Bayley, W. S., 1897, The Marquette iron-bearing district of Michigan: U. S. Geol. Survey Mon. 28, 608 p.
99. Van Hise, C. R., and Leith, C. K., 1911, The geology of the Lake Superior region: U. S. Geol. Survey Mon. 52, 641 p.
100 Vinogradov, A. P., Komlev, L. V., Danilevich, S. I., Savonenko, V. G., Tugarinov, A. I., and Filippov, M. S., 1960, Absolute geochronology of the Ukrainian Pre-Cambrian, in Pre-Cambrian stratigraphy and correlations: 21st Internat. Geol. Congr. (Norden), Copenhagen, Pt. IX, p. 116–132.
101. Vinogradov, A. P., and Tugarinov, A. I., 1961, Geochronology of the Precambrian: Geochemistry, no. 9, p. 787–800 (translation of Geokhimia).
102. Wagner, P. A., 1928, The iron deposits of the Union of South Africa: Union of South Africa, Dept. Mines and Industries, Geol. Survey Mem. 26, 264 p.
103. White, D. A., 1954, The stratigraphy and structure of the Mesabi Range, Minnesota: Minnesota Geol. Survey Bull. 38, 92 p.
104. Whitehead, T. H., Anderson, W., Wilson, V., and Wray, D. A., 1952, The Liassic ironstones: Geol. Survey Great Britain Mem., 211 p.
105. Wilson, A. F., Compston, W., and Jeffery, P. M., 1961, Radioactive ages from the pre-Cambrian rocks of Australia: New York Acad. Sci., Annals, v. 91, p. 514–520.
106. Wolff, J. F., 1917, Recent geological developments on the Mesabi iron range: Am. Inst. Min. Eng. Trans., v. 56, p. 142–169.
107. Woolnough, W. G., 1937, Sedimentation in barred basins, and source rocks of oil: Am. Assoc. Petroleum Geologists Bull., v. 21, p. 1101–1157.
108. Woolnough, W. G., 1941, Origin of banded iron deposits—a suggestion: Econ. Geol., v. 36, p. 465–489.

22

Reprinted from Petrologic Studies: A Volume to Honor A. F. Buddington, Geological Society of America, New York, 1962, pp. 447–477

Model for the Evolution of the Earth's Atmosphere

Heinrich D. Holland

Princeton University, Princeton, New Jersey

ABSTRACT

The history of the earth's atmosphere has been divided into three stages. The first covers the earliest part of earth history, prior to the formation of the core; during this time native iron was probably present in the upper part of the mantle. Volcanic gases ejected during the first stage probably contained a large amount of hydrogen, and the atmosphere was highly reduced. Carbon was probably present in the form of methane, nitrogen may have been present as ammonia, and free oxygen was absent in all but nonequilibrium quantities.

The second stage of atmospheric history began when iron was removed from the upper mantle. Volcanic gases became much less reducing and probably approached rather rapidly the present oxidation state of Hawaiian volcanic gases. The atmosphere responded to this change largely in terms of the chemistry of atmospheric carbon. Carbon dioxide took the place of methane and began to participate in the processes of weathering and chemical precipitation. Free oxygen was probably absent during the second stage except in trace, nonequilibrium quantities.

The third stage began when the rate of production of oxygen by photosynthesis exceeded the rate needed to oxidize injected volcanic gases completely. During this, the present, stage oxygen has probably accumulated at a roughly constant rate.

The earth's core was probably formed quite early, and the first stage of the atmosphere probably lasted less than 0.5 b.y. (billion years). The second stage may have continued until about 1.0 b.y. ago. The presence of detrital uraninite in mid-Precambrian sediments suggests that the second stage lasted at least until 2.0 b.y. ago, and the oxygen requirements of late Paleozoic insects suggest that a fairly high oxygen pressure obtained 0.3 b.y. ago.

INTRODUCTION

During the last decade our understanding of the evolution of the earth's atmosphere has been advanced very materially by the contributions of Urey (1952a; 1952b; 1959) and Rubey (1951; 1955). The hypothesis of a strongly reducing atmosphere during at least the earliest part of earth history has been established by Urey. The hypothesis of the gradual accumulation of the ocean and of excess volatiles in the hydrosphere and atmosphere has been virtually established by Rubey. Nevertheless, a good deal of uncertainty remains concerning the composition of the atmosphere between the earliest period and the late Paleozoic. Urey has pointed out that a highly reducing state may have prevailed during much of geologic time. The evidence for this suggestion came from the work of Thode *et al.* (1953) on the fractionation of sulfur isotopes in the Precambrian. Since that time Ault and Kulp (1959) have shown that isotopic fractionation of sulfur extends very far into the Precambrian, so that the basis for Urey's proposal has been removed.

Rubey (1955) on the other hand maintains that the excess volatiles now present at the earth's surface do not contain enough hydrogen to permit a methane-hydrogen atmosphere during geologic time. This evidence is not at all conclusive. If a methane-hydrogen atmosphere existed in the past, the free hydrogen and that bound to carbon would since have escaped. Rubey's excess volatiles must therefore be regarded as residual excess volatiles.

We are therefore left with only scant evidence for the chemistry of the atmosphere during much of geologic time. This paper is an attempt to present a unified model for the evolution of the atmosphere. The model must be regarded as a first approximation. However, it brings together a number of mutually supporting lines of evidence and explains some otherwise puzzling data.

FIRST STAGE OF THE ATMOSPHERE

There seems little question now that the constituents of the earth's atmosphere have been largely, if not wholly, evolved from the interior of the earth. The strong-

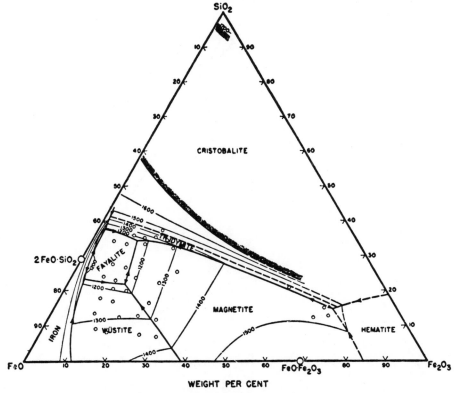

Figure 1. Phase diagram for the system $FeO-Fe_2O_3-SiO_2$. Circles represent compositions, as shown by analysis, of mixtures at liquidus temperatures. Light lines are liquidus isotherms at 100° C intervals, and heavy lines are boundary curves, with arrows pointing in the directions of falling temperatures. The tridymite-cristobalite boundary is dashed. Medium lines with stippling on one side indicate limit of two-liquid region (Muan, 1955).

est evidence for this is the very much smaller abundance of the rare gases in the earth than in average cosmic matter (Brown, 1952), silicon being the standard for comparison. The chemistry of the earliest atmosphere was therefore dependent largely on the nature of volcanic gases ejected at that time, and on the rate of escape of hydrogen from the upper atmosphere. There is still considerable controversy concerning the temperature during and soon after accretion of the earth (Urey, 1959). It seems likely, however, that iron was present in the outer parts of the earth. Sufficient melting must then have taken place to permit the accumulation of iron at the center of the earth. The oxidation state of magmas generated while iron was still present in the outer part of the mantle was buffered by the pair Fe^0-Fe^{+2}. We can obtain a first approximation of the partial pressure of oxygen in a vapor phase in equilibrium with magmas produced by partial melting in the upper mantle from Muan's (1955) data on the system FeO-Fe_2O_3-SiO_2 and from Muan and Osborn's (1956) data on the system MgO-FeO-Fe_2O_3-SiO_2.

Two of Muan's (1955) diagrams have been reproduced in Figures 1 and 2. Melts in equilibrium with solid iron, fayalite, and either tridymite or wüstite near

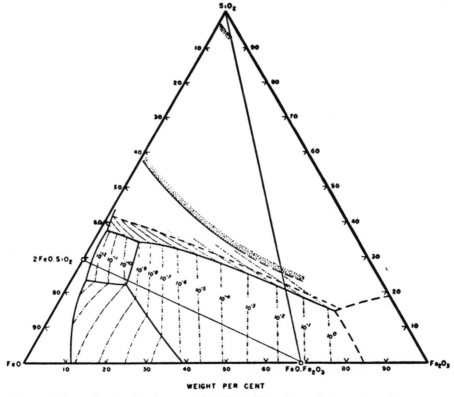

Figure 2. Phase diagram for the system FeO-Fe_2O_3-SiO_2. Heavy lines are boundary curves, and dash-dot lines are lines of equal O_2 pressures for points on liquidus surface. Lines with stippling on one side indicate limit of two-liquid region (Muan, 1955).

1200° C are in equilibrium with a vapor phase having an oxygen pressure of about $10^{-12.5}$ atm. The oxygen pressure defines the ratio of p_{H_2O} to p_{H_2} and of p_{CO_2} to p_{CO} since

$$(H_2)_g + \tfrac{1}{2}(O_2)_g \rightleftarrows (H_2O)_g \tag{1}$$

$$K_1 = \frac{p_{H_2O}}{p_{H_2} \cdot p_{O_2}^{1/2}}$$

and

$$(CO)_g + \tfrac{1}{2}(O_2)_g \rightleftarrows (CO_2)_g \tag{2}$$

$$K_2 = \frac{p_{CO_2}}{p_{CO} \cdot p_{O_2}^{1/2}}$$

Values of these equilibrium constants are listed in the Appendix. At 1200° C

$$K_1 = 10^{5.89} = \frac{p_{H_2O}}{p_{H_2} \cdot p_{O_2}^{1/2}} .$$

Thus, when $p_{O_2} = 10^{-12.5}$ atm

$$\frac{p_{H_2O}}{p_{H_2}} = 0.44 .$$

Also

$$K_2 = 10^{5.47} = \frac{p_{CO_2}}{p_{CO} \cdot p_{O_2}^{1/2}}$$

and thus

$$\frac{p_{CO_2}}{p_{CO}} = 0.17 .$$

In gases in equilibrium with such a melt, hydrogen is therefore roughly twice as abundant as water, and CO more than five times as abundant as CO_2. Methane would normally be a minor constituent. Its pressure would be determined by the equilibrium

$$(CO_2)_g + 4(H_2)_g \rightleftarrows (CH_4)_g + 2(H_2O)_g , \tag{3}$$

for which the equilibrium constant, K_3, is

$$K_3 = \frac{p_{CH_4} \cdot p_{H_2O}^2}{p_{CO_2} \, p_{H_2}^4} .$$

At 1200° C

$$K_3 = 10^{-3.15} = \left(\frac{p_{CH_4}}{p_{CO_2}}\right)\left(\frac{p_{H_2O}}{p_{H_2}}\right)^4 \cdot \frac{1}{p_{H_2O}^2}$$

$$\frac{p_{CH_4}}{p_{CO_2}} = 10^{-3.15} \cdot \frac{1}{(0.44)^4} \cdot p_{H_2O}^2$$

$$= 1.9 \times 10^{-2} p_{H_2O}^2 .$$

TABLE 1. COMPARISON OF RUBEY'S (1951) "EXCESS VOLATILES" AND EATON AND MURATA'S (1960) AVERAGE COMPOSITION OF HAWAIIAN VOLCANIC GASES

	Rubey (1951) (gm per 1000 gm H_2O)	Eaton and Murata (1960) (gm per 1000 gm H_2O)
Total C as CO_2	55	370
Cl	18	2.5
N	2.5	25
S	1.3	155
H	0.6	0.8
B, Br, Ar, F, etc.	0.2	1.1

The total pressure on ejection was probably greater than 1 and less than 10 atmospheres. Methane was therefore probably much less abundant than CO_2.

Rubey (1951) has estimated the total amount of water and the total amount of carbon as CO_2 which have been ejected from volcanoes and hot springs. If the early volcanoes were "average" in this respect

$$R = \frac{p_{H_2O}}{p_{CO} + p_{CO_2} + p_{CH_4}} \approx 43.$$

It is of interest to note that in Hawaiian gases today (Eaton and Murata, 1960)

$$R = \frac{p_{H_2O}}{p_{CO} + p_{CO_2} + p_{CH_4}} \approx 6.6.$$

Sulfur and its compounds were undoubtedly present in these early volcanic gases. In Hawaiian volcanic gases today SO_2 is the dominant molecule containing sulfur. Just as CO_2 and CO are more abundant relative to water in these gases than in the total "excess volatiles," so the rate of sulfur ejection is abnormally high relative to water. Rubey's (1951) results and the data of Eaton and Murata (1960) have been recast in Table 1 to point out these differences. It is difficult to assess the extent to which the Hawaiian samples are representative of volcanic gases through time. If they are representative, a large fraction of water now in the hydrosphere must have reached its present position by a route other than volcanic ejection. It may be that volatiles released at depth from cooling intrusives lose much of their sulfur as sulfides and much of their carbon as carbonates. On the other hand, there is no evidence that the very large quantities of sulfide and carbonate demanded by this alternative are present in the earth's crust. At present it therefore seems likely that the samples of Hawaiian gases studied by Eaton and Murata (1960) were abnormally enriched in compounds of carbon and sulfur.

Regardless of the ratio of sulfur and sulfur compounds to water in the earliest gases, H_2S should have been the dominant molecular species. The equilibrium constant K_4 for the reaction

$$(H_2S)_g + 2(H_2O)_g \rightleftharpoons (SO_2)_g + 3(H_2)_g \qquad (4)$$

at 1200° C is

$$K_4^{1200°C} = \frac{p_{SO_2} \cdot p_{H_2}^3}{p_{H_2S} \cdot p_{H_2O}^2} = 10^{-3.29};$$

thus

$$\frac{p_{SO_2}}{p_{H_2S}} \approx 10^{-3.29} \cdot (0.44)^3 \cdot p_{H_2O}$$

$$\approx 10^{-4.36} \cdot p_{H_2O}.$$

At all reasonable values of the water vapor pressure, H_2S would therefore have far exceeded SO_2 in abundance. The equilibrium constant K_5 for the reaction

$$(H_2)_g + \frac{1}{2}(S_2)_g \rightleftharpoons (H_2S)_g \tag{5}$$

at 1200° C is

$$K_5^{1200°C} = \frac{p_{H_2S}}{p_{H_2} \cdot p_{S_2}^{1/2}} = 10^{+0.61}.$$

Since hydrogen was probably the most abundant constituent of these gases, p_{H_2} would have been in the neighborhood of 1 to 10 atm. Therefore

$$p_{S_2} \approx 10^{-2} \cdot p_{H_2S}^2.$$

The H_2S pressure was probably well below 1 atm if the sulfur content of Hawaiian gases can be taken as a guide, so that normally p_{S_2} would have been smaller than p_{H_2S}.

We can conclude then, that volcanic gases in equilibrium with iron, fayalite, and either tridymite or wüstite at an oxygen pressure of $10^{-12.5}$ atm at 1200° C would consist largely of H_2, H_2O, CO, and H_2S. Carbon dioxide and sulfur molecules would be present in minor amounts, and the partial pressure of CH_4 and SO_2 would be very small.

Equivalent data for the system MgO-FeO-Fe_2O_3-SiO_2 are not available, but the effect of MgO on the oxygen pressure of a gas in equilibrium with a melt and solid iron would probably be well below one order of magnitude. As an indication that this is the case, consider the oxygen pressure at the ternary invariant point magnetite-fayalite-tridymite in the system FeO-Fe_2O_3-SiO_2 and at the corresponding quaternary invariant point magnesioferrite-olivine-pyroxene-tridymite in the system MgO-FeO-Fe_2O_3-SiO_2. The ternary point lies at 1140° C and at an oxygen pressure of $10^{-9.0}$ atm (Muan, 1955). At this temperature and p_{O_2}

$$\frac{p_{H_2O}}{p_{H_2}} = K_1^{1140°C} \cdot p_{O_2}^{1/2} = 10^{6.30} \times 10^{-4.50} = 10^{+1.80}.$$

The quaternary invariant point lies at 1255° C and at an oxygen pressure of $10^{-7.1}$ atm (Muan and Osborn, 1956). At this temperature and p_{O_2}

$$\frac{p_{H_2O}}{p_{H_2}} = K_1^{1255°C} \cdot p_{O_2}^{1/2} = 10^{5.67} \times 10^{-3.55} = 10^{+2.02}.$$

Thus volcanic gases ejected under these two sets of conditions have nearly the same ratio of water pressure to hydrogen pressure.

At temperatures above 1300° C the presence of magnesium affects the ratio

p_{H_2O}/p_{H_2} considerably. This effect is probably due to the presence of large amounts of MgO in the spinel phase.

Undoubtedly, the presence of other components will modify gas compositions. We cannot determine the magnitude of these effects at present. They are probably small, because the concentration of metal oxides other than FeO, Fe_2O_3, MgO, and SiO_2 in a homogeneous earth based on a chondrite model is only about 8 per cent (Urey and Craig, 1953). The present oxidation state of volcanic gases can be explained reasonably well on the basis of data on the system MgO-FeO-Fe_2O_3-SiO_2 alone. It seems probable, then, that the composition of gases in equilibrium with a melt in the system FeO-Fe_2O_3-SiO_2 and with solid iron, fayalite, and either tridymite or wüstite, is not very different from the composition of volcanic gases in the pre-core stage of the earth.

If the temperature structure of the pre-core (first stage) atmosphere was similar to the present-day structure, much of the water contained in these gases would condense on cooling. The fate of H_2, CO, CO_2, H_2S, and CH_4 would depend in large part on the escape rate of hydrogen from the upper atmosphere, on the rate of reaction of CO and CO_2 with H_2, and on the rate of reaction of H_2S with iron silicates. Reaction rates in the upper atmosphere are fast today, and it seems very unlikely that it would have taken more than 1 year to approach chemical equilibrium in the pre-core atmosphere. Such reaction rates would have been much faster than H_2 escape from the exosphere. Thus equilibrium in the reactions

$$(CO)_g + 3(H_2)_g \rightleftharpoons (CH_4)_g + (H_2O)_g \tag{6}$$

and

$$(CO_2)_g + 4(H_2)_g \rightleftharpoons (CH_4)_g + 2(H_2O)_g \tag{7}$$

would have determined the final distribution of carbon among the three compounds CO, CO_2 and CH_4. The value of K_6 at 25° C is

$$K_6^{25°C} = 10^{24.90} = \frac{p_{CH_4} \cdot p_{H_2O}}{p_{CO} \cdot p_{H_2}^3}.$$

Thus

$$\frac{p_{CH_4}}{p_{CO}} = 10^{24.90} \cdot \frac{p_{H_2}^3}{p_{H_2O}}.$$

The water pressure at 25° C will have approximately the value in equilibrium with an ocean. Thus at 25° C for all values of p_{H_2} greater than ca. 10^{-6} atm, the ratio p_{CH_4}/p_{CO} would be greater than 1. The same will be true of the ratio p_{CH_4}/p_{CO_2}.

The hydrogen pressure in this primeval atmosphere is difficult to calculate. However, broad limits can be set on its value. The total amount of water evolved during geologic time (Rubey, 1951) is about 1.6×10^{24} gm. The mean rate of evolution is therefore about 3.6×10^{14} gm/year. Since most of the water today is in the ocean, recycling through the sedimentary-metamorphic-igneous sequence has presumably not been of great importance. If evolution of volcanic gases took place at an average rate, the hydrogen would have been evolved approximately at the rate of

$$\frac{3.6 \times 10^{14}}{0.44} \times \frac{2}{18} = 9 \times 10^{13} \text{ gm/year.}$$

The actual rate was probably no smaller than this. It was probably no greater than this by more than a factor of 10; otherwise all the water in the oceans would have appeared in less than 5×10^8 years. Thus hydrogen evolution took place at a rate of about $10^{14.5\pm0.5}$ gm/year. In these gases

$$p_{H_2} \gg 3\, p_{CO} + 4\, p_{CO_2}.$$

Therefore the production of methane would not decrease markedly the hydrogen available for escape. The amount of hydrogen in the atmosphere at any time after equilibrium had been established would have been defined by the equation

$$N_{H_2} = \frac{1}{\lambda_{H_2}} \cdot \frac{dN_{H_2}}{dt},$$

where dN_{H_2}/dt equals the rate of introduction, and

$$\lambda_{H_2} = \frac{0.693}{(T_{1/2})_{H_2}}$$

where $(T_{1/2})_{H_2}$ is the half-life for hydrogen escape from the atmosphere. Today $(T_{1/2})_{H_2}$ must be less than 5×10^5, the escape half-life of He^3 (Damon and Kulp, 1958). It is probably greater than 10^3 years (Spitzer, 1952). It is very difficult to estimate the temperature structure in the early atmosphere of the earth. Zabriskie (1960, Ph.D. thesis, Princeton Univ., Princeton, N. J.) has discussed the similar problem of the present atmosphere of Jupiter. It seems unlikely, however, that $(T_{1/2})_{H_2}$ was less than 10^3 or more than 10^6 years in the early atmosphere. Thus the minimum weight of H_2 in the atmosphere would have been

$$(N_{H_2})_{\min} \approx \frac{10^3}{0.693} \times 10^{14} \approx 1.4 \times 10^{17} \text{ gm,}$$

and the maximum weight

$$(N_{H_2})_{\max} \approx \frac{10^6}{0.693} \times 10^{15} \approx 1.4 \times 10^{21} \text{ gm.}$$

These values correspond to a minimum value for p_{H_2} of 2.7×10^{-5} and a maximum of 0.27 atmosphere; it seems unlikely that the hydrogen pressure ever exceeded one atm.

If the hydrogen pressure was close to the lower limit during this period of earth history, then CO_2 was already an important part of the atmosphere. This can be seen from the relationships in Figure 3. Line I represents the relationship between p_{CH_4} and p_{H_2} in equilibrium with graphite at 25° C. Its position is determined by the equilibrium

$$\langle C \rangle_c + 2(H_2)_g \rightleftharpoons (CH_4)_g$$

$$K_8^{25°C} = \frac{p_{CH_4}}{p_{H_2}^2} = 10^{8.90}. \tag{8}$$

Line II represents the equilibrium at 25° C between p_{CO_2} and p_{H_2} in equilibrium with graphite and water-saturated air.

$$\langle C \rangle_c + 2(H_2O)_g \rightleftharpoons (CO_2)_g + 2(H_2)_g$$

$$K_9^{25°C} = \frac{p_{H_2}^2 \cdot p_{CO_2}}{p_{H_2O}^2} = 10^{-11.01}. \tag{9}$$

It can be shown that p_{CO} is always much less than 10^{-6} atm.

If now p_{H_2} is 3×10^{-6} atm, p_{CH_4} is nearly equal to p_{CO_2}, and the atmosphere is in equilibrium with graphite. If the prevailing hydrogen pressure had an intermediate value of, say, 10^{-3} atm, methane would be the dominant gas. Under these conditions, the methane pressure would be determined by the total amount of $CO + CO_2 + CH_4$ evolved up to that time. The maximum pressure might be obtained by converting all C and CO_2 now in the crust to methane. If we again take Rubey's (1951) figure for the total amount of C as CO_2,

$$(p_{CH_4})_{max} \approx \frac{920 \times 10^{20} \times 16}{44 \times 5.1 \times 10^{21}} \approx 6.6 \text{ atm.}$$

This is obviously a very strong maximum.

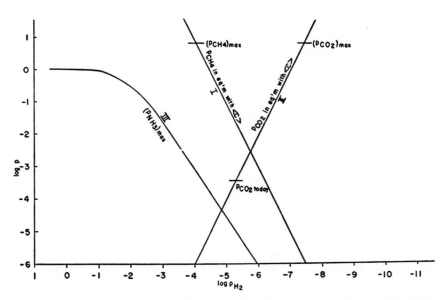

Figure 3. Relationship between H_2 pressure and the pressure of CH_4, Co_2, and NH_3 during the first stage of the earth's atmosphere

Nitrogen would be evolved from volcanoes, together with the compounds already discussed. The reaction with hydrogen to ammonia

$$(N_2)_g + 3(H_2)_g \rightleftharpoons 2(NH_3)_g$$

$$K_{10}^{25°C} = \frac{p_{NH_3}^2}{p_{N_2} \cdot p_{H_2}^3} = 10^{6.83} \tag{10}$$

depends on the hydrogen pressure and on the total amount of nitrogen evolved. The maximum ammonia pressure at a given p_{H_2} increases with the total amount of nitrogen in the atmosphere. Since the atmosphere has served nearly as a simple accumulator for nitrogen during geologic time, the maximum amount of ammonia at any time is defined by p_{H_2} and $(p_{N_2} + p_{NH_3}) = 0.8$ atm. Curve III has been constructed under this assumption. It follows that ammonia may have been, but was probably not, a major constituent of the earliest atmosphere.

The H_2S pressure was probably determined largely by the balance between the rate of ejection from volcanoes and the rate of reaction of atmospheric H_2S with iron oxides and silicates to form pyrite or pyrrhotite. In this sense the situation of H_2S is similar to that of CO_2 in the present atmosphere.

The maximum H_2S pressure can be calculated like that of methane, on the basis of Rubey's estimate of "excess volatiles":

$$(p_{H_2S})_{max} \approx \frac{22 \times 10^{20} \times 34}{5.1 \times 10^{21} \times 32} \approx 0.46 \text{ atm.}$$

As in the case of methane, this is a very strong maximum. It seems most unlikely that p_{H_2S} was ever greater than 0.05 atm.

In discussing the pressure of methane, of nitrogen plus ammonia, and of hydrogen sulfide we have to set some value on the amount of degassing during this earliest stage of earth history. The time span was probably less than 0.5×10^9 years, because in the parent body (or bodies) of meteorites fractionation into silicate and metal phase apparently took place faster than this. The very large release of gravitational energy during settling of the metallic phase makes such a time span appear reasonable. If the rate of degassing was nearly average, less than 10 per cent of the total volume of volatiles would have evolved during this time. However, the rate of evolution of volatiles may have been much faster than the mean rate, so that little can be said on this basis concerning the total quantity of volatiles evolved during the pre-core period.

One line of evidence seems to offer hope of setting some limits to speculation concerning this matter. Kokubu, Mayeda, and Urey (1961) have recently confirmed the result that the deuterium-hydrogen ratio in the hydrosphere is about 5 per cent greater than that in volcanic waters. The explanations offered by them seem inadequate. To account for the observed effect by assuming a large amount of water dissociation followed by protium loss involves the production of embarrassingly large quantities of oxygen and disagrees with Urey's (1959) calculations on the rate of photodissociation of water and the subsequent escape of hydrogen. To account for the effect by assuming the presence of surface water "at the beginning" seems out of harmony with the data on rare gas abundances. On the other hand, the observed effect would follow reasonably logically from the model of volcanic gases and atmospheric chemistry proposed here for the pre-core stage of the atmosphere.

Hydrogen evolving from volcanoes would escape from the atmosphere. Unless the temperature in the atmosphere was very high, the deuterium escape rate would have been smaller than that of protium by at least a factor of 10 and possibly by a factor of 100. There would therefore have been a tendency to build up in the atmosphere a deuterium-protium ratio well in excess of that in volcanic gases. At equilibrium, and neglecting all other factors, the ratio of deuterium to protium in the atmosphere would have been equal to the ratio of deuterium to protium in volcanic gases, multiplied by their respective escape half-lives. However, such a ratio could not be maintained at present surface temperatures since water of isotopic composition equal to that of volcanic water would be cycled through the atmosphere. The amount of water passing annually through the atmosphere would have been determined by the surface area of the oceans at the time.

We can then write down the equations controlling the amount of deuterium, D, and protium, P, in the atmosphere and in the oceans at this time. If x_1 is the number of moles of protium in the atmosphere, x_2 the number of moles of deuterium in the atmosphere, y_1 the number of moles of protium in the oceans, and y_2 the number of moles of deuterium in the oceans, then

$$\frac{dx_1}{dt} = k_1 - \lambda_1 x_1 + \beta M \left(\frac{x_2}{x_1} - \frac{y_2}{y_1} \right)$$

$$\frac{dx_2}{dt} = k_2 - \lambda_2 x_2 - \beta M \left(\frac{x_2}{x_1} - \frac{y_2}{y_1} \right)$$

$$\frac{dy_1}{dt} = K_1 - \beta M \left(\frac{x_2}{x_1} - \frac{y_2}{y_1} \right)$$

$$\frac{dy_2}{dt} = K_2 + \beta M \left(\frac{x_2}{x_1} - \frac{y_2}{y_1} \right)$$

where

k_1 = number of moles of protium injected into the atmosphere per year
k_2 = number of moles of deuterium injected into the atmosphere per year
K_1 = number of moles of protium combined as water injected into the atmosphere per year
K_2 = number of moles of deuterium combined as water injected into the atmosphere per year
λ_1, λ_2 = escape rate of protium and deuterium from the atmosphere per year
M = number of moles of hydrogen circulated annually as water through the atmosphere
β = probability that a water molecule circulating through the atmosphere attains isotopic equilibrium with the atmosphere.

Since the D/P ratio in volcanic gases is very small, the effect of the exchange term on the protium content of the atmosphere and oceans can be neglected. Thus at equilibrium

$$x_1 \approx \frac{k_1}{\lambda_1}$$

and

$$y_1 \approx K_1 t;$$

the value of x_2 depends on the relative magnitude of the term $\lambda_2 x_2$ and of the exchange term

$$\beta M\left(\frac{x_2}{x_1} - \frac{y_2}{y_1}\right).$$

An approximate relationship between these terms can be obtained from a consideration of the maximum permissible difference between the isotopic composition of the atmosphere and ocean, since

$$\left(\frac{x_2}{x_1} - \frac{y_2}{y_1}\right) < \frac{k_2}{\beta M}.$$

Now

$$k_2 \approx k_1/6700 \approx \frac{10^{14.5 \pm 0.5}}{2 \times 6700} \approx 0.7 \times 10^{10.5 \pm 0.5}.$$

The total mass of water circulating annually through the atmosphere today is 4×10^{20} gm, corresponding to 2.2×10^{19} moles of hydrogen. The mass was almost certainly smaller in the pre-core stage of the earth. Perhaps a reasonable lower limit would be 10^{16} moles/year. Today the exchange in the upper atmosphere would probably be essentially complete. Mixing between the upper and lower atmosphere today is reasonably fast, so that β might well be between 0.01 and 1. In the absence of an ozone layer, dissociation due to ultraviolet rays would have been more important in the troposphere than today, so that the value of β may well have been greater during the first stage of the atmosphere. If we use the present-day value of M and an estimated value of 0.1 for β, we obtain

$$\frac{k_2}{\beta M} \approx \frac{10^{10}}{2 \times 10^{19} \times 10^{-1}}$$

$$\approx 0.5 \times 10^{-8}.$$

But

$$\frac{y_2}{y_1} \approx 1.4 \times 10^{-4}.$$

Therefore, under present-day conditions the difference

$$\left(\frac{x_2}{x_1} - \frac{y_2}{y_1}\right)$$

would be a very small fraction of y_2/y_1 ; i.e., the isotopic composition of the atmosphere would be essentially identical with that of the oceans. Only if the product βM was less than 1/1000 of the present value would any appreciable difference in the isotopic composition of these reservoirs have been possible.

Today isotopic exchange would therefore probably overshadow deuterium escape as a means of removing deuterium from the atmosphere. It follows that, as protium escapes from the atmosphere, deuterium enters the oceans, so that the D/P ratio in the atmosphere remains nearly the same as in the ocean reservoir. When the ratio λ_1/λ_2 is large, nearly all the deuterium enters the oceans. A steady state will thus be established in which the D/P ratio in the oceans is determined by the D/P ratio and by the ratio p_{H_2O}/p_{H_2} in volcanic gases. If, under these conditions

$$\frac{p_{H_2O}}{p_{H_2}} = 0.44,$$

then

$$\left(\frac{D}{P}\right)_{oceans} \approx \left(\frac{D}{P}\right)_{volcanic\ gases} \times \left(1 + \frac{1}{0.44}\right)$$

$$\approx 3.3 \left(\frac{D}{P}\right)_{volcanic\ gases}$$

At present, only a very small amount of hydrogen is ejected from volcanoes. This situation has probably prevailed for a long time. Any "heavy" water from the pre-core stage has, therefore, been diluted with water of volcanic composition. The amount of dilution required to reduce the difference between ocean water and volcanic water is shown in graphical form in Figure 4. If the ratio p_{H_2O}/p_{H_2} in

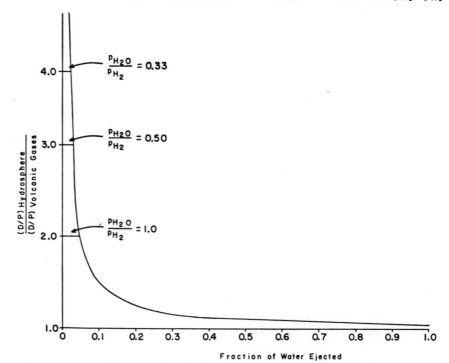

Figure 4. Ratio of the deuterium-protium ratio in the hydrosphere to that in volcanic gases as a function of the size of the hydrosphere

pre-core volcanic gases was 1.0, the fraction of water ejected at that time was about 5 per cent of the total amount now present in the hydrosphere. If the ratio p_{H_2O}/p_{H_2} was 0.50, the total amount ejected would have been only 2.5 per cent of the total amount now present. If the ratio had been 0.33, only 1.6 per cent of the present water would have been ejected in the pre-core stage.

There are far too many uncertainties in the propositions concerning the early atmosphere to permit us to take these results too seriously. If exchange between ocean water and atmospheric hydrogen were incomplete, then the fraction of water ejected in the pre-core stage would have been larger than the calculated value. Until more data become available on the probable temperature structure of the pre-core atmosphere, the matter must remain in an unsatisfactory state. What we can conclude is that the difference between the deuterium content of the hydrosphere and of volcanic gases could be explained satisfactorily in terms of the proposed mechanism.

If the proposed mechanism and atmospheric conditions are correct, approximately 1–5 per cent of the total water now present in the hydrosphere was ejected during this high-hydrogen stage. If the remainder of the water has been ejected at a nearly uniform rate, an age effect on the D/H ratio of ancient waters would probably be observed only in rocks more than 2 b.y. old. On the other hand, a strong age effect should be observed if a large fraction of the present hydrosphere has accumulated since the beginning of the Paleozoic Era.

SECOND STAGE

During the pre-core first stage, the chemistry of the atmosphere was proably determined in large part by the high percentage of hydrogen in volcanic gases. After the formation of the core, essentially all metallic iron was probably removed from at least the upper part of the mantle.

If we again take the system $MgO-FeO-Fe_2O_3-SiO_2$ as our model, the crystalline phases in the upper mantle would probably have become olivine, pyroxene, and magnetite. Other phases were probably present as well, and at depths of more than a few tens of kilometers polymorphic transformations would probably have altered the prevailing mineralogy. Nevertheless, it seems worth while to assess the consequences of a simplified model based on Muan and Osborn's (1956) data on the system $MgO-FeO-Fe_2O_3-SiO_2$. Magmas formed in the upper mantle by partial fusion would have an oxidation state defined by equilibrium between olivine, pyroxene, and magnetite:

$$olivine + O_2 \rightleftarrows pyroxene + magnetite.$$

None of the phases would be pure. If we take Urey and Craig's (1953) average for the composition of chondritic meteorites as the best estimate of the composition of the mantle, the olivine would have had a composition near $Fo_{75}Fa_{25}$ and the pyroxene the approximate composition of $En_{75}Fs_{25}$. The magnetite would have contained at least some MgO, Cr_2O_3, and TiO_2 in solid solution. A precise statement concerning the oxidation state of gases in equilibrium with these mineral phases is not yet possible. It does seem unlikely, however, that p_{O_2} would differ markedly from $10^{-7.1}$ atm, its value at the quaternary invariant point in the system $MgO\text{-}FeO\text{-}Fe_2O_3\text{-}SiO_2$, if melting started near $1255°$ C.

At this temperature and oxygen pressure

$$\frac{p_{H_2O}}{p_{H_2}} \approx 105$$

and

$$\frac{p_{CO_2}}{p_{CO}} \approx 37.$$

In these gases water would therefore be much more abundant than hydrogen, and CO_2 much more abundant than CO. Methane should be virtually absent. SO_2 would be the dominant sulfur species. If the total amount of melting in the mantle has been small, the chemistry of volcanic gases might be expected to remain nearly constant at this composition with time. Table 2 compares the composition of volcanic gases predicted on the basis of this model with the composition of average volcanic gases from Kilauea (Eaton and Murata, 1960). The temperature of the associated lavas was between *ca.* $1150°$ C and $1200°$ C. The agreement is exceedingly good and supports the validity of the simplified model.

The model also demands that the oxidation state of volcanic gases has not changed appreciably with time. Perhaps the best indication of the oxidation state of the gases can be obtained from the Fe^{+3}/Fe^{+2} ratio in magmas (Kennedy, 1948). The safest indication of this ratio is probably the Fe^{+3}/Fe^{+2} ratio in the chilled borders of intrusive rocks. Table 3 contains analyses of chill-facies samples from Triassic diabases of New Jersey, U. S. A., and of Nipissing diabase from Ontario, Canada, 1.8 ± 0.3 b.y. old. There is no significant difference between the oxidation state of these samples, and of Hawaiian tholeiitic basalt (Eaton and Murata, 1960) and therefore none between the oxidation state of the gases that would have been evolved from extrusive phases of these units. Green and Poldervaart (1955) have pointed out the lack of a time effect on the Fe_2O_3 content of basaltic lavas, so that the hypothesis of reasonably constant oxidation state of volcanic gases

TABLE 2. ACTUAL AND PREDICTED OXIDATION STATE OF AVERAGE HAWAIIAN VOLCANIC GASES

Average Hawaiian volcanic gases (volume per cent)		Ratios found	Ratios predicted
H_2O	79.31	$\dfrac{p_{H_2O}}{p_{H_2}} = 137$	105
H_2	0.58		
CO_2	11.61	$\dfrac{p_{CO_2}}{p_{CO}} = 31$	37
CO	0.37		
SO_2	6.48		
S_2	0.24		
H_2S	—		

TABLE 3. COMPOSITION OF THE CHILLED BORDERS OF SOME DIABASIC INTRUSIVE ROCKS

	Palisades diabase (H. H. Hess, unpub.)					Rocky Hill diabase (H. H. Hess, unpub.)	Nipissing diabase (Hriskevich, unpub.)	
	Pl 1	Pl 2	Pl 3	Pl 6	Pl 9	RH 2	312	166
SiO_2......	52.11	51.98	52.41	51.74	52.14	52.21	51.89	51.98
Al_2O_3.....	14.41	14.26	13.91	14.31	14.51	14.59	14.79	14.69
Fe_2O_3.....	1.25	1.28	1.02	1.35	1.70	1.46	1.60	0.85
FeO.......	8.91	8.96	9.10	8.90	8.51	8.75	7.95	8.70
MgO......	7.68	7.94	8.48	7.74	7.35	7.35	7.54	7.50
CaO.......	10.58	10.56	11.10	10.11	10.26	10.79	10.93	11.46
Na_2O.....	2.04	1.93	1.93	2.07	2.07	2.02	1.74	1.83
K_2O.......	0.73	0.80	0.54	1.11	0.86	0.63	0.89	0.51
H_2O^+......	0.77	0.86	0.26	1.08	1.01	0.60	1.29	1.35
H_2O^-......	0.10	0.11	0.13	0.22	0.25	0.16	0.05	0.08
P_2O_5......	0.13	0.14	0.13	0.13	0.15	0.14	0.06	0.06
TiO_2......	1.14	1.14	1.08	1.15	1.16	1.17	0.62	0.62
Cr_2O_3.....	—	—	—	—	—	—	—	—
MnO......	0.18	0.18	0.18	0.18	0.20	0.17	0.18	0.18
NiO.......	—	—	—	—	—	—	—	—
CO_2.......							0.22	0.04
	100.03	100.14	100.27	100.09	100.17	100.04	99.75	99.85
	G. Kahan, analyst						H. Baadsgaard, analyst	

emitted during all but the pre-core stage of the earth's history appears to be tenable.

Thus, the chemistry of the atmosphere must have been controlled in large part by the composition of such volcanic gases after the formation of the earth's core. This should have been true at least until the appearance of organisms capable of photosynthesis. The time when life processes became important for atmospheric chemistry is not known with any degree of certainty, but mineralogical evidence cited below suggests that free oxygen appeared more recently than 2.0 b.y. ago. If this is correct, there was a long period of time between the formation of the core and the appearance of free oxygen during which the chemistry of the atmosphere was controlled by the injection of Hawaiian-type volcanic gases. This period of atmospheric history is here called the "second stage".

The composition of the atmosphere during the second stage can be defined reasonably well. There is ample evidence that between 2 and 3 b.y. ago water was present at the earth's surface as standing bodies, and the assumption of temperatures similar to those of the present seems justified. Water from volcanoes would then join the hydrosphere, and the water content of the atmosphere would depend in large measure on the surface temperature. Since the amount of hydrogen was very much smaller than that required to reduce CO and CO_2 to methane, carbon dioxide would have remained to participate in the weathering process much as it is doing today. The concentration of CO_2 in the atmosphere would have been controlled by the kinetics of weathering and would probably not have risen to values much above its present value. Carbonates of calcium and magnesium would have precipitated from the oceans.

CO is unstable in such an atmosphere in all but trace amounts. Numerous reactions probably took place in its destruction, but two deserve special mention:

$$(CO)_g + (H_2)_g \rightleftharpoons \langle C \rangle_c + (H_2O)_g \tag{11}$$

and

$$(CO)_g + (H_2O)_g \rightleftharpoons (CO_2)_g + (H_2)_g. \tag{12}$$

Equation (11) is important at higher hydrogen pressures than equation (12). The crossover point is defined by the two equilibrium constants:

$$K_{11} = \frac{p_{H_2O}}{p_{H_2} \cdot p_{CO}}$$

$$K_{12} = \frac{p_{CO_2} \cdot p_{H_2}}{p_{CO} \cdot p_{H_2O}}.$$

When equilibrium is established in both reactions at 25° C,

$$\frac{K_{11}^{25°C}}{K_{12}^{25°C}} = \frac{10^{16.00}}{10^{5.00}} = \frac{p_{H_2O}^2}{p_{CO_2} \cdot p_{H_2}^2}$$

$$p_{H_2} = \frac{p_{H_2O}}{p_{CO_2}^{1/2}} \cdot 10^{-5.50}$$

If we use

$$p_{H_2O} = 3 \times 10^{-2} \text{ atm}$$

$$p_{CO_2} = 3 \times 10^{-4} \text{ atm}$$

then

$$p_{H_2} = 10^{-5.25} \text{ atm.}$$

At hydrogen pressures greater than $10^{-5.25}$ atm, CO and H_2 will react to form graphite and water. A slow shower of graphite may therefore have been associated with volcanic eruptions during a part of the second stage of atmospheric evolution.

One of the principal uncertainties in predictions concerning the chemistry of the atmosphere during the second state is introduced by the assumption concerning the SO_2 content of volcanic gases. If these gases contained as much SO_2 as Hawaiian gases, then p_{SO_2} would have been more than ten times greater than p_{H_2}. Hydrogen would have been used up by reaction (13):

$$(SO_2)_g + 3(H_2)_g \rightleftharpoons (H_2S)_g + 2(H_2O)_l; \tag{13}$$

at 25° C

$$K_{13}^{25°C} = 10^{+36.4} = \frac{p_{H_2S}}{p_{SO_2} \cdot p_{H_2}^3}$$

and

$$p_{H_2} = 10^{-12.1} \left(\frac{p_{H_2S}}{p_{SO_2}}\right)^{1/3}.$$

Since p_{H_2S} would have been smaller than p_{SO_2}, the hydrogen pressure would have been less than $10^{-12.1}$ atm. (Fig. 5). At this H_2 pressure CO would react with water to form CO_2 and H_2, and the small amount of hydrogen thus generated would be used up in reacting with SO_2. Therefore if the SO_2 content of volcanic gases was similar to that of the Hawaiian gases, hydrogen was essentially absent from the atmosphere during the second stage.

On the other hand, it seems much more likely that the ratio of p_{H_2O} to p_{SO_2} in volcanic gases during this stage was equal to the value demanded by Rubey's (1951) "excess volatiles". Then p_{SO_2} would have been less than one tenth of p_{H_2}. Under such circumstances the rate of escape of H_2 from the atmosphere probably was the most important single factor in determining p_{H_2} in the atmosphere.

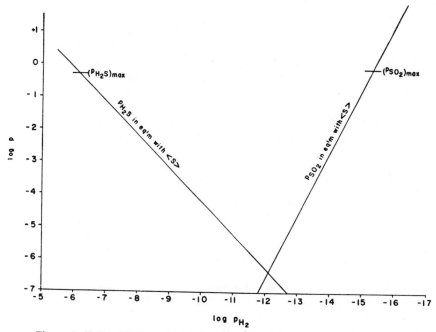

Figure 5. H_2S and SO_2 pressure as a function of the H_2 pressure during the second stage of the atmosphere

We can calculate the hydrogen pressure in the atmosphere on this basis. As before we have

$$N_{H_2} = \frac{1}{\lambda_{H_2}} \cdot \left(\frac{dN_{H_2}}{dt} \right).$$

If the escape half-life was 10^4 years, the rate of introduction of water vapor $10^{14.5}$ gm/year, and the ratio p_{H_2O}/p_{H_2} in volcanic gases equal to 100, then

$$p_{H_2} \approx \frac{10^4}{0.693} \cdot \frac{10^{14.5}}{9 \times 100} \cdot \frac{1}{5.1 \times 10^{21}} \text{ atm}$$

$$\approx 1 \times 10^{-6} \text{ atm.}$$

A variety of reactions involving CO and SO_2 probably modified the effect of escape on the hydrogen pressure. But it seems very likely that hydrogen was an extremely minor constituent of the atmosphere. Its actual partial pressure may therefore have been influenced strongly—as it is today—by the persistence of nonequilibrium conditions.

Photodissociation of water vapor in the upper atmosphere would have affected the hydrogen content of the atmosphere. Kuiper (1952, p. 312) finds that the present rate of oxygen production due to dissociation followed by hydrogen escape is about 2×10^{12} gm per year, an amount nearly sufficient to oxidize the calculated annual amount of hydrogen ejected from volcanoes. Recently Urey (1959) has again looked at this problem and concludes that Kuiper's figure is much too high. The photodissociation of water vapor followed by hydrogen escape has apparently affected the oxidation state of the atmosphere only very slightly.[1]

Oxidation of sediments was already taking place rather early in the Precambrian. At 25° C in the presence of hydrogen pressures less than $10^{-7.5}$ atm, the reaction

$$2 \langle Fe_3O_4 \rangle_c + (H_2O)_l \rightleftharpoons 3 \langle Fe_2O_3 \rangle_c + (H_2)_g \qquad (14)$$

proceeds to the right. It can be doubted that this reaction would proceed at a geologically significant rate in the absence of free oxygen. Possibly, trace quantities of H_2O_2 produced by the effect of ultraviolet radiation on surface waters catalyzed the reaction.

The SO_2 would undoubtedly have participated in oxidizing reactions. In a very schematic way we could summarize these reactions by the equations:

$$11 \langle Fe_2SiO_4 \rangle_c + 4(SO_2)_g \rightleftharpoons 10 \langle Fe_2O_3 \rangle_c + 2 \langle FeS_2 \rangle_c + 11 \langle SiO_2 \rangle_c \qquad (15)$$

and

$$\langle CaSiO_3 \rangle_c + (SO_2)_g + (H_2O)_l \rightleftharpoons (Ca^{+2})_{aq} + (SO_4^{-2})_{aq} + \langle SiO_2 \rangle_c + (H_2)_g. \qquad (16)$$

The equilibrium pressure of SO_2 demanded by these equilibria is less than 10^{-10} atm. Undoubtedly, however, p_{SO_2} would have been greater than this, just as the

---

[1] This conclusion might require some modification if the increased intensity of ultraviolet radiation due to the absence of an ozone layer during the first and second stage resulted in a very much larger amount of H_2O dissociation in the lower parts of the atmosphere.

value of p_{CO_2} is greater today than the pressure in equilibrium with calcium silicates and calcium carbonate.

The most abundant constituent of the atmosphere during the second stage would probably have been nitrogen, which has been accumulating in the atmosphere somewhat like argon. Its abundance must have increased with time but could not have exceeded its present value. A probable composition for the atmosphere during the second stage is presented in Table 4. The total pressure was almost certainly less than 1 atmosphere.

Available geologic evidence suggests that the atmosphere was sufficiently oxidized to permit formation of calcite in the earliest known sediments (Armstrong, 1960). The oxidation state was such that hematite was the stable oxide of iron in equilibrium with the atmosphere at least as early as 2.0 ± 0.2 b.y. ago in Huronian sediments (James, 1951; 1954) and probably as early as Keewatin time (Klinger, 1956), more than 2.5 b.y. ago (Goldich *et al.*, 1958). In order to appear, neither hematite nor calcite demands the presence of free oxygen, but the stability of hematite demands that the hydrogen pressure must have fallen very low. Both minerals are, of course, stable under present atmospheric conditions as well, so that their presence merely defines a lower limit for the degree of oxidation of the atmosphere during mid-Precambrian time.

The upper limit for the oxidation state of the atmosphere 2.0 to 2.5 b.y. ago is not so well defined. An indication that free oxygen was not present in the atmosphere comes from the apparent persistence of uraninite in detrital sediments formed during this period. Today uraninite, UO_2, normally weathers rapidly to a variety of hydrated oxides of U^{+6}. Oxidation is often followed by solution in surface waters and transport as a complex ion of U^{+6}. There is evidence that this sequence of events was not characteristic of the weathering of uraninite in the early Precambrian. Ramdohr (1958) has shown that the uranium-gold deposits of the Dominion Reef and Witwatersrand systems in South Africa, at Blind River in Canada, and at Serra de Jacobina in Brazil had a placer origin, and that they were later partially recrystallized during a period of metamorphism. Ramdohr's

TABLE 4. SOME PROBABLE LIMITS FOR THE COMPOSITION OF THE ATMOSPHERE DURING THE SECOND STAGE

Component	Pressure range (atm)
N_2	> 0.01; < 0.8
H_2O	3×10^{-3} to 3×10^{-2}
CO_2	*ca.* 3×10^{-4}
Ar	$> 10^{-4}$; $< 10^{-2}$
Ne	$> 10^{-6}$; $< 10^{-4}$
He	$> 10^{-6}$; $< 10^{-4}$
CH_4	$> 10^{-6}$; $< 10^{-4}$
NH_3	trace
SO_2	trace
H_2S	trace
O_2	nonequilibrium trace

conclusions, based mainly on evidence from ore microscopy, reinforce the conclusion reached by Liebenberg (1955; 1958) on similar grounds for the South African gold-uranium deposits. Liebenberg (1958) concludes that "there is no doubt as to the origin of the uraninite grains, the available evidence showing that the uraninite is one of the detrital minerals of a heavy-mineral suite which was deposited simultaneously with the other components of the conglomerates". Nel (1960) has recently summarized the pertinent field evidence and rules out a hydrothermal origin for the South African gold-uranium deposits.

Recently, Mair, Maynes, Patchett, and Russell (1960) have presented isotopic data on the age of monazite, zircon, uraninite, pyrite, pyrrhotite, sericite, and feldspar of the Blind River uranium deposits. They report that the apparent age of the uraninite (*ca.* 1.7 b.y.) is much less than that of the associated monazite and zircon (*ca.* 2.5 b.y.) and could be identical with the time of formation of the sediments. It is possible, however, that the later metamorphism of the sediments recorded by Mair *et al.* (1960) has reduced the apparent age of the uraninite from an original value in excess of 2.0 b.y. Goldich *et al.* (1958) suggest that an event 1.7 ± 0.1 b.y. ago marked the end of the Huronian period. This is supported by a 1.62 b.y. K/Ar age for a fresh biotite from sheared basal Huronian quartzite (Porter Township, between Blind River and Sudbury; R. K. Wanless, personal communication, 1960). Aldrich and Wetherill (1960) report a Rb/Sr age of 2.23 b.y. and a K/Ar age of 1.99 b.y. for pre-Cobalt biotite from Coleman Township, Ontario. These results indicate that Bruce Series sedimentation took place between 2.2 and 1.7 b.y. ago, and that the apparent age of the uraninite has been affected by post-depositional metamorphism.

A similar instance is recorded in the careful study by de Villiers, Lurger, and Nicolaysen (1958) of galenas from the Witwatersrand System. They suggest very strongly that the 2.1 b.y. age found for the uraninites from this system was reduced from an initial value equal to or greater than 2.5 b.y. during a "chemical rejuvenation" of the ores 2.1 b.y. ago. This event did not produce a gross change in the morphology of the uraninite grains. Despite the arguments of Davidson (1957), it seems to me that the evidence for a sedimentary origin of these deposits is overwhelming. The only reasonable explanation for the observed difference in the behavior of uraninite seems to be that the oxygen content of the atmosphere during the formation of these deposits was much lower than at present.

Uraninite is an extremely rare mineral in sediments today. Only two occurrences have been described (Steacy, 1953; Zeschke, 1960); neither is similar to those of the South African uraninite deposits, but Zeschke's (1960) observations show that rivers can transport uraninite in the form of minute grains.

If we could assume that the uraninite of the Witwatersrand type was in equilibrium with the atmosphere during weathering and transportation, we could set an upper limit to the prevailing oxygen pressure in the atmosphere. The system uranium-oxygen has been studied most recently by Grønvold (1955) and by Blackburn (1958) and Blackburn *et al.* (1958). At high temperatures UO_2, U_3O_8, and UO_3 are the stable oxides of uranium. Below 300° C several phases in the range $UO_{2.20}$ and $UO_{2.40}$ can be produced, but these may be metastable and are of no particular concern here. The boundary of importance is that between the stability field of U_3O_8 and UO_3 as defined by the reaction

$$2 \langle U_3O_8 \rangle_c + (O_2)_g \rightleftharpoons 6 \langle UO_3 \rangle_c \tag{17}$$

and by the equilibrium constant

$$K_{17} = \frac{1}{p_{O_2}}.$$

The value of this equilibrium constant can be calculated from the data of Blackburn *et al.* (1958). The older data of Coughlin (1954) are included in Table 5. The agreement between the two sets of data for UO_3 is satisfactory; the discrepancy for U_3O_8 is quite large. The values of p_{O_2} calculated from the two sets of data are shown in Table 6.

Both sets of data show that U_3O_8 can be in equilibrium with the atmosphere only when essentially no free oxygen is present. However, the rate of oxidation of U_3O_8 to UO_3 at oxygen pressures somewhat above the equilibrium pressure may be slow enough at 25° C to permit the accumulation of uraninite in placers. The values of p_{O_2} in Table 6 are therefore not definite maximum values for p_{O_2} during the formation of detrital uraninite deposits. They show, however, that U_3O_8 can oxidize to UO_3 even in the presence of minute traces of oxygen, and this suggests the virtual absence of atmospheric oxygen during the deposition of these sediments.

Davidson (1957) has argued against the possibility of a low oxygen pressure during the accumulation of the Witwatersrand System. He has pointed out that hematite occurs in the same assemblage of rocks and that this implies the availability of free atmospheric oxygen during sedimentation. The argument is not

TABLE 5. THERMOCHEMICAL DATA FOR THE FORMATION OF U_3O_8 AND UO_3
FROM U AND O_2 AT 25° C

	Blackburn *et al.* (1958)		Coughlin (1954)	
	$-\Delta H°$	$-\Delta S°$	$-\Delta H°$	$-\Delta S°$
U_3O_8	846.6	150.9	851.0	159.7
UO_3	290.1	59.3	290.5	60.0

TABLE 6. VALUES OF THE OXYGEN PRESSURE AT THE STABILITY BOUNDARY U_3O_8-UO_3
CALCULATED FROM THE DATA OF BLACKBURN *et al.* (1958) AND COUGHLIN (1954)

T (° C)	p_{O_2}	
	Blackburn	Coughlin
0	$10^{-25.8}$	$10^{-23.9}$
10	$10^{-24.5}$	$10^{-22.8}$
20	$10^{-23.3}$	$10^{-21.7}$
30	$10^{-22.1}$	$10^{-20.7}$

valid, because the equilibrium constants of reactions (14) and (17) show that there is a range of oxygen pressure (10^{-72} to 10^{-21}) in which U_3O_8 and Fe_2O_3 are stable together.

The conclusion that atmospheric oxygen was virtually absent during mid-Precambrian time could be checked by studying the oxidation state of other metals. Manganese is a suitable example since three oxidation states are represented in manganese minerals. Data for oxygen pressures at the stability boundary of lower oxides and silicates of manganese have been published by Muan (1959) and by Hahn and Muan (1960). Klingsberg and Roy (1959) have studied some of the reactions in the Mn-O-H system and have reported recently (Klingsberg and Roy, 1960) on the univariant equilibrium curves between Mn_3O_4 and Mn_2O_3 and between Mn_2O_3 and β-MnO_2 (pyrolusite). Goldsmith and Graf (1957) have studied the decomposition of rhodochrosite.

If Klingsberg and Roy's (1960) data are extrapolated to 25° C, the boundary between the stability field of bixbyite and pyrolusite lies at an oxygen pressure of about 10^{-3} atm. This figure may be uncertain by one or two orders of magnitude. Nevertheless, the bixbyite-pyrolusite boundary lies at the highest oxygen pressure of any of the mineral pairs so far considered and should prove a sensitive indicator of lower oxygen pressures in the Precambrian. Unfortunately Precambrian manganese deposits have normally been recrystallized so thoroughly by subsequent metamorphism that the original mineralogy has been obscured. The relationships in the system Mn-O-C-H are extremely pertinent to the problem of the history of the earth's atmosphere, and a further search for suitable, unmetamorphosed deposits seems amply warranted.

Several possible oxygen barometers can also be built around the oxidation of S^{-2} to SO_4^{-2}. For example, the equilibrium between galena and anglesite

$$\langle PbS \rangle_c + 2(O_2)_g \rightleftharpoons \langle PbSO_4 \rangle_c \tag{18}$$

depends on the prevailing oxygen pressure. The equilibria between sphalerite and zinkosite ($ZnSO_4$) or goslarite ($ZnSO_4 \cdot 7H_2O$) and the equilibrium between pyrrhotite and melanterite ($FeSO_4 \cdot 7H_2O$) are similarly a function of the oxygen pressure, although the partial pressure of water also influences the stability of the hydrates.

Kellogg and Basu (1960) have recently re-examined the thermodynamic properties of the system Pb-S-O. The free energy of reaction for the oxidation of one mole of galena to anglesite at 25° C according to these authors is -172.1 kcal, and the equilibrium p_{O_2} at this temperature is therefore 10^{-63} atm. In the presence of liquid water p_{H_2} would be $10^{-10.1}$ atm. For calculating the equilibrium between sphalerite and zinkosite we can take the data of Kubaschewski and Evans (1955), since the redetermination of the free energy of formation of ZnS by Curlook and Pidgeon (1958) is in reasonable agreement with that of the earlier compilation. A calculation similar to that for the galena-anglesite equilibrium shows that sphalerite and zinkosite are in equilibrium at 25° C at an oxygen pressure of 10^{-59} atm, and in equilibrium with water at a p_{H_2} of $10^{-12.1}$ atm.

These results show that magnetite oxidizes to hematite at a lower oxygen pressure than galena to anglesite or sphalerite to zinkosite, but that galena is oxidized to anglesite and sphalerite to zinkosite before U_3O_8 is oxidized to UO_3. Perhaps this explains the presence of a normal "hydrothermal" assemblage of minerals in the uranium-gold ores of Witwatersrand type. Galena and sphalerite may well have been present as detrital minerals in these sediments. Rounded pebbles of pyrite are certainly common, and there may well have been other sulfides which are today oxidized rapidly at the outcrop of ore deposits. During metamorphism these sulfides have been partially or wholly recrystallized. I believe that the hypothesis of the virtual absence of oxygen during much of the Precambrian would solve the mineralogic problem of these deposits that has been posed by the proponents of a hydrothermal origin. The search for detrital sulfides in less metamorphosed sedimentary rocks more than 1.5 b.y. old should prove fruitful in this connection.

The evidence thus suggests that between about 1.8 and 2.5 b.y. ago the atmosphere had a composition within the range predicted in Table 4 for the second stage of the evolution of the earth's atmosphere. There are no geologic data to check the proposal that the second stage started much before 2.5 b.y. ago. The fact that detrital uraninite persisted during mid-Precambrian time suggests that oxidation due to the photodissociation of water vapor followed by hydrogen escape was not sufficiently rapid to permit the accumulation of free oxygen in the atmosphere. This observation also suggests that the combined oxidative effect of the photodissociation of water vapor and of photosynthesis was not sufficiently pronounced to yield appreciable amounts of free atmospheric oxygen. This does not, of course, demand that life originated more recently than 1.8 b.y. ago but only that the rate of oxygen production was insufficient to overcome the reducing effect of the introduction of volcanic gases and of the loss of oxygen due to the oxidation of surface materials. The end of the second stage of atmospheric evolution can be set as the time when the combined effect of photodissociation of water vapor and photosynthesis first produced free oxygen in equilibrium amounts. If the evidence from detrital uraninites is valid, this took place more recently than about 1.8 b.y. ago.

THIRD STAGE

The beginning of the third stage of the history of the atmosphere has been defined as the time when the combined rate of oxygen production due to photosynthesis and photodissociation of water vapor exceeded the rate of consumption of oxygen by the oxidation of volcanic gases, organic remains, and sediments and rocks exposed to weathering. The accumulation of free oxygen has been contingent on production of a continuing surplus of oxygen, and the present amount of free oxygen must have accumulated very gradually. At the moment, we can set only rather broad limits on the variation of p_{O_2} with time since the beginning of the third stage.

A balance sheet for the production of atmospheric oxygen and for its present distribution is a convenient starting point. The most recent compilation of this kind (Hutchinson, 1954) must now be modified and can be expanded (Table 7). Oxygen production by photosynthesis followed by incomplete decay of organic

TABLE 7. PRODUCTION AND USE OF OXYGEN

Total estimated production
By photosynthesis in excess of decay . . . . 181×10^{20} gm
By photodissociation of water vapor followed by hydrogen escape . .   1

Total . . . . 132×10^{20} gm

Total estimated use
Oxidation of ferrous iron to ferric iron during weathering   14
Oxidation of S^{-2} to SO_4^{-2} during weathering   12
Oxidation of volcanic gases
  CO to CO_2   10
  SO_2 to SO_3   11
  H_2 to H_2O . . . . <150

Total . . . . $<197 \times 10^{20}$ gm

Free in atmosphere $\sim 12 \times 10^{20}$ gm

matter appears to be much more important than oxygen production by photodissociation of water vapor followed by hydrogen escape. The estimate for the former source comes from Rubey's (1951) estimate of the total amount of organic carbon in sediments. The estimate for photolysis of water vapor is taken from Urey's (1959) analysis.

The estimate of the use of oxygen in the oxidation of Fe^{+2} to Fe^{+3} is taken from Hutchinson's (1954) table. The oxygen consumption in the oxidation of S^{-2} to SO_4^{-2}, CO to CO_2, SO_2 to SO_3, and H_2 to H_2O is based on Rubey's (1951) estimate of total excess C, excess S, and H_2O and on the oxidation state of volcanic gases predicted for all but the first stage of the earth's atmosphere. The consumption of oxygen in the oxidation of H_2 has been computed assuming no loss of H_2 from the upper atmosphere, and therefore must approximate an upper limit.

Despite its many shortcomings, Table 7 points out important relationships, particularly the relatively small amount of oxygen now in the atmosphere. Further, it seems that the oxidation of H_2 to H_2O has been a major user of oxygen and that the oxidation of Fe^{+2}, S^{-2}, CO, and SO_2 has played a rather subordinate role.

Production and use can be balanced in Table 7 by permitting some hydrogen escape from the upper atmosphere, but no reliable figure for the mean percentage of hydrogen escape can be obtained because the estimate both of the total amount of organic carbon in sediments and of the H_2O/H_2 ratio in volcanic gases is uncertain by a factor of 2. The data do however rule out the idea that the oceans have been produced by the oxidation of hydrogen captured by the earth from interplanetary space. About $12,000 \times 10^{20}$ gm of oxygen would have been needed for this purpose.

The time dependence of the oxygen content of the atmosphere during the third stage is difficult to estimate accurately, chiefly because the rate of accumulation of oxygen was determined by the small difference between two much faster rates.

Nevertheless, there is evidence that the oxygen content of the atmosphere increased essentially linearly with time. The rate of oxygen production by photosynthesis in excess of plant decay during a particular period is reflected in the carbon content of sediments deposited during that period. There seems to be no well-defined trend of the carbon content in sediments of similar origin (Pettijohn, 1943). This suggests that the rate of oxygen production by this mechanism has been roughly constant during the third stage. Since there is no good reason to suspect variations in the rate of photodissociation of water vapor in the upper atmosphere during the third stage, the total rate of oxygen production during this time was probably reasonably constant.

The rate of consumption of oxygen in the past was determined largely by the intensity of volcanic activity. This in turn was related to the intensity of orogenic activity. No definitive data on the time dependence of the intensity of orogenic activity on a world-wide basis are available. Certainly the intensity has fluctuated with time, but it seems unlikely that there has been a continuing trend in this rate for a large part of the third stage. Perhaps the best indication of this is the reasonably steady rate of precipitation of calcium and magnesium carbonates, at least since the beginning of the Paleozoic Era. The rate of precipitation of these carbonates is directly related to the rate of volcanic activity. The time lag between changes in volcanic activity and carbonate sedimentation is geologically insignificant, so that a precise determination of past volcanic activity is possible, at least in principle, if the contribution of CO_2 from weathering is known.

The hypothesis of a rate of oxygen production essentially constant when taken over periods on the order of 100 million years seems attractive. The duration of the third stage is not known and is difficult to assess. Certainly the oxygen pressure in the atmosphere must have been an appreciable fraction of its present value when land animals evolved during late Paleozoic time. The oxygen demands of the large Carboniferous insects cannot be determined accurately, but it seems most unlikely that p_{O_2} could have been less than one tenth of its present value during their lifetime. Unfortunately this restriction on the possible range of p_{O_2} is not particularly helpful in limiting our thinking on the time of origin of the third stage. If stage three started 1.5 b.y. ago, p_{O_2} would have been about 80 per cent of its present value during Carboniferous time; if the start of stage three was only 0.5 b.y. ago, p_{O_2} would have been about 50 per cent of its present value during Carboniferous time, provided the oxygen pressure increased at a linear rate. It seems to me that the beginning of stage three cannot at present be bracketed more closely than between 1.8 and 0.5 b.y. ago.

During the accumulation of free oxygen, the nitrogen, neon, argon, krypton, and xenon content of the atmosphere must have approached its present value gradually. The other constituents should have been present in about their present concentration.

SUMMARY, A CRITIQUE AND SOME IMPLICATIONS

The history of the atmosphere has been divided into three stages. During the first, the input into the atmosphere consisted largely of volcanic gases in equilibrium with magmas generated in a mantle which still contained metallic iron. The partial pressure of oxygen in such gases was probably about $10^{-12.5}$ atm, and the temperature on ejection from volcanoes on the order of 1200° C. Tnese gases consisted largely of H_2, H_2O, and CO, with minor amounts of N_2, CO_2, and H_2S. Hydrogen was probably the most important constituent. On cooling, the CO and CO_2 reacted with H_2 to form methane. Nitrogen may have reacted to form ammonia, provided the rate of escape of hydrogen from the atmosphere was sufficiently slow to permit the build-up of an appreciable hydrogen pressure. If surface temperatures were similar to those at present, nearly all the water condensed, and the main constituents of the atmosphere were CH_4 and H_2 (Table 8).

The duration of the first stage was determined by the time that elapsed between the accretion of the earth and the removal of metallic iron from the upper mantle. It seems unlikely that this process took more than 0.5 b.y. After the removal of the iron phase, crystallization in the upper mantle probably proceeded until olivine, pyroxene, and magnetite were among the mineral phases. Subsequent melting and volcanism was accompanied by the ejection of gases much more highly oxidized that those of the first stage. If the available data on the system MgO-FeO-Fe_2O_3-SiO_2 apply to this situation, the oxidation state of these gases was very similar to that of Hawaiian volcanic gases today. It seems likely, therefore, that the oxidation state of volcanic gases has not changed a great deal since the beginning of the second stage.

Water was the dominant component in these volcanic gases; CO_2, CO, H_2, SO_2, and N_2 were minor constituents. The atmosphere during the second stage contained largely N_2 with minor amounts of CO_2 and H_2O (Table 8). Free oxygen was not present in equilibrium amounts, since the rate of oxygen production by

TABLE 8. SUMMARY OF DATA ON THE PROBABLE CHEMICAL COMPOSITION OF THE ATMOSPHERE DURING STAGES 1, 2, AND 3

	Stage 1	Stage 2	Stage 3
Major components $P > 10^{-2}$ atm	CH_4 H_2 (?)	N_2	N_2 O_2
Minor components $10^{-2} < P < 10^{-4}$ atm	H_2 (?) H_2O N_2 H_2S NH_3 Ar	H_2O CO_2 Ar	Ar H_2O CO_2
Trace components $10^{-4} < P < 10^{-6}$ atm	He	Ne He CH_4 NH_3 (?) SO_2 (?) H_2S (?)	Ne He CH_4 Kr

photodissociation was less than that required to oxidize volcanic gases introduced into the atmosphere.

The second stage ended when oxygen input exceeded oxygen use; that is, when the combined rate of oxygen production by photosynthesis, followed by burial of organic carbon with sediments, and by photodissociation of water vapor, exceeded the rate of use in oxidizing surface rocks and the products of volcanoes and hot springs. The time of transition from the second to the third stage is not yet well determined. Rather scant evidence from the mineralogy of Precambrian sedimentary uranium deposits suggests that free oxygen in appreciable amounts was not present in the atmosphere *ca.* 1.8 b.y. ago. The oxygen requirements of Paleozoic land animals suggest that the oxygen content of the atmosphere at the end of the Paleozoic Era was a large fraction of its present value. Thus the second stage probably ended between 0.5 and 1.8 b.y. ago.

During the third stage oxygen gradually accumulated in the atmosphere. It is difficult to be sure of the time function describing the oxygen pressure. The increase may well have been roughly linear, since there is little evidence that the rate of volcanic activity and the rate of burial of carbon with sediments differed markedly in successive time intervals of 100 m.y.

The validity of the proposed model for the development of the atmosphere depends on the validity of the assumptions used in its construction. Chief among these are the assumptions that the data on the system MgO-FeO-Fe_2O_3-SiO_2 can be applied directly to melting phenomena in the upper mantle, and that the temperature structure of the atmosphere has always been similar to that of the present day. Phase studies at high pressures and in systems approaching more closely the composition of the upper mantle are needed to assess the validity of the first assumption.

It should soon be possible to check the second assumption, at least in semiquantitative fashion, by applying to this problem the results of recent data on the chemistry of the upper atmosphere of the earth, and possibly data yet to come on the temperature structure of the atmosphere of Mars, Venus, and the major planets.

If the proposed model is nearly correct, a highly reducing atmosphere existed for only a small fraction of earth history. If the processes leading to the first forms of life required such a highly reducing environment, then life on earth must be a very ancient phenomenon. There seems to be no evidence at present to suggest that life did not originate more than 3 b.y. ago.

The proposed model suggests that the chemistry of the oceans has not changed a great deal since the beginning of the second stage in early Precambrian time. This model suggests a history very different from that proposed by Krynine (1960) and is in essential agreement with that proposed by Rubey (1951).

ACKNOWLEDGMENTS

The growth of the ideas in this paper has been stimulated by discussions with members of the staff at Princeton. In particular, I thank Professor A. G. Fischer for guidance regarding biologic and paleontologic matters, Professor H. H. Hess for suggestions regarding the oxidation state of magmas, and Professor E. Sampson for help with the section on uranium deposits. Professor A. F. Buddington's comments were always helpful, and this paper is dedicated to him with deep affection.

REFERENCES CITED

Aldrich, L. T., and Wetherill, G. W., 1960, Rb-Sr and K-A ages of rocks in Ontario and northern Minnesota: Jour. Geophys. Research, v. 65, p. 337–340

Armstrong, H. S., 1960, Marbles in the "Archean" of the southern Canadian shield: 21st Internat. Geol. Cong. (Norden) Rept., pt. 9, p. 7–20

Ault, W. U., and Kulp, J. L., 1959, Isotopic geochemistry of sulphur: Geochim. et Cosmochim. Acta, v. 16, p. 201–235

Blackburn, P. E., 1958, Oxygen dissociation pressures over uranium oxides: Jour. Phys. Chemistry, v. 62, p. 897–902

———— Weissbart, J., and Gulbransen, E. A., 1958, Oxidation of uranium dioxide: Jour. Phys. Chemistry, v. 62, p. 902–908

Brown, H., 1952, Rare gases and the formation of the earth's atmosphere, p. 258–266 in Kuiper, G. P., Editor, The atmospheres of the earth and planets, 2d ed.: Chicago, Univ. Chicago Press, 434 p.

Coughlin, J. P., 1954, Contributions to the data on theoretical metallurgy. XII. Heats and free energies of formation of inorganic oxides: U. S. Bur. Mines Bull. 542, 80 p.

Curloook, W., and Pidgeon, L. M., 1958, Determination of the standard free energies of formation of zinc sulfide and magnesium sulfide: Am. Inst. Min. Met. Eng. Trans., v. 212, p. 671–676

Damon, P. E., and Kulp, J. L., 1958, Inert gases and the evolution of the atmosphere: Geochim. et Cosmochim. Acta, v. 13, p. 280–292

Davidson, C. F., 1957, On the occurrence of uranium in ancient conglomerates: Econ. Geology, v. 52, p. 668–693

de Villiers, J. W. L., Burger, A. J., and Nicolaysen, L. O., 1958, The interpretation of age measurements on the Witwatersrand uraninite: 2d United Nations Internat. Conf. on the Peaceful Uses of Atomic Energy Proc., v. 2, p. 237–238

Eaton, J. P., and Murata, K. J., 1960, How volcanoes grow: Science, v. 132, p. 925–938

Goldich, S. S., Nier, A. O., Krueger, H. W., and Hoffman, J. H., 1958, K-A dating of Precambrian iron formations (Abstract): Am. Geophys. Union Trans., v. 39, p. 516

Goldsmith, J. R., and Graf., D. L., 1957, The system CaO-MnO-CO₂: Solid-solution and decomposition relations: Geochim. et Cosmochim. Acta, v. 11, p. 310–334

Green, J., and Poldervaart, A., 1955, Some basaltic provinces: Geochim. et Cosmochim. Acta, v. 7, p. 177–188

Grønvold, F., 1955, High-temperature x-ray study of uranium oxides in the UO₂-U₃O₈ region: Jour. Inorg. and Nuclear Chem., v. 1, p. 357–370

Hahn, W. C., Jr., and Muan, A., 1960, Studies in the system Mn-O: the Mn₂O₃-Mn₃O₄ and Mn₃O₄-MnO equilibria: Am. Jour. Sci., v. 258, p. 66–78

Hutchinson, G. E., 1954, The biochemistry of the terrestrial atmosphere, p. 371–433 in Kuiper, G. P., Editor, The earth as a planet: Chicago, Univ. Chicago Press, 751 p.

James, H. L., 1951, Iron formation and associated rocks in the Iron River district, Michigan: Geol. Soc. America Bull., v. 62, p. 251–266

———— 1954, Sedimentary facies of iron-formation: Econ. Geology, v. 49, p. 235–293

Kellogg, H. H., and Basu, S. K., 1960, Thermodynamic properties of the system Pb-S-O to 1100° K: Am. Inst. Min. Met. Eng. Trans., v. 218, p. 70–81

Kennedy, G. C., 1948, Equilibrium between volatiles and iron oxides in igneous rocks: Am. Jour. Sci., v. 246, p. 529–549

Klinger, F. L., 1956, Geology of the Soudan mine and vicinity, p. 120–134 in Schwartz, G. M., Editor, Precambrian of northeastern Minnesota: Geol. Soc. America Guidebook, 235 p.

Klingsberg, C., and Roy, R., 1959, Stability and interconvertibility of phases in the system Mn-O-OH: Am. Mineralogist, v. 44, p. 819–838

—————— 1960, Solid-solid and solid-vapor reactions and a new phase in the system Mn-O: Am. Ceram. Soc. Jour., v. 43, p. 620–626

Kokubu, Nobuhide, Mayeda, T., and Urey, H. C., 1961, Deuterium content of minerals, rocks and liquid inclusions from rocks: Geochim. et Cosmochim. Acta, v. 21, p. 247–256

Krynine, P. D., 1960, Primeval ocean (Abstract): Geol. Soc. America Bull., v. 71, p. 1911

Kubaschewski, O., and Evans, E. L., 1955, Metallurgical thermochemistry, 2d ed., London, Pergamon Press, 410 p.

Kuiper, G. P., 1952, Planetary atmospheres and their origin, p. 306–405 in Kuiper, G. P., Editor, The atmospheres of the earth and planets: Chicago, Univ. Chicago Press, 434 p.

Liebenberg, W. R., 1955, The occurrence and origin of gold and radioactive minerals in the Witwatersrand system, the Dominion reef, the Ventersdorp contact reef and the Black reef: Geol. Soc. South Africa Trans., v. 58, p. 101–227

—————— 1958, The mode of occurrence and theory of origin of the uranium minerals and gold in the Witwatersrand ores: 2d United Nations Internat. Conf. on the Peaceful Uses of Atomic Energy Proc., v. 2, p. 379–387

Mair, J. A., Maynes, A. D., Patchett, J. E., and Russell, R. D., 1960, Isotopic evidence on the origin and age of the Blind River uranium deposits: Jour. Geophys. Research, v. 65, p. 341–348

Muan, A., 1955, Phase equilibria in the system $FeO-Fe_2O_3-SiO_2$: Jour. Metals, v. 7, p. 965–976

—————— 1959, Stability relations among some manganese minerals: Am. Mineralogist, v. 44, p. 946–960

Muan, A., and Osborn, E. F., 1956, Phase equilibria at liquidus temperatures in the system $MgO-FeO-Fe_2O_3-SiO_2$: Am. Ceram. Soc. Jour., v. 39, p. 121–140

Nel, L. T., 1960, The genetic problem of uraninite in the South African gold-bearing conglomerates: 21st Internat. Geol. Cong. (Norden) Rept., pt. 15, p. 15–25

Pettijohn, F. J., 1943, Archean sedimentation: Geol. Soc. America Bull., v. 54, p. 925–972

Ramdohr, P., 1958, Die Uran- und Goldlagerstätten Witwatersrand-Blind River District-Dominion Reef-Serra de Jacobina: Erzmikroskopische Untersuchungen und ein geologischer Vergleich: Deutsche Akad. Wiss. Berlin Abh., Klasse für Chemie, Geologie und Biologie, No. 3, 35 p.

Rubey, W. W., 1951, Geologic history of sea water. An attempt to state the problem: Geol. Soc. America Bull., v. 62, p. 1111–1148

—————— 1955, Development of the hydrosphere and atmosphere, with special reference to probable composition of the early atmosphere, p. 631–650 in Poldervaart, Arie, Editor, Crust of the earth: Geol. Soc. America Special Paper 62, 762 p.

Spitzer, L., Jr., 1952, The terrestrial atmosphere above 300 km, p. 211–247 in Kuiper, G. P., Editor, The atmospheres of the earth and planets, 2d ed.: Chicago, Univ. Chicago Press, 434 p.

Steacy, H. R., 1953, An occurrence of uraninite in a black sand: Am. Mineralogist, v. 38, p. 549–550

Thode, H. G., Macnamara, J., and Fleming, W. H., 1953, Sulphur isotope fractionation in nature and geological and biological time scales: Geochim. et Cosmochim. Acta, v. 3, p. 235–243

Urey, H. C., 1952a, On the early chemical history of the earth and the origin of life: Nat. Acad. Sci. Proc., v. 38, p. 351–363

—————— 1952b, The planets, their origin and development: New Haven, Yale Univ. Press, 245 p.

—————— 1959, The atmosphere of the planets: Handbuch der Physik, v. 52, p. 363–418

Urey, H. C., and Craig, H., 1953, The composition of the stone meteorites and the origin of the meteorites: Geochim. et Cosmochim. Acta, v. 4, p. 36–82

Zeschke, G., 1960, Transportation of uraninite in the Indus River, Pakistan: South Africa Geol. Soc. Trans. 63 (preprint)

APPENDIX

VALUES FOR SOME EQUILIBRIUM CONSTANTS

$T(° C)$	$\log K_1$	$\log K_2$	$\log K_3$	$\log K_4$	$\log K_5$
25	40.07	45.08	19.93	−33.91	12.92
900	8.16	8.02	−1.65	−5.32	1.45
1000	7.28	7.03	−2.23	−4.54	1.13
1100	6.54	6.19	−2.72	−3.87	0.85
1200	5.89	5.47	−3.15	−3.29	0.61
1300	5.32	4.83	−3.54	−2.78	0.40
1400	4.83	4.28	−3.85	−2.34	0.21

23

Reprinted from *Nature*, **212**(5067), 1225–1226 (1966)

PLANETARY SCIENCE

Primitive Atmosphere of the Earth

THIS communication reports the results of calculations on the temperature structure in a model of the primitive atmosphere, which suggest that during the early history of the Earth a reducing atmosphere could be stable against gravitational escape for as long as 10^9 yr. These results may help partly to resolve the considerable amount of controversy in the literature about the composition of the atmosphere of the Earth 3×10^9–4×10^9 yr ago when the first living organisms are thought to have been synthesized.

Oparin[1] has argued that the atmosphere was reducing in character and suggested that the first organic compounds may have been produced under such conditions. Miller and Urey[2] have presented arguments in favour of such an atmosphere, suggesting methane, ammonia, nitrogen and hydrogen as the major components of the atmosphere at that time; the atmosphere is a remnant of the primordial gases that the Earth probably acquired at the time of its condensation out of the solar nebula. Miller[3] has demonstrated that such a mixture of gases, when exposed to electric discharges in the laboratory, actually produce amino-acids.

Abelson[4], on the other hand, rejects the methane–ammonia atmosphere hypothesis, mainly on the grounds that the Earth seems to have completely lost its primordial atmosphere. This important conclusion is based on the observation of Brown[5] that noble gases like neon, argon-36, krypton and xenon are almost completely absent from the Earth. It has therefore been argued that if xenon-131 escaped from the Earth, the whole atmosphere must have been lost at the same time. It has been suggested that the present atmosphere and hydrosphere of the Earth originated through outgassing from the interior, mainly through the exhalation of volcanic gases[6,7]. Today, water and carbon dioxide are the principal gases emanating from volcanoes. Using these arguments, Abelson suggests that the composition of the primitive atmosphere of the Earth was dominated by carbon monoxide, carbon dioxide, nitrogen and hydrogen. Radiation interacting with this mixture would produce hydrocyanic acid, and would eventually yield amino-acids and other substances of biological interest.

Although it is widely accepted that the present atmosphere of the Earth is a result of outgassing, it is not certain if the composition of the volcanic gases has remained the same throughout the history of the Earth. Holland[7] has presented a model for the primitive atmosphere, which he derived from outgassing from the interior. This includes the novel suggestion that the gases which emanated from volcanoes were highly reducing in character because in the early history of the Earth free iron was still present in large quantities in the crust and upper mantle. The composition of the atmosphere at that time would be mainly methane, with small amounts of hydrogen, ammonia and water vapour. The stability of this atmosphere would have depended on the amount of free hydrogen that was available to prevent the methane from being oxidized to carbon dioxide.

Hydrogen is the lightest of all elements, and therefore escapes from the gravitational field of the Earth relatively rapidly. The amount of hydrogen in the atmosphere depends partly on the temperature of the exosphere of the Earth, and partly on the rate of diffusion of this gas from the lower to the upper atmosphere.

We have attempted to calculate the exospheric temperature for the model suggested by Holland. It is assumed that the upper atmosphere is in conductive equilibrium, and that the following equation describes the energy balance:

$$\varkappa \frac{dT}{dz} = \int_z^\infty Q \, dz - \int_z^\infty R \, dz$$

where $\varkappa$ is the thermal conductivity of the gas which varies as $\sqrt{T}$, Q is the amount of energy available for heating, and R is the thermal emission by the atmospheric constituents.

If we integrate this equation using the observed value of Q, the theoretically derived rates of emission R, and a $\varkappa$ value for the present composition of the upper atmosphere (3×10^3 ergs cm^{-1} sec^{-1} deg^{-1} at $0°$ C), we find the average exospheric temperature of the Earth to be about $1,500°$ K, in agreement with what is observed. In the Holland model of the primitive atmosphere, where methane is the principal constituent, however, the situation is very different. The upper atmosphere will mainly contain free hydrogen with a thermal conductivity of 17.3×10^3 ergs cm^{-1} sec^{-1} deg^{-1} (at $0°$ C). This is higher by about a factor of five than the conductivity of a nitrogen–oxygen mixture. Our calculations indicate that mainly because of the high conductivity, and also as a result of the high emission rates, R, in the infra-red of methane and its dissociation products, the exospheric temperature for this primitive model atmosphere would be considerably less than the present value of $1,500°$ K and would lie in the range of $500°$–$900°$ K, depending on the various amounts of methane and hydrogen assumed in the model. Preliminary calculations of escape indicate that for an exospheric temperature of $600°$ K, a methane atmosphere, with small amounts of ammonia, nitrogen and hydrogen, could be stable against gravitational escape for as long as 10^9 yr.

These results imply that the atmosphere required for the synthesis of amino-acids, in the Miller experiment, could have been present on the Earth during the first 10^9 yr.

S. I. Rasool

Institute for Space Studies,
Goddard Space Flight Center, NASA,
New York, and
Department of Meteorology and Oceanography,
New York University,
New York.

W. E. McGovern

Department of Meteorology and Oceanography,
New York University,
New York.

Received November 4, 1966.

[1] Oparin, A. I., *The Origin of Life* (Dover Publications, New York, 1953).

[2] Miller, S. L., and Urey, H. C., *Science*, **130**, 245 (1959).

[3] Miller, S. L., *Science*. **117**, 528 (1953).

[4] Abelson, P. H., *Proc. U.S. Nat. Acad. Sci.*, **55**, 1365 (1966).

[5] Brown, H., "Rare Gases and the Formation of the Earth's Atmosphere", in *The Atmospheres of the Earth and Planets* (edit. by Kuiper, G.), 258 (University of Chicago Press, 1952).

[6] Rubey, W. W., "Development of the Hydrosphere and Atmosphere, with Special Reference to Probable Composition of the Early Atmosphere", in *Crust of the Earth* (edit. by Poldervaart, A.), 631 (Geological Society of America, New York, 1955).

[7] Holland, H. D., in *Petrologic Studies: a Volume in Honor of A. F. Buddington* (edit. by Engel, A. E., James, H., and Leonard, B. F.), 447 (Geological Society of America, New York, 1962).

24

Reprinted from *Jour. Atmos. Sci.*, **22**(3), 225 (1965) with the permission of the
American Meteorological Society.

On the Origin and Rise of Oxygen Concentration in the Earth's Atmosphere[1]

L. V. BERKNER AND L. C. MARSHALL

Southwest Center for Advanced Studies, Dallas, Texas

(Manuscript received 23 November 1963, in revised form 3 March 1965)

ABSTRACT

This study adopts the basic premise that the Earth was without a primordial atmosphere and that its secondary atmosphere has arisen primarily from local heating and volcanic action associated with continent building. Since no oxygen can be derived in this way, the initial formation of oxygen from photochemical dissociation of water vapor is found to provide the primitive oxygen in the atmosphere. Because of the Urey self-regulation of this process by shielding H_2O vapor with O_2, O_3, and CO_2, primitive oxygen levels cannot exceed $O_2 \sim 0.001$ present atmospheric level (P.A.L.). The analysis of photochemistry of the atmospheric constituents is made possible by measurements of solar radiation with space vehicles and the now excellent data on uv absorption. The rates of oxidation of lithospheric materials are examined in this primitive atmosphere and, because of active species of oxygen present, found adequate to make unnecessary the usual assumption of high oxygenic levels in the pre-Cambrian eras to account for such lithospheric oxides. The appearance of an oxygenic atmosphere awaits a rate of production that exceeds O_2 photodissociation and loss.

The rise of oxygen from the primitive levels can only be associated with photosynthetic activity, which in turn depends upon the range of ecologic conditions at any period. Throughout the pre-Cambrian, lethal quantities of uv will penetrate to 5 or 10 meters depth in water. This limits the origin and early evolution of life to benthic organisms in shallow pools, small lakes or protected shallow seas where excessive convection does not bring life too close to the surface, and yet where it can receive a maximum of non-lethal but attenuated sunlight. Life cannot exist in the oceans generally and pelagic organisms are forbidden. Atmospheric oxygen cannot rise significantly until continental extensions and climatic circumstances combine to achieve the necessary extent of this protected photosynthesis, over an area estimated at 1 to 10 per cent of present continental areas.

When oxygen passes ~ 0.01 P.A.L., the ocean surfaces are sufficiently shadowed to permit widespread extension of life to the entire hydrosphere. Likewise, a variety of other biological opportunities arising from the metabolic potentials of respiration are opened to major evolutionary modification when oxygenic concentration rises to this level. Therefore, this oxygenic level is specified as the "first critical level" which is identified by immediate inference with the explosive evolutionary advances of the Cambrian period (-600 m.y.). The consequent rate of oxygen production is expedited.

When oxygen passes 0.1 P.A.L., the land surfaces are sufficiently shadowed from lethal uv to permit spread of life to dry land. This oxygenic level is specified as the "second critical level" and by immediate inference is identified with the appearance and explosive spread of evolutionary organisms on the land at the end of the Silurian (-420 m.y.).

Subsequently, oxygen must have risen rapidly to the Carboniferous. Because of the phase lag in the process of decay, the change of atmospheric oxygen may have fluctuated as a damped saw-toothed oscillation through late Paleozoic, Mesozoic, and even Cenozoic times in arriving at the present quasi-permanent level.

The physical and evolutionary evidence concerning the development of the Earth appears fully to support such a model, which removes the so-called "puzzle" of the Cambrian evolutionary explosion and of certain subsequent radical evolutionary advances. In particular, consideration of the rise of oxygen permits a view of the history of the Earth in a rather new and more advanced perspective.

It seems reasonable that the atmosphere of Mars appears similar to the above model of the Earth's in its primitive state. Life on Mars would therefore be subject to the same restrictive ecology as existed on Earth during Archaeozoic or early Proterozoic eras.

[1] This research was supported in part by the National Science Foundation under grant GP-768, and by the United States Weather Bureau under grant Cwb 10531. This paper is based on a report submitted to the NSF dated 26 November 1963.

25

Reprinted from *Space Life Sci.*, **2**, 5–17 (1970)

THE HISTORY OF ATMOSPHERIC OXYGEN

M. G. RUTTEN

Geological Institute of the State University of Utrecht, The Netherlands

(Received 6 June, 1969)

Abstract. A primeval anoxygenic terrestrial atmosphere having been postulated on astronomical grounds, experiments using simulated conditions have shown that the formation of 'organic' molecules by abiogenic processes will proceed freely in such an environment.

Atmospheric oxygen will at first be limited to 0.001 PAL through the Urey mechanism which inhibits further dissociation of water above this level. All atmospheric oxygen exceeding this level must be biogenic and produced by photosynthesis. Molecular fossils prove its existence 2.7 billion years ago. Sedimentary ores, notably pyrite sands of gold-uranium reefs and banded iron formations, attest to the existence of an atmosphere with 'little' oxygen up to 1.8 billion years ago. Geochemistry does not, however, supply us with data as to the level of oxygen at that time. The Pasteur Point, on the other hand, at which microbes change from fermentation to respiration and vice versa, is a powerful regulating factor situated at 0.01 PAL of free oxygen.

It is postulated that the primeval atmosphere of Lower and Middle Precambrian was limited to this level of free oxygen. At this level pre-life – the formation of 'organic' compounds through inorganic processes – still exists. Pre-life and early life therefore were coexistent for two billion years at least, and were able to influence each other over all this time. The primeval atmosphere was definitely superseded by an oxygenic one about 1.45 billion years ago, but the level of 0.1 PAL of free oxygen was only reached during the Ordovician, 0.4–0.5 billion years ago.

1. Introduction

A primeval anoxygenic atmosphere has been postulated on astronomical grounds by Bernal (1951), Urey (1952) and on geochemical grounds by Abelson (1966). This idea agreed well with concepts of biologists postulating the origin of life through natural causes. They had come to the conclusion that such origin was only possible in an anoxygenic – or 'anaerobic', as it was mistakenly called – environment (Oparin, 1938, 1964). In experiments using the simulated conditions of anoxygenic environment, water and a mixture of the gases thought to have made up the primeval atmosphere, 'organic' molecules have since been synthesized *in vitro* (Calvin, 1965; Oró, 1965) by inorganic processes. These 'organic' molecules form the building materials of living matter. Although not every step in the formation of more complicated molecules of living matter has as yet been imitated, the overall result is impressive indeed. Moreover, such syntheses have been successful in quite a variety of simulated anoxygenic environments. These range from hot and dry – simulating a volcano in eruption – to more normal and aqueous – simulating lakes and oceans. The sources of energy applied in the experiments have varied too, and include electric sparking, ultraviolet irradiation and simple heat.

These experiments prove conclusively that, once given a primeval anoxygenic atmosphere and a hydrosphere containing liquid H_2O, 'organic' molecules will freely be formed through inorganic processes; leading to a sort of pre-life containing the building stones out of which early life developed. Nor may we visualize such a formation as a rare event. Instead it must have been a common and ubiquitous process, taking place continuously in a wide variety of terrestrial environments.

2. Early Geological History

From the early geological history of the earth – i.e., from the Early and Middle Precambrian in the timetable of Goldich *et al.* (1961) – sediments are known which attest the presence of an atmosphere with 'little' oxygen, and which no longer occur during later history, i.e., during Late Precambrian and Phanerozoic. The two best known types are the pyrite sands of the gold-uranium reefs (Ramdohr, 1958; Rutten, 1962) and the banded iron formations (Govett, 1966). The pyrite sands contain sulphides, such as pyrite, FeS, and originally the mineral uraninite, UO_2.* These have not undergone any oxidation, either during the weathering of the parent rocks from which the elements forming the sandstones were derived, or during the subsequent transportation and sedimentation, although they must have been in contact with the atmosphere during most of this time. The main original mineral of the banded iron formations, on the other hand, is magnetite, Fe_3O_4. Although this is the least oxidized form of iron, this still attests to some oxidation by the contemporary atmosphere during the ancient sequence of weathering-transportation- sedimentation.

It may seem strange at first hand that during the same geological period sediments could form, some of which are not oxidized at all, while others show signs of some oxidation. The apparent discrepancy can be explained by differences in speed of the ancient sequences of weathering-transportation-sedimentation, which are dependent upon the time of formation of these sediments in relation to the orogenetic cycle. The successive orogenetic cycles into which the earth's history can be divided are each made up by three periods characterized by marked differences in the rate of crustal movements and accompanying processes. These are the *geosynclinal-*, the *orogenetic-* and the *post-orogenetic* periods. During the first period of each successive orogenetic cycle relatively quiet circumstances prevail. This means that on the rising parts of the crust erosion is able to base-level the continents, and no mountains can develop. In the sedimentary basins, on the other hand, sedimentation normally keeps pace with subsidence, and is able to fill most basins up to sea level, a feature found both in the very slightly subsiding epicontinental basins and in the more strongly subsiding geosynclinal basins. During the second period the earlier geosynclinal basins are transformed into fold belts. During the post-orogenetic period strong vertical crustal movements occur due to the re-establishment of the isostatic equilibrium which lead to rapid erosion, transportation and sedimentation.

The pyrite sands belong to the post-orogenetic periods of various orogenetic cycles. Their material has been weathered mechanically from the parent rocks and undergone rapid transportation and sedimentation. This has led to relatively quick burial and

---

* See Schidlowski (1966b) for a full description of the alterations the uraninite has undergone during the later history of these rocks.

exclusion from further contact with the contemporary atmosphere. The banded iron formations, on the other hand, were formed during geosynclinal periods. Their material was leached by chemical weathering from the parent rocks in base-leveled continents and presumably was also precipitated chemically in lakes (Hough, 1958), during which the iron ions had ample time to come into contact with the contemporary atmosphere, and become partially oxidized, even if there was but little oxygen present.

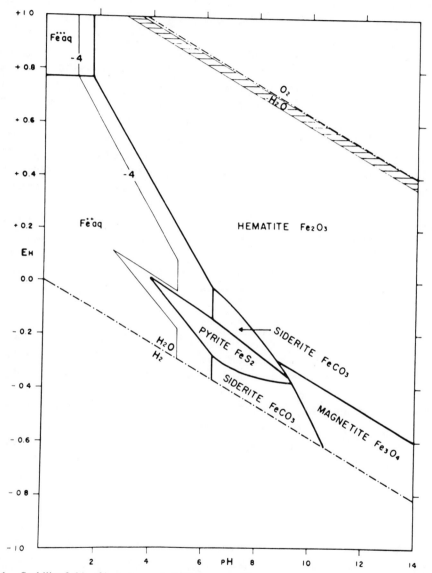

Fig. 1. Stability fields of iron compounds in water at 25° and 1 atm pressure. (From Garrels, 1960.) The barred area in the upper part of the picture indicates the shift in the boundary between the fields of O_2 and H_2O between 1 PAL O_2 (upper line) and 0.01 PAL O_2 (lower line). It does not affect the stability field of hematite (courtesy Dr. W. C. Kelly).

It follows that the state of oxidation of sediments depends not only on the level of oxygen in the contemporary atmosphere, but on the balance between the rate of oxidation for a given mineral and the speed of the weathering-transportation-erosion sequence. It must even be stated that if the minerals of sediments were allowed to reach equilibrium with the atmosphere, all sedimentary minerals, even those deposited under atmospheres with extremely little oxygen, would be completely oxidized. For it follows from Figure 1 that even a reduction of the oxygen level of the present (PAL), to one percent of the present (0.01 PAL) would not affect the stability fields of the iron minerals.

It follows that in all our considerations on the state of oxidation of ancient sediments we must use the kinetics of the processes of oxidation versus weathering, transportation and sedimentation. The equilibrium conditions, so diligently studied by the inorganic geochemists, can offer no help at all. We have no data, either for the speed of oxidation prevailing during, for instance, the formation of the pyrite sands, or for the speed of weathering, transportation and sedimentation. So it further follows that these ancient sediments, different though they are from all younger sediments, can give us no clue as to the exact level of free oxygen in the contemporary atmosphere. All we can say is that the so-called primeval anoxygenic atmosphere was an atmosphere with 'a small amount' of free oxygen.

The Blind River uranium deposits of Ontario, dated provisionally at 1.8 billion years (Derry, 1960) can be accepted as the youngest known specimen of ancient sediments deposited under the primeval atmosphere. The Dala Sandstone of central Sweden, dated at 1.4 billion years (Priem et al., 1968) can be provisionally accepted as the oldest red bed indicating the existence of an oxygenic atmosphere.

3. Self-Regulating Mechanisms

In search of other indications as to the level of free atmospheric oxygen in the early atmosphere, we come across two important self-regulating mechanisms, named respectively after H. C. Urey and Louis Pasteur, and regulating the oxygen level at about 0.001 PAL and 0.01 PAL. The Urey mechanism depends on inorganic atmospheric processes, mainly on the freezing of water vapour in the so-called cold trap at around 10 km altitude in the atmosphere. The Pasteur mechanism, on the other hand, is purely biological and depends on the change of metabolism found in microbes, which turn from fermentation to respiration when enough free oxygen becomes available.

In addition to the existence of the cold trap in the atmosphere, the Urey mechanism depends on the circumstance that the same wave lengths of ultraviolet sun rays which dissociate water and form free oxygen are also used by that same oxygen to form ozone (Berkner and Marshall, 1966). Hence there is a competition for the use of this part of the sunlight, and the more free oxygen there is in the atmosphere, the less sunlight is available for further dissociation of water. The critical level, at which there is so much oxygen that no more dissociation of water will take place is strongly influenced by the distribution of water vapour and oxygen in the atmosphere. Oxygen is distributed exponentially throughout the atmosphere, as are most gases, and not affected by the cold trap. Consequently this gas rises far higher into the atmosphere than water vapour and is able to blanket the lower atmosphere, even at the very low level of 0.001 PAL, in such a way that no further dissociation of water will take place.

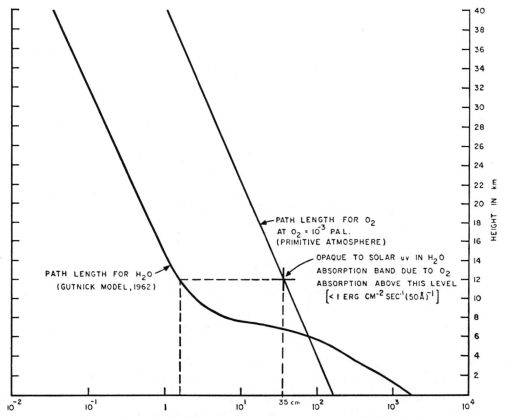

Fig. 2. Integrated path lengths of water vapour at PAL and of oxygen at 0.001 PAL. From Berkner and Marshall (1966). At the integrated path length of 35 cm for oxygen, the H₂O adsorption band is already filtered out at an altitude of 12 km in the atmosphere, and does not reach any more the bulk of the water vapour distributed lower down in the atmosphere. Note the difference in distribution between water vapour and oxygen, due to the cold trap which does not affect oxygen.

As Berkner and Marshall (1966) insist, this is an important level, which cannot be broken by any inorganic process. The only possible ways in which it could be broken are:
(1) by some unknown extraterrestrial influence,
(2) by supernatural intervention, or
(3) by biogenic production of free oxygen.

Taking up the last possibility as the most plausible, we must be aware that organic photosynthesis in its first stages will accomplish no more than replace the oxygen produced earlier by the inorganic dissociation of water. The Urey mechanism operates on the total of atmospheric oxygen, regardless of its genesis. Only when biogenic production of oxygen is strong enough to compensate for all of the oxygen previously formed by inorganic dissociation of water will the Urey level be broken.

In the Pasteur mechanism it is known that, given the same amount of original material, respiration yields more than ten times the amount of energy as fermentation. Those microbes who can change from fermentation to respiration will therefore find their metabolism much more rewarding in the latter case. It is not known why respiration evidently ceases below 0.01 PAL of atmospheric oxygen. Presumably it has to do with the kinetics of the process, either with the rate of diffusion of oxygen through water or with the rates of one or more of the enzyme reactions which together make up respiration. An interesting field of research would seem to lie wide open here. For our purposes the important point is that the Pasteur level seems to operate at about the same level of free oxygen in a varied and non-related group of microbes which are classed together as the facultative aerobes. We may therefore assume that this mechanism is controlled by some – as yet unknown – general physicochemical threshold(s) and is not related to a single genetic group of microbes. We may therefore assume that it has been operative in the past too, in which case we had better speak of 'facultative respirators' than of facultative aerobes.

4. Interpretations

In my interpretation of the Pasteur level – which is diametrically opposed to that of Berkner and Marshall (1965) – it forms a second important self-regulating mechanism. For, once early photosynthetic life had produced enough free oxygen to reach this level, respiration could develop. Of course, there will have been a time lag between the point when the oxygen level reached 0.01 PAL for the first time and the development of perhaps a primitive form of respiration. But, in agreement with Berkner and Marshall, we may postulate that the possibility of respiration offered such an immense advantage to early life that it must have developed in what was, geologically speaking, a short time. Once respiration had developed, the facultative respirators present at that time would switch from fermentation to respiration every time the level of free oxygen reached 0.01 PAL. If it should again drop below that level because the consumption of oxygen by respiration was higher than the production by organic photosynthesis, the organisms would revert to fermentation, and so on. A feedback mechanism will therefore have operated, regulating the level of free oxygen at 0.01 PAL.

We may now define the primeval anoxygenic atmosphere as an atmosphere containing free oxygen up to 0.01 PAL. We may date it by the occurrence of the ancient sediments laid down in contact with such an anoxygenic atmosphere.

The Pasteur level was eventually broken when organic photosynthesis was able to produce so much oxygen that respiration (+ oxidation of surface minerals + other oxygen losses) could no longer consume all the oxygen released. Presumably this has been due to the development of a better way of photosynthesis. For instance by the

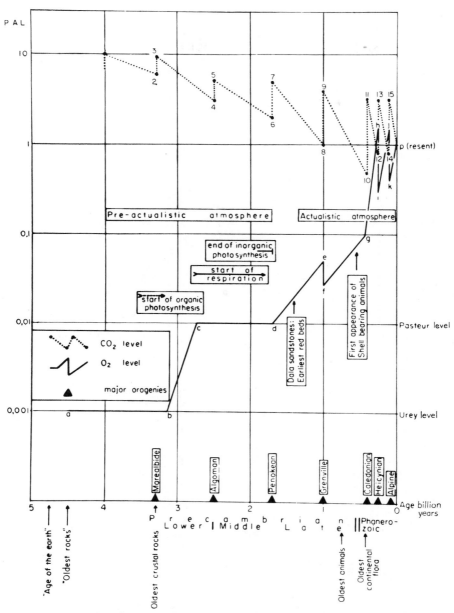

Fig. 3. The rise and fall of atmospheric oxygen and carbon dioxide in the history of the earth. For the history of carbon dioxide, compare Rutten (1966). In the early history of the earth, the level of oxygen was maintained at the Urey level at 0.001 PAL (*a-b*). At some unknown time organic photosynthesis started (see box) and at some later time it was able to break the Urey level (*b*). It is assumed that, once the Urey level was broken, expansion of photosynthesis was rapid. Moreover, all things being equal, life will expand exponentially, and so will oxygen production. This is indicated by the straight line *b-c* in the semi-logarithmic representation of this figure. It is assumed that during the deposition of the partly oxidized Soudan Iron Formation (over 2.7 billion years old) the Pasteur level was reached, and maintained until the time of deposition of the Blind River beds (1.8 billion years ago) as represented by the horizontal line *c-d*. Shortly thereafter the Pasteur level must have been broken, for in the Dala Sandstones (1.45 billion years old) there is already indication for far stronger oxidation. The oxygen level might have oscillated due to the effects of the Grenville orogeny, a point not taken up in this paper. It reached 0.1 PAL during the Ordovician 0.45 billion years ago, when the first continental flora gave evidence that life had 'conquered the land' (*g*). It will have overshot PAL during the Upper Carboniferous (*h*), and oscillated around PAL up to the present.

ascendancy of photolithotrophy over the earlier photo'organo'trophy. Or by the development of the eucaryotic cell (Sagan, 1967), in which photosynthesis is located in specialized organelles, and can be much more effective than in procaryotic cells. But here everybody's guess is as good as mine, because we really have no idea through what process or processes the Pasteur level has been broken.

We may now draw up a very schematic model for the history of atmospheric oxygen as seen in Figure 3.

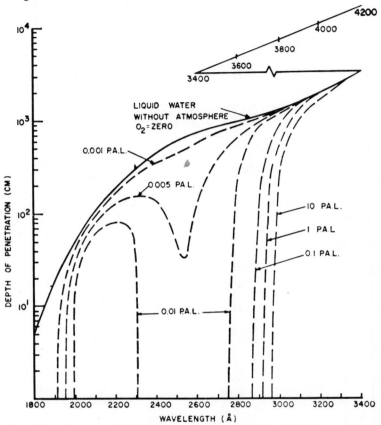

Fig. 4.   Depth of penetration of UV sunlight in water for atmospheres with a level of free oxygen of 0.001, 0.01, 0.1, 1 and 10 PAL. (From Berkner and Marshall, 1965.)

Assuming the model of Figure 3 to represent a working hypothesis, a conclusion of prime importance can be drawn, i.e., that the oxygen level of the primeval atmosphere was maintained by the Pasteur mechanism at 0.01 PAL for at least a billion years, from over 2.7 billion years to 1.8 billion years ago. At that level the shorter UV sunlight still penetrates the atmosphere, but is filtered out by a layer of 1 m water (see Figure 4).

Taking in account the time life had already evolved, but had not yet produced enough free oxygen to produce the Pasteur level (a point somewhere between *a* and *b* in Figure 3), this means that on the continents, and in shallow pools and rivulets,

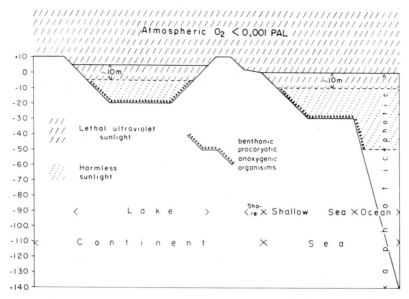

Fig. 5. Schematic representation of various environments at the surface of the earth under a primeval atmosphere of about 0.001 PAL free oxygen. Lethal UV sunlight strikes the continents and penetrates up to 10 m in water, forming the 'organic' material of pre-life. Life, still thought to be procaryotic, is benthonic in lakes and along the coasts of the oceans, being confined to a depth between 10 m and 50 m, the latter being the approximate base of the photic zone.

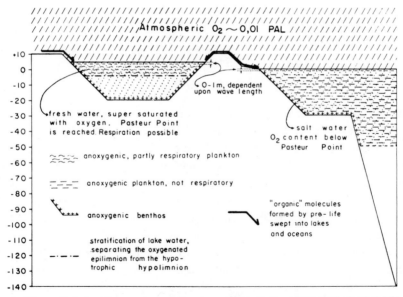

Fig. 6. As Figure 5, but under an anoxygenic atmosphere of about 0.01 PAL. Lethal UV sunlight still reaches the continents, but is filtered out already in 1 m of water. Planktonic life is now thought to be possible. As fresh water contains more oxygen than salt water, when in contact with atmospheres of the same level of O_2, it is thought that respiration will first develop in lakes.

inorganic formation of 'organic' materials still proceeded, whereas in lakes and oceans and possibly also in pores in the soil, early life could already exist. Hence, pre-life and early life have been contemporaneous for 2 billion years at least!

Based upon the admittedly rickety scaffolding of the meager data and bold assumptions mentioned above, we may now draw up a schematic evaluation of the environment of life during the three main levels of free atmospheric oxygen of 0.001 PAL, 0.01 PAL and 0.1 PAL respectively (Figures 5–7).

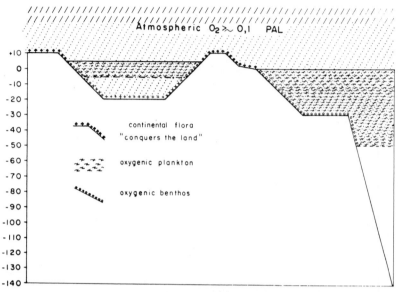

Fig. 7. As Figure 5, but under an oxygenic atmosphere of about 0.1 PAL. Lethal UV sunlight is extinguished already in the atmosphere. The flora 'conquers the land'.

5. Conclusions

To close our speculations on the history of atmospheric oxygen, we might look for a moment at its repercussions for the origin of life. The first meaningful graphic representation of this problem was given by Pirie (1959), in a figure reproduced here as Figure 8. In his famous 'hourglass' stress is laid on the major differences between pre-life and life. The former shows chemical diversity combined with structural simplicity, the latter (bio)chemical uniformity with structural and morphological complexity. The origin of life is, however, represented by a single point only, situated at the waist of the hourglass.

From a comparison with the evolution of life, as known to us from the later history of the earth, I argued in 1962 that pre-life has probably shown many and independent evolutionary trends towards life, from which eventually a single, or a number of related, pathways would actually have to led to life (Figure 9).

Now that we have admitted the regulatory influence of the Pasteur mechanism at about 0.01 PAL free oxygen, the situation has, however, changed still more. At this

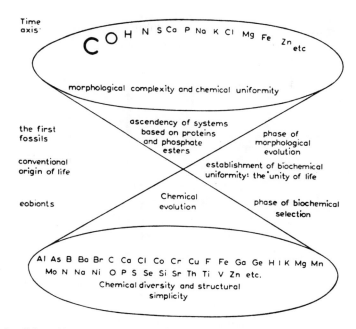

Fig. 8.   Schematic representation of the origin of life on earth. (From Pirie, 1959.)

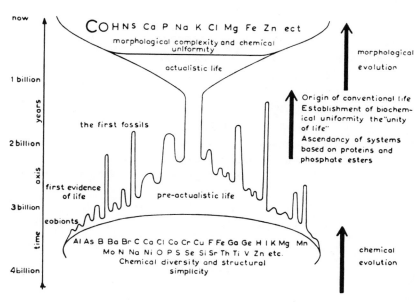

Fig. 9.   Modified version of Pirie's representation of the origin of life, stressing the fact that pre-life will have known many evolutionary trends, and that the transition from pre-life into life must have taken up considerable time. (From Rutten, 1962.)

level pre-life will, as we saw, exist contemporaneously with early life. The evolutionary trends in pre-life will thus continue to evolve parallel to those of early life. Over the period of 2 billion years transitions from pre-life to early life may occur, as indicated in Figure 10. If, by some coincidence, one or more of these later transitions happens

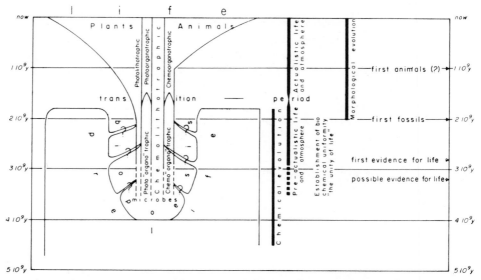

Fig. 10.   Schematic representation of the origin of life on earth, taking into consideration the regulatory effect of the Pasteur mechanism at about 0.01 PAL free atmospheric oxygen. Pre-life will have been coexistent with early life for 2 billion years at least, and biopoesis – the transition from pre-life to life – can have taken place over and over again during this period.

to be more successful than earlier ones, the newly formed life will have an advantage over the life already developed and this might lead to extermination of life which had known a successful development at an earlier date.

Although we do not know, and probably never will know, how the transition or transitions from pre-life to life actually took place, we now at least have sufficient time available for a trial and error period in which it could be accomplished. For me as a geologist, who always needs time in quantity when visualizing natural processes, this is a more gratifying situation than that in Pirie's hourglass in which the transition from pre-life to life is indicated to have happened almost instantaneously.

The tentative history of atmospheric oxygen has been drawn up according to terrestrial data. It is thought, however, that any planet on which life capable of organic photosynthesis has developed, will show a comparable history of its atmospheric oxygen.

References

Abelson, P. H.: 1966, 'Chemical Events on the Primitive Earth', *Proc. Natl. Acad. Sci. (U.S.)* **55**,1365.
Barghoorn, E. S. and Tyler, S. A.: 1965, 'Microorganisms from the Gunflint Chert', *Science* **147**, 563.
Berkner, L. V. and Marshall, L. C.: 1965, 'On the Origin and Rise of Oxygen Concentration in the Earth's Atmosphere', *J. Atmosph. Sci.* **22**, 225.

Berkner, L. V. and Marshall, L. C.: 1966, 'Limitation on Oxygen Concentration in a Primitive Planetary Atmosphere', *J. Atmosph. Sci.* **23**, 133.

Bernal, J. D.: 1951, *The Physical Basis of Life*, Routledge & Paul, London.

Calvin, M.: 1965, 'Chemical Evolution', *Proc. Roy. Soc.* A **288**, 441.

Derry, D. R.: 1960, 'Evidence of the Origin of the Blind River Uranium Deposits', *Econ. Geology* **55**, 906.

Garrels, R. M.: 1960, *Mineral Equilibria*, Harper, New York.

Goldich, S. S., Nier, A. O., Baadsgaard, H., Hoffman, J. H. and Krueger, H.: 1961, 'The Precambrian Geology and Geochronology of Minnesota', *Minnesota Geol. Surv., Bull.* **41**, Univ. Minnesota Press.

Govett, G. J. S.: 1966, 'Origin of Banded Iron Formations', *Bull. Geol. Soc. Amer.* **77**, 1191.

Hough, J. L.: 1958, 'Fresh-Water Environment of Deposition of Precambrian Banded Iron Formations', *Jour. Sed. Petrology* **28**, 414.

Oparin, A. I.: 1938, *The Origin of Life*, 2nd ed., Dover Publications Inc, New York.

Oparin, A. I.: 1964, *The Chemical Origin of Life*, Amer. Lecture Ser., Springfield, Mass.

Oró, J.: 1965, 'Investigation of Organo-Chemical Evolution', in *Current Aspects of Exobiology* (ed. by G. Mamikunian and M. H. Briggs), Pergamon, London.

Pirie, N. W.; 1959, '*Chemical Diversity and the Origins of Life. Origin of Life on Earth*, IUB Symposium Series, Vol. I, Pergamon Press, London.

Priem, H. N. A., Mulder, F., Boelrijk, N. A. I. M., Hebeda, E. H., Verschure, F. H. and Verdurmen, E. A. T.: 1968, 'Geochronological and Paleomagnetic Investigations in Southern Sweden', *Phys. Earth Planet. Interiors* **1**, 373.

Ramdohr, P.: 1958, 'Die Uran- und Goldlagerstätten Witwatersrand, Blind River District, Dominion Reef, Serra de Jacobina: Erzmikroskopische Untersuchungen und ein geologischer Vergleich', *Abhandl. Deutsch. Akad. Wiss., Berlin, Kl.f. Chem., Geol. u. Biol.* 1958, **3**.

Rutten, M. G.: 1962, *The Geological Aspects of the Origin of Life on Earth*, Elsevier, Amsterdam.

Rutten, M. G.: 1966, 'Geologic Data on Atmospheric History', *Palaeogeography, etc.* **2**, 47.

Rutten, M. G.: 1969, 'Sedimentary Ores of the Early and Middle Precambrian and the History of Atmospheric Oxygen', in *Sedimentary Ores; Proc. 15th Inter-Univ. Congr., 1957, Leicester*, p. 187.

Sagan, L.: 1967, 'On the Origin of Mitosing Cells', *J. Theoret. Biol.* **14**, 225.

Schidlowski. M.: 1966b, 'Beiträge zur Kenntnis der radioaktiven Bestandteile der Witwatersrand-Konglomerate. II: Brannerit und "Uranpecherzgeisterzgeister"', *N.Jb. Mineral. Abh.* **105**, 310.

Urey, H. C.: 1952, *The Planets*, Yale Univ. Press, New Haven, Conn.

26

Reprinted from Am. Jour. Sci., 272, 537–548 (June 1972)

A WORKING MODEL OF THE PRIMITIVE EARTH*

PRESTON CLOUD

Department of Geological Sciences, University of California,
Santa Barbara, California 93106

ABSTRACT. Knowledge of the sedimentary and biological record of pre-Paleozoic time, scanty though it still is, suggests a working model of early Earth history consisting of four successive modes (or Eons) punctuated by "events" in which surface conditions on the Earth changed relatively rapidly. The four modes are:

1. An interval (Hadean) for which no certain record has been discovered on Earth, though it has on the moon — ended 3.6 to 3.5 aeons ago by a world-wide thermal event, possibly cosmic as the moon may also have been affected.

2. An interval (Archean) characterized by rocks mainly or entirely of the greenstone–graywacke–granodiorite suite, relatively low in potassium and little differentiated, by the earliest suggestions of life, and by thin crust and no large continental cratons —ended about 2.6 aeons ago by the beginning of extensive cratonization.

3. An interval (Proterophytic) characterized by the formation of large continental cratons and upon them a gamut of differentiated sedimentary and igneous rock types, and by the oxidation of vast amounts of iron by photo-autotrophic procaryotic microorganisms, although the atmosphere remained reducing. This terminated around 1.9 aeons ago with the development of advanced oxygen-mediating enzymes that permitted the evasion of oxygen to the atmosphere and probably led quickly to the development of eucaryotes, mitosis, and sex.

4. An interval (Proterozoic in a restricted sense) characterized by an increasingly oxidizing atmosphere and oxidized sediments, especially red beds, and by an abundance of unicellular eucaryotes, but without differentiated multicellular animal life (Metazoa) —ended about 0.68 aeons ago by the appearance of an ozone layer in the atmosphere, by the onset of Metazoa, and perhaps by a climax of continental glaciation.

INTRODUCTION

Recent advances in geochronology, geochemistry, biogeology, microbiology, electron microscopy, and sedimentology converge with the increasingly genetic focus of pre-Paleozoic geology to illuminate the evolution of the primitive Earth with growing clarity. Scanty though it still is, we now have a presumptive record of life going back to the oldest sedimentary rocks, and one that we can unequivocally relate to known living organisms for at least the last 2 aeons (years $\times$ 10^9). Broad trends in crustal evolution are now also clear, and some of these can be related to biospheric and atmospheric evolution. The larger need is for a consistent working model that will integrate present knowledge and well-reasoned inference about the interdependent variables so as to focus on the central problems and predict future directions of advance. Such a model, under study for some years now, is here outlined—not with any thought of finality, but as a framework for discussion and a focus for observation.

RATIONALE

Because time is continuous and without natural subdivisions, it becomes necessary in all historical science to identify events or broad

* In preparing this paper, I have drawn heavily on the data and ideas of others, especially the participants of the Third Penrose Conference at Laramie, Wyoming (September 1970, Cloud 1971) and those who preceded me on the program of the Geological Society of America at which this paper was presented (November 2, 1971). The research on which this paper is based was supported by NSF Grant Nos. GB-7851 and GB-23809 and by NASA Grant No. NGR-05-010-035. This is Contribution #27, Biogeology Clean Lab, University of California, Santa Barbara, California 93106

modalities that set off one part of the sequence from preceding and following parts so as to bring out historical trends. One doesn't simply
divide human history or crustal evolution into units of equal length in
years without evidence for a true cyclicity of events, which has yet to be
demonstrated for geologically long intervals at a global scale.

Nevertheless, an historical system must be calibrated as well as
punctuated, and this requires some kind of time sequence and nomenclature. The best sequence we can have is one that reflects real time
quantitatively, as do the radiometric numbers which in recent years have
contributed so importantly to the emergence of an interregional pre-
Paleozoic stratigraphy. At the same time, it is important to discriminate
between crustal evolution and the system with which we attempt to cali-
brate it, and I will do that here.

In speaking of crustal evolution, I include not only the historical
sequence of the rocks themselves but also the evidence contained within
them of the interrelated evolutions of biosphere, atmosphere, hydrosphere, and chemosphere. Indeed, what is known or reasonably inferred
about each of these lines of evolution places constraints on what may
justifiably be inferred about each of the others. When all such evidence
is considered together and found to converge, a powerful fabric of
credibility results, even where individual lines of evidence may be less-
than-compelling.

A difficulty in trying to reconstruct pre-Paleozoic history has been
our disproportionately greater level of information about crystalline
rocks, in contrast to those of demonstrably sedimentary origin, and our
slowness in learning to recognize the sedimentary origin of some crystalline rocks. Our best radiometric numbers are on intrusive rocks, and,
although they cluster in most interesting ways, it now seems clear that
near or seemingly correlatable intrusive events took place over different
time intervals in different regions, that some of the most important ones
may have been scores of millions of years between emplacement and final
cooling, and that some numbers represent episodes of rejuvenation
rather than initial crystallization.

For the punctuation of our geologic continuum, we seek trend-re-
lated events that have affected the entire Earth over relatively short intervals of time and left recognizable signatures in the rock sequences
of the globe. Such attributes are more likely to result from events in
atmospheric, climatic, or biologic evolution than in plutonic evolution
and hence should be more characteristic of the sedimentary record than
of the igneous or metamorphic record, although the latter must be included in any meaningful global assessment.

MAIN TRENDS IN CRUSTAL EVOLUTION

What broad modal trends and events stand out in the history of
the primitive Earth, defining that history as beginning with the accumulation of the Earth as a solid body and ending with the appearance of
differentiated multicellular animal life—whose elaboration comprises
the stuff of conventional historical geology? And how can we array these

modalities and events into a consistently interrelated and testable system that gives us some feeling for the broad aspects of crustal evolution?

Four broad modes can be discerned, succeeded by a fifth, the Phanerozoic, representing geologically "modern" times. The oldest of the pre-Phanerozoic modes comprises the first aeon of Earth-history, from the time of accumulation of the Earth about 4.6 aeons or more ago to that of the oldest confidently dated rocks, about 3.5 to 3.6 aeons ago.[1] These oldest firmly dated rocks are generally gneisses and schists which appear to represent still older parent rocks in which radiometric clocks have been reset by some global and perhaps cosmic thermal event. We do not really know the age of deposition or original nature of these "basement" rocks but only that they are very old and that some are metavolcanic or metasedimentary, or both.

Indeed the only rocks we know with certainty to be older than about 3.6 aeons are a few that have been returned from the moon. On the moon these rocks are called pre-Imbrian. If and when the occurrence of rocks of this age is confirmed on Earth (and some believe it is already confirmed), we might find it preferable to call them Hadean, implying their basal position in the geologic sequence; or, perhaps, Selenian, in token of their lunar affinities.

The second great modality, and the oldest to be represented by unequivocally dated terrestial rocks, is commonly designated as Archean. These rocks, roughly 3.6 to 2.6 aeons old, are "granitic" (commonly granodioritic) and gneissic complexes, ordinarily with relatively low ratios of potassium to sodium (Ronov and Migdisov, 1971; also A. E. J. and Celeste Engel, oral commun., September 1970), unless K-metasomatized; and enclosed or embedded in belts of greenstone and volcaniclastic graywacke-like sediments that are analogous in their implication of mantle sources and immature weathering to the volcanic-sedimentary complexes of modern island-arcs. Although similar rocks of younger age are common, their association, to the near-exclusion of other types (except ultramafics, charnokites, and other granulite facies rocks), seems to be a universal feature of this second major phase of Earth history. Archean folded terranes commonly, and perhaps generally, consist of contiguous subsiding belts without intervening anticlinal structures, often associated with ultramafics. Over most of the Archean earth, the crust appears to have been thin, cratonization weak, and roughly linear subsiding belts common. A few places, such as in west central Arizona, show essentially Archean styles of rock associations of much younger age. Such sequences must be viewed in the larger context of interregional geology.

The third big modal trend in early crustal evolution is an interval of marked cratonization, during which plutonic and sedimentary rocks alike became increasingly sialic and cratonal. It shows an abundance of

---

[1] Black and others (1971) have now published *preliminary* evidence suggesting an age approaching 4 aeons on amphibolitic gneisses from West Greenland; but, as these authors are careful to point out, the basic data can be interpreted in several ways, one of which implies an age not much greater than 3.6 aeons.

relatively potassic granites and gneisses and plenty of clean quartzites and arkoses. Many of its platform deposits remain undeformed to the present day. This interval extended from about 2.6 to around 1.9 aeons ago. The ratio of potassium to sodium in granites and other rocks of this age is, as a rule, greater than that of older rocks. The oldest thick and extensive carbonate rocks occur near the top of this sequence. It terminates with a great episode of banded iron formation and is followed by the oldest red beds of consequence. Its sediments, although cratonal, are not subaerially oxidized. Rocks of this age have been designated by a confusing variety of terms, most commonly Lower Proterozoic (although Lower Proterozoic conventionally includes somewhat younger rocks as well). Inasmuch, however, as evolution of primitive plants (blue-green algae and bacteria) rather than animals is what is biologically important about this interval, we might cut the Gordian Knot of conflicting terminology by calling it Proterophytic and reserving Proterozoic for pre-Paleozoic rocks younger than about 1.9 aeons— an interval most of which is included in the Proterozoic of all classifications that employ that term. Although the Proterozoic, so defined, is not known to contain animal fossils, Protozoan remains are to be expected in its upper parts, and long-standing usage, although not strictly logical, does deserve to be conserved as far as practicable.

The fourth and last great modality observed, the Proterozoic proper (or Upper Proterozoic, if Proterophytic doesn't appeal), includes an aggregation in which oxidized sediments are conspicuous, and which is generally similar to Phanerozoic rocks except for the rarity of sedimentary calcium sulfate, the preponderance of dolomite over limestone, and the absence of Metazoan fossils, tracks, burrows, or after-death imprints. It includes rock types characteristic of the older sequences as well but is distinguished by the addition of oxidized cratonal sediments. It began around 1.9 aeons ago, when thick and extensive continental red beds first appeared, and lasted until about 680 m.y. ago, when the late pre-Paleozoic ice ages were waning, and the first authentic Metazoa are known.

These broad subdivisions of primitive crustal evolution can be recognized on every continent that has a good pre-Paleozoic sequence. They are also reflected, if somewhat unevenly, in the clustering of radiometric ages of plutonic rocks. But they must be integrated with other evidence to create geologic history and to discern potential turning points in that history.

In what remains of this paper, I shall attempt to relate some of these other elements to the broad features of crustal evolution suggested above.

INTERRELATION WITH ATMOSPHERIC AND BIOSPHERIC EVOLUTION

Source of atmospheric oxygen.—Consider the atmosphere, in particular its anomalous 21 percent of oxygen. No plausible model of atmospheric origin provides either a primary or a juvenile source of free oxygen. Whether or not Earth accumulated with a residual· atmosphere, the very low ratios of the noble gases in our present atmosphere, as compared with cosmic abundances, tell us that it has descended from one that accumulated as a consequence of the volcanic outgassing through time of originally occluded gases. Oxygen, not being among such gases, must have arisen through some secondary process. Plausible sources include the photolytic fission of the water molecule with escape of hydrogen from Earth's gravity field and the photosynthetic combination of CO_2 and H_2O, with sedimentary segregation of carbon (plus relatively minor additions from rock weathering). Both processes occur, and each has undoubtedly contributed some of the oxygen now in atmosphere and hydrosphere, and combined with other elements in sedimentary sulfates, ferric oxides, and other oxygen sinks. One way to assess the problem is to see whether a geochemical balance exists between oxygen and carbon in the present atmosphere, hydrosphere, and lithosphere. Unfortunately, we cannot make a similar test for oxygen and hydrogen, because no way has been found to measure independently the mass of hydrogen that has escaped Earth's gravity field, or, conversely, that has been added to our atmosphere by the solar wind.

Fortunately, numbers are available for oxygen and carbon that were compiled quite independently and for a different purpose. These are shown in figure 1, which was prepared from the data of W. W. Rubey

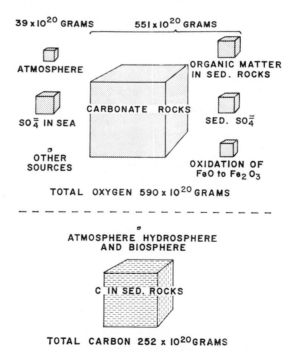

Fig. 1. Oxygen-carbon geochemical balance.

(1951), modified only to allow for larger deposits of ferric oxides than were then known. These numbers show an unexpectedly close geochemical balance between the combining equivalents of carbon and oxygen, with just enough carbon left over to account for the conversion of about 6×10^{20} g, or 6 geograms, of volcanic carbon monoxide to carbon dioxide (which seems low for consumption of oxygen by reduced volcanic gases). Of course, this estimate does not allow for the recycling of crustal materials beneath drifting continents, but that should not affect ratios, and it is these that are critical. The oxygen budget also presumably should include some O_2 resulting from rock weathering and some from outer atmosphere dissociation of H_2O— estimated respectively as 29 geograms and 89 to 100 geograms by G. D. Nicholls (Manchester, England, oral commun., 5 January 1972). Both these estimates, however, are only small fractions of the total O_2 budget, the physical dissociation estimate is very uncertain, and the consumption of O_2 in the conversion of volcanic CO to CO_2 is probably larger than suggested above—perhaps by some tens of geograms, which do not show in the estimated oxygen budget. Thus the near-balance observed between C and O_2 in figure 1 implies that photosynthesis was the main source of free oxygen and that rock-weathering and non-biological photolysis was comparatively unimportant.

Now it is time to see how the broad aspects of primitive crustal evolution may be integrated (fig. 2).

Hadean Time.—We have known that the solid parts of the solar system originated somewhat more than 4.5 aeons ago ever since the classic demonstration by Patterson (1956) that the isotopes of meteoritic and oceanic leads defined an isochron of that age ("primordial lead"), and the number has been revised only slightly upward. But the oldest unequivocally dated minerals on Earth are only 3.55 aeons old (Goldich, Hedge, and Stern, 1970), and they approach a similar age on several continents. It looks as though a thermal event of global scope had reset all the radiometric time-keeping systems about then. Available estimates indicate that this could not have been the heat of planetary accumulation and probably not radiogenic heating, because those processes should have had their main effects much earlier. The finding of similar numbers on the moon suggests that there may have been a relation, as I suggested before the lunar numbers were known (Cloud, 1968a)—perhaps a near approach of moon to Earth; or maybe an episode of bombardment by large meteorites, whose kinetic energy, converted to heat, produced magma pools and thermal metamorphism; or other cosmic event.

Archean and Proterophytic Time.—I suppose that if there had been an older atmosphere, hydrosphere, or biosphere, it would not have survived this thermal event, and that the beginning of a well-defined record of primitive Earth evolution dates from after that time. At least we know, from the evidence of thick and extensive sedimentary deposits approaching this age in eastern South Africa and Swaziland, that there

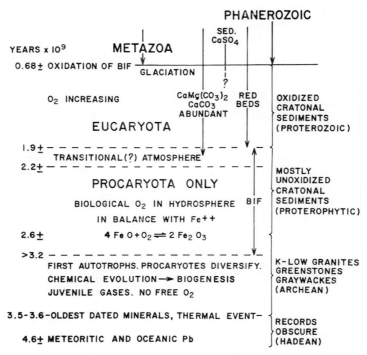

Fig. 2. Evolution of the primitive Earth.

was a substantial atmosphere and hydrosphere, and probably a biosphere, shortly after this 3.5 to 3.6 aeon event, whatever it was. The most recent number we have is a Rb–Sr isochron age of 3.38 aeons on Onverwacht sediments by the M.I.T. geochronology group (Hurley and others, 1971).

The volcanism and outgassing associated with the heating of Earth's outer shell at that time probably gave rise to a very large fraction of all the gases that ever reached the surface of the Earth—perhaps 10 to 30 percent of the volume of the present atmosphere and hydrosphere. An atmosphere of juvenile gases (CO, CO_2, H_2O, N, HCl, and probably some H, NH_3, and minor gases), in the absence of free oxygen, provides ideal circumstances for the chemical synthesis of large organic molecules, leading to the origin of life. In the absence of an ozone screen, high-energy UV radiation would impinge on Earth's surface waters, and organic compounds produced could accumulate without serious threat of oxidation. Many important steps in chemical evolution leading toward a living cell have been repeated in the laboratory by biochemists and biophysicists, experimenting with likely models of the primitive atmosphere—the common and essential feature of the successful experiments being the absence of free oxygen. Although a living cell has not as yet resulted from such experiments, the only serious remaining question (and the quite unanswerable one in the face of a variety of conceivable routes) is, "exactly what pathway did lead to the evolution of the first living organism?"

The first living cells, in any case, almost certainly inhabited an oxygen-free environment and fed on organic substances of non-biological origin or on one another. They were, technically speaking, anaerobic heterotrophs. But evolution would not have gone very far without the emergence of an organism that could manufacture its own food products, using sunlight or chemical energy to activate the process—an autotroph. When that happened, we should expect to see a diversification of the morphologically simple and essentially asexual forms of life we designate as Procaryota to signify their rudimentary nuclear structure and cellular organization.

An interesting question is: "What happened when the first of these early autotrophs became an oxygen-releasing photo-autotroph—oxygen being poisonous to all forms of life in the absence of suitable oxygen-mediating enzymes?" Unless it simultaneously evolved a full complement of advanced oxygen-mediating enzymes, it would have been in grave need of some chemical oxygen-acceptor in the surrounding physical environment. Water-splitting photo-autotrophs which today lack oxygen defenses survive by attaching their oxygen waste-products to the sulfide ion, converting it to sulfate, or to other suitable and available reduced ions—but we do not see in the older geologic record the large deposits of the oxidized products of these substances that we would expect to see if they had then been important oxygen sinks.

Instead we see vast deposits of banded iron formation (BIF) of near global distribution in sedimentary rocks older than about 1.8 to 1.9 aeons. This mainly siliceous BIF (especially between about 1.9 to 2.2 aeons ago) is commonly thick, of great lateral extent, relatively rich in hematite, most commonly low in phosphorus, and primarily a chemical blanket-deposit from open waters of broad extent. Excepting rare, mostly restricted, and often earthy younger deposits, commonly with discontinuous banding and characteristically associated with volcanism and narrow, subsiding, euxinic trenches, typical BIF is not known from rocks younger than about 1.9 aeons.

The extensive, time-limited, blanket deposits of hematitic and magnetitic BIF pose a geochemical dilemma. How could the iron have been transported over such large areas had the atmosphere (and hydrosphere) been oxidizing, and how could it have been precipitated in the absence of free oxygen? The dilemma is resolved by calling on ferrous iron in solution to serve as an oxygen acceptor for the early photosynthesizing microorganisms. In the absence or rarity of free O_2 in the atmosphere and hydrosphere, ferrous iron would be freely transported in solution wherever waters moved. Biological O_2 production in the hydrosphere in fluctuating balance with a sink of ferrous iron would account for its precipitation as insoluble ferric oxide, as well as for the banding and varying facies of BIF (these being products of varying levels of local oxidation and distance from sources of oxygen). I must note that there are interesting and perhaps significant differences between Archean and Proterophytic BIF that sometimes lead to differences of opinion about origin—but that story, interesting though it is, would be discursive here. I emphasize instead only that the presence of thick and extensive BIF of the Algoma and Lake Superior types, and the essential absence of red beds, characterizes a long episode of sedimentation and biological evolution that ends with the third major modality of crustal evolution, the Proterophytic.

The microorganisms of this time (and life then was wholly microbiological) were probably limited to deep-planktonic and stromatolite-building protonucleate and non-mitosing cells—the procaryotes. Since, moreover, all living mitosing or eucaryotic cells are either obligate aerobes or dependent on obligate aerobes, it seems likely that eucaryotes could not have evolved until a suitable level of atmospheric oxygen arose. The BIF, the absence of substantial red beds, the presence of widely distributed detrital uraninite and pyrite, and the probable nature of the biosphere converge to imply that the atmosphere anterior to about 1.9 aeons ago contained only trivial and transient amounts of free oxygen.

The advent of advanced oxygen-mediating enzymes, releasing the photosynthesizers of the time from dependence on the ferrous iron sink, was more than a major development in biochemical evolution. It necessarily had pronounced, geologically near-instantaneous, global responses that should be observable in both stratigraphic and biologic evolution. The previously iron-dependent photosynthesizers would now be freed of such dependence. They could flourish throughout all parts of the hydrosphere that were both within the photic zone and sheltered from UV in the range of 2400 to 2600 angstroms. Ferrous iron would be swept out of solution in a final episode of blanketing ferric oxides and other oxygen-sinks would be wiped out. Oxygen could begin to accumulate in the atmosphere provided (and as rapidly as) some of its carbonaceous by-product was sequestered. With high-energy UV still impinging on Earth's surface, highly reactive ozone and atomic oxygen, as well as molecular oxygen, would be available for the oxidation of reduced substances then abundant at the surface of the Earth and in the atmosphere.

Proterozoic Time (restricted).—Evidence that the oldest major episode of atmospheric oxidation began about 1.8 to 1.9 aeons ago is seen in the continental red beds of this age on most continents. The youngest blanket deposits of BIF seem to be older than the oldest prominent red beds. The carbon counterpart of this oxygen influx may be seen in the roughly 1.8 to 1.9 aeon metacoals or shungites of Soviet Karelia and Siberia as well as in volumetrically more important carbonaceous sediments of this age elsewhere.

Continuing refinement of the evidence, especially in the Canadian Shield, suggests that there may have been a transitional interval of perhaps as much as 300 m.y. between the almost completely anoxygenous atmosphere of older pre-Paleozoic time and the freely (but not highly) oxygenous atmosphere in which the first important red beds formed— the occurrence of thin red siltstones and shales in the Cobalt Group and younger Huronian beds has been summarized by Roscoe (1969, p. 81-82). But the massive onset of thick and extensive red beds and the termination of the blanket deposits of BIF are near enough in time to suggest a common relationship to the emergence of a substantially oxygenous atmosphere about 1.9 aeons ago. Eucaryotes may have evolved (the evidence is equivocal) at about the same time. Thus we seem to have an event of relatively short duration and global extent that separates long trends in atmospheric, biospheric, and lithospheric evolution.

I suggest that this is the kind of event we have been looking for, that it is probably real if not yet precisely placed in time, and that we should use it to delimit major divisions of pre-Paleozoic crustal evolution. Whether we choose to call these divisions Proterophytic and Proterozoic, Lower and Upper Proterozoic, or something else, is of no substantive concern, although eventually geologists will have to submerge these inconsequential differences and agree on names if they want to communicate clearly.

The beginning of substantial evasion of O_2 from hydrosphere to atmosphere thus marks the beginning of our fourth main pre-Paleozoic evolutionary package—the terminal sequence of dominantly cratonal and oxidized sediments. The oxygen did not come all at once, however. To assert, as some have done, that all the oxygen now in the atmosphere could arise abruptly following the onset of photosynthesis is to fail to recognize the reversible nature of the photosynthetic reaction (basically $CO_2 + H_2O \rightleftarrows CH_2O + O_2$), such that oxygen can accumulate no faster than a chemically equivalent mass of carbon is sequestered (nor until the oxygen sinks of the time are filled). We see suggestions of the gradual accretion of O_2 in these younger pre-Paleozoic rocks in the continuing abundance of carbonaceous sediments, in the succession of red beds, and in the abundance and wide distribution of carbonate rocks of this age.

Proterozoic-Phanerozoic transition.—We come now to a convergence of events in biospheric, lithospheric, and perhaps atmospheric evolution which, I believe, logically marks the termination of primitive Earth evolution and the beginning, geologically speaking, of essentially modern times. I take this as the logical transition from Proterozoic to Phanerozoic, that is from pre-Paleozoic to Paleozoic times. Of course it is not a sharp line, any more than any other geological "boundary" of more than regional extent; but it seems to me to be an interval of profound transition in evolutionary style that deserves serious attention in establishing local boundaries of convenience.

I refer to that convergence of events to be seen in the onset of Metazoan life, the appearance of sedimentary sulfate as a common rock type, the waxing and waning of extensive mainly late pre-Paleozoic glaciations,[2] and the consequent lowering of water tables and oxidative enrichment of the BIF protores that we see especially in North America. The wasting of the late pre-Paleozoic (and locally early Paleozoic) glaciers perhaps also gave rise to the initial epicontinental floods of Phanerozoic time.

These several developments are consistent with the suggestion that there may have been a noticeable increase in atmospheric oxygen around 680 m.y. ago[3]—perhaps from around 1 percent present atmospheric level, where an effective ozone screen appears (for example Berkner and Marshall, 1965), and a large increase of photosynthetic populations is conceivable to substantially more than that. One reason for supposing that this may have been the case is the fact that, of the two essential preconditions for the emergence of a Metazoan grade of life, that of a eucaryotic level of cellular evolution had long previously been realized. The most probable explanation for the relatively abrupt appearance of the Metazoa at this time (Cloud, 1968b) is, thus, the triggering effect of the attainment of a sufficient level of free oxygen, as originally suggested by Nursall (1959) and later by Berkner and Marshall (1965). The relatively rapid diversification implied during basal Paleozoic or Ediacarian time (an interval of perhaps 100 m.y. or more) becomes comprehensible if one visualizes a polyphyletic origin, with relatively uncontested adaptive radiation into previously unoccupied ecologic niches.

That such an emergence and adaptive radiation of the Metazoa should define the base of the Paleozoic (and, of course, the Phanerozoic) would hardly be questioned if it were widely accepted as a reality. I think it probably is a reality and that, as the evidence is critically reviewed by others, it will become generally accepted as such. It is, of course, reinforced by other aspects of crustal evolution already alluded to. The oxidative enrichment of the banded iron formation and the relatively abrupt onset of thick and extensive sedimentary sulfates would be logical consequences of the suggested increase in oxygen. The glaciation (if it is real and nearly synchronous) could have been triggered by a reduction in CO_2 (and the greenhouse effect) related to the suggested increase in O_2.

It is also a prediction of the mechanism suggested that there should be substantial deposits of carbon of late pre-Paleozoic and early Paleozoic age. It is of interest that the fine clastics associated with tillites and dropstone breccias of this age are often relatively carbonaceous. Ronov and Migdisov (1971, p. 176, fig. 9) have also documented a marked in-

[2] These glaciations, sometimes taken to imply an improbable global freezing, may reflect instead an episode of relatively rapid drift of continents over the polar regions or a clustering of continental crust around one or both poles. They are not critical to the model, but they cannot be ignored, and, if true, they would be consistent with it.

[3] Stratigraphic and radiometric data elaborated by Evans, Ford, and Allen (1968) imply that this age, earlier given by the author as 640 m.y. from Soviet data, must be increased to the number here given. Geochemical similarities between fossiliferous volcanigenic sediments and associated dated intrusives indicate near contemporeity of sediments and intrusives dated at 680 m.y. in the British Midlands.

crease in the carbon content of sedimentary rocks of the Russian and North American platforms beginning at about this point in geologic time.

The needed testing of this and other predictive aspects of the model is a task for the future. So is its refinement and local subdivision into conventional time and time-rock units. But I venture the judgments: (1) that the four major subdivisions suggested reflect real modal trends; (2) that they deserve to be designated by formal names comparable to the now generally accepted fifth and final major division of geologic time, the Phanerozoic Eon; and (3) that, over the next few decades, we shall see the emergence of a more detailed historical geology anterior to the Phanerozoic, in which modern principles of sedimentology and bio-geology will play an important part, along with geochronology and the more traditional tools of pre-Paleozoic geology.

REFERENCES

Black, L. P., Gale, N. H., Moorbath, S., Parkhurst, R. J., and McGregor, V. R., 1971, Isotopic dating of very early Precambian amphibolite facies gneisses from the Godthaab District, West Greenland: Earth and Planetary Science Letters, v. 12, p. 245-259 (North Holland Publishing Co.).

Berkner, L. V., and Marshall, L. C., 1965, History of major atmospheric components: Natl. Acad. Sci. (U.S.A.) Proc., v. 53, no. 6, p. 1215-1225.

Cloud, Preston, 1968a, Atmospheric and hydrospheric evolution on the primitive earth: Science, v. 160, p. 729-736.

———— 1968b, Pre-metazoan evolution and the origins of the Metazoa, *in* Drake, E. T., ed., Evolution and Environment: New Haven, Ct., Yale Univ. Press, p. 1-72.

———— 1971, The Third Penrose Conference—The Precambrian: Geotimes, v. 16, no. 3, p. 13-18.

Evans, A. M., Ford, T. D., and Allen, J. R. L., 1968, Precambrian rocks, *in* Sylvester-Bradley, P. C., and Ford, T. D., ed., The geology of the East Midlands: Leicester Univ. Press, p. 1-19.

Goldich, S. S., Hedge, C. E., and Stern, T. W., 1970, Age of the Morton and Montevideo gneisses and related rocks, southwestern Minnesota: Geol. Soc. America Bull., v. 81, p. 3671-3695.

Hurley, P. M., and others, 1971, Ancient age of the Middle Marker Horizon, Onverwacht Group, Swaziland Sequence, South Africa: M.I.T.-1381-19, 19th Ann. Progress Rept. to U.S. Atomic Energy Comm., p. 1-4.

Nursall, J. R., 1959, Oxygen as a prerequisite to the origin of the Metazoa: Nature, v. 183, p. 1170-1172.

Patterson, C. C. 1956, Age of meteorites and the earth: Geochim. et Cosmochim. Acta, v. 10, p. 230-237.

Ronov, A. B., and Migdisov, A. A., 1971, Geochemical history of the crystalline basement and the sedimentary cover of the Russian and North American Platforms: Sedimentology, v. 16, p. 137-185.

Roscoe, S. M., 1969, Huronian rocks and uraniferous conglomerates: Canada Geol. Survey Paper 68-40, 205 p.

Rubey, W. W., 1951, Geologic history of sea water: Geol. Soc. America Bull., v. 62, p. 1111-1148.

Erratum: On page 542, line 5, the numbers should read 73×10^{20} g or 73 geograms (P. Cloud, personal communication).

IV
Organic Geochemistry—
Carbon Chemistry of
Precambrian Rocks

Editor's Comments on Papers 27 Through 38

Organic geochemistry has been applied to understand the occurrence, distribution, fate, and significance of organic compounds found in Precambrian rocks. Serious geochemical considerations of these old rocks began in the early 1960s, when it was realized that Precambrian organic materials might provide clues to Precambrian life. Previously, most organic geochemistry was practiced in petroleum companies, where there was little interest or incentive for exploration in rocks of Precambrian age. With the advent of programs for the exploration of space and the increasing interest within these programs regarding the origin of life, new efforts in organic geochemistry were directed to the most ancient rocks on earth. Twelve papers are included here to attempt to highlight the progress that has been made. The early results were received with interest as compounds of biological significance were extracted from these ancient rocks (Eglinton and Calvin, 1967). Examples are given of chemical discoveries in rocks ranging in age from 1 to 3 billion years. Critical evaluations of some of these results have suggested, however, that many of the organic molecules, especially those discovered in the oldest rocks, probably have little to do with the biology at the time these oldest rocks were formed, but rather represented some sort of natural contamination.

Hoering (Paper 27) was one of the first investigators to apply techniques of organic geochemistry to a variety of Precambrian samples. By means of oxidation, pyrolysis, reduction, and solvent extraction he obtained results from which he concluded that the carbon in these samples was originally in the form of kerogen, the state of most organic matter in geologically younger sedimentary rocks. His evidence strongly supported the idea that biological activity existed very early in the Precambrian. His paper marks the beginning of numerous subsequent studies.

During the next 10 years a number of papers appeared in which various classes of organic compounds were reported to be present in Precambrian rocks. For example, hydrocarbons were found in shales 1 billion years old (Meinschein et al., 1964; Eglinton et al., Paper 28; Barghoorn et al., Paper 29), in cherts 2 billion years old (Oró et al., Paper 30) and in cherts 3 billion years old (Hoering Paper 31; Meinschein, 1967; Oró and Nooner, 1967; MacLeod, 1968; Han and Calvin, 1969). Other classes of organic compounds, such as steranes (Burlingame et al., 1965), fatty acids (Van Hoeven et al., 1969; Han and Calvin, 1969), porphyrins (Barghoorn et al., Paper 29; Kvenvolden and Hodgson, 1969; Rho et al., 1973), and amino acids (Schopf et al., Paper 32) were found in association with Precambrian sedimentary rocks. Most of these results were thought to have some significance with regard to Precambrian life.

At exactly the same time Meinschein et al. (1964) and Eglinton et al. (Paper 28) published papers announcing the discovery of *n*-alkanes and isoprenoid hydrocarbons in 1-billion-year-old Nonesuch Shale from the White Pine area of Michigan. These papers reported for the first time the finding of the "biological markers," pristane and phytane, in Precambrian sedimentary rocks. The presence of these hydrocarbons was believed to reflect evidence for life in Precambrian times. The paper by Meinschein et al. (1964) is not included here because it was succeeded the following year by a more detailed report (Barghoorn et al., Paper 29). This paper contains an extensive description of the geology, paleobiology, and chemistry of the Nonesuch Shale. The paper is notable for its comprehensive descriptions of the organic geochemical measurements that were made. Particularly significant is the description of the discovery of porphyrins found in the Nonesuch Shales but not in the associated crude oils. The evidence presented in this paper argues strongly that the organic compounds are indeed indigenous to this Precambrian shale and are the products from photosynthetic processes operative during late Precambrian times.

With the successful discovery of indigenous organic compounds in Precambrian rocks 1 billion years old, it was natural to extend these kinds of analyses to older rocks. Oró et al. (Paper 30) examined 2-billion-year-old Gunflint chert, and they found both *n*-alkanes and the isoprenoid hydrocarbons, pristane and phytane. These discoveries were believed to provide a chemical characterization of Precambrian life and extended the record of life to 2 billion years ago.

The occurrence of biological-type alkanes in rocks 3 billion years old was first reported by Hoering (Paper 31). Fig Tree shale from the Swaziland Supergroup of South Africa was shown to contain *n*-alkanes ranging from C_{18} to C_{31}. The isotopic compositions of the extractable organic material and insoluble polymer, kerogen, were remarkably similar. Although this similarity could be interpreted to indicate that the alkanes are indigenous to the Fig Tree shale and, therefore, likely to be as old as the shale, Hoering argues that the agreement in isotopic compositions may be fortuitous. Since this paper was published there has been increasing skepticism with regard to the significance, relative to Precambrian life, of the organic molecules found in the oldest sedimentary rocks. To be assured that the organic molecules are not only indigenous to the rocks but also as old as the rocks is extremely difficult.

In 1968 Schopf et al. (Paper 32) reported finding amino acids in fossiliferous cherts—Bitter Springs, Gunflint, and Fig Tree, which are approximately 1, 2, and 3 billion years old, respectively. The authors believed that the amino acids were indigenous to all the samples studied and suggested that, from geological and geochemical considerations, the compounds date from the time the rocks were originally deposited; however, the skepticism concerning the significance of organic molecules in very ancient sediments, voiced first about hydrocarbons by Hoering (Paper 31), can now be applied to amino acids. Abelson and Hare (1969) concluded after a series of experiments that Gunflint chert contains small amounts of amino acids of recent origin but insignificant amounts of very ancient amino acids. Concurrent with the work of Abelson and Hare (1969), Kvenvolden et al. (Paper 33) applied a newly developed gas-chromatographic technique to determine the ratio of optical isomers of the amino acids in the Fig Tree chert. The results showed that amino acids were present in only the **L** configuration. This finding suggested that the amino acids are of recent origin because, were they ancient in origin, they would generally be expected to occur as racemic mixtures.

In order to try to cope with the problem of understanding the significance of organic molecules found in association with Precambrian rocks, Hoering (Paper 34) developed criteria for judging the suitability of specimens used in studies of organic geochemistry applied to Precambrian sediments. He lists various ways in which rocks can be contaminated, both by natural, geological circumstances and by laboratory practices. This paper began to dampen the enthusiasm of many investigators who were trying to discover clues to early Precambrian biochemistry through organic geochemical studies. By 1970 the difficulties of interpreting the Precambrian organic geochemical record were generally recognized. Smith et al. (Paper 35) examined the extractable organic matter in three Precambrian cherts and compared their results with the extractable organic material from cherts of Phanerozoic age. They found only traces of hydrocarbons and fatty acids that could be considered indigenous to the samples. No evidence could be found that these compound dated from the time of sedimentation of the rocks. They conclude "that unless chemical criteria indicative of a 'Precambrian origin' can be established, the extractable constituents of very ancient sediments will provide little or no interpretable evidence of early biological processes."

Two reviews summarize the evidence for early life provided by organic geochemical studies. Essentially all literature concerned with the organic geochemistry of Precambrian sediments is cited. In his consideration of evolutionary events prior to the origin of vascular plants, Schopf (Paper 36) treats the subject of "chemical fossils" found throughout the entire Precambrian record; it is only the geochemical part of his total paper that is included here. He divides the geochemical evidence into "uncertain" and "probable" occurrences. Notably lacking are "certain occurrences." Although it is recognized that the geochemical evidence is most difficult to interpret, much of the evidence is consistent with the existence of life during much of Precambrian time. Kvenvolden (Paper 37) has also reviewed the organic geochemistry of the Precambrian but has confined his considerations to the earliest part of the record, which exceeds 3 billion years in age. He concludes that although it is possible to show that extracted organic materials are indigenous to a sediment and are not contamination, it is difficult, if not impossible, to establish that the formation of the organic material and the time of deposition of the sediment were contemporaneous. The problem is especially evident in the study of very early Precambrian rocks where the amounts of extractable organic materials are small and often approach the limits of detection of the analytical methods used.

Although it is commonly difficult to ascertain the significance of small concentrations of organic molecules recovered from very ancient rocks, the insoluble, polymeric organic material called kerogen offers a potentially valuable source of information about early life. Early Precambrian kerogen has been examined by techniques that involve hydrogenation (Hoering, 1964), ozonolysis (Nagy and Nagy, 1969), pyrolysis (Scott et al., 1970; Simmonds et al., 1969), and carbon isotopic abundance measurements (Hoering, Papers 31 and 34; Schopf et al., Paper 40; Oehler et al., Paper 38). The work by Oehler et al., (Paper 38) represents the first systematic study of carbon isotopic abundances in the Swaziland Supergroup, the oldest sedimentary record known at this time. From their work they suggested that a discontinuity in the early Precambrian carbon isotopic record reflected a major event in biological evolution. Further studies of this isotopic record are, of course, necessary to establish confidence in any interpretation regarding Precambrian biochemistry. It seems apparent that if any interpretable chemical record of early Precambrian life exists, it most likely resides in the kerogen of the earliest sedimentary rocks.

Organic geochemistry as a branch of geochemistry is relatively young, having its beginnings about 1936. Two important contributors to the field in recent years have been Thomas C. Hoering and Geoffrey Eglinton.

Hoering is a physical chemist who was born in 1925 in Illinois and received his doctorate at Washington University at St. Louis. He taught at the University of Arkansas for seven years and has been on the staff at the Carnegie Institution of Washington since 1959. His work on stable isotope geochemistry and the geochemistry of organic substances is well known. The group at Carnegie Institution composed of T. C. Hoering, P. E. Hare, and P. H. Abelson has been the source of many ideas and important research in organic geochemistry.

Eglinton was trained in natural-products chemistry and now leads the Organic Geochemistry Unit in the School of Chemistry, University of Bristol, England. Before going to Bristol he was on the staff of the University of Glasgow. Born in Wales in 1927 and educated in Manchester, he was first recognized for his work on the structure of natural products, particularly leaf waxes. He emphasized the use of infrared spectrometry and the gas chromotographic separation of complex mixtures. He has applied these techniques in his more recent interests in the use of organic chemistry for the solution of geological problems. He has been honored for his contributions to understanding the carbon chemistry of the moon.

References

Abelson, P. H., and P. E. Hare (1969) Recent amino acids in the Gunflint chert: *Carnegie Inst. Washington Yearbook 67*, 208–210.

Burlingame, A. L., P. Haug, T. Belsky, and M. Calvin (1965) Occurrence of biogenic steranes and pentacyclic triterpanes in an Eocene shale (52 million years) and in an early Precambrian shale (2.7 billion years): a preliminary report: *Proc. Natl. Acad. Sci. (U.S.)*, **54**, 1406–1412.

Eglinton, G., and M. Calvin (1967) Chemical fossils: *Sci. Am.*, **216**, 32–43.

Han, J., and M. Calvin (1969) Occurrence of fatty acids and aliphatic hydrocarbons in a 3.4 billion-year-old sediment: *Nature*, **224**, 576–577.

Hoering, T. C. (1964) The hydrogenation of kerogen from sedimentary rocks with phosphorus and anhydrous hydrogen iodide: *Carnegie Inst. Washington Yearbook 63*, 258–262.

Kvenvolden, K. A. and G. W. Hodgson (1969) Evidence for porphyrins in early Precambrian Swaziland System sediments: *Geochim. Cosmochim. Acta*, **33**, 1195–1202.

MacLeod, W. D., Jr. (1968) Combined gas chromatography–mass spectrometry of complex hydrocarbon trace residues in sediments: *Jour. Gas Chrom.*, **6**, 591–594.

Meinschein, W. G. (1967) Paleobiochemistry—1967: *McGraw-Hill Yearbook of Science and Technology*, pp. 283–285.

Meinschein, W. G., E. S. Barghoorn, and J. W. Schopf (1964) Biological remnants in a Precambrian sediment: *Science*, **145**, 262–263.

Nagy, B., and L. A. Nagy (1969) Early Precambrian Onverwacht microstructures: possibly the oldest fossils on earth: *Nature*, **223**, 1226–1227.

Oró, J., and D. W. Nooner (1967) Aliphatic hydrocarbons in Precambrian rocks: *Nature*, **213**, 1082–1085.

Rho, J. H., A. J. Bauman, H. G. Boettger, and T. F. Yen (1973) A search for porphyrin biomarkers in Nonesuch shale and extraterrestrial samples: *Space Life Sci.*, **4**, 69–77.

Scott, W. M., V. E. Modzeleski, and B. Nagy (1970) Pyrolysis of early Precambrian Onverwacht organic matter ($>3 \times 10^9$ years old): *Nature*, **225**, 1129–1130.

Simmonds, P. G., G. P. Shulman, and C. H. Stembridge (1969) Organic analysis by pyrolysis–gas chromatography–mass spectrometry: a candidate experiment for the biological exploration of Mars: *Jour. Chrom. Sci.*, **7**, 36–41.

Van Hoeven, W., J. R. Maxwell, and M. Calvin (1969) Fatty acids and hydrocarbons as evidence of life processes in ancient sediments and crude oils: *Geochim. Cosmochim. Acta*, **33**, 877–881.

Additional Suggested Readings

Belsky, T., R. B. Johns, E. D. McCarthy, A. L. Burlingame, W. Richter, and M. Calvin (1965) Evidence of life processes in a sediment two and a half billion years ago: *Nature*, **206**, 446–447.

Brooks, J., and G. Shaw (1971) Evidence for life in the oldest known sedimentary rocks—the Onverwacht Series chert, Swaziland System of southern Africa: *Grana*, **11**, 1–8.

Eglinton, G., P. M. Scott, T. Belsky, A. L. Burlingame, W. Richter, and M. Calvin (1966) Occurrence of isoprenoid alkanes in a Precambrian sediment: in *Advan. Org. Geochem., 1964*, edited by G. D. Hobson and M. C. Louis: Pergamon Press, Elmsford, N.Y., pp. 41–74.

Hoering, T. C. (1967) The organic geochemistry of Precambrian rocks: in *Researches in Geochemistry*, Vol. 2, edited by P. H. Abelson: Wiley, New York, pp. 87–111.

Meinschein, W. G. (1965) Soudan formation: organic extracts of early Precambrian rocks: *Science*, **150**, 601–605.

Palacas, J. G., F. M. Swain, and F. Smith (1960) Presence of carbohydrates and other organic compounds in ancient sedimentary rocks: *Nature*, **185**, 234.

Swain, F. M., A. Blumentals, and N. Prokopovich (1958) Bitumens and other organic substances in Precambrian in Minnesota: *Am. Assoc. Petrol. Geol. Bull.*, **42**, 173–189.

Swain, F. M., J. M. Bratt, and S. Kirkwood (1970) Carbohydrates from Precambrian and Cambrian rocks and fossils: *Geol. Soc. Am. Bull.*, **81**, 499–504.

Welte, D. H., and G. Knetsch (1970) Organischer Kohlenstoff und die Entwicklung der Photosynthese auf der Erde: *Naturwiss.*, **57**, 17–23.

Welte, D. H., and A. Wurm (1967) Das Problem der fruhesten organischen Lebensspuren: *Naturwiss.*, **54**, 325–329.

27

Reprinted from *Carnegie Inst. Washington Yearbook 62*, 184–187 (1963)

The Isolation of Organic Compounds from Precambrian Rocks

T. C. Hoering

The ultimate fate of most organic materials is to be oxidized to carbon dioxide. Some organic substances escape this fate by being buried in sediments. A few remain as compounds similar to those of the original living cells. Thus amino acids, carbohydrates, fatty acids, and pigments have been found in sedimentary rocks. However, the majority of the organic compounds in rocks have been converted to an insoluble substance known as kerogen.

At 25°C organic substances are unstable with respect to decomposition into methane, carbon dioxide, and graphite. At this temperature reactions leading toward these products require times as long as billions of years. The so-called "graphite" of Precambrian sedimentary rocks may contain intermediate molecules in the chemical pathways of the decomposition of kerogen.

It is the purpose of this work to consider the chemical nature of the carbon of Precambrian sedimentary rocks and to see whether any recognizable organic compounds can be isolated from it. Any such organic compounds need not bear much resemblance to the chemical components of living cells, but as the nature and transformations of kerogen are gradually understood they may give some insight into the existence and the nature of Precambrian life.

The reduced carbon of Precambrian rocks is reminiscent of high-rank anthracite coal, and therefore some of the techniques for the elucidation of coal structure were employed. The reactions used included (*a*) oxidation and recovery

184

of aromatic and aliphatic acids, (b) thermal pyrolysis and isolation of aliphatic and olefinic hydrocarbons, (c) reduction with anhydrous hydrogen iodide and identification of saturated hydrocarbons, (d) solvent extraction followed by spectroscopy of the extracts.

For experimental simplicity, much of the work was done on massive graphite of Precambrian age. The samples include the following:

1. Michigami coal from the Iron River formation of northern Michigan. It has been described by Tyler, Barghoorn, and Barret (1957). Samples were collected by E. S. Barghoorn and P. H. Abelson.

2. Anthroxolite from Sudbury, Ontario, Canada (Thompson, 1956). The samples were collected by P. H. Abelson.

3. Graphitic material from the Soudan iron mine, Oliver Mining Company, Soudan, Minnesota. Samples were collected by F. L. Klinger.

4. Carbon leader from the Main Reef series, Transvaal, South Africa. Sample was donated by P. Ramdohr.

Some work was done also on the finely dispersed carbon of the Gunflint chert, the Bulawayan limestone, and the Transvaal dolomite. These rocks are described in another section of the writer's report.

Oxidation of coal by alkaline potassium permanganate is a well known reaction. The products are a mixture of benzene polycarboxylic acids (Holly and Montgomery, 1956). Figure 70 is a drawing of a paper chromatogram of the aromatic acids isolated from the oxidation of the carbonaceous material from the Soudan iron mine. The acids on the chromatogram appeared as dark blue and fluorescent spots when viewed under ultraviolet light. The ultraviolet adsorption spectrum of an aromatic acid from one of the spots of the chromatogram is shown in figure 71; it is typical of this class of compounds. Through a comparison of the rate of migration of known substances on paper chromatograms, the presence of benzenetricarboxylic, benzenetetracarboxylic, and benzenepentacarboxylic

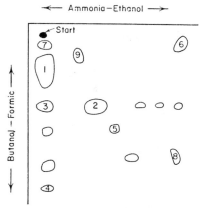

Fig. 70. Paper chromatogram of the aromatic acids from the oxidation of Michigami coal. A mixture of 1 part of coal with 1.6 parts of KOH was refluxed with excess $KMnO_4$ for 24 hours. The solution was acidified, treated with SO_2, and evaporated to dryness. The solids were extracted with diethyl ether. The extract was separated by two-dimensional paper chromatography according to the procedure of Germain (1959). The separated acids gave a deep blue color or a bright fluorescence when viewed under ultraviolet light. A comparison of R_f values and colors under ultraviolet light, with known acids, indicated the presence of benzene polycarboxylic acids.

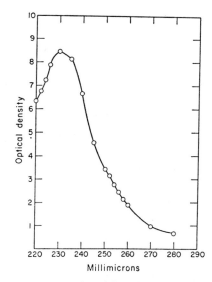

Fig. 71. The ultraviolet adsorption spectrum of an aromatic acid from the oxidation of Michigami coal. The spot numbered 2 in the paper chromatogram shown in figure 70 was eluted with dilute sodium hydroxide, and the ultraviolet adsorption spectrum was taken.

acids was indicated. Benzenehexacarboxylic acid (mellitic acid) was identified in all samples, but as this compound can be made from purely inorganic graphite its presence is of little significance to this research. The oxidation products were also examined for low-molecular-weight aliphatic acids, but only acetic acid was identified.

The pyrolysis of coals is a well studied process. Figure 72 is a tracing from the gas-liquid chromatography separation of

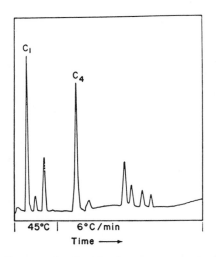

Fig. 72. Gas-liquid chromatogram of the hydrocarbons from the pyrolysis of Michigami coal. Samples of graphite were pyrolyzed in a vacuum, and the gases were pumped off for chemical analysis. The temperature was raised gradually. The gases given off below 250°C were due to adsorbed air. At 300°C hydrocarbon gases began to be evolved, and above 600°C molecular hydrogen was observed. The gases were transferred to a temperature-programmed gas-liquid chromatograph and separated with a 6-foot silicone rubber packed column. This figure shows a typical chromatogram with gases from methane through pentane being observed.

the hydrocarbons derived from the heating of Michigami coal in a vacuum. All the hydrocarbons from methane through pentane have been identified.

Thermal pyrolysis of coal is very destructive to any structure and does not give much insight into the nature of the organic substances present. The chemical reduction and liquefaction of coal may be more informative. Figure 73 shows a typical mass spectrometric analysis of the saturated hydrocarbons obtained from the action of anhydrous hydrogen iodide on Michigami coal. A mixture of hydrocarbons from methane through hexane is indicated. The orders of magnitude of the yields of hydrocarbons liberated by pyrolysis and reduction range from 10 to 100 parts per million of starting rock.

The exhaustive extraction of coal by basic solvents such as pyridine has long been a means of isolating organic substances. Infrared adsorption spectra of organic molecules are very specific for the types of chemical bonds contained in them. Figure 74 is an infrared adsorption spectrum of the organic substances extracted from the carbonaceous material of the Transvaal dolomite by pyridine. The presence of methylene groups ($-CH_2-$) is the most conspicuous feature of this spectrum.

The chance of contamination is an ever-present danger in the search for trace amounts of organic substances in material that has had such a long history as the rocks studied in this work. Contamination in the chemical reagents and water used for the work can be tested by running the appropriate blanks. Airborne dust or pollen is another source of contamination. By using a number of different procedures and by looking for a number of different organic substances, we can hope to decide whether laboratory contamination is a problem. Natural contamination of the rock during the long period from its deposition in the Precambrian to the present is much more difficult to evaluate.

The results obtained so far support the premise that the so-called "graphite" of Precambrian rocks was originally kerogen. If so, it is of interest to ask whether the organic compounds that formed this kerogen were the product of living cells or whether they could represent abio-

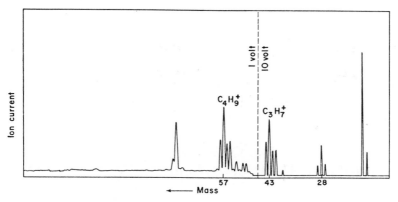

Fig. 73. The mass spectrum of the hydrocarbons from the treatment of Precambrian "graphite" with anhydrous hydrogen iodide. Ten grams of Michigami coal was placed in a bomb, and 50 grams of anhydrous hydrogen iodide was distilled in. The bomb was heated to 180°C for 16 hours. Substances volatile at 100°C were distilled off the reaction mixture, and the iodine and hydrogen iodide were removed. The gases were separated into fractions by gas-liquid chromatography, and various fractions were admitted into the mass spectrometer for analysis. This figure shows the mass spectrum of a mixture of hydrocarbons obtained in this manner from Michigami coal. A mixture of hydrocarbons from methane through pentane is shown.

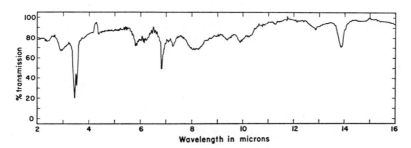

Fig. 74. Infrared adsorption spectrum of pyridine extract from Precambrian Transvaal dolomite. The "graphite" from the Transvaal dolomite was exhaustively extracted in a Soxhlet apparatus with pyridine. The pyridine was evaporated, and the resulting oil was pressed into a KBr pellet. The sharp adsorptions at 3.4–3.5, 6.8–6.9, and 14.0 microns are characteristic of isolated methylene (–CH₂–) groups. The broad adsorptions at 5.5–5.6 microns are suggestive of substituted aromatic hydrocarbons.

logically produced organic compounds from a period that preceded terrestrial life. A number of the rocks studied have textures that are generally described as due to colonial algae. The carbon isotope studies reported by the writer here indicate that photosynthesis was occurring during the time of their formation.

Thus the evidence is in favor of the existence of biological activity very early in the Precambrian era.

28

Reprinted from *Science*, **145**, 263–264 (July 17, 1964)

Hydrocarbons of Biological Origin
from a One-Billion-Year-Old Sediment

GEOFFREY EGLINTON, P. M. SCOTT, TED BELSKY,
A. L. BURLINGAME, and MELVIN CALVIN

Abstract. *The isoprenoid hydrocarbons, phytane ($C_{20}H_{42}$) and pristane ($C_{19}H_{40}$), are present in the oil seeping from the Precambrian Nonesuch formation at the White Pine Mine, Michigan. Gas-liquid chromatography and mass spectrometry provide the isolation and identification procedures.*

Two experimental methods are now being used to study the origin of terrestrial life and the time of its first appearance. The "primitive atmosphere" experiments (*1*) demonstrate that a wide variety of small molecules of biological significance can be formed in the laboratory from mixtures of extremely simple substances like methane, ammonia, and water. Geologists and geochemists examine ancient sediments for fossil organisms and determine the chemical nature of the imprisoned organic matter (*2–4*). Earlier than about 600 million years ago well-defined morphological remains are scanty, and generally are difficult to relate conclusively to specific living things (*5*). A firm correlation between the morphological evidence and the organic matter present in the same rock would permit a systematic search for chemical evidence of early life in the ancient sediments. Certain classes of organic compounds—the alkanes (*6*), the long-chain fatty acids (*7*), and the porphyrin pigments (*8*)—show promise as biological markers since they are evidently stable for long periods of time under geologic conditions. These compounds are valid as biological markers only insofar as they cannot be synthesized in significant proportions by abiogenic means. For this reason "primitive atmosphere" experiments play an important role. The range of compounds based on the isoprenoid subunit is particularly useful, for here we have a high degree of structural specificity coupled with a widespread distribution in na-

ture. Thus pristane (2,6,10,14-tetramethylpentadecane) and other isoprenoid hydrocarbons have been isolated from crude petroleums of moderate ages (Mesozoic and Paleozoic) in concentrations vastly greater than those anticipated for individual branched alkanes in a thermally derived mixture (*9*). Pristane is a known constituent of living things—zooplankton (*10*), fish and whale oils (*11*), wool wax (*12*), and marine sponges (*13*)—but the original source of the mineralized material may be the phytol portion of chlorophyll degraded either biogenically or abiogenically (*9, 14*). There is every prospect that the isoprenoid hydrocarbons, and the related alcohols and acids, will be useful biological markers.

We now report the isolation and identification of phytane (2,6,10,14-tetramethylhexadecane) and pristane in the oil which seeps in small quantities from the Precambrian Nonesuch shale in Michigan. This rock (*15*) is of Keweenawan age and is in the region of 1 billion (10^9) years old. The identification of these hydrocarbons augurs well for the extension of such analyses to even older Precambrian formations.

We established the conditions for the isolation by processing suitable model mixtures. Thus, the Linde molecular 5 Å sieve (composed of 1.6-mm pellets, dried at 200°C and 10^{-3} mm-Hg, and used at a ratio of 20:1 by weight to the alkanes) quantitatively removed (*16*) the normal isomers from a benzene solution of alkanes from tobacco wax (*17*); the analysis was performed by gas-liquid chromatography at 230°C (3 percent SE-30 silicone gum on Gas Chrom Z, 100-120 mesh, the column being 170 cm × 3 mm). Even after prolonged reflux lasting up to 72 hours, the 2-methyl and 3-methyl substituted *n*-paraffins (C_{25}—C_{35}) were quantitatively retained in the solvent and

the necessary solvent washings. A second sieving treatment of the hydrocarbons was found to be unnecessary. The same fractionation quantitatively removed *n*-heptacosane from a mixture containing cholestane, pristane, and squalane. In each case the *n*-alkanes were readily recovered from the sieve by heating in *n*-hexane or by dissolution of the sieve in dilute HF.

In another experiment an oil shale (*18*) from the Green River Formation (Eocene age, about $60 × 10^6$ years) at Rifle, Colorado, was extracted for several hours with *n*-hexane under reflux and the extract placed on a washed activated alumina column (the particle size being 2 to 44 μ). The initial hexane eluate contained only the alkane fraction which was then subjected to the sieving process, followed by gas-liquid chromatographic analysis (3 percent SE-30 column, programmed from 100° to 300°C at 6°C per minute). The distribution of the *n*-alkanes closely paralleled that reported for this shale by Cummins and Robinson (*18*), with the marked dominance of the odd-carbon number alkanes, especially C_{27}, C_{29}, C_{31}, so characteristic of most plant (*19*) and relatively young sediments (*6*). Again, we confirm these workers' prior findings (*18*) that the lower molecular weight range of the branched and cyclic alkane fraction is mainly composed of phytane, pristane, and other terpenoids, since gas-liquid chromatographic fractions collected in capillary tubes from 6-mm columns displayed the appropriate mass spectrometric characteristics (*20*). Thus, the biological history of this Cenozoic rock is evident from both the very uneven distribution of the *n*-alkanes and from the presence of large proportions of isoprenoid alkanes.

We treated a small sample (4.1 g) of

280

the black viscous oil which oozes from the Precambrian Nonesuch shale (*15*) using the procedures described and found that the members of the *n*-alkane series range from C_{11} to about C_{35}. The distribution reaches a maximum around C_{19} and there is a very slight, but quite definite, predominance of odd-numbered members. The branched-cyclic fraction (2.5 g), which still contains small amounts of aromatic hydrocarbon, is very complex, but is characterized by several prominent peaks in the region corresponding in retention time to that between *n*-C_{14} and *n*-C_{18} on the silicone gum column. We collected amounts of the order of a milligram for each of these peaks by successive sampling on a 6-mm column and then by rechromatographing these at 90°C upon a strongly polar substrate—5 percent tetracyanoethylated pentaerythritol on Gas-Chrom RA, 80–100 mesh, the column being 170 cm × 6 mm—that provided a particularly effective separation. Two of the fractions so obtained when chromatographed on a variety of substrates (Apiezon 'L', fluorosilicone QF-1, silicone gum SE-30, Carbowax-20M, and tetracyanoethylated pentaerythritol) gave single peaks of the same retention times as phytane and pristane. These identifications were then confirmed by direct comparison on the mass spectrometer (*20*).

The mass spectrometric data on incompletely separated fractions show that other compounds with isoprenoid skeletons are present; these fractions may represent geologic breakdown products from the more abundant phytane and pristane. Studies on a capillary column (with Apiezon 'L', 45 m × 0.25 mm at 170°C) of cuts taken from the packed columns show that the branched fraction is an extremely complex mixture: even so, the phytane and pristane comprise approximately 0.6 and 1.2 percent, respectively, of the branched-cyclic fraction.

If one accepts the presence of these hydrocarbons as evidence of life in Precambrian times, there remains the question of the relation between the oil and the rock. We believe that the oil is indigenous to the rock, since we have found an almost identical pattern for the normal and the branched-cyclic alkane fractions (about 3 mg total) isolated from carefully washed (water, HF, and benzene-methanol) and pulverized Nonesuch shale (18 g) taken

from the so-called "marker bed" situated above the stratum from which the oil had been collected. No method for the dating of ancient organic matter exists as yet, so that some doubt must remain concerning the precise age of this oil; however, the geologic evidence (*15*) favors the viewpoint that the organic matter and the associated copper are sedimentary or early diagenetic in origin.

References and Notes

1. N. H. Horowitz and S. L. Miller, *Fortschr. Chem. Org. Naturstoffe* **20**, 423 (1962).
2. P. H. Abelson, *ibid.* **17**, 379 (1959); P. E. Cloud, Jr., and P. H. Abelson, *Proc. Natl. Acad. Sci. U.S.* **47**, 1705 (1961).
3. I. A. Breger, Ed., *Organic Geochemistry* (Macmillan, New York, 1963).
4. U. Colombo and G. D. Hobson, Eds., *Advances in Organic Geochemistry* (Macmillan, New York, 1964).
5. E. S. Barghoorn and S. A. Tyler, *Ann. N.Y. Acad. Sci.* **108**, 451 (1963); J. S. Harrington and P. D. Toens, *Nature* **200**, 947 (1963); C. G. A. Marshall, J. W. May, C. J. Perret, *Science* **144**, 290 (1964); A. G. Vologdin, *The Oldest Algae in the USSR* (Academy of Sciences of the U.S.S.R., Moscow, 1962).
6. W. G. Meinschein, *Space Sci. Rev.* **2**, 653 (1963).
7. P. H. Abelson, T. C. Hoering, P. L. Parker, in U. Colombo and G. D. Hobson (*4*, p. 169); J. E. Cooper and E. E. Bray, *Geochim. Cosmochim. Acta* **27**, 1113 (1963).
8. H. N. Dunning, in I. A. Breger (*3*), p. 367.
9. J. G. Bendoraitis, B. L. Brown, L. S. Hepner, *Anal. Chem.* **34**, 49 (1962); for other papers, J. J. Cummins and W. E. Robinson, *J. Chem. Eng. Data* **9**, 304 (1964).
10. M. Blumer, M. M. Mullin, D. W. Thomas. *Science* **140**, 974 (1963).
11. B. Hallgren and S. Larsson, *Acta Chem. Scand.* **17**, 543 (1963); G. Lambertsen and R. T. Holman, *ibid.*, p. 281, and references cited therein.
12. J. D. Mold, R. K. Stevens, R. E. Means, J. M. Ruth, *Nature* **199**, 283 (1963).
13. W. Bergmann, in Breger (*3*), p. 534.
14. E. G. Curphey, *Petroleum, London* **15**, 297 (1952).
15. W. S. White and J. C. Wright, *Econ. Geol.* **49**, 675 (1954); W. S. White, *ibid.* **55**, 402 (1960).
16. J. G. O'Connor, F. H. Burow, M. S. Norris, *Anal. Chem.* **34**, 82 (1962).
17. J. D. Mold, R. K. Stevens, R. E. Means, J. M. Ruth, *Biochemistry* **2**, 605 (1963).
18. J. J. Cummins and W. E. Robinson, *J. Chem. Eng. Data* **9**, 304 (1964).
19. G. Eglinton and R. J. Hamilton, in *Chemical Plant Taxonomy*, T. Swain, Ed. (Academic Press, London, 1963), p. 187.
20. Mass spectra were determined on a modified CEC 21-103C mass spectrometer (for details of modifications and performance. see F. C. Walls and A. L. Burlingame, *Anal. Chem.*, submitted) equipped with heated glass inlet system operated at 200°C. All spectra were determined at ionizing voltage of 70 ev. ionizing current at 10 μa and 160 to 180 volts per stage on the multiplier.
21. This work supported by NASA (grant 101-61) and by the AEC. We thank Dr. P. E. Cloud, Jr., of the University of Minnesota and the owners and geologists of the White Pine Copper Mine, White Pine, Ontanagon County, Michigan, for providing the samples and for the information pertaining to them. We thank Dr. W. E. Robinson, Bureau of Mines, Laramie, Wyoming, for providing the Colorado oil shale.
* NATO fellow, 1962–64.

15 May 1964

Reprinted from *Science*, **148**(3669), 461–472 (1965)

Paleobiology of a Precambrian Shale

Geology, organic geochemistry, and paleontology are applied to the problem of detection of ancient life.

Elso S. Barghoorn, Warren G. Meinschein, J. William Schopf

Little more than a decade ago serious effort to invade the paleobiological "desert" of the Precambrian was limited to a few devoted, hopeful, and persistent paleontologists employing with ingenuity the traditional techniques of field observation and laboratory microscopy. The cumulative efforts of nearly a century of study of rocks underlying the Cambrian fossiliferous "boundary" have resulted in a small but impressive body of evidence indicating that life had originated and diversified at a time far more remote in geologic history than the lowermost Paleozoic. Unfortunately much of this earlier evidence was equivocal and based on alleged metazoan megafossils or stromatolitic structures secured from partially or highly metamorphosed sediments. It is understandable that an attitude of scepticism has traditionally prevailed in the assessment of Precambrian fossils and their acceptance as bona fide evidence of organisms. On the other hand, geologists and biologists have, somewhat inconsistently, long accepted the view, on grounds of sheer logic, that life must have existed long before the appearance of the diversified "parade" of metazoans and multicellular algae in the lower Paleozoic.

Several factors appear to have been responsible for retarding progress in the understanding of the Precambrian record of life. Among these are: (i) preoccupation with the problem of discovering megascopic invertebrate fossil organisms and a concomitant lack of attention to, or interest in, microfossils or

Dr. Barghoorn is professor of botany and curator of Paleobotanical Collections of the Botanical Museum at Harvard University, Cambridge, Massachusetts; Dr. Meinschein is research associate at Esso Research and Engineering Company, Linden, New Jersey; Mr. Schopf is a graduate student in the department of biology at Harvard University.

the fine structure of potentially fossiliferous sedimentary rocks, such as cherts; (ii) lack or paucity of chronological control in the Precambrian, such as is now provided by radiogenic dating; and (iii) lack or inaccessibility of corroborative evidence such as is now available through application of paleobiochemical techniques and the sensitive instruments of organic chemical analysis.

It might be postulated that organisms early in the history of living things would be unicellular, or at least simply organized multicellular aggregates, and of microscopic size. Concrete evidence for this assumption has been obtained in recent years through the discovery of a varied assemblage of microorganisms showing definitive and three-dimensional morphology in sediments almost 2000 million years old (*1*). The lithologic environment of preservation of these organisms is a cryptocrystalline chalcedonic chert (Gunflint chert), the incompressibility of which permitted retention of three-dimensional form. Present evidence and continuing study of the paleochemistry of the Gunflint rocks indicate the presence of relatively complex organic molecules directly associated with the structurally preserved organic remains (*1*). The Gunflint assemblage of organisms establishes, at least temporarily, a minimal time for the duration of multicellular life. Between Gunflint time and the beginning of the lower Cambrian, however, there is an extreme paucity of coherent evidence of organic activity and of fossils (*2*). This article deals with various aspects of the geology, organic geochemistry, and paleontology of one of the more interesting intermediate links in the continuity of life—with evidence from rocks approximately 1000 million years old.

Geology

The genesis of the economically important copper ore present in the basal portion of the Precambrian Nonesuch shale formation, one of the several cupriferous deposits of northern Michigan, has been the subject of considerable controversy (*3–8*). In part because of this controversy, and in part because of the economic importance of the formation, many of the petrological, structural, and inorganic geochemical aspects of the basal portion of this ancient deposit have been investigated (*3, 6, 7, 9, 10*). In a preliminary report we stated that the Nonesuch shale appears to be the oldest formation in which crude oil, optically active alkanes, porphyrins, and the isoprenoid hydrocarbons pristane and phytane are known to occur, and, in addition, that what appear to be fossil microorganisms of several types occur in the formation (*11*). The preservation of this varied and unique assemblage of paleontological and organic geochemical evidence of early life must be directly related to the depositional, diagenetic, tectonic, and thermal history of the formation, a history which would include the period when copper minerals were emplaced in this deposit.

Geologic setting. The White Pine area, located in the northeast half of Township 50 N, Range 42 W, Ontonagon County, northern Michigan, is situated just southeast of the Porcupine Mountains (see Fig. 1). The Nonesuch formation, as exposed at White Pine, is a dominantly gray, well-bedded siltstone and shale, easily differentiated from the red-to-brown coarser-grained rocks of the formations above and below. The cupriferous zone comprises the basal 7½ to 9 meters of the Nonesuch formation and includes also the uppermost 1½ meters of the underlying middle Keweenawan Copper Harbor conglomerate (*3*). The biological remnants present at White Pine occur in the cupriferous zone. Both the Nonesuch shale and the overlying Freda sandstone are of late Keweenawan age. The radiogenic age of the Nonesuch formation, based on rubidium-strontium ratios in whole-rock samples from the White Pine area, has recently been determined as 1046 ± 46 million years, by S. Chaudhuri and G. Faure (*12*). Figure 2 represents a generalized columnar section of these upper Precambrian formations as they occur in the White Pine area.

The Keweenawan strata of northern

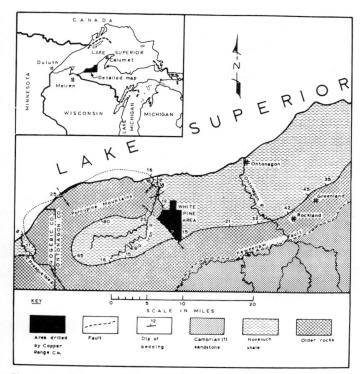

Fig. 1. Index map showing location of White Pine area, altitude and location of the base of the Nonesuch shale, and presumed subsurface areal extent of the formation. [After White and Wright (*3*)]

Michigan and northern Wisconsin lie on the southern flank of the Lake Superior basin and dip gently northwest and north toward the center of the basin at most places (*3*). The structure of the Nonesuch shale in the vicinity of White Pine is dominated by the White Pine fault (see Fig. 1). In addition to this major fault, which has a horizontal displacement of several hundred meters (*3*), many small-scale, steeply dipping strike-slip faults disrupt the strata in the vicinity of White Pine.

Distribution of organic material. The cupriferous zone in the White Pine area has been divided, from oldest to youngest, as follows: Lower sandstone, Parting shale, Upper sandstone, and Upper shale (*3*). Mineralogically, with the exception of the copper minerals which were probably emplaced after burial but prior to lithification (*3, 6*), the shales and sandstones of the cupriferous zone appear to be similar to most other shales, silts, and sands. Figure 3 represents a detailed columnar section of this zone as it occurs near White Pine.

The Parting shale and the Upper shale units of the cupriferous zone contain sporadically distributed vugular structures which often contain small amounts of crude oil; liquid hydrocarbons also occur, associated with fractures in these strata. The organization of asphaltic material present within the interstices of the Parting shale unit, as observed in thin section in transmitted light, is typified by alternating, sharply delineated layers and spheroids of reddish and light amber material interpreted as residual petroleum exhibiting varying degrees of devolatilization (Fig. 4, parts 1, 2, and 4). This organic material is frequently arranged in an undulating, concentric manner about a mineral center, often chalcocite. The significance of these structures with regard to the paleontological differentiation between biogenic morphological organization at or above the microscopic level ("fossils"), abiogenic morphological organization ("pseudofossils"), and biogenic indicators at the molecular level ("chemical fossils") is discussed in a later section.

Finely disseminated carbonaceous material, locally but scantily present in the two sandy units (*3, 13*), is relatively abundant and found throughout the shale units of the cupriferous zone. In general, the vertical and horizontal distribution of copper minerals within the zone, predominantly chalcocite and native copper, appears to vary in direct proportion to the amount of organic matter present (*3, 9*). In the case of the Lower sandstone unit this spatial relationship appears to be specific at a microscopic level—that is, particles of native copper are commonly surrounded by, and in intimate contact with, anthraxolitic organic material (*14*).

The direct association of copper minerals with finely disseminated carbonaceous matter, anthraxolitic organic material, and partially or wholly devolatilized asphaltic organic residues within the cupriferous zone tends to substantiate the hypothesis, presented by other investigators (*4, 6*), that the precipitation of copper minerals in the basal Nonesuch shale was facilitated by the presence of organic material. Inasmuch as lateral and vertical distribution of the copper minerals strongly suggests their diagenetic emplacement (*3, 6, 9*), a syngenetic or early diagenetic origin for the organic material may be inferred. The possibility of petroleum migration from younger rocks is negated by consideration of the great thickness of coarse-grained, permeable sediments devoid of organic matter directly overlying the Nonesuch formation, and by the nearly complete absence of crude oil in the 180 meters of essentially similar silts and sands directly overlying the basal oil-bearing cupriferous zone.

Summary of geologic history. The geologic history of the cupriferous zone of the Nonesuch shale in the White Pine area may be briefly summarized as follows:

1) Pre-Keweenawan, basic and acidic, volcanic and plutonic igneous rocks to the south, with a minor contribution from sedimentary and metamorphic strata, produced particles which were carried north from their terrain of rather low relief to a proximate coarse-sand delta by a fluvial system. The gradual aqueous inundation of the delta and associated sand plain and the simultaneous decrease in the average particle size of the detritus deposited at White Pine resulted in the deposition of highly organic, laminated silts and shales in shallow-water, peri-

odically reducing environments present in apparently disconnected depressions of the delta. Further inundation resulted in the deposition of plant remains within a laterally continuous, fairly homogeneous siltstone unit. During the deposition of this unit the local basin reached a state of maximum submergence; this was followed by a gradual subsidence of water level, culminating in exposure of the strata to subaerial weathering and the end of the first sedimentary cycle.

2) Following this diastem and an associated readjustment of drainage patterns in the delta, the depositional cycle was repeated. Minor authigenic alteration of the less stable minerals of the sediment took place (9), followed by burial, partial degradation of the organic material, and the accumulation of petroleum. Prior to lithification, copper was introduced and probably distributed by connate water (6). Subsequently the sediment compacted and the synchronous chromatographic segregation of organic molecular types took place. At some time following lithification the formation was locally faulted and folded. This was apparently associated with both the uplift of the anticlinal Porcupine Mountains and the genesis of the White Pine fault. Some copper minerals were probably redistributed during this tectonic activity (3–8). Nevertheless, metamorphism has been minimal, and the formation has had a mild thermal history. Further deposition, uplift, and the subsequent erosion of much of the overburden has exposed the present formation.

Paleobiology

Ranges of variabilities of animate and inanimate materials have not been completely defined. In dealing with the general problem of recognizing evidence of ancient life, one is forced to consider what similarities between biological remnants and existing organisms are required to ensure that abiogenic entities may not be mistaken for biologically produced substances. Although an absolute evaluation of the reliability of "fossil evidence" cannot be made, historical and theoretical considerations provide a basis for limited appraisals of the accuracies and potentials of various methods used for detecting the remains of former living things.

Traditionally, a fossil is defined as a morphologically organized entity or remnant of a preexistent organism, ir-

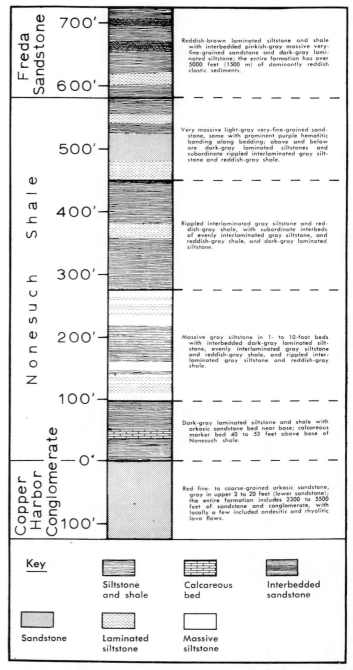

Fig. 2. Columnar section of Nonesuch shale formation in the White Pine area, Michigan. [After White and Wright (*3*)]

3

respective of its size or completeness. This concept of a fossil has been extended in recent decades by increasing the discrimination of the level of observation, with the result that at present it is possible in many cases to determine biologically produced structures in organic residues at the molecular or even the submolecular level. Isotopic measurements have demonstrated that biological processes are selective at the atomic level, and purely inorganic manifestations of biological activity, such as certain carbonates, phosphates, and free sulfur, for example, are known. Thus, the term *fossil* in its broadest sense may embrace *any* evidence of ancient life, and the reliability of identifications of biological remnants, as well as the information that can be gained from their study, may be increased by the use of a variety of analytical and observational methods. It is for this reason that the results of organic chemical and isotopic analyses, in addition to classical geological and traditional micropaleontological observations, have been used in this investigation of biological remnants in the Nonesuch formation.

Paleobiochemical Experiments

Samples. Siltstone and crude oils were obtained from the Parting shale subdivision of the Nonesuch shale formation at White Pine, Michigan (*15*).

Solvents. The solvents used were of reagent grade, and the solvents used in the alkane analyses were distilled and tested prior to use, as described elsewhere (*16*).

Porphyrin isolation and characterization. Soxhlet extractions were carried out on three batches of siltstone pulverized to pass 200 mesh. The total weight of sample was 215 grams, and a solution of 2 volumes of benzene in 1 volume of methanol was used as the extracting solvent. Each batch of siltstone was extracted for 24 hours. The solvent was removed by evaporation under vacuum at 65°C. The residue (0.2 g) was redissolved in a minimum volume of chloroform (about 5 ml), and asphaltenes were precipitated by the addition of 50 milliliters of hot isooctane. The asphaltene precipitate was removed by vacuum filtration. Additional asphaltenes were separated from the filtrate

by repeating these asphaltene removal steps, and the asphaltene-free residue was dissolved in a minimum volume of isooctane for subsequent chromatographic fractionation.

A column (2 by 30 cm) was filled with 125 grams of activated Merck alumina (reagent grade) in isooctane. The asphaltene-free residue (0.18 g) was placed on this column, and elution was accomplished by successive additions of 200 milliliters of each of the various solvents and solvent solutions in ratios (by volume) as follows: isooctane; isooctane and benzene (1:1); benzene; benzene and diethyl ether (3:1; 1:1; 1:3); diethyl ether; diethyl ether and methanol (1:1); and methanol.

Separate eluates were collected as each solvent solution passed onto the top of the column. Visual spectra, obtained with a Perkin-Elmer 202 visible ultraviolet spectrophotometer, indicated that, although some vanadyl porphyrins were present in the ether-methanol eluate, most of these porphyrins were concentrated in the benzene-ether eluates. Ultraviolet spectra were also obtained at this time. Infrared absorption spectra of these eluates were measured on a Perkin-Elmer 137 Infracord spectrophotometer.

Petroleum from the Nonesuch formation was also investigated for evidence of the presence of porphyrins. Three grams of crude oil were dissolved in an excess of benzene, and the suspended rock particles were removed by vacuum filtration. Benzene was removed by evaporation under vacuum at 65°C. Asphaltene precipitation was accomplished by means of the chloroform-isooctane method described above; this separation was carried out three times. The asphaltene-free oil was placed on a chromatography column prepared as described. Elution and collection of eluates was accomplished by the successive addition of 200-milliliter volumes of solvent solutions of the kinds and in the ratios specified above. Visible spectra obtained by the method utilized in identification of vanadyl porphyrins from the siltstone unit of the formation yielded no evidence of the presence of porphyrins in this crude oil. This isolation procedure was repeated with a petroleum sample weighing 14 grams, and again no evidence of the presence of porphyrins was detected.

To complete the study, the asphaltene portion of the crude oil was investigated for evidence of the presence of porphyrins. The asphaltene precipitates from the two oil samples were com-

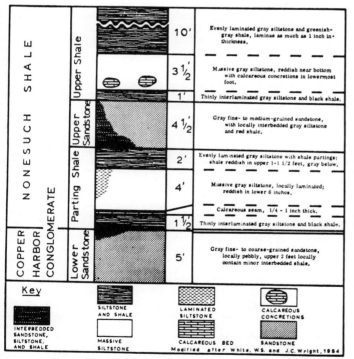

Fig. 3. Detailed columnar section of cupriferous zone of the Nonesuch shale in the White pine area. [After White and Wright (*3*)]

bined, redissolved in benzene, and placed on an alumina chromatography column (2 by 30 cm; Merck activated alumina, reagent grade) packed in benzene. Except for the fact that no isooctane-containing solvent solutions were used, elution, collection, and spectral analysis were accomplished as described above. As in the case of the asphaltene-free portion of the oil from the Nonesuch formation, no porphyrins were detected in the asphaltene constituents.

Alkane analysis. Descriptions of the procedures employed in extraction, removal of solvents or sample recovery (*17*), and liquid-solid phase chromatography, in alkane analysis, have appeared elsewhere (*16–18*). A 10-gram sample of siltstone and 10 grams of crude oil were separately analyzed. Mass spectrometric, ultraviolet, and infrared spectroscopic analyses were made of alkanes from the siltstone and the crude oil (*18*).

The optical activities of the alkanes from the crude oil were measured on an O. C. Rudolph and Sons instrument [model 80 CSPZA (Calcite Optics)] with a No. 600 mercury light source, a 3-millimeter micro device, and a simm angle of about 2 to 3 degrees. Measurements were made (in a 100-mm tube with 6-mm bore) on a solution (0.396 g/ml) of the alkanes in spectral-grade isooctane (*19*).

Gas-liquid chromatography analyses of the alkanes were made by means of five capillary columns (J. Bishop and Company type 316 stainless steel, 0.025 by 3000 cm). One of these columns was coated with 10-percent solution of Apiezon L; this column was prepared by the Barber-Colman Company. Another column was coated with a 10-percent solution of Apiezon N, and the remaining three columns were coated with one or another of three different liquid-solid chromatographic fractions of Apiezon L. These fractions were separated on silica gel (ratio for gel to Apiezon L, 100:1); n-heptane, benzene, and a solution of benzene and methanol (1:1, by volume) were used successively as eluants. In each case the ratio of eluant to silica gel was 1.5:1 by volume. The concentrations of the various Apiezon-L fractions in the benzene used for coating the columns are shown in Table 1.

Gas-liquid chromatography analyses of reference hydrocarbons and of the Nonesuch alkanes were made with the five capillary columns described above, and the analyses presented in Figs. 5, 6, and 7 were made with the column

supplied by the Barber-Colman Company. These chromatograms were obtained with a Barber-Colman model 10 instrument equipped with a temperature programmer and a modified inlet system. The modification consisted of mounting the inlet against the detection cell block and placing additional insulation about the inlet system. Temperature was controlled to increase 1°C per minute (from 25° to 300°C) for conditioning the columns and 5°C per minute (from 70° to 300°C or 320°C) for the analyses. The argon-inlet pressure on the columns was 30 pounds per square inch (2.04 atm). The split flow was 35 milliliters per minute.

Operation conditions were as follows: inlet temperature, 450°C; microionization detector (radium sulfate source), 1000 volts; amplifier sensitivity setting, 10; attenuation, 2 to 16; range, 1×10^7; chart speed, 38 centimeters per minute.

Results and discussion: Porphyrins. The visible spectrum (in chloroform) of the porphyrin aggregate isolated in the benzene-ether eluates from the siltstone of the Parting shale unit is given in Fig. 8. This aggregate constitutes about 45 parts of the original rock sample per million. The peak configuration and their locations (411, 534, and 573 μ) are typical of vanadyl porphyrins isolated from crude oil and bituminous strata (*20*). Although most of the porphyrins were isolated in the benzene-ether eluates, the presence of minor amounts (about 5 parts per million) of vanadyl porphyrins in the more polar ether-methanol eluate suggests that at least two types of porphyrins are present. The infrared spectrum of the porphyrin aggregate from the benzene-ether eluates, given in Fig. 9, shows a well-defined bifurcated absorption band at about 5.8 microns, indicating the presence of two carbonyl groups. Although careful chromatographic techniques seldom produce a sample with porphyrin content higher than 50 percent (*21*), the intensity of these peaks and their presence in the fractions eluted by the relatively weakly polar

benzene-ether solvent solutions suggests that they are related to the dominant species of the aggregate and that they contribute polarity to a rather large and otherwise nonpolar molecule. Analyses of the visible and ultraviolet spectra indicate that vanadium-porphyrin complexes are the dominant members of this aggregate. The chromatographic characteristics of these rather large and otherwise nonpolar molecules are probably the result of polarity contributed by attached carbonyl groups.

The limit of porphyrin detection in the procedure used in this investigation is about 0.05 part per million. For this reason it might be suggested that the apparent absence of porphyrins in the crude oil of the Nonesuch formation is due to insufficient detection sensitivity. However, the two portions of asphaltene-free oil placed on the alumina column were 14 and 65 times, respectively, the volumes of analogous material obtained from the extracted siltstone. The evidence therefore suggests that there has been a very nearly complete segregation of porphyrins between the siltstone and the petroleum phases of the formation. The presence of vanadyl porphyrins within the fine-grained strata, together with their absence in the petroleum present in the associated vugular structures of the Parting shale unit, establishes the fact that the porphyrins have had limited mobility within the formation and indicates that they are indigenous to the siltstone and not secondarily emplaced.

The vanadyl porphyrins indigenous to this ancient formation appear to be the oldest respiratory pigment derivatives yet discovered. Although it might be postulated that they represent degradation products of hemoglobin or other animal pigments, any such animals must have ultimately relied on plants as a source of oxygen and as a food supply. This consideration and the presence of fragments of plant tissue within the strata indicate that the vanadyl porphyrins of the Nonesuch formation constitute the oldest presently

Table 1. Identical retention temperatures (in degrees $\pm$ 1°C) of the peaks for reference pristane and phytane and for pristane and phytane in the Nonesuch alkanes.

Compound	Apiezon L column supplied by Barber-Colman	10% Solution of Apiezon N	8% Solution of n-heptane eluate of Apiezon L	0.5% Solution of benzene eluate of Apiezon L	0.7% Solution of benzene-methanol eluate of Apiezon L
Pristane	193	184	186	137	133
Phytane	208	198	199	154	144

5

286

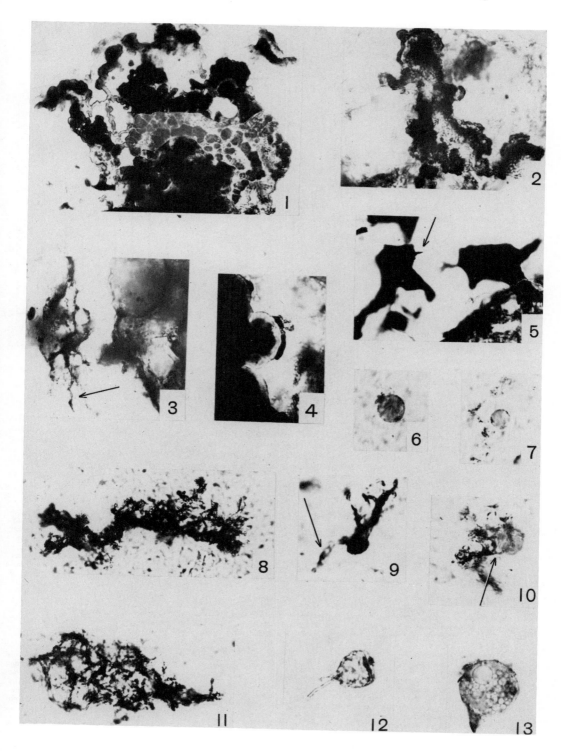

known direct evidence of the photosynthetic mechanism.

The presence of porphyrins within the cupriferous zone of the Nonesuch shale has important bearing on the origin of the economically important copper minerals of the deposit. Sales (4) and Joralemon (5) have suggested that all copper minerals of the basal Nonesuch were emplaced by magmatic hydrothermal solutions emanating from the White Pine fault at a late period in the history of the formation. White and Wright (3) and White (6) have suggested that most of the copper minerals of the zone were deposited through diagenetic processes, involving relatively low temperatures, prior to lithification. Thus, the time and temperature of copper emplacement are important factors in evaluating these two conflicting theories.

As early as 1934 it was recognized that preservation of porphyrins is dependent upon mild geothermal conditions (22). Calculations based on the activation energy required for the thermal degradation of porphyrins indicate that these chemical fossils could not have persisted longer than about 100 years in a lithologic environment subjected to temperatures of 250°C, for longer than about 11.5 days (10^6 seconds) at temperatures of 350°C, or longer than about 100 seconds at temperatures of 500°C (23). The presence of porphyrins within the basal Nonesuch is consistent with a low-temperature origin for the copper minerals of the deposit and renders highly unlikely any proposed mechanism of emplacement involving widespread high temperatures, such as the epigenic hydro-

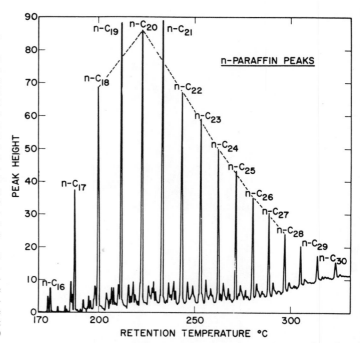

Fig. 5. Gas-liquid chromatographic analysis of alkanes in the *n*-heptane eluate from silica-gel columns of Nonesuch crude oil.

thermal theory would probably require.

Results and discussion: Hydrocarbons. Crude oils are generally thought to be of biological origin, and extensive investigations have established the identities and similarities in the C_{15} and C_{30} alkanes that are common to crude oils, sedimental extracts, and biological lipids (18, 24). Data in Table 2 make

it possible to characterize the crude oil and benzene extract from the Nonesuch siltstone. These data, in conjunction with results of other analyses, show that both the Nonesuch samples contain higher concentrations of alkanes and lower concentrations of aromatic hydrocarbons than an average crude oil. Differences in the results reported in

Fig. 4 (left). Physical organization of organic matter in Parting shale unit, Nonesuch shale formation, White Pine, Michigan (all photographs taken in transmitted light). Part 1. Representative of category 2 in text. Composite photograph showing concentric layers of alternating reddish-brown and translucent amber asphaltic-like organic matter interpreted as residual petroleum exhibiting varying degrees of devolatilization. (In the photograph the reddish-brown layers are dark, the translucent layers lighter.) Central area, superimposed from photograph taken at lower focal depth, shows radially oriented condensation cracks. (Thin section; about × 1275.) Part 2. Representative of category 2 in text. Asphaltic-like organic material, mammillary in outline, showing well-defined outer layer of residual petroleum and associated solid spheroidal body (at upper left). (Thin section; about × 1275.) Part 3. Representative of category 3 in text. Anastomosing aggregate of filamentous plant fragments oriented parallel to bedding planes of siltstone; arrow points to cell wall of nonseptate filament. (Thin section; about × 1275.) Part 4. Representative of category 2 in text. Globular structure showing broad, sharply delineated layers of asphaltic-like organic material in varying degrees of devolatilization; inner layer (light in photograph) exhibits poorly defined radial condensation cracks. (Thin section; about × 1275.) Part 5. Representative of category 1 in text. Angular anthraxolitic organic matter occurring as interstitial filling between mineral grains; arrow points to organic material extruded at grain boundaries, suggesting a precompaction origin for this material. (Thin section; about × 1025.) Part 6. Solid organic spheroidal body, probably derived from disaggregation of asphaltic-like organic residues (see parts 1 and 2), about 7 μ in diameter. (Maceration; about × 1325.) Part 7. Hollow, hyaline, spherical organic body, about 5 μ in diameter, associated with poorly defined plant filament. (Maceration; about × 1325.) Part 8. Massive sheet of plant tissue composed of filamentous cellular residues. (Maceration; about × 675.) Part 9. Ovoid solid organic bleb in juxtaposition with branched and unbranched plant filaments; arrow points to filament showing septated cell-like divisions approximately 1 μ wide. (Maceration; about × 1600.) Part 10. Aggregate of filamentous plant fragments in juxtaposition with solid spheroidal organic body; arrow points to branching filament. (Maceration; about × 1325.) Part 11. Sheet of plant tissue composed of occasionally branching, anastomosing filaments. (Maceration; about × 1025.) Part 12. Filamentous plant fragment associated with subangular solid organic particle having a fossil-like appearance. (Maceration; about × 1025.) Part 13. Hollow, hyaline spherical organic body, 42 μ in diameter, containing an internal cluster of hollow spherical nodes ranging in diameter from 1 to 13 μ; origin of this pseudofossil structure apparently involves an emulsification or alveolation process. (Maceration; about × 500.)

7

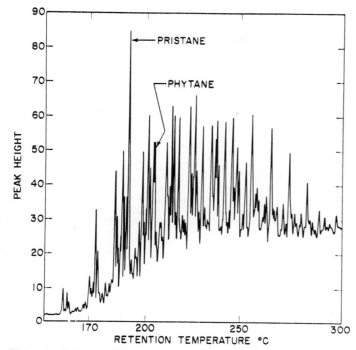

Fig. 6. Gas-liquid chromatographic analysis of the *n*-paraffin-free, first *n*-heptane eluate of Nonesuch alkanes from the alumina column. *n*-Paraffins were removed from this eluate by means of molecular sieves.

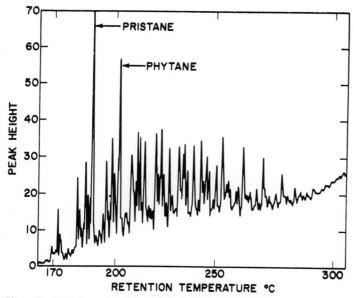

Fig. 7. Gas-liquid chromatographic analysis of the *n*-paraffin-free, first *n*-heptane eluate of Nonesuch alkanes from the alumina column, after the addition of reference pristane and phytane to this eluate.

Table 2 for the Nonesuch oil and extracts are not significant, and infrared, ultraviolet, and mass spectra of these chromatographic fractions of the Nonesuch samples further confirm their equivalence. Compositionally, the Nonesuch samples resemble a highly paraffinic petroleum of the type frequently produced in Pennsylvania. Oils from older sediments are usually more paraffinic than oils from younger sediments (*25*).

Organisms are uniquely capable of synthesizing carbon compounds which possess optical activity (*26*). Alkanes from living organisms (*26*), recent sediments (*24, 27*), and crude oils (*28*) rotate polarized light in a clockwise or positive direction. Values for the optical activities of alkanes from these sources and for the Nonesuch crude oil (*11, 19*) are presented in Table 3. It is apparent from these values that the Nonesuch alkanes have optical rotations comparable in direction and magnitude to those of other biological and sedimental alkanes.

A gas chromatogram of the alkane fraction from the Nonesuch crude oil is presented in Fig. 5. All the large peaks in this chromatogram are produced by *n*-paraffins. In Fig. 5, the tops of successive peaks produced by *n*-paraffins of even carbon number (n-C_{18} to n-C_{24}) have been joined by dotted lines. It is noteworthy that the peaks for *n*-paraffins of odd carbon number extend above the lines which join their even-carbon-number homologs. Biological and sedimental alkanes frequently contain more odd- than even-carbon-number *n*-paraffins (*24, 29*). Although this "odd carbon preference" is not pronounced in the alkanes from some organisms (*30*) and from most ancient sediments, the slightly greater abundances of odd- than of even-carbon-number *n*-paraffins in sedimental alkanes have been reported as evidence that these compounds were made either by formerly living organisms or from biological acids and alcohols (*24, 29*). The "odd carbon preference" of the Nonesuch *n*-paraffins is greater than that of the *n*-paraffins from some organisms (*30*) and from most crude oils, even some oils that are younger than Paleozoic (*29, 31*).

Branched-chain alkanes and n-C_{21} and smaller-molecular *n*-paraffins are concentrated in the first fraction of sedimental alkanes that are eluted by *n*-heptane from alumina columns (*32*), and *n*-paraffins are removed from mixtures of alkanes by molecular sieves

8

(33). A gas chromatogram of this first *n*-heptane eluate of the Nonesuch alkanes, obtained after the removal of *n*-paraffins, is shown in Fig. 6. Peaks labeled "pristane" and "phytane" in this chromatogram are chromatographically equivalent to pristane and phytane as established in Fig. 7. The heights of the pristane and phytane peaks of Fig. 7 had been increased by the addition of reference compounds. The chromatographic equivalence of the reference and Nonesuch pristane and phytane were established on five gas-liquid chromatography columns, each coated with a different substrate. Retention temperatures for reference pristane and phytane and the pristane and phytane peaks in the Nonesuch alkanes are given in Table 1 for these five columns. Mass spectra of the branched-chain concentrate from the Nonesuch alkanes contained large peaks at mass 183. In the mass spectra of pristane and phytane, large peaks at mass 183 are assumed to be caused by 2,6,10-trimethyldecyl ions (34).

Polycyclic alkanes and *n*-C_{25} and larger molecule *n*-paraffins are concentrated in carbon tetrachloride eluates of sedimental alkanes from alumina. Mass spectra of the polycyclic alkane concentrates from the Nonesuch alkanes have large peaks at mass 372, 218, 217, and 149; such peaks are characteristic of parent sterol hydrocarbons (16). Similar polycyclic alkanes are apparently found ubiquitously in biological and sedimental alkanes (18, 35).

Geological evidence, the occurrence of plant fragments within the formation, and the distribution of porphyrins discussed above clearly indicate that biological materials in the Nonesuch formation were not derived from younger sediments. The pristane, phytane, optically active alkanes, *n*-paraffins, and steranes in the Nonesuch crude oil and extract, therefore, are apparently indigenous to sediments which were deposited approximately 1 billion years ago. Identities or similarities in the structures and optical activities of the Nonesuch alkanes and of modern biological alkanes suggest that certain hydrocarbons can be preserved for long periods of geologic time in some sediments. Additional support for the view that alkanes may retain their structures for nearly 1 billion years in some sedimentary environments may be deduced from the distribution of *n*-paraffins in the Nonesuch formation. Random cleavages of carbon bonds in *n*-paraffins would produce

Table 2. Data from silica gel chromatographic analyses of material from the Nonesuch formation.

Material	Percentage eluted by			
	n-Heptane (alkanes)	Carbon tetrachloride (alkanes and aromatics)	Benzene (aromatics)	Methanol (nonhydrocarbons)
Nonesuch samples:				
Crude oil	61.2	22.2	12.7	3.9
Extract	61.5	20.3	15.4	2.8
Average crude oil*	47.2	12.1	32.4	8.3

* Average for 110 crude oils.

equal amounts of odd- and even-carbon-number molecules. The "odd carbon preference" of the Nonesuch *n*-paraffins indicates that these compounds are not degradation products.

The pristane, phytane, and porphyrins in the Nonesuch sediments may be derived from chlorophyll. Reaction pathways by which chlorophyll may have been converted into the types of porphyrins which appear widely distributed in ancient sediments have been experimentally defined (36), and pristane and phytane contain carbon skeletons that are present in phytol, which appears in an ester substituent in chlorophyll. Phytol has been commonly suggested as a precursor of the pristane and phytane in sediments (34, 37). However, pristane is a constituent of various organisms (38), and phytane has been tentatively identified in the bacterium *Vibrio ponticus* (30).

Since living organisms synthesize pristane and apparently phytane, and the production of these compounds

from phytol has not been accomplished by abiotic reactions that might occur in sediments, pristane and phytane are probably less reliable indicators of the preexistence of photosynthetic organisms than porphyrins are. The metabolic processes of living organisms suggest that complementary information about the biosynthetic pathways of ancient organisms may be obtained through consideration of the porphyrins and alkanes present in the Nonesuch sediments.

Chlorophyll absorbs the solar energy that photosynthetically reduces CO_2 to carbohydrates in green plants. Carbohydrates are a source of energy and precursors for the biosynthesis of other biological compounds. Hexoses, derived from carbohydrates, are converted in living cells into pyruvates, which in turn participate both in the Krebs cycle and in lipid biosynthesis (39). The biosynthesis of the pyrrole rings in porphyrins has been traced from succinate, a compound of the Krebs cycle (40). Succinate condenses with glycine, and the glycine carboxyl is then cleaved to form δ-aminolevulinic acid, which cyclizes to form the pyrrole rings of porphyrins (21).

Because alkanes are minor and chemically inert products of plants and animals, the metabolism of alkanes has received less attention than has that of most other organic compounds. However, alkanes are structurally and isotopically similar to the acids and alcohols in biological fats and waxes (18, 21). Furthermore, direct metabolic relationships between *n*-paraffins and fatty acids have been established. *n*-Paraffins are converted by Ω oxidation into fatty acids in the livers of certain verte-

Table 3. Optical activities of alkanes.

Sample	Measured rotation (deg)	α_D^{20} (deg)	α_{5460}^{35} † (deg)	α_{5780}^{35} ‡ (deg)	Reference
Biological					(26)
Spirogyra	+1.28				
Hydrodictyon reticulatum	+1.0				
Vibrio ponticus	+0.94				
Recent sediments:					
Gulf of Mexico		+0.514			(24)
Atlantic Coast of France		+3.3			(27)
Crude oils					(28)
Pennsylvania		+0.25			
Rodessa		+0.79			
Mid-Continental No. 1		+0.93			
Mid-Continental No. 2		+1.20			
California		+2.73			
Nonesuch alkanes			+0.541	+1.569	(11, 19)

* α_D^{20} Means rotation measured at 20°C in light source of wavelength of the D line of sodium, 5890 Å. † α_{5460}^{35} Means rotation measured at 35°C in light source of wavelength 5460 Å. ‡ α_{5780}^{35} Means rotation measured at 35°C in light source of wavelength 5780 Å.

9

290

Table 4. The δC^{13} determinations in Nonesuch shale.

Sample	δC^{13}
CaCO₃	− 3.2
CaCO₃	− 2.1
"Kerogen"	−28.9
Petroleum	−29.7

brates (*41*). Thus, information about the structural (*24, 39, 40*) and isotopic (*21, 41*) selectivities of biological processes and the limited data on the metabolism of *n*-paraffins provide substantial reasons for presuming that the alkane constituents of plant and animal lipids are biosynthesized either from acetates, as are the acids and alcohols, or from the acids and alcohols (*18*).

Lipid biosyntheses are defined in even greater detail than is the biosynthesis of porphyrins (*39, 42*). Acetates, made from pyruvates, are the precursors of malonates and mevalonates, which serve, respectively, in the production of straight- and branched-chain acids, alcohols, and hydrocarbons. Squalene, a branched-chain C_{30} alkene, is in turn an intermediate in steroid syntheses (*39*). Equilibria exist between hexoses, pyruvates, acetates, and the tricarboxylates in living cells. Amino acids, building units of proteins, as well as pyrrole rings are made from the tricarboxylic acids participating in the Krebs cycle (*39*).

In essence, pyruvates are apparently intimately involved in the production of most biological compounds, and alkanes and porphyrins are probably products of the two major biosynthetic pathways from pyruvates. One pathway leads through acetates to lipids, and the other, through the Krebs cycle to the syntheses of amino acids and the porphyrins. The presence in the Nonesuch formation of porphyrins and alkanes that resemble porphyrins in geologically young sediments and alkanes in living organisms suggests that life which existed in Precambrian times was metabolically similar to existing life.

Paleontology and Physical Organization of Organic Matter

As previously noted, organic matter is widely disseminated throughout the cupriferous zone of the formation. The organic matter, as seen in thin sections of the rock, exhibits three basic patterns of organization.

1) Irregular, often highly angular discrete or anastomosing opaque bodies occurring as interstitial fillings between mineral grains (Fig. 4, part 5).

2) Globular spheroidal masses, translucent amber to reddish-brown, often mammillary in outline and commonly exhibiting radial cracks or checks which are oriented at right angles to the interface or to the surface (Fig. 4, parts 1, 2, and 4).

3) Threads, or aggregates of threads, of organic matter, often tightly anastomosed and "shreddy" in outline (Fig. 4, part 3).

These filaments vary in color from nearly hyaline to dense brown, and they vary markedly in length, organization, and degree of curvature. Bodies in categories 1 and 2 seem most logically interpretable as structures which have resulted from diffusion of liquid organic material through the porosities of the mineral matrix prior to compaction and from subsequent devolatilization and condensation. Material in category 3 is interpreted as being fragments of plant tissue, ranging from triturated filamentous structures 1 to 15 microns long to aggregates exceeding 100 microns in length.

The putative plant fragments have been rigorously examined with a view to determining whether they are in fact original plant tissue, as opposed to pseudomorphs or secondary aggregates of organic matter unrelated to primary structure. It should be noted that the poor preservation of these filamentous bodies makes their interpretation difficult. However, careful resolution of the structure of these fragments shows the following features characteristic of plant microfossils, which form the bases for our interpretation of their organismal origin: (i) filamentous or laminar form with discrete cell walls and occasional septations; (ii) prevailing orientation of filaments and laminae parallel to the bedding planes; (iii) tendency of the filament aggregates to anastomose; (iv) occasional appearance of true branching; (v) absence of birefringence in polarized light; and (vi) optical differentiation from organic matter of categories 1 and 2 present in the same rock.

Although it seems probable that these threads or filaments constitute remnants of photosynthetic organisms, the great age and poor preservation of these plant fossils makes it impossible to assign a possible phylogenetic position with any degree of certainty. Gross morphology, the depositional environment, and evolutionary considerations would be consistent with an assumed algal or fungal affinity for these forms, but it would probably be unwarranted to base such an assignment on the available evidence.

Preparations of organic residues from the Nonesuch shale were secured by dissolution of the rock in mixtures of hydrochloric and hydrofluoric acids, followed by washing and centrifugation in heavy liquid (ZnBr, specific gravity 2.3). Slurries of the organic fraction recovered in this manner were mounted on clean slides in a medium of purified glycerine jelly. Numerous rounded or subangular solid blebs of organic matter were observed distributed among masses of plant-tissue fragments, often in juxtaposition with the fragments (Fig. 4, parts 6, 9, 10, 12). It is apparent that many of the rounded or subangular solid particles have resulted from disaggregation of the spheroidal or mammillary asphaltic residues observed in thin sections of the shale. Plant fragments, observed in macerations far more clearly than in thin sections, consist of fragments of filaments, aggregates of filaments, and occasional sheets of tissue which appear to show cellular residues (Fig. 4, parts 8 and 11). It is not possible to demonstrate the presence of spores with any assurance. The pseudosporelike bodies seen in Fig. 4, parts 9 and 10, are actually solid blebs of organic material observed in juxtaposition with, or superimposed on, filaments of plant residues. The distinction between what appear to be organized sporelike entities and true plant structures is not easy to make. Our interpretation that the pseudospores are disaggregated particles of condensed spheroidal organic matter and not true microfossils is based on the facts that they are solid, often subangular, and optically readily differentiated from the filamentous or laminar masses, which can be shown to possess lumena in various degrees of occlusion (Fig. 4, part 9). The fact that the solid spheroidal bodies are commonly in physical contact with the filamentous, branched or unbranched plant fragments complicates their paleontological interpretation.

An additional complexity in interpreting the microstructure of the organic residues results from the occasional presence, in macerations, of small organic bodies in the form of perfect and apparently hollow spheres of various sizes. These range in diameter from approximately 1 micron to more than 40 microns. They are gen-

10

291

erally hyaline in appearance, but they possess a distinct refractive outer layer, which renders them bubble-like as seen in optical section. In a few cases these spherical bodies possess an internal cluster of smaller bodies of disparate size but perfectly spherical outline (Fig. 4, part 13). These spherical objects are not included in the three major categories of visible organic matter since they are of infrequent occurrence and of highly problematical origin. Their superficial resemblance to spores or related structures of biological origin is notable. However, their variable size, their unusually refractive membranes or walls, and the occasional occurrence of contained alveolar masses renders this interpretation questionable. Although no simple explanation is suggested for the mechanism of formation of these spherical bodies, their appearance, structural features, and variable size suggest phenomena involving emulsification or some mechanism for alveolation of the organic matter either during its emplacement in the shale or possibly during sample preparation by the acid treatment or heavy-liquid separation procedures. Scrupulous care in sample treatment and slide preparation eliminates the possibility that these structures were introduced as laboratory contaminants.

Accessory evidence from isotope studies. The presence of the metal-porphyrin complexes in the Nonesuch shale, probably originating from alteration products of chlorophyll, provides a reasonable basis for postulating a photosynthetic system. One auxiliary line of evidence for this interpretation is of much interest—that is, comparison of the isotopic ratios of the biologically fractionated elements carbon and sulfur.

As shown by the early studies of Nier and Gulbransen (*43*), Wickman (*44*), Rankama (*45*), and Craig (*46*), the ratio of the stable carbon isotopes C^{12} and C^{13} in biogenically synthesized carbon compounds consistently exhibits an enrichment of the lighter isotope C^{12} with respect to the ratio of inorganic terrestrial C^{12} and C^{13} in the atmosphere, in the hydrosphere, and in nonbiogenic sedimentary carbonates. The terrestrial ratio of the two isotopes C^{12} and C^{13} is approximately 90:1 in inorganic environments in which the photosynthetic system does not operate. Reduction of CO_2 during the photosynthetic reaction tends to selectively concentrate the lighter isotope relative to the heavy isotope, with the resulting

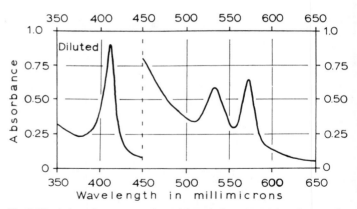

Fig. 8. Visual absorption spectrum of vanadyl porphyrin recovered from benzene-ether eluates of nonasphaltic extractable constituents of Parting shale unit, Nonesuch formation. Approximately 2.5 mg of chloroform per milliliter.

enrichment of C^{12} in biogenic organic matter. As Craig (*46*) expresses it, terrestrial organic matter (that is, organic matter photosynthetically produced and incorporated into the biosphere) and carbonate rocks constitute two well-defined groups in which the carbonates are richer in C^{13}. Organic matter of fossil origin exhibits a preferential enrichment of C^{12}, presumably arising from its initial formation through photosynthesis. However, it has been shown by studies (*47*) subsequent to those of Craig that the fractionation of the two isotopes is biochemically complex (*46*). Although complications exist pertaining to the carbon reservoir of the terrestrial environment and the rate of cycling of carbon, it is evident that ancient organic matter which shows a substantial enrichment of the lighter carbon isotope relative to inorganic carbon of the same deposit is most reasonably interpreted as biogenic in

origin. In the case of the Nonesuch shale the stable-carbon-isotope ratios for both the inorganic carbonate carbon and the reduced (organic) carbon have been determined by T. C. Hoering (*48*). These values are expressed in units of δC^{13}, the difference between the C^{13} content of the sample and that of an arbitrary standard, according to the equation

$$\delta C^{13} = \frac{(C^{13}/C^{12})x - (C^{13}/C^{12})s}{(C^{13}/C^{12})} \times 1000$$

in which x is the sample and s indicates the National Bureau of Standards isotope reference sample No. 20 (Solenhofen limestone). The δC^{13} determinations are shown in Table 4.

When the data are considered in conjunction with our evidence of the presence of pristane, phytane, and metal-porphyrin complexes in the organic matter of the Nonesuch formation, it appears reasonable to infer that

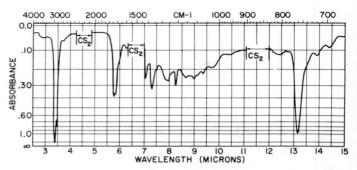

Fig. 9. Infrared absorption spectrum of vanadyl porphyrin aggregate recovered from benzene-ether eluates of nonasphaltic extractable constituents of Parting shale unit, Nonesuch formation. Spectrum solvent is CS_2.

11

the carbon-isotope fractionation and enrichment of C^{12} constitute corroborative evidence of a photosynthetic system operating in the primary biosynthesis in the Nonesuch basin.

The situation with respect to sulfur isotopes in the Nonesuch shale is comparable to that for the carbon isotopes as indicators of biological processes in Nonesuch time. Of the four stable sulfur isotopes found in nature, only two, S^{32} and S^{34}, are quantitatively important. The light isotope, S^{32}, is preferentially reduced by sulfate-reducing bacteria. Thus, an enrichment of the heavier isotope, S^{34}, relative to a standard of abiotic origin, has been proposed as evidence of the biogenic nature of sedimentary sulfides. In the sulfur isotope investigations of the Nonesuch formation the arbitrary standard used was the Cañon Diablo meteorite ($S^{32}/S^{34} = 22.21$) [49]. The mean isotopic ratio for oceanic sulfates, relative to the standard, is $S^{32}/S^{34} = 21.76$, whereas the mean value for sulfides of sedimentary origin is 22.49 [50]. Wiese [9] reports that the mean S^{32}/S^{34} value, relative to the Cañon Diablo meteorite, for ten samples of chalcocite from various horizons of the Parting shale unit of the Nonesuch formation is 22.16; the spread of values is from 21.93 to 22.39. Although all ratios from the Nonesuch shale are intermediate between the mean value for oceanic sulfates and the mean value for sedimentary sulfides, the wide range of isotopic values has been interpreted by Jensen, according to Wiese [9], as being indicative of biogenically fractionated sulfur.

Summary

Investigations have been made of crude oil, pristane, phytane, sterane-type and optically active alkanes, porphyrins, microfossils, and the stable isotopes of carbon and of sulfur found in the Nonesuch shale of Precambrian age from Northern Michigan. These sediments are approximately 1 billion years old. Geologic evidence indicates that they were deposited in a nearshore deltaic environment. Porphyrins are found in the siltstones but not in the crude oils of the Nonesuch formation—evidence that these chemical fossils are adsorbed or absorbed and immobile. This immobility makes it highly unlikely that these porphyrins could have moved from younger formations into the Nonesuch sediments, and the widely disseminated particulate organic matters and fossils in this Precambrian shale are certainly indigenous.

The thermal stability of the Nonesuch porphyrins adds support to the concept of low-temperature emplacement of the copper minerals in the deposit. The concatenation of geologic, geochemical, and micropaleontological evidence strongly indicates that the organic matter, including the crude oil of the Nonesuch deposits, is an indigenous product of primary photosynthetic processes operating in Nonesuch time [51].

References and Notes

1. S. A. Tyler and E. S. Barghoorn, *Science* 119, 606 (1954); E. S. Barghoorn, in "Treatise on Marine Ecology and Paleoecology," *Geol. Soc. Am. Mem.* 67 (1957), vol. 2, pp. 75–86; ——— and S. A. Tyler, *Ann. N.Y. Acad. Sci.* 108, 451 (1963); ———, *Science* 147, 563 (1965).
2. M. F. Glaessner, *Biol. Rev. Cambridge Phil. Soc.* 37, 467 (1962); M. G. Rutten, *The Geological Aspects of the Origin of Life on Earth* (Elsevier, New York, 1962).
3. W. S. White and J. C. Wright, *Econ. Geol.* 49, 675 (1954).
4. R. H. Sales, *ibid.* 54, 947 (1959).
5. I. B. Joralemon, *ibid.*, p. 1127.
6. W. S. White, *ibid.* 55, 402 (1960).
7. R. H. Carpenter, *ibid.* 58, 643 (1963).
8. I. B. Joralemon, *ibid.*, p. 1345; J. R. Rand, *ibid.* 59, 160 (1964); E. L. Ohle, *ibid.* 57, 831 (1962).
9. R. G. Wiese, Jr., thesis, Harvard University (1960).
10. C. K. Leith, R. J. Lund, A. Leith, *U.S. Geol. Survey Profess. Paper 184* (1935); B. S. Butler and W. S. Burbank, *U.S. Geol. Survey Profess. Paper 144* (1929).
11. W. G. Meinschein, E. S. Barghoorn, J. W. Schopf, *Science* 145, 262 (1964).
12. S. Chaudhuri (Ohio State University), personal communication. The methods and results of age determinations of the Nonesuch shale will appear in the 1964 *M.I.T. Annual Progress Report* (in press), in an article by S. Chaudhuri and G. Faure.
13. K. Nishio, *Econ. Geol.* 14, 324 (1919).
14. S. Hamilton, personal communication.
15. We thank the officers and geologists of the White Pine Copper Company for the samples of the rock and petroleum used for this study, and for the information pertaining to them.
16. W. G. Meinschein and G. S. Kenny, *Anal. Chem.* 29, 1153 (1957).
17. W. G. Meinschein, B. Nagy, D. J. Hennessy, *Ann. N.Y. Acad. Sci.* 108, 553 (1963).
18. W. G. Meinschein, *Bull. Am. Assoc. Petrol. Geologists* 43, 925 (1959).
19. Optical activities of Nonesuch alkanes were measured by B. Poczik of O. C. Rudolph and Sons.
20. H. N. Dunning, J. W. Moore, A. T. Myers, *Ind. Eng. Chem.* 46, 2000 (1954); H. N. Dunning, in *Organic Geochemistry*, I. A. Breger, Ed. (Pergamon, New York, 1963), pp. 367–430; G. W. Hodgson, N. Ushijima, K. Taguchi, I. Shimad, *Sci. Rept. Tohoku Univ. Fifth Ser.* 8, 483–513 (1963); M. Blumer and G. S. Omenn, *Geochim. Cosmochim. Acta* 25, 81 (1961).
21. R. Park and H. N. Dunning, *Geochem. Cosmochim. Acta* 22, 99 (1961).
22. A. Treibs, *Ann. Chem.* 510, 42 (1934).
23. P. H. Abelson, in *Researches in Geochemistry*, P. H. Abelson, Ed. (Wiley, New York, 1959), pp. 79–103.
24. P. V. Smith, *Bull. Am. Assoc. Petrol. Geologists* 38, 377 (1954); E. G. Baker, *Science* 129, 871 (1959); ———, *Geochim. Cosmochim. Acta* 25, 81 (1961); W. G. Meinschein, *Space Sci. Rev.* 2, 653 (1963); R. L. Martin, J. G. Winters, J. A. Williams, *Nature* 199, 110 (1963).
25. H. M. Smith, H. N. Dunning, H. T. Rall, J. S. Ball, *Proc. Am. Petrol. Inst. Sect. III* 39, 443 (1959).
26. T. S. Oakwood and R. W. Stone, *Fundamental Research on Occurrence and Recovery of Petroleum 1944–45* (Lord Baltimore Press, Baltimore, Md., 1946), p. 100.
27. M. Louis, *Rev. Inst. Franc. Petrole Ann. Combust. Liquides* 19, 277 (1964).
28. M. R. Fenske, F. L. Carnaham, J. N. Beston, A. H. Casper, A. R. Rescorla, *Ind. Eng. Chem.* 34, 638 (1942).
29. A. C. Chibnall and S. H. Piper, *Biochem. J.* 28, 2008 (1934); N. P. Steven, E. E. Bray, E. D. Evans, *Bull. Am. Assoc. Petrol. Geologists* 40, 975 (1956); E. E. Bray and E. D. Evans, *Geochim. Cosmochim. Acta* 22, 2 (1961); W. G. Meinschein, *ibid.*, p. 58.
30. L. S. Ciereszko, D. H. Attaway, C. B. Koons, *Vapor Pressure* 33, No. 3, 59 (1963); W. G. Meinschein, quarterly report, NASA contract No. NASw508, 1 July 1964, with Esso Research and Engineering.
31. W. G. Meinschein, *Life Sciences and Space Research* (COSPAR, Paris, in press).
32. E. D. Evans, G. S. Kenny, W. G. Meinschein, E. E. Bray, *Anal. Chem.* 29, 1858 (1957).
33. J. C. O'Connor, F. H. Burow, M. S. Morris, *ibid.* 34, 52 (1962).
34. J. G. Bendoraitus, B. L. Brown, R. S. Hepner, *ibid.*, p. 49; J. J. Cummins and W. E. Robinson, *J. Chem. Eng. Data* 9, 304 (1964).
35. W. G. Meinschein, quarterly report, NASA contract No. NASw508, 1 October 1963, with Esso Research and Engineering.
36. G. W. Hodgson, B. Hitchon, R. M. E. Lofson, B. L. Baker, E. Peake, *Geochim. Cosmochim. Acta* 19, 272 (1960); G. W. Hodgson and E. Peake, *Nature* 191, 766 (1961).
37. G. Eglinton, P. M. Scott, T. Belsky, A. L. Burlingame, M. Calvin, *Science* 145, 263 (1964).
38. N. A. Sorensen and J. Mehlum, *Acta Chem. Scand.* 2, 140 (1948); J. Pliva and N. A. Sorensen, *ibid.* 4, 846 (1950); B. Hallgren and S. Larson, *ibid.* 17, 543 (1963); J. D. Mold, R. K. Stevens, R. E. Means, J. M. Ruth, *Nature* 199, 283 (1963).
39. J. Bonner, *Plant Biochemistry* (Academic Press, New York, 1950); C. H. Werkman and P. W. Wilson, *Bacterial Physiology* (Academic Press, New York, 1951); G. E. W. Wolstenholme and M. O'Connor, *Ciba Foundation Symposium on the Biosynthesis of Terpenes and Sterols* (Churchill, London, 1959).
40. D. Shemin, *Ciba Foundation Symposium on Porphyrin Biosynthesis and Metabolism* (Churchill, London, 1955).
41. R. Park and S. Epstein, *Geochim. Cosmochim. Acta* 21, 110 (1960); *Plant Physiol.* 36, 133 (1961).
42. R. O. Brady, *Proc. Nat. Acad. Sci. U.S.* 44, 993 (1958); S. J. Wakil and J. Ganguly, *J. Am. Chem. Soc.* 81, 2597 (1959); J. D. Brodie, G. Wasson, J. W. Porter, *J. Biol. Chem.* 238, 1294 (1963).
43. A. O. Nier and E. A. Gulbransen, *J. Am. Chem. Soc.* 61, 697 (1939).
44. F. E. Wickman, *Geochim. Cosmochim. Acta* 2, 243 (1952).
45. K. Rankama, *Isotope Geology* (Pergamon, London, 1954), pp. 181–234.
46. H. Craig, *Geochim. Cosmochim. Acta* 3, 53 (1953).
47. W. D. Rosenfeld and S. R. Silverman, *Science* 130, 1658 (1959); P. H. Abelson and T. C. Hoering, *Proc. Nat. Acad. Sci. U.S.* 47, 623 (1961); S. R. Silverman, in *Isotopic and Cosmic Chemistry*, H. Craig, S. L. Miller, G. J. Wasserburg, Eds. (North-Holland, Amsterdam, 1964), pp. 92–102.
48. T. C. Hoering, personal communication.
49. M. L. Jensen, *Econ. Geol.* 53, 598 (1958).
50. W. V. Ault and J. L. Kulp, *ibid.* 55, 73 (1960).
51. We acknowledge the assistance and helpful cooperation of John Trammell of the White Pine Copper Company; H. Carillo and K. C. Klein of Esso Research and Engineering; Max Blumer of the Woods Hole Oceanographic Institution; Robert Wiese, Jr., of Mount Union College; Mrs. Dorothy D. Barghoorn, Harvard University; and Gilbert Hillman of Harvard College. This work was sponsored in part by financial assistance under NASA contract No. NASw508 with Esso Research and Engineering and NSF grants G18858 and G19727 to Harvard University.

12

Reprinted from *Science*, **148**(3666), 77–79 (1965)

Hydrocarbons of Biological Origin in Sediments about Two Billion Years Old

JOHN ORÓ, D. W. NOONER, A. ZLATKIS,
S. A. WILKSTRÖM, AND E. S. BARGHOORN

Abstract. *Normal paraffins in the C_{16} to C_{32} range and the saturated isoprenoid hydrocarbons, pristane and phytane, have been found in chert from the Gunflint iron formation (1.9×10^9 years old) of the north shore of Lake Superior. The distribution of n-alkanes shows two maxima, one at about C_{18} to C_{19} and the other at about C_{22} with a minimum occurring at C_{20} to C_{21}. No predominance of odd- to even-carbon-number alkanes is observed within the C_{16} to C_{32} range. The results agree with micropaleontological observations made on the Gunflint chert and provide a chemical characterization of Precambrian life existing about two eons ago.*

Chert from the Gunflint iron formation is perhaps the oldest Precambrian sedimentary rock known containing, in a three-dimensional matrix, well-preserved microfossils with the morphology of algae and simple fungi. Other microfossils found less frequently in the chert differ morphologically from all known existing plant and animal microorganisms (*1–3*). The organic residues of these microfossils yield small amounts of aliphatic and aromatic hydrocarbons (*2, 3*), and they contain hydrocarbons with 20 or more carbon atoms (*2*). The large abundance of microfossils resembling blue-green algae (*1–3*) suggests that chlorophyll degradation products such as phytane and pristane could also be present in the organic residues.

On the basis of these observations and suggestions we have analyzed the Gunflint chert for high molecular weight paraffins with the aim of establishing their identity, determining their molecular weight distribution, and finding out if hydrocarbons of definitely biological origin such as phytane and pristane are present (*4*).

A specimen of solid, black chert (No. C 6353 from the collection of E. S. Barghoorn) weighing approximately 150 g was cut and broken into several major pieces. Several thin sections were prepared from one of these pieces and examined microscopically. The most abundant microfossils appeared to be blue-green algae (*5*), an observation in agreement with those of Barghoorn and associates (*1–3*). The paraffinic hydrocarbons from the chert were analyzed by gas chromatography and by a combination of gas chromatography and mass spectrometry (*6, 7*). Only the necessary additional information pertaining to the treatment of the chert samples and extraction of the organic matter is described here. The organic matter was extracted from the chert, in an all-glass Soxhlet-type apparatus, with a mixture of benzene and methanol (3:1) (*8*) after the sample had been either demineralized with hydrofluoric acid (*9*) or pulverized (*10*). After the extraction, the organic residue was separated on a silica-gel column into four fractions (*10*). Only the first fraction which was eluted from the column with *n*-heptane and contained the paraffinic hydrocarbons was used in the analysis. The other fractions (the carbon tetrachloride, benzene, and methanol eluates) were retained for subsequent analysis.

Three samples of chert were treated as follows. Chert sample No. 1 (22.9 g) was demineralized, without prior removal of possible surface contamination, and the residue (230 mg) was extracted. Chert sample No. 2 (22.8 g) was pulverized, without prior removal of possible surface contamination and extracted. Chert sample No. 3 (10.8 g) was taken from an inside piece to eliminate contamination (*11*), pulverized, and extracted.

The residues from the *n*-heptane eluates, which approximated 0.1 mg hydrocarbon from a 20-g sample of

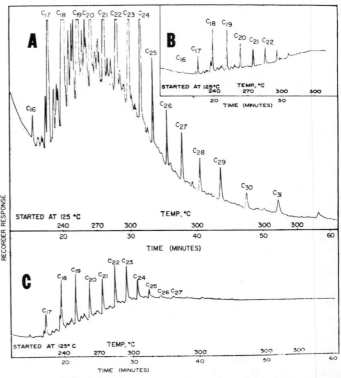

Fig. 1. Gas chromatographic separation of alkanes from the Gunflint chert. Nitrogen pressure, 2550 g/cm². Programmed from 125° to 300°C at 5.8°C per minute. Small split. Barber-Colman Series 5000 apparatus equipped with flame detector. *A*, Chert sample No. 1 (residue from HF treatment). About one-half of the *n*-heptane eluate was injected. Attenuation, 30. *B*, Same as *A* except about one-tenth of the hydrocarbons was injected. Attenuation, 100. *C*, Chert sample No. 2 (untreated with HF, pulverized). About one-sixth of the *n*-heptane eluate was injected. Attenuation, 300.

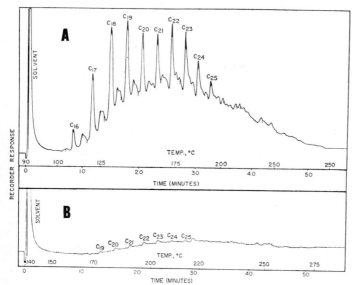

Fig. 2. Gas chromatographic separation of alkanes from the Gunflint chert with a packed column (F-60). Glass tubing, 1.8 m × 4 mm, containing 1 percent F-60 silicone oil (12) on 80 to 100 mesh acid washed silanized Gas Chrom P. Helium pressure, 1400 g/cm². No split. Mass spectrometer-gas chromatograph combination (6). *A*, Chert sample No. 2 (untreated with HF, pulverized). About one-third of the *n*-heptane eluate was injected. Attenuation, 100. Programmed from 90° to 250°C at approximately 3.5°C per minute. *B*, Surface washings from chert. About one-half of the *n*-heptane eluate was injected. Attenuation, 30. Programmed from 140° to 275°C at approximately 2.5°C per minute.

tion of high-molecular-weight alkanes which are only present in relatively small amounts. Although only *n*-alkanes up to C_{32} are shown, small peaks for *n*-$C_{33}H_{68}$ and *n*-$C_{34}H_{70}$ were also observed in the chromatogram. This chromatogram also shows that many, probably isomeric, hydrocarbons are present which contribute to the base line hump and form the small incompletely resolved peaks between the normal alkanes. The peak immediately preceding that of the C_{17} *n*-alkane corresponds to pristane as judged by its retention time. The phytane peak which would immediately precede the C_{18} *n*-alkane peak is off scale and can not be seen in Fig. 1*A*, but its presence is obvious in Fig. 1*B*, where the chromatogram was made from a smaller sample of the same benzene solution.

Figure 1*C* shows a chromatogram of the paraffinic hydrocarbons obtained from chert sample No. 2. The distribution of hydrocarbons observed in the aforementioned two different treatments (demineralization and pulverization) is essentially the same. Some of the small differences may be caused by loss of the more volatile hydrocarbons during the evaporation procedure. The hydrocarbons appear to be distributed in two major groups of *n*-alkanes, one with a maximum at C_{18} to C_{19} and the other at C_{22} showing a minimum at C_{20} to C_{21} (13). This distribution can be interpreted as indicating that the hydrocarbons were derived from at least two major biosynthetic pathways or two major classes of organisms. No predominance of alkanes with an odd number of carbon atoms (C-odd) over those with an even number of carbon atoms (C-even) is observed in these hydrocarbons. Although, at first, this may appear strange for hydrocarbons of biological origin, the nonpreponderance of C-odd to C-even has been shown in the alkanes from a number of living and fossilized organisms (14).

Figure 2*A* shows the analysis of chert sample No. 2 on a packed column attached to an Atlas CH4 mass spectrometer (6, 15). Confirmation of the identification of the C_{16} through C_{25} *n*-alkanes was afforded by the mass spectra of the individual components as they emerged from the column. The short lines at the top of the peaks indicate where the mass spectra were taken. The column used did not resolve pristane and phytane,

chert, were dissolved in benzene and samples were taken for chromatography (Figs. 1, 2, and 3). The chromatograms shown in Fig. 1 were obtained with an 0.03-cm capillary column coated with Apiezon L (12). Figure 1*A* shows a chromatogram of the alkanes in sample No. 1 which was demineralized by hydrofluoric

acid. As judged by the retention times of individual hydrocarbons and hydrocarbon mixtures of known identity the major peaks in this and subsequent chromatograms corresponded to normal alkanes. The *n*-alkanes shown in Fig. 1*A* have 16 to 32 carbon atoms. The large sample taken and the high sensitivity of the test permitted detec-

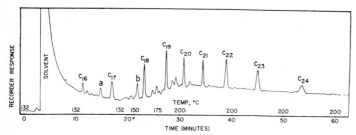

Fig. 3. Gas chromatographic separation of alkanes from the Gunflint chert with a capillary column (Polysev). Stainless steel tubing, 75 m × 0.08 cm, coated with Polysev (12). Nitrogen pressure, 700 g/cm². Isothermal at 132°C for 18 minutes, then programmed at 5.8°C/min at 200°C. Small split. Barber-Colman Series 5000 apparatus equipped with a flame detector. Chert sample No. 3 (untreated with HF; pulverized inside sample). About one-tenth of the *n*-heptane eluate was injected. Attenuation, 30.

but, with standard samples, these two hydrocarbons had retention times essentially identical to *n*-heptadecane and *n*-octadecane respectively. Evidence for the presence of pristane and phytane was obtained since the mass spectra for the peaks which gave cracking patterns of C_{17} and C_{18} *n*-alkanes also had small maxima at m/e (mass/charge) = 183 ($C_{13}H_{27}^+$). This indicated that the peaks consisted primarily of the *n*-alkane with a small admixture of pristane in peak C_{17} and phytane in peak C_{18}.

To check the effects of possible surface contamination (brought about by handling or by contemporary organisms) on the results shown in Figs. 1 and 2*A*, a large sample of chert was thoroughly washed with a total of 50 ml of benzene-methanol (3:1) mixture. The solution was then handled exactly as the chert extracts, and the alkane fraction was chromatographed. Figure 2*B* shows that contamination accounts for less than 1 percent of the hydrocarbons recovered from the chert (*16*).

The inside piece of chert (sample No. 3) was washed with benzene-methanol (3:1) before being pulverized, and the extract was chromatographed in a capillary column coated with Polysev (*12*) which very efficiently separates pristane, phytane, and other isomers from the *n*-alkanes (Fig. 3). This chromatogram also shows the bimodal distribution of alkanes characteristic of other samples from the Gunflint chert. It also gives convincing evidence for the presence of pristane and phytane because peaks *a* and *b* have retention times identical to that of pristane and phytane standards. The paraffinic hydrocarbons from the inside piece of chert were also analyzed on another capillary—stainless steel (0.03 cm × 90 m)—coated with Carbowax 20M terminated with terephthalic acid (*12*)—and programmed from 125° to 200°C at the rate of 5°C per minute. This system resolves pristane and phytane from the other alkanes. Again peaks with retention times corresponding to these two isoprenoid hydrocarbons were observed.

The carotenoid pigments and the phytol portion of the chlorophyll molecule have been mentioned as possible precursors of the isoprenoid hydrocarbons found in petroleum (*17*). To these may be added the substituted napthoquinones and tocopherols (vitamins E and K, respectively), ubiqui-

nones and plastoquinones. Because the organic material of the chert is derived for the most part from algae we believe that the pristane and phytane in this sediment came from the phytyl alcohol of chlorophyll. Conceivably these two saturated isoprenoid hydrocarbons could also have been derived from the other possible precursors (substituted quinones) by diagenetic processes. However, if these biological quinones were present in sufficient quantities to produce pristane and phytane, then higher homologs of these branched compounds should also have been detected. The same reasoning can be used to show that these isoprenoid hydrocarbons could not have been produced abiotically. Independent evidence for the biological origin of the Gunflint organic matter has been provided recently by measurement of the C^{13}/C^{12} ratio of this matter and its comparison with the C^{13}/C^{12} ratio of the carbonate fraction in the same rock (*3*). These studies have shown that the degree of C^{13} depletion of the Gunflint organic compounds is similar to that found in contemporary organic compounds produced by photosynthesis (*3*).

From the foregoing chromatographic and mass spectrometric studies, we conclude that (i) the Gunflint chert contains pristane and phytane, (ii) the chemical evidence of life (presence of pristane and phytane) correlates with the morphological evidence provided by fossilized organisms, and (iii) the Gunflint chert also contains normal paraffinic hydrocarbons ranging approximately from C_{16} to C_{32}, the distribution being bimodal but showing no predominance of C-odd over C-even alkanes. As a corollary of (iii) and of other recent observations (*14*) it should be added that an odd- to even-carbon-number preference in the distribution of normal alkanes may be a sufficient but is not a necessary indication of biological origin.

References and Notes

1. S. A. Tyler and E. S. Barghoorn, *Science* 119, 606 (1954); E. S. Barghoorn, in *Current Aspects of Exobiology*, M. H. Briggs and G. Manikunian, Eds. (Pergamon Press, London, 1965), in press.
2. E. S. Barghoorn and S. A. Tyler, *Ann. N.Y. Acad. Sci.* 108, 451 (1963).
3. ——, *Science* 147, 563 (1965).
4. Evidence for the presence of phytane and pristane in the one-billion-year-old Precambrian Nonesuch formation has been presented by G. Eglinton, P. M. Scott, T. Belsky, A. L. Burlingame, M. Calvin, *Science* 145, 263 (1964), and by W. G. Meinschein, E. S. Barghoorn, J. W. Schopf, *ibid.*, p. 262.
5. A number of spore-like bodies and a few structures of more complex morphology were also observed. These results are being prepared for publication.
6. R. Ryhage, *Anal. Chem.* 36, 759 (1964).
7. J. Oró, D. W. Nooner, S. A. Wikström, *Science* 147, 870 (1965); *J. Gas. Chromatography*, in press.
8. W. G. Meinschein, Quarterly Rept., contract No. NASw-508, 1 April 1964; W. G. Meinschein and G. L. Kenny, *Anal. Chem.* 29, 1153 (1957).
9. We thank R. Louden for demineralizing the chert with hydrofluoric acid.
10. The chert was pulverized in a Carver press test cylinder. Material that passed through a 100-mesh sieve was extracted.
11. A piece of the chert was sawed and ground with Carborundum tools until at least 1 cm thickness of all the surface rock had been removed.
12. Apiezon L (a high temperature grease) and Polysev [*m*-bis *m*-(phenoxyphenoxy)-phenoxybenzene] were obtained from Applied Science Laboratories, Inc., State College, Pennsylvania. The F-60 silicone oil was obtained from Dow Corning Corporation, Midland, Michigan. Carbowax 20M (terminated with terephthalic acid) was obtained from Wilkens Instrument and Research, Inc., Walnut Creek, California.
13. The maximum at C_{18} to C_{19} may be caused in part by the evaporation of the more volatile hydrocarbons during the isolation procedure.
14. J. Oró, D. W. Nooner, S. A. Wikström, unpublished; W. G. Meinschein, Quarterly Rept., contract No. NASw-508, 1 July 1964; B. Pasby, B. S. Cooper, D. W. Hood, Abstracts of papers presented at the 1964 Annual Meetings of the Geological Society of America, 19–21 November 1964, at Miami Beach, Florida, p. 149; N. P. Stevens, E. E. Bray, E. D. Evans in *Habitat of Oil*, L. G. Weeks, Ed. (American Association of Petroleum Geologists, Tulsa, Oklahoma, 1958), p. 779; L. S. Ciereszko, D. H. Attaway, M. A. Wolf, Petroleum Research Fund, 8th Annual Report (1963), p. 33.
15. We thank E. C. Horning, Lipid Research Center, Department of Biochemistry, Baylor University College of Medicine, Houston, Texas, for making the gas chromatograph-mass spectrometer combination available to us.
16. The small peaks in Fig. 2*B* correspond to *n*-alkanes showing a unimodal distribution with maximum at about C_{23}. This can be observed more clearly by injecting a much larger sample of the paraffinic fraction obtained from the chert surface washings. It is not yet certain whether these traces of alkanes are derived from the lipids of human hands or from other sources.
17. J. C. Bendoraitis, B. L. Brown, L. S. Hepner, *Anal. Chem.* 34, 49 (1962).
18. This work was presented at the meeting of the Meteoritical Society, Arizona State University, Tempe, 30 October 1964, and the Group for the Analysis of Carbon Compounds in Carbonaceous Chondrites, Washington, D.C. We thank Dr. W. G. Meinschein for a sample of paraffins containing pristane, phytane, and squalane; Dr. M. Blumer for a sample of phytane and P. Simmonds for valuable assistance with the gas chromatographic analyses. Supported in part by NASA grant, NsG-257-62.
* Permanent Address: Laboratory for Mass Spectrometry, Karolinska Institutet, Stockholm, Sweden.

28 January 1965

3

31

Reprinted from *Carnegie Inst. Washington Yearbook 64*, 215–218 (1965)

THE EXTRACTABLE ORGANIC
MATTER IN PRECAMBRIAN ROCKS
AND THE PROBLEM OF CONTAMINATION

T. C. Hoering

Sedimentary rocks of all ages contain reduced carbon. In severely metamorphosed formations the carbon is graphitic. Very old rocks can be found, however, that have experienced a mild thermal history, and any organic matter originally resident in them could be expected to have survived with only moderate degradation. We have assembled a good collection of old, carbonaceous, lightly metamorphosed rocks and have sought to identify some of the organic matter that can be extracted from them. We have found that sedimentary rocks even as old as 3,000 m.y. contain petroleumlike substances that can be extracted into organic solvents. The amounts of such material in Precambrian rocks are generally small, but there are several exceptions. We have found that a rock of the Nonesuch shale of upper Michigan with an age of 1,100 m.y. contains 3,100 ppm extractable organic matter. The McMinn shale of the Northern Territory of Australia, whose age is about 1,600 m.y., has 1,500 ppm of the weight of the rock as benzene-soluble material. Typically, however, the amounts are less than 80 ppm.

We have analyzed the extracts for straight-chained hydrocarbons in the range of $C_{14}H_{30}$–$C_{30}H_{62}$ and have found some of these compounds in all of them. Compounds of this type are known to result from transformation of biologically produced organic matter. Before we can discuss the biogeochemical significance of finding this class of compounds in rocks of great age we must ask if they are indigenous.

Some of the ways that a rock may become contaminated can be evaluated. Inadvertent introduction of refined petroleum products is a potential hazard. The molecular weight distribution of the normal alkanes in them, however, is distinctly different from those in rock extracts. The problem of contamination during chemical processing can be assessed by running blanks and control experiments. A much more likely source of contamination of Precambrian rocks is the migration of petroleum fluids in the earth. Hedberg (1964) has reviewed the literature on the migration of oil. Although the mechanism and mode of such transport are not well understood, it is clear that petroleum can move for long distances and may appear in rocks of all kinds. Occurrences of petroleum in such unlikely places as igneous and metamorphic rocks are known.

The occurrence of biological-type alkanes in a rock older than 3,000 m.y. is described below, and a method of evaluating organic contamination is discussed. The Barberton area of eastern Transvaal in the Republic of South Africa contains some of the world's oldest rocks. Rocks of the Swaziland system have been dated by Nicolaysen as older than 3,000 m.y. This system contains the lower Onverwacht series of lava flows and the upper

Fig Tree series, consisting of pelitic rocks and great thicknesses of banded cherts, banded iron stones, and carbonaceous shales. The shales in some places have suffered surprisingly little metamorphism. Samples of them were collected in the Montrose Gold Mine.

An experiment was designed to determine the location of the extractable organic matter in the shale. The rocks were crushed into coarse pieces approximately ½ inch in size, and extracted in a Soxhlet extractor with 70 per cent benzene, 30 per cent methanol. The solvent was evaporated slowly under vacuum, and 13 ppm of extracted organic matter was obtained. The rock pieces were then crushed to −6, +30 mesh size and extracted for 24 hours to give 30 ppm extractables. The crushed rock was then ball milled to 3–10 micron-sized particles and extracted to give 15 ppm. The ball-milled rock was demineralized with hydrofluoric and hydrochloric acid to give a kerogen concentrate. This was dried and extracted to give 6 ppm soluble matter based on the weight of the original rock. It appears that an appreciable amount of the extractable organic matter is on surfaces and grain boundaries that are easily accessible to the solvent.

The extracts were combined, and a nonpolar fraction was isolated by silica-gel chromatography using heptane as the eluant. The straight-chained hydro-carbons were isolated by forming adducts with urea. They were recovered and analyzed by gas-liquid chromatography. The chromatogram is shown in Fig. 77. Although the individual compounds are easily detected, they each represent only about one part in 10^7 of the original rock. This assemblage of compounds is typical of hydrocarbons formed in association with biologically produced organic matter of all ages.

In most sedimentary rocks the bulk of the organic matter exists as an insoluble high polymer called kerogen. It is unusual to find a shale with more than a few per cent of its organic matter in a form that is soluble in benzene. Kerogen cannot migrate far, so it is of interest to compare the C^{13}/C^{12} ratio of the soluble and insoluble organic matter in a series of rocks. The rocks chosen had a wide range of ages and types. In some of them there is abundant extractable material, which is very likely related to the kerogen. Some of the Precambrian rocks contain only small amounts of extractable material and the relation of the extract to the kerogen is less certain.

Samples of finely ball-milled rocks were extracted in a Soxhlet extractor with benzene-methanol mixture, and the solvent was evaporated. The residual rock was dried and treated with hydrochloric acid to remove carbonates. The two fractions were quantitatively combusted

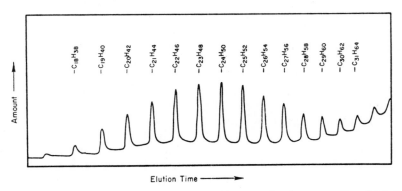

Fig. 77. Normal alkanes from Fig Tree shale of the Swaziland system. F and M Corporation Model 1609 hydrogen flame detector, 100-foot by 0.016-inch column coated with Apiezon L grease, 30 psi helium. Temperature programming from 100°C–300°C.

to carbon dioxide and analyzed in the mass spectrometer. The results, shown in Table 30, are expressed in parts-per-thousand difference in the C^{13}/C^{12} ratio compared with a standard (National Bureau of Standards isotope reference sample 20, limestone from Solenhofen).

$$\delta C^{13} = \frac{(C^{13}/C^{12})_x - (C^{13}/C^{12})_s}{(C^{13}/C^{12})_s} \times 1,000$$

where x refers to the sample and s to the standard.

In samples B through I, where there was an abundant amount of soluble material, the isotope ratios are within about two parts per thousand of each other. Sample A, the recent sediment, shows a relationship between the two fractions that is similar to that found between the "lipide" and "nonlipide" portions of living cells. In the Precambrian samples J, K, and L, where the extractable portion is less than 80 ppm, there is a considerable discrepancy. Sample M, the

shale from the Fig Tree series of the Swaziland system described earlier in this report, shows agreement.

To see how uniformly the carbon isotopes were distributed in a sample of rock extract, samples B, D, H, and I were fractionated by silica-gel chromatography. The sample was eluted successively with heptane, carbon tetrachloride, benzene, and methanol. Each fraction was combusted to carbon dioxide, and analyzed. There was little difference in the carbon isotope ratios between the four classes of compounds in every case. Thus a discrepancy cannot be explained by preferential loss of different portions of the organic matter in the rock.

The chemical pathways leading from biological organic matter to kerogen and extractable organic matter in rocks are not known. If the system has been closed since compaction of the rock, however, a consistent relationship would be expected between the carbon isotope ratios

TABLE 30. Carbon Isotope Ratios in the Extractable and Insoluble Organic Matter in Rocks

Rock	Location	Age	δC^{13} Insoluble	δC^{13} Soluble
A. Unconsolidated sediment	Aransas Bay, Tex.	Recent	−18.40	−22.01
B. Oil shale	Green River formation, de Beque, Colo.	Eocene	−28.80	−27.57
C. Shale	Chattanooga formation, Tenn.	Devonian	−27.46	−27.48
D. Shale	Woodford formation, Okla.	Mississippian	−27.82	−28.26
E. Shale	Alum formation, Sweden	Cambrian	−27.43	−26.19
F. Lignite coal	Fort Union formation, N. Dak.	Paleocene	−22.76	−23.60
G. Bituminous coal	Pittsburgh, Pa.	Pennsylvanian	−22.34	−22.84
H. Shale	Nonesuch formation, Mich.	Precambrian (1,100 m.y.)	−28.15	−28.14
I. Shale	McMinn formation, Northern Territory, Australia	Precambrian (1,600 m.y.)	−30.71	−30.59
J. Carbonaceous slate	Soudan formation, Minn.	Precambrian (2,500 m.y.)	−34.81	−25.00
K. Limestone	Transvaal system, South Africa	Precambrian (2,000 m.y.)	−38.21	−25.01
L. Shale	Ventersdorp system, South Africa	Precambrian (2,100 m.y.)	−36.86	−25.78
M. Shale	Swaziland system, South Africa	Precambrian (3,000 m.y.)	−26.94	−27.55

of the two, and this is what has been found empirically.

It is then necessary to understand the discrepancy in samples J, K, and L. The simplest explanation could be the migration of organic matter from an outside source. The values of δC^{13} found for the extractables in these rocks fall in the range of younger petroleums.

One possible explanation is that a process has removed or destroyed organic matter in a way that fractionated isotopic molecules. If this process preferentially removed C^{12}-containing compounds, the residue would be enriched in C^{13} compared with the insoluble organic matter. The agreement in the case of the Fig Tree shale could be fortuitous since the value of δC^{13} of -27.55 found for the extractable organic matter falls in the range of most younger petroleums.

One must conclude that it is difficult to say whether the extractable organic matter in a very ancient rock is of the same age as the rock. Other criteria will have to be employed before it can be said that the n alkanes in the Fig Tree shale are diagnostic of biologically produced organic matter.

32

Reprinted from *Proc. Natl. Acad. Sci. (U.S.)*, **59**(2), 639–646 (1968)

AMINO ACIDS IN PRECAMBRIAN SEDIMENTS: AN ASSAY

By J. William Schopf, Keith A. Kvenvolden, and Elso S. Barghoorn

DEPARTMENT OF BIOLOGY, AND SOCIETY OF FELLOWS, HARVARD UNIVERSITY, CAMBRIDGE;
EXOBIOLOGY DIVISION, AMES RESEARCH CENTER, MOFFETT FIELD, CALIFORNIA;
AND DEPARTMENT OF BIOLOGY, HARVARD UNIVERSITY

Communicated November 16, 1967

Recent studies have documented the occurrence of organically and structurally preserved algae, bacteria, and other microorganisms of less certain affinities, in sediments of Early,[1, 2] Middle,[3–7] and Late Precambrian age.[8, 9] Many of these organisms are morphologically comparable to or identical with living thallophytes.[9, 10] Investigations of hydrocarbons,[11–13] porphyrins,[12] fatty acids,[14] and carbon isotopic ratios[4, 15] in ancient sediments suggest basic biochemical similarities between Precambrian and extant organisms. To extend these studies, we have analyzed three fossiliferous Precambrian black cherts, ranging from about 1 to more than 3 billion years in age, for the presence of fossil amino acids.

Our choice of amino acids for this preliminary survey is based on the following considerations: (1) Polypeptides, composed of amino acids, perform the indispensable role of mediating biochemical reactions in all known organisms; (2) analyses of Phanerozoic sediments and fossils[16, 17] and laboratory studies of reaction kinetics[16–18] have established that many amino acids are geochemically stable; (3) sensitive methods of amino acid analysis by cation-exchange chromatography[19] and by gas chromatography[20] have been developed.

Materials and Methods.—Samples: *Black cherts from fossiliferous localities of the Fig Tree Series* (Early Precambrian, South Africa), *the Gunflint Iron Formation* (Middle Precambrian, Ontario, Canada), *and the Bitter Springs Formation* (Late Precambrian, Northern Territory, Australia) were analyzed for free and combined amino acids. All three cherts are dense, black, waxy, and somewhat lustrous on freshly broken conchoidal faces, and finely and rather irregularly laminated. They are composed of about 98% cryptocrystalline quartz, and contain 0.3–0.8% organic matter; they appear to be chemical sediments of primary, rather than secondary, origin. Each of the deposits has had a mild thermal history. *Bacteriumlike microfossils[1] and algalike organic spheroids[2] in the Fig Tree cherts*, having a minimum radiogenic age of 3.1 billion years,[21] constitute the oldest structurally preserved evidence of life now known from the geologic record. Pristane, phytane, and *n*-alkanes have been reported from the Fig Tree sediments,[2, 22] and the $C^{13}:C^{12}$ ratio of the indigenous organic matter,[15] as well as the geology and mineralogy of the sedimentary sequence,[2, 10] suggests the presence of photosynthetic organisms. The Fig Tree chert analyzed for amino acids was collected at the same locality and horizon from which microfossils have been reported.[23] *The Gunflint Iron Formation* has a radiogenic age of approximately 1.9 billion years.[4] The fossiliferous chert analyzed in the present study was collected near Schreiber, Ontario,[24] at the type-locality for many of the twelve species of microscopic plants recognized in the well-documented Gunflint microbiota.[3–7] Pristane, phytane, and normal paraffins have been reported from these cherts,[13] as have $C^{13}:C^{12}$ ratios for both organic and inorganic carbon.[4] *Black cherts of the Bitter Springs Formation*, from the Amadeus Basin of central Australia, contain at least 30 species of organically preserved microorganisms;[8–10] chert from this fossiliferous locality[25] was investigated for amino acids. Radiogenic age determinations indicate that these sediments are approximately 1 billion years old.[9]

Reagents and analytical controls: The commercially prepared chemicals used were of reagent grade; prior to use, all reagents were tested for amino acid content. Reagent grade ammonium hydroxide and hydrochloric acid were found to contain four amino acids (Ser, Gly, a-Ala, Asp)[26] in concentrations of 10^{-8} to 10^{-9} M/liter. The NH$_4$OH used in these analyses was prepared from gaseous ammonia and triple-distilled water; no amino acids were detected in this reagent at a sensitivity of about 10^{-10} M/liter. Distillation of reagent grade hydrochloric acid reduced the amino acid content by an order of magnitude, and only three amino acids (Ser, Gly, Asp) were detected. The quantities of these amino acids which could have been introduced during hydrolysis vary from more than 2 to over 4 orders of magnitude less than those detected in the chert extracts. All other reagents contained no detectable amino acids at a sensitivity of about 10^{-10} M/liter. An analytical blank (acid-cleaned glass beads), using commercially prepared ammonium hydroxide and hydrochloric acid, was carried through the entire analytical procedure. Trace amounts ($<10^{-10}$ M/gm) of serine, glycine, a-alanine, and aspartic acid were detected. Using freshly prepared NH$_4$OH and distilled HCl, a second analytical blank (acid-cleaned quartz sand) was treated in the same manner as the rock samples, except that pulverization was omitted. These extracts were free of amino acids at a detection sensitivity of about 10^{-11} M/gm.

Additional techniques were adopted to avoid introduction of contamination: objects coming in contact with the samples following hydrofluoric acid treatment (e.g., metal tongs, pulverizing equipment, glassware, etc.) were washed in a chromic-sulfuric acid solution and distilled water prior to use; filters were pre-extracted with hot 0.5 N NH$_4$OAc, 6 N HCl (where appropriate), and hot H$_2$O, and were handled with acid-cleaned forceps; disposable plastic gloves, changed frequently, were worn throughout.

Preparation of powdered chert: A hand-sized sample (500–800 gm) of chert from each of the three formations was mechanically cleaned, and immersed in a chromic-sulfuric acid solution overnight. The dried sample was broken into pieces which were fragmented in a jaw crusher, and the resulting $^1/_8$- to $^1/_2$-in. fraction was immersed in 25% hydrofluoric acid until a total loss in dry weight of approximately 10% was measured ($^1/_2$–2 hr). After being thoroughly washed in triple-distilled water, the chalky-gray, etched chert fragments were ground for 2–6 min in a precleaned "shatter-box" disc pulverizer,[27] and the smaller than 115-mesh fraction (350–700 gm) was separated and stored in acid-cleaned bottles. At least 3 aliquots (50–100 gm each) of powdered chert from each sample were separately analyzed.

Ammonium acetate leach: In a typical analysis, 100 gm of $<$115-mesh chert powder was mixed on a magnetic stirrer with 400 ml 0.5 N ammonium acetate for $^1/_2$ hr at 80°C.[28] The Teflon-coated stirring bar, with adhering iron filings derived from the pulverizer, was removed and the solution filtered through a Buchner funnel. The powdered chert was leached with 100 ml hot 0.5 N NH$_4$OAc, followed by leaching with 500 ml hot triple-distilled water. The combined filtrate and leachate, containing free amino acids, was dried by evaporation under vacuum at 80°C, and the remaining ammonium acetate was removed by sublimation.

HCl hydrolysis: The leached powder was placed in a 250-ml Erlenmeyer flask to which 100 ml distilled 6 N hydrochloric acid was added. The flask was partially evacuated, sealed, and placed in a constant temperature oven at 105°C for 22 hr.[29] The greenish HCl hydrolysate was filtered through a Buchner funnel, and the residual powder was washed with 250 ml of hot triple-distilled water. The combined HCl hydrolysate and washings, containing combined amino acids, was dried by evaporation under vacuum.

Desalting procedure: The ammonium acetate leachate and the HCl hydrolysate (pH adjusted to about 5.5), washed, respectively, onto the top of 100- and 200-cc chromatography columns packed with Bio-Rad AG 50W-X8 cation-exchange resin, were deionized by flushing with 2-column vol of triple-distilled water; amino acids were collected in 4-column vol of 2 N ammonium hydroxide.[28] The NH$_4$OH eluants were dried by evaporation under vacuum at 80°C. Typically, 4 ml of triple-distilled water was added to the dry eluants, and the solutions were each divided into three fractions representing $^1/_4$, $^1/_4$, and $^1/_2$ of the total extracts. These six fractions were dried by evaporation under vacuum at 60°C, and stored at -5°C.

Identification of amino acids: Determination of amino acid content was carried out on a modified amino acid analyzer fitted with a microcolorimeter, using cation-exchange chromatography and spectrophotometry of ninhydrin-reacted products.[19] As is shown in Figure 1, all amino acids reported, with the exception of lysine and ornithine, are well resolved in this system. Dissolved in pH 2.2 citrate buffer, $1/4$ of each extract was applied to the analyzer to determine acidic and neutral amino acid content, and $1/4$ was used for determination of basic amino acids. The remaining half of each extract was investigated by gas chromatography. N-trifluoroacetyl *n*-butyl (N-TFA) ester derivatives of amino acids in the chert extracts were prepared according to the method of Gehrke and co-workers.[20] Operating conditions for gas chromatography of these derivatives are given in Figure 3.

Results.—Free amino acids: Relatively large amounts of glycine were de-

TABLE 1. *Amino acids in ammonium acetate leachates.*

Sample	Age $(\times 10^9 \text{ yr})$	Glycine (nM/gm)	a-Alanine (nM/gm)	Total free amino acids (nM/gm)
Bitter Springs chert	*ca.* 1.0	18.1	0.3	18.4
Gunflint chert	1.9	9.8	—	9.8
Fig Tree chert	>3.1	5.9	—	5.9

tected in the ammonium acetate leachate of each of the three chert samples; *a*-alanine was present in minor amounts in the Late Precambrian Bitter Springs chert. The quantities listed in Table 1 represent average values based on amino acid analyzer chromatograms from at least three separate analyses, with a maximum deviation of 11 per cent for the Bitter Springs chert, 9 per cent for the Gunflint chert, and 10 per cent for the Fig Tree chert. Confirmatory identification of glycine and alanine in these extracts was obtained by gas chromatography of the N-TFA ester derivatives, based on both comparative retention times, and on coinjection of the samples with the appropriate authentic derivatives.

Combined amino acids: Figure 1 shows amino acid analyzer chromatograms obtained from HCl hydrolysates of the three Precambrian cherts. The quanti-

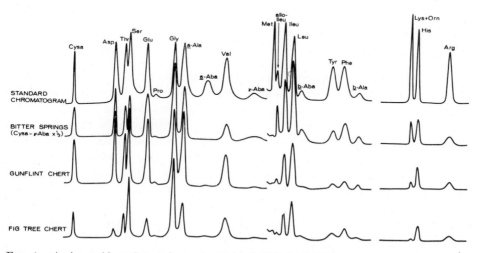

FIG. 1.—Amino acid analyzer chromatograms of HCl hydrolysates of three Precambrian cherts. Abbreviations for amino acids are listed in Fig. 2.

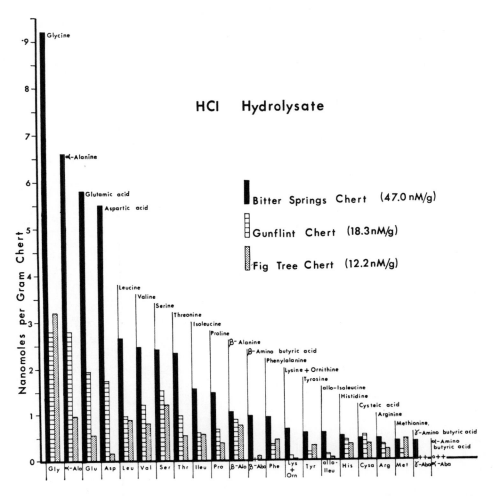

FIG. 2.—Histogram showing distribution of amino acids in HCl hydrolysates of the Bitter Springs chert (*ca.* 1.0×10^9 years old), the Gunflint chert (1.9×10^9 years old), and the Fig Tree chert ($>3.1 \times 10^9$ years old).

ties shown in Figure 2 represent average values based on triplicate analyses of each sample, with a deviation of 7 ± 3 per cent for the Bitter Springs amino acids, 8 ± 4 per cent for those from the Gunflint hydrolysate, and 10 ± 4 per cent for the amino acids of the Fig Tree chert. Identification of 11 of the amino acids detected in these hydrolysates (Gly, *a*-Ala, Glu, Asp, Leu, Val, Ser, Thr, Ileu, Pro, *b*-Ala) was confirmed by gas chromatography of their N-TFA ester derivatives. A gas chromatogram of the derivatives from the Fig Tree hydrolysate is shown in Figure 3. Because of limitations on sample size, the identification of several amino acids (Phe, allo-Ileu, Met, amino butyric acids) could not be confirmed by this technique. Derivatives of other amino acids (Lys, Orn, Tyr, His, Arg) would not be detected under the conditions employed.[20, 30]

Discussion.—The interpretation of these results is dependent on four basic considerations: (1) Are the reported compounds correctly identified? (2)

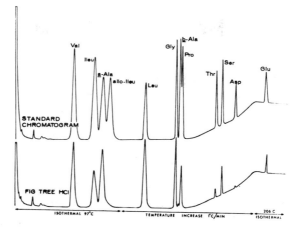

FIG. 3.—Gas chromatogram of N-TFA ester derivatives of amino acids in HCl hydrolysate of Early Precambrian Fig Tree chert. Abbreviations for amino acids are listed in Fig. 2. Operating conditions: Perkin-Elmer F-11 flame-ionization G.C.; Carbowax 20M, 0.02 in. I.D., 50 ft; N_2 = 7 ml/min, H_2 = 10 ml/min; chart speed = 1 in./10 min.

Are these compounds indigenous to the rock samples analyzed? (3) Are the organic compounds syngenetic with original sedimentation? (4) Are the compounds biogenic? Although only three Precambrian sediments have been investigated in this preliminary study, and the data are therefore somewhat limited, these questions can be answered, at least tentatively, in the affirmative.

Identification: The identification of amino acids in extracts of the three Precambrian cherts is based on amino acid analyzer chromatograms, in part confirmed by gas chromatography of the N-TFA ester derivatives. This appears to be the first report of gas chromatographic identification of amino acids from geologic materials. Solubility properties of these compounds, and their chromatographic characteristics on desalting columns, are similarly characteristic of amino acids. Identification of these compounds seems well established.

Indigenousness: In analyzing sedimentary rocks for low concentrations of amino acids, care must be taken to avoid contamination from two major sources: (a) sample contamination from contemporary plant fragments, microorganisms, and other organic matter on surfaces of the rock sample; and (b) laboratory contamination from hands, reagents, glassware, etc. Sample contamination was minimized by immersing the chert specimens in a sulfuric-chromic acid solution, followed by coarse crushing and treatment with hydrofluoric acid. Scrupulous care was taken to avoid laboratory contamination. The success of these techniques is evidenced by the results obtained by analyses of analytical blanks.

The indigenous nature of the amino acids detected is further supported by the following five considerations: (1) The absence of contamination from fingerprints is evidenced by the detection of only two amino acids (Gly and Ala) in the ammonium acetate leachates; fingerprints normally contain at least 17 free amino acids including citrulline,[31] a nonprotein amino acid not detected in the chert extracts. (2) The absence of hand contamination is further indicated by a lack of similarity between the amino acid content of the chert hydrolysates and the distribution reported from hydrolyzed hand rinses,[31–33] and by the detection of several nonprotein amino acids (b-Ala, allo-Ileu, amino butyric acids) not present in human hands. (3) Triplicate analyses of each rock sample show quantitative and qualitative consistency, and the three cherts differ in amino acid content. This distribution is unrelated to the order of analysis: the samples were analyzed sequentially in order of decreasing geologic age, increasing geologic age, and a third sequence in which the sample of intermediate age was analyzed first. (4) Analyses of both 50- and 100-gm aliquots of powdered chert demonstrate that the quantities of amino acids detected vary in direct proportion to the amount of sample extracted. (5) Amino acids have been detected in different samples of the Bitter Springs chert at two different laboratories.[34]

Syngenesis: Although the evidence indicates that the amino acids are almost certainly indigenous to the chert samples, it is somewhat more difficult to establish that these compounds are syngenetic with original sedimentation, and are therefore the same age as the sediments in which they have been detected. This problem arises because techniques are unavailable for direct dating of the organic fraction of ancient rocks. Geologic and geochemical considerations, however, suggest that the amino acids detected are of great geologic age, and most probably were emplaced at the time of original sedimentation.

Because of their relatively high degree of impermeability and incompressibility, primary cherts are excellent sediments for organic geochemical studies. The petrology and mineralogy of the three cherts analyzed indicates that they are essentially unmetamorphosed, and none has been recrystallized. On the basis of petrologic and geologic evidence, it has been established that most, and presumably all, of the organic matter in these deposits was emplaced prior to lithification, at the time of original sedimentation.[2, 4, 9]

Chemical evidence suggests that the amino acid distribution in these sediments has been geochemically altered since original deposition. Of particular significance is the occurrence of several nonprotein amino acids (*b*-Ala, allo-Ileu, amino butyric acids), reported to be products of the geochemical degradation of more common amino acids.[18, 35] The presence of only two free amino acids, glycine and *a*-alanine, seems attributable to their inherent geochemical stability, and to the fact that they are degradation products of less stable amino acids.[18, 35] Although only thermally stable amino acids were detected in the ammonium acetate leachates, several less stable compounds (e.g., Thr, Met, Ser, Arg) were freed from the cherts by HCl hydrolysis. This distribution is consistent with other studies which suggest that geochemical stability may be enhanced by chelation with an inorganic or organic phase.[36, 37]

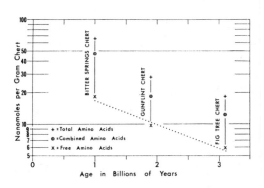

FIG. 4.—Semi-log plot of total, combined, and free amino acid content of the three Precambrian cherts investigated. Note the approximately exponential decrease (*dotted line*) in free amino acid content with increasing age.

As is shown in Figure 4, the quantity of amino acids detected is inversely related to the radiometric age of the cherts. Although plotted in terms of nanomoles of amino acids per gram of chert, a similar trend would result if the data were plotted as nM/gm organic matter, since the three sediments contain quite similar amounts of organic material. If the three cherts initially contained comparable quantities of amino acids, this trend could be interpreted as resulting from gradual chemical degradation, and since the sediments have had similar geologic histories, a comparison of the residual amino acid content would indicate relative geologic age. The apparently exponential decrease in free amino acid content could be interpreted as resulting primarily from the degradation of glycine by a first-order irreversible reaction.[17, 18] Based on the preliminary data available, such an interpretation would be premature. Data from other Precambrian cherts, and from additional samples of the three sediments investigated, are needed to establish that the observed trend is not merely fortuitous.

Biogenicity: Complex organic compounds have been abiotically synthesized under conditions similar to those postulated for the primitive earth.[38] Few criteria are available to enable discrimination between biologically and abiotically produced organic compounds, and amino acids are readily formed by abiotic means. Although it might be suggested, therefore, that the amino acids here reported from Precambrian rocks are of nonbiological origin, the available evidence seems indicative of biogenicity.

Perhaps the most telling argument for the biological origin of these compounds is the fact that they are associated with organically preserved microfossils.[1-9] It may also be noted that the distribution of amino acids in the chert extracts differs from typical distributions reported from abiotic syntheses,[38] and that all of the amino acids detected are either known from extant organisms or are apparently geochemical derivatives of biological amino acids. New techniques for gas chromatographic analysis of amino acid isomers[39] may provide additional evidence relating to the origin of these apparently biogenic compounds.

Summary and Conclusions.—Three fossiliferous Precambrian cherts, approximately 1, 2, and 3 billion years in age, were analyzed for the presence of amino acids. Small but measurable quantities of one or two free amino acids (10^{-8} to 10^{-9} M/gm), and 21 or 22 combined amino acids (10^{-8} M/gm) were detected in triplicate analyses of each sediment. Determination of amino acid content is based on amino acid analyzer data, and the identification of 11 amino acids was confirmed by gas chromatography of the N-TFA ester derivatives. Several lines of evidence indicate that the amino acids are indigenous to the rock samples investigated, and geologic and geochemical considerations strongly suggest that these compounds date from the time of original sedimentation.

In general, there appear to be few qualitative differences in amino acid distribution between the three Precambrian cherts, but the quantity of amino acids detected per gram of sediment decreases with increasing geologic age. Although the stereoisomeric nature of the amino acids has not been determined, their association with organically preserved microorganisms seems indicative of a biological, rather than an abiotic, origin. These preliminary findings are consistent with earlier studies which indicated that biochemically complex organisms were in existence more than 3.1 billion years ago,[1, 2, 10] and they suggest that the amino acid composition of living systems has not changed significantly since very early in biological history.

We thank Dr. Egon T. Degens, Woods Hole Oceanographic Institution, for suggestions relating to analytical procedure and for allowing J. W. S. to carry out preliminary analyses in his laboratory. We also thank Dr. Cyril Ponnamperuma for making available the facilities of the Ames Research Center to J. W. S., and Dr. Sam Aronoff, Iowa State University, for helpful advice regarding laboratory techniques.

* This work supported in part by NSF grant GA-1115 to Harvard University.

[1] Barghoorn, E. S., and J. W. Schopf, *Science*, **152**, 758 (1966).

[2] Schopf, J. W., and E. S. Barghoorn, *Science*, **156**, 508 (1967).

[3] Tyler, S. A., and E. S. Barghoorn, *Science*, **119**, 606 (1954).

[4] Barghoorn, E. S., and S. A. Tyler, *Science*, **147**, 563 (1965).

[5] Schopf, J. W., E. S. Barghoorn, M. D. Maser, and R. O. Gordon, *Science*, **149**, 1365 (1965)

[6] Cloud, P. E., Jr., *Science*, **148**, 27 (1965).

[7] Cloud, P. E., Jr., and H. Hagen, these Proceedings, **54**, 1 (1965).

[8] Barghoorn, E. S., and J. W. Schopf, *Science*, **150**, 337 (1965).

[9] Schopf, J. W., *J. Paleontol.*, in press.

[10] Schopf, J. W., in *McGraw-Hill Yearbook of Science and Technology* (1967), p. 46.

[11] Johns, R. B., T. Belsky, E. D. McCarthy, A. L. Burlingame, P. Haug, H. K. Schnoes, W. Richter, and M. Calvin, *Geochim. Cosmochim. Acta*, **30**, 1191 (1967).

[12] Barghoorn, E. S., W. G. Meinschein, and J. W. Schopf, *Science*, **148**, 461 (1965).

[13] Oró, J., D. W. Nooner, A. Zlatkis, S. A. Wikström, and E. S. Barghoorn, *Science*, **148**, 77 (1965).

[14] Hoering, T. C., and P. H. Abelson, *Carnegie Institution of Washington Year Book 64* (1965), p. 218.

[15] Hoering, T. C., *Carnegie Institution of Washington Year Book 64* (1965), p. 215.

[16] Jones, J. D., and J. R. Vallentyne, *Geochim. Cosmochim. Acta*, **21**, 1 (1960).

[17] Abelson, P. H., in *Organic Geochemistry*, ed. I. A. Breger (New York: Pergamon Press, 1963), p. 431.

[18] Vallentyne, J. R., *Geochim. Cosmochim. Acta*, **28**, 157 (1964).

[19] Moore, S., and W. H. Stein, *J. Biol. Chem.*, **211**, 893 (1954).

[20] Gehrke, C. W., and D. L. Stalling, *Separ. Sci.*, **2**, 101 (1967).

[21] Ulrych, T. J., A. Burger, and L. O. Nicholaysen, *Earth Planet. Sci. Letters*, **2**, 179 (1967).

[22] Oró, J., and D. W. Nooner, *Nature*, **213**, 1082 (1967).

[23] Black chert facies in upper third of Fig Tree Series, exposed by road cutting 100 m northwest of surface opening to Daylight Mine (Barbrook Mining Co.), 28 km east-northeast of Barberton, South Africa. Collected by E. S. B., February 1965.

[24] Lower Algal Chert Member of Gunflint Iron Formation, exposed on northern shore of Lake Superior, 6.4 km west of Schreiber, Ontario. Collected by J. W. S. and E. S. B., August 1966.

[25] Black chert facies in middle third of Bitter Springs Formation, exposed on south slope of ridge about 1.6 km north of Ross River Tourist Camp (Love's Creek Homestead), 64 km east-northeast of Alice Springs, Northern Territory, Australia. Collected by E. S. B., April 1965.

[26] Abbreviations for amino acids are listed in Fig. 2.

[27] T-250 Laboratory Disc Mill, Angstrom, Inc., Chicago, Ill.

[28] Degens, E. T., and J. H. Reuter, in *Advances in Organic Geochemistry* (New York: Pergamon Press, 1964), p. 377.

[29] Hydrolysis of this type, in the presence of oxygen, results in the quantitative oxidation of cystine to cysteic acid and the partial destruction of methionine.[31]

[30] Pollock, G. E., *Anal. Chem.*, **39**, 1194 (1967).

[31] Hare, P. E., *Carnegie Institution of Washington Year Book 64* (1965), p. 232.

[32] Oró, J., and H. B. Skewes, *Nature*, **207**, 1042 (1965).

[33] Hamilton, P. B., *Nature*, **205**, 284 (1965).

[34] Analyses here reported were carried out at the Ames Research Center; preliminary analyses of the Bitter Springs chert were carried out at the Woods Hole Oceanographic Institution, through the courtesy of Dr. E. T. Degens.

[35] Hare, P. E., and R. M. Mitterer, *Carnegie Institution of Washington Year Book 65* (1966), p. 362.

[36] Degens, E. T., J. M. Hunt, J. H. Reuter, and W. E. Reed, *Sedimentology*, **3**, 199 (1964).

[37] Degens, E. T., *Geochemistry of Sediments* (Englewood Cliffs, N.J.: Prentice-Hall, 1965), 342 pp.

[38] Ponnamperuma, C., and N. W. Gabel, *Space Life Sci.*, in press.

[39] Pollock, G. E., and V. I. Oyama, *J. Gas Chromatog.*, **4**, 126 (1966).

33

Reprinted from *Nature*, **221**(5176), 141–143 (1969)

Optical Configuration of Amino-acids in Pre-Cambrian Fig Tree Chert

Optically active amino-acids have been found in a sample of Pre-Cambrian Fig Tree chert at least three thousand million years old

by

KEITH A. KVENVOLDEN, ETTA PETERSON

and

GLENN E. POLLOCK

Exobiology Division, Ames Research Centre, NASA, Moffett Field, California 94035

THE occurrence and fate of amino-acids in the biosphere and lithosphere have intrigued numerous investigators during the past few years[1-4]. We are interested in amino-acids for the information they may provide concerning the origin and evolution of early life on Earth. Recent work from this laboratory[5] showed the presence of about twenty-two amino-acids in ancient cherts which range in age from one to three billion years. As a continuation of the previous work, the optical configuration of the amino-acids associated with the oldest chert sample has now been determined by gas chromatography. (Chert is a microcrystalline sedimentary rock composed chiefly of SiO_2.) The chert sample examined here came from the Fig Tree Series of the Swaziland System at the Daylight Mine area of eastern Transvaal, South Africa. Its minimum age of about $3 \cdot 1 \times 10^9$ years[6] places it among the oldest sedimentary rocks known. Recent studies of Fig Tree chert have revealed the presence of specific organic compounds such as *n*-alkanes[7], isoprenoid hydrocarbons[8,9] and amino-acids[5], as well as microfossils[6,8].

Great care was taken in processing the chert sample for amino-acids in order to avoid contamination. Amino-acids could not be detected in procedural blanks. Analysis of the chert sample followed the previously published procedure[5].

Amino-acid analyser chromatograms showed that glycine and a trace of α-alanine were present in the chert as free amino-acids. Bound amino-acids in the HCl

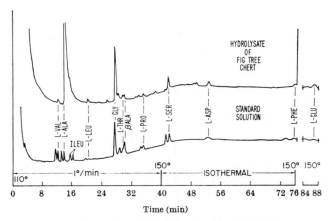

Fig. 1. Gas chromatogram of N-trifluoroacetyl-(+)-2-butyl esters and O-acetyl-N-trifluoroacetyl-(+)-2-butyl esters of amino-acids in the hydrolysate from Fig Tree chert and in the racemic standard. Abbreviations for amino-acids are listed in the text. Column: 0·02 inch × 150 feet, *UCON* 50 *HB* 2000. Perkin Elmer 881 gas chromatograph with flame ionization detector. He = 10 ml./min. Temperature programme from 110° to 150° C at 1° C/min.

hydrolysate were glycine (Gly), serine (Ser), threonine (Thr), leucine (Leu), α-alanine (α-Ala), valine (Val), proline (Pro), aspartic acid (Asp), glutamic acid (Glu), β-alanine (β-Ala), phenylalanine (Phe) and isoleucine (Ileu). These amino-acids are the same as the twelve most abundant ones discovered earlier in rock from the same area[5]. No analysis was made for basic amino-acids. The concentration of amino-acids found in this work is less than about 2 nmoles/g of chert.

The optical configuration of the bound amino-acids was determined by gas chromatography[10]. For this work, N-trifluoroacetyl-(+)-2-butyl ester derivatives of the amino-acids were used[11]. In the case of hydroxy amino-acids, the hydroxyl group was converted to an acetyl instead of a trifluoroacetyl group to give O-acetyl-N-trifluoroacetyl-(+)-2-butyl esters[12]. Previous work on this gas chromatographic method[11] established that the order of peak elution for diastereomers of racemic amino-acids is D(+), followed by L(+). (D and L refer to amino-acid, and (+) to alcohol.) Gas chromatography of the derivatives was carried out on two capillary columns, each coated with a different liquid phase.

Figs. 1 and 2 show the gas chromatograms of the derivatives of bound amino-acids in the hydrolysate from Fig Tree chert. The lower portion of each figure contains a chromatogram of a mixture of twelve derivatized racemic amino-acids. The amino-acids were weighed out, except for leucine, in proportion to the twelve most abundant amino-acids found earlier[5], and were derivatized

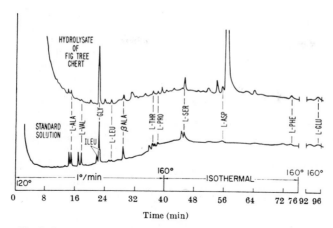

Fig. 2. Same as Fig. 1, except column: 0·02 inch × 150 feet, *XE* 60; temperature programme from 120° to 160° C at 1° C/min.

in the same manner as the sample. The resulting standard chromatograms show two peaks for every derivatized amino-acid containing an asymmetric carbon atom except for the aspartic acid derivatives which are poorly resolved. Glycine and β-alanine, having no asymmetric centres, produce one peak each. For the Fig Tree sample, the retention times of the predominant amino-acid derivatives correspond to those of the L(+) diastereomers, the D(+) forms being present in very low concentrations or in concentrations below the level of accurate detection. The presence of isoleucine could not be confirmed by gas chromatography because of interference from other peaks. Unidentified peaks appear in both chromatograms, but, except for isoleucine, these peaks usually do not interfere with peaks attributed to amino-acid derivatives. Standard solutions of amino-acid derivatives were co-injected with the sample preparations, and augmentation of the expected peaks occurred.

In the earlier work[5] several lines of evidence strongly suggested that the amino-acids were indigenous to the Fig Tree chert and dated from time of original sedimentation. If this is true, the finding of predominantly L amino-acids in these rocks indicates that biological processes as we know them were active three billion years ago because, with few exceptions[13], amino-acids in modern living systems have the L configuration. Small contributions to the amino-acids from abiotic or prebiotic processes, which probably would yield racemic products, cannot be excluded by our results. Chert apparently has provided an extremely stable environment in which racemization reactions have been greatly inhibited.

From another point of view, the finding of optically

active amino-acids in this chert may suggest that the amino-acids are much younger than the rock itself. Contamination in our laboratory is unlikely because of the stringent precautions and the absence of amino-acids in the numerous blanks that have been run. Entry of amino-acids into the rocks at some time subsequent to lithification may be possible. These amino-acids could have been carried in solution or could have been components of micro-organisms which have entered the rock. Although these cherts are dense and apparently impermeable, discrete microfractures are evident in hand samples. In processing the chert, however, efforts were made to break the rock along planes of weakness, such as microfractures, so that these surfaces could be cleaned. The work of Hare and Abelson[14] has shown that complete racemization of amino-acids occurred in *Mercenaria* shells as young as Miocene—about 11 million years. On the basis of this work, one might expect that amino-acids three billion years old should also be racemic. Final interpretation of the significance of optically active amino-acids in Fig Tree chert will require further laboratory study.

We thank the Economic Geology Research Unit, University of the Witwatersrand, and Dr I. A. Breger for the chert sample.

Received September 26; revised November 4, 1968.

[1] Abelson, P. H., *Researches in Geochemistry*, 1 (John Wiley, New York, 1959).

[2] Briggs, M. H., *NZ J. Geol. Geophys.*, **4**, 387 (1961).

[3] Degens, E. T., Reuter, J. H., and Shaw, K. N. F., *Geochim. Cosmochim. Acta*, **28**, 45 (1964).

[4] Hare, P. E., and Abelson, P. H., *Carnegie Inst. Wash. Yearbook*, **63**, 267 (1964).

[5] Schopf, J. W., Kvenvolden, K. A., and Barghoorn, E. S., *Proc. US Nat. Acad. Sci.*, **59**, 639 (1968).

[6] Schopf, J. W., and Barghoorn, E. S., *Science*, **156**, 508 (1967).

[7] Hoering, T. C., *Carnegie Inst. Wash. Yearbook*, **64**, 215 (1965).

[8] Barghoorn, E. S., and Schopf, J. W., *Science*, **152**, 758 (1966).

[9] Oro, J., and Nooner, J. W., *Nature*, **212**, 1082 (1967).

[10] Charles, R., Fischer, G., and Gil-Av, E., *Israel J. Chem.*, **1**, 234 (1963).

[11] Pollock, G. E., and Oyama, V. I., *J. Gas. Chromatog.*, **4**, 126 (1966).

[12] Pollock, G. E., and Kawachi, A. H., *Anal. Chem.*, **40**, 1356 (1968).

[13] Stanier, R. Y., Doudoroff, M., and Adelberg, E. A., *The Microbial World*, second ed. (Prentice Hall, New Jersey, 1963).

[14] Hare, P. E., and Abelson, P. H., *Carnegie Inst. Wash. Yearbook*, **66**, 526 (1968).

34

Reprinted from *Carnegie Inst. Washington Yearbook 65*, 365–372 (1967)

CRITERIA FOR SUITABLE ROCKS IN PRECAMBRIAN ORGANIC GEOCHEMISTRY

T. C. Hoering

Successful study of the organic geochemistry of Precambrian rocks requires well-preserved material with sufficient indigenous organic chemicals. This report describes experiments designed to guide the discovery of such rocks. Criteria are proposed for deciding whether organic molecules are indigenous. Some possible sources of contamination are identified.

Only a few organic-rich rocks older than 600 million years (m.y.) are known that have had a mild thermal and chemical history. Outstanding among them is the 1100-m.y.-old Nonesuch shale of Michigan. Reports of organic molecules in this formation have been presented in the past 4 years. Continued advances in the organic geochemistry of the Precambrian depend on additional examples of a wide range of ages.

The organic structure of the carbon in most sedimentary rocks that escape weathering has been destroyed by metamorphism. In extreme cases, it has been converted to crystalline graphite. Most Precambrian sedimentary rocks contain only small amounts of organic chemicals. The quantities appear to be such that they could have been introduced from contamination.

Operationally, the organic matter in a rock can be divided into two fractions by solvent extraction. The portion soluble in a benzene-methanol mixture has some properties of a petroleum. This fraction is attractive for experimental work because the molecules are of sufficiently low molecular weight to be separated and analyzed. However, this fraction is susceptible to contamination by younger organic matter.

The solvent insoluble fraction (kerogen) is more difficult to investigate. Since it is not mobile, however, it is likely to be of the same age as the host rock. The same chemical environment has affected both fractions and the presence of soluble organic matter in a rock together with severely metamorphosed kerogen would be suspect.

The experimental procedure used in this laboratory for the solvent extraction of rocks is as follows: The surfaces are mechanically and chemically cleaned. The rock is crushed and ball-milled to micron-sized particles. The powder is extracted with redistilled benzene-methanol mixture (7/3 by volume) in a 500-watt ultrasonic generator. The mixture is stirred and extracted for 4 to 6 hours, during which time the solvent reaches the boiling point. After centrifugation the solution is treated with metallic copper to remove sulfur. It is then filtered, and the solvent is carefully evaporated under vacuum. The extract is recovered and weighed.

Three Precambrian sedimentary rocks with extractable organic matter greater than 100 ppm have been found. Later it will be shown that in addition to the

TABLE 27. Precambrian Sedimentary Rocks with Greater than 100 Parts per Million Extractable Organic Matter

Rock	Location	Age, million years	% Organic Carbon	Parts per Million Extractable
Calcareous shale, McMinn formation, Roper River series	Northern Territory, Australia	1600	1.04	1500
Shale, Muhos formation, Jotnian series	Finland	1200	0.41	304
Carbonaceous lens, shale, Jeerinah formation, Fortescue series	Western Australia	2000	80.5	250

The portions of the total organic carbon and the extractable organic matter are figured on the weight of the carbonate-free rock.

Nonesuch shale the kerogen in these rocks gives indications of being unmetamorphosed. They are described in Table 27. On the other hand, about twenty Precambrian rocks with less than 100 ppm extractable organic matter have been examined. The kerogen in them appears to have been altered. Some indication of possible "background levels" of organic contamination is given by the fact that the extraction of a red Precambrian granite yielded 20 ppm organic matter. Solvent blanks are about 3 ppm.

Experience in this laboratory shows that a level of about 100 ppm is a lower limit at which organic geochemistry can be done confidently. Rocks with greater amounts than this satisfy the criteria for suitable specimens that will be proposed later. Rocks with smaller amounts than this fail to satisfy one or more of them.

The hydrocarbon fractions from a number of Precambrian rocks have been analyzed by gas-liquid chromatography with capillary columns. Silica gel chromatography with a pentane eluate was used to separate the hydrocarbon from the rock extracts.

From Fig. 93 it is clear that the normal alkanes from the Nonesuch shale are in marked excess over all the possible isomers and that they predominate over the thermodynamically more stable isoalkanes and cycloalkanes. That the major peaks are n-alkanes is proved by separating them as their urea adducts or by absorbing them on molecular sieve 5A and repeating the gas-chromatographic analysis.

Figure 94 shows results of the hydrocarbon analysis on an extract of the 3000-m.y.-old Fig Tree shale. Although there are some differences in detail, there is a predominance of normal alkanes. The amounts are very small, however, each n-alkane representing only about 1 part in 10 million of the original rock. This is in the range where contamination must be carefully evaluated. As we shall see below, some questions are raised on whether these n-alkanes are actually indigenous.

In contrast, Fig. 95 shows the gas chromatogram of the high molecular weight hydrocarbons produced by treating a cleaned, high-carbon cast iron with hydrochloric acid. The sample was fractionated and analyzed identically to the previous two. Although high molecular weight hydrocarbons are formed by this abiological reaction, there is no excess of normal alkanes. Probably the mixture consists of isoalkanes and cycloalkanes.

Living organisms have the ability to synthesize preferentially straight-chained organic compounds. The fatty acids of cells are one expression of this. Although fatty acids of cells consist primarily of molecules with an even number of carbon atoms, there is a geochemical mechanism for converting them to normal alkanes with a smooth distribution of carbon numbers. The presence of an excess

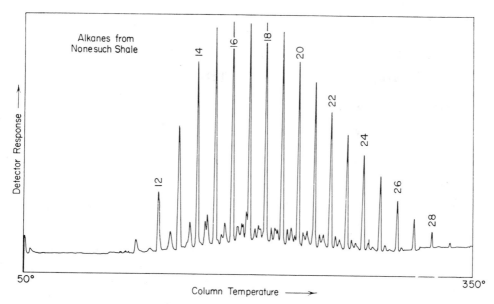

Fig. 93. Gas chromatogram of the hydrocarbon fraction of the extractable organic matter from the Nonesuch shale. Analyses made on a Wilkens model 600 gas chromatograph with a hydrogen flame detector. A 200-foot by 0.018-inch capillary column coated with Apiezon L grease was used. The helium pressure was 30 psi gauge. The column temperature was programmed from 50° to 350°C during the analysis; 200 μg of hydrocarbons were injected.

amount of indigenous normal alkanes may be taken as evidence that they have been derived from the products of living organisms. The problem is to show that they are indigenous.

Last year's report contained results on the C^{13}/C^{12} ratios of the soluble and insoluble organic matter in sedimentary rocks. The isotope ratios of the two fractions agreed to within 2 parts per 1000 for rocks in which the soluble organic matter was indigenous. Table 28 provides further data on the rocks listed in Table 27. There is agreement in the first three cases. Although the Jeerinah shale gave amounts of extractable organic matter above the level expected from contamination, and its kerogen seemed to be unaltered by the pyrolysis test given below, results from this rock are suspect, since there is a discrepancy in the carbon isotope ratios.

Table 29 includes results of carbon isotope ratio measurements on several rocks with less than 100 ppm extractables. Discrepancies in the ratios from the two fractions are explained as being due to contamination by younger organic matter. The values of δC^{13} in the soluble fraction are typical for petroleum of marine origin. The agreement in the case of the Fig Tree shale could be fortuitous.

Previous reports give results on the low-temperature pyrolysis of unmetamorphosed kerogens. Abundant amounts of low molecular weight, saturated hydrocarbons in the range of ethane through heptane are produced by such treatment. The examination of rocks by this technique can give valuable information on the state of preservation of kerogens. The following simple experiments investigated this in Precambrian rocks.

A few grams of untreated powdered rocks were heated under vacuum on the inlet of a mass spectrometer. The evolved gases were collected in a trap cooled by liquid nitrogen, and the noncondensed gases were pumped off. After several hours of outgassing at 110°C, the temperature was increased to 150°C and the gases

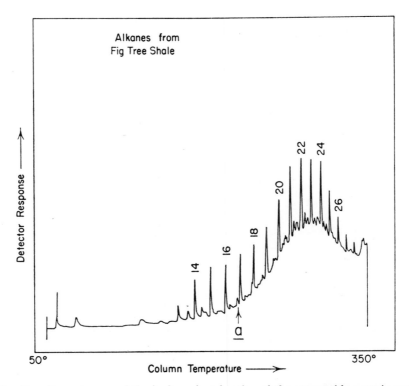

Fig. 94. Gas chromatogram of the hydrocarbon fraction of the extractable organic matter from the Fig Tree shale. The analysis was made on a Perkin-Elmer model 880 gas chromatograph. A 200-foot by 0.010-inch capillary column coated with Apiezon L grease was used. The helium pressure was 50 psi gauge. The column temperature was programmed from 50° to 350°C during the analysis; 100 μg of hydrocarbons were injected.

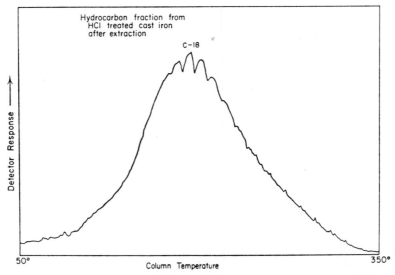

Fig. 95. Gas chromatographic analysis of the hydrocarbon fraction produced by hydrochloric acid on iron carbide. Conditions for analysis same as in Fig. 93.

TABLE 28. C^{13}/C^{12} Ratios in the Soluble and Insoluble Organic Matter of Four Unmetamorphosed Precambrian Rocks

Rock	Location	δC^{13} Soluble Organic Matter	δC^{13} Insoluble Organic Matter
Shale, Nonesuch formation	Michigan, U.S.A.	−28.14	−28.15
Shale, McMinn formation, Roper River series	Northern Territory, Australia	−30.59	−30.71
Shale, Muhos formation, Jotnian series	Finland	−27.51	−28.71
Shale, Jeerinah formation, Fortescue series	Western Australia	−24.11	−36.50

The units used are defined as follows:

$$\delta C^{13} = \frac{(C^{13}/C^{12})_x - (C^{13}/C^{12})_s}{(C^{13}/C^{12})_s} \times 1000$$

where the subscript x refers to the sample and the subscript s refers to a standard carbon. The standard is National Bureau of Standards isotope reference material no. 20, limestone from Solenhofen, Bavaria.

TABLE 29. C^{13}/C^{12} Ratios in the Soluble and Insoluble Organic Matter of Some Metamorphosed Precambrian Rocks

Rock	Location	δC^{13} Soluble Organic Matter	δC^{13} Insoluble Organic Matter
Shale, Soudan formation	Michigan, U.S.A.	−25.00	−34.81
Carbon leader, Witwatersrand system	South Africa	−27.36	−35.22
Shale, Fig Tree series, Swaziland system	South Africa	−27.55	−26.94
Shale, Ventersdorp system	South Africa	−25.78	−36.86
Carbonaceous limestone, Transvaal system	South Africa	−25.01	−38.21

The units of δC^{13} are shown in Table 28.

were collected for several hours. The trap was then warmed, and the gases were cycled over phosphorous pentoxide and solid potassium hydroxide. The gases were admitted to the mass spectrometer, and the mass spectrum was scanned from masses 12 to 200. The temperature was increased in increments of 50°C up to 300°C, and the gases were analyzed after each increase. By this sensitive method about 10^{-8} grams of volatile hydrocarbons can be detected.

Figure 96 shows the mass spectrum of volatile hydrocarbons generated from the Muhos shale described in Table 27. This assemblage of hydrocarbons is typical of those from unmetamorphosed kerogens of all ages. The ions at masses 29, 43, 57, 85, and 99 are due to hydrocarbon fragments with the empirical formula C_nH_{2n+1}, where n has values from 2 to 7. These ions can only come from saturated hydrocarbons of the empirical formula C_nH_{2n+2}.

Figure 97 shows the gases from the Fig Tree shale obtained in the same way. The amount of gas is about a factor of 10 less. Ions at masses 44, 28, 18, and 17 are from carbon dioxide, nitrogen, carbon monoxide, and water that were incompletely removed. The ions at masses 92, 91, 78, 52, 51, and 50 come from benzene, toluene, and substituted aromatic compounds. Barely a trace of the ions could have come from saturated hydrocarbons. Since metamorphism removes hydrogen from organic compounds and converts it to aromatic substances, and eventually

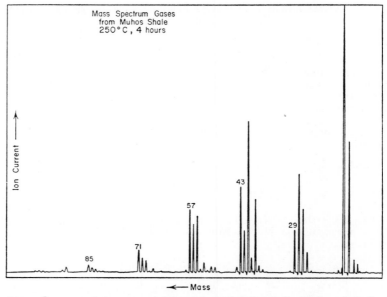

Fig. 96.   Mass spectrum of the gases produced by pyrolysis of the Muhos shale for 12 hours at 250°C. Samples were analyzed in a 6-inch, 60°-sector-field, single-focusing mass spectrometer. The electron energy was 70 volts. The accelerating voltage was 2000 volts, and the spectrum was taken by linearly varying the magnetic field.

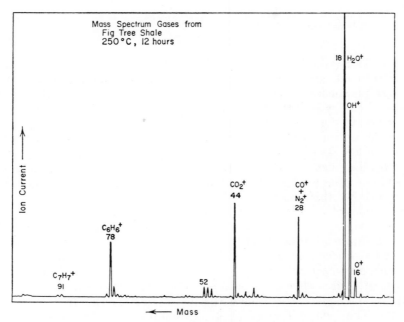

Fig. 97.   Mass spectrum of gases from the Fig Tree shale.

to graphite, it appears that the kerogen in the Fig Tree shale has been severely altered. Some questions are then raised concerning the hydrocarbons shown in Fig. 94. How did they escape destruction when the kerogen seems to be so completely aromatized? The pyrolysis of the other rocks in Table 29, exhibiting a discrepancy in the carbon isotope ratios of the soluble and insoluble fractions, gave predominantly aromatic hydrocarbons.

Many of the Precambrian sedimentary rocks that appear to be metamorphosed contain free sulfur. This element plays an important role in the destruction of organic matter. It is a good dehydrogenating agent, and organic sediments exposed to it at low temperatures for geologically short periods suffer great changes. Samples of the McMinn shale (Table 27), which contain little free sulfur, were pyrolyzed with and without added sulfur. The presence of this element produced a marked change in the total amount and distribution of the hydrocarbons that were evolved. The ratio of aromatic to saturated hydrocarbons increased sharply and the amounts of hydrocarbons decreased by a factor of 10. These effects became noticeable at a temperature as low as 200°C.

We recognize now several criteria for judging whether a Precambrian sedimentary rock is suitable for organic geochemical studies: (1) Its geological setting, mineral assemblages, and chemical composition should indicate that it has not been severely metamorphosed. (2) It should contain an amount of extractable organic matter above the level expected from contamination. (3) The C^{13}/C^{12} ratios of the soluble and insoluble organic matter should agree to within 2 parts per 1000. (4) The kerogen should evolve mainly saturated hydrocarbons on pyrolysis.

A Precambrian rock can be contaminated in several ways by younger organic matter: The migration of petroleum fluids in the earth was mentioned in last year's report. Soil organic matter from percolating ground waters may get into a rock.

Vast quantities of geologically younger petroleum products are a part of our everyday environment. Some of it appears in unlikely places and could find its way into rocks being studied. For example, the heating system of this laboratory contains a dust filter that consists of a metal screen that is continually bathed with a light petroleum oil. During the heating season, all laboratory air is saturated with these oils. Fortunately, the normal alkanes in it are of a lower molecular weight range than is normally analyzed for in rock extracts. The presence of this contamination has been noticed in several cases.

It is common practice to wrap rock specimens in paper for shipment and storage. Many papers contain oil products from their manufacturing processes. Newspapers are a rich source of oils. Printer's ink is oil based. The extraction of a newspaper yielded 2% of its weight as material soluble in benzene-methanol mixture, 0.7% of which was hydrocarbons. Dust, collected from corners of the laboratory, contained 1.6% by weight of soluble organic matter, of which 0.51% was hydrocarbons. The assemblage of normal alkanes in it is similar to that of a rock extract.

A sample of the 2500-m.y.-old Soudan shale of Minnesota, which had been stored in newspapers for 2 years, yielded 70 ppm of soluble organic matter. The kerogen in it yielded mainly aromatic hydrocarbons on pyrolysis. It contains an abundance of free sulfur. Mineralogical evidence points to a period of exposure of 300° to 400°C. The C^{13}/C^{12} ratios of the soluble and insoluble organic matter show a discrepancy of 10 parts per 1000. The carbon

TABLE 30.   Carbon Isotope Ratios in the Soudan Shale

Fraction	δC^{13}
Kerogen	−34.81
Soluble organic matter in rock	−25.00
Soluble organic matter from wrapping paper	−25.87

The units used are defined in Table 28.

isotope ratio of the soluble fraction is in the range found for more recent petroleums.

The hydrocarbons from the enclosing newspapers were isolated and compared with those found in the rock. They both contain normal alkanes in the region from $C_{16}H_{34}$ to $C_{30}H_{62}$, but those in the rock have a slightly higher proportion of hydrocarbons in the lower molecular weight range. Distillation of hydrocarbons from the paper into the rock is possible and would lead to a fractionation. The vapor pressure of n-hexadecane is 0.9×10^{-3} mm Hg at 25°C. Table 30 shows the carbon isotope ratios of the two rock fractions and of the newspaper extract. Although difficult to prove, it appears that contamination has occurred in this case.

Reprinted from *Geochim. Cosmochim. Acta*, **34**, 659–675 (1970) with permission of Microforms International Marketing Corporation as exclusive copyright licensee of Pergamon Press journal back files.

Extractable organic matter in Precambrian cherts*

J. W. SMITH,† J. WILLIAM SCHOPF and I. R. KAPLAN

Institute of Geophysics and Planetary Physics, and Department of Geology
University of California, Los Angeles 90024

(*Received 5 November* 1969; *accepted in revised form* 19 *January* 1970)

Abstract—The concentrations of hydrocarbons and fatty acids, and the ratios of the stable isotopes of carbon and sulfur were determined for Precambrian cherts from the Gunflint Iron-Formation, the Paradise Creek Formation, and the Bitter Springs Formation. All three cherts are known to contain organically preserved microfossils. For comparison, studies were conducted on two fossiliferous Phanerozoic cherts (from the Rhynie Chert Beds and Serian Volcanic Formation) of comparable origin and geologic history. The highest concentrations of n-alkanes, pristane and phytane, and saturated and unsaturated fatty acids were generally recovered from the untreated surfaces of the samples; these compounds are primarily, and probably entirely, of recent origin. Extremely small concentrations (a few ppb) of similar compounds were extracted from interior portions of the Precambrian samples; although in part apparently indigenous to the sediments, these compounds are not demonstrably syngenetic with original sedimentation, and the major portion of these extracts also appears to be of relatively recent origin. Permeability and porosity measurements conducted on separate rock samples from the collection on which the organic studies were made, showed the presence of microfractures that could allow the passage of ground water under a pressure gradient. In the absence of chemical criteria firmly establishing the syngenetic nature of extracted organic constituents, such studies of Precambrian sediments may only provide ambiguous evidence of early biochemical processes.

INTRODUCTION

PALEONTOLOGICAL evidence suggests that Precambrian time was characterized by gradually accelerating biological evolution (SCHOPF, 1969); the detection of chemical fossils in early sediments should serve to augment the morphological fossil record, possibly yielding evidence for the time of origin of major biochemical innovations. To data, however, organic geochemical studies have provided little evidence of early evolutionary development. In fact, no major differences between the extractable components of Precambrian and younger rocks have been demonstrated and, therefore, no specific chemical tests for "Precambrian origin" now exist.

The yields of extracts from Precambrian sedimentary rocks vary considerably. In the work here reported, the total yields of isolated material rarely exceed 0·1 ppm; it may be noted that these yields are two to three orders of magnitude lower than those previously reported from Precambrian cherts, including the well-known Gunflint chert (ORO *et al.*, 1965; VAN HOEVEN *et al.*, 1969) which we have here reinvestigated. Clastic sediments, however, may contain much greater amounts of extractable organic matter, and yields as great as 1500 ppm have been reported from Precambrian shales (HOERING, 1967). Since mobile fluids may migrate along bedding planes in shales, constituting a potential source of *in situ* contamination by materials of younger geological age, we have restricted this investigation to highly indurated, and relatively impermeable, carbonaceous cherts. We have studied five sediments

---

* Publication No. 803, Institute of Geophysics and Planetary Physics, UCLA.

† Permanent address: CSIRO, Division of Mineral Chemistry, P.O. Box 175, Chatswood, NSW, Australia.

of varying geological age known to contain well-preserved plant fossils in an attempt to relate the extractable constituents of these cherts to the biota extant during their deposition.

Pristane, phytane, normal paraffins and fatty acids have been previously reported from chert of the Gunflint Iron Formation by Oro *et al.* (1965) and Van Hoeven *et al.* (1969). Amino acids were isolated by Schopf *et al.* (1968) and Abelson and Hare (1968), and C^{13}/C^{12} ratios were measured by Hoering (1967) on the organic constituents. Amino acids were also identified in extracts from black cherts of the Bitter Springs Formation (Schopf *et al.*, 1968).

Several approaches have been used in the present study to differentiate between Precambrian organic matter and contaminants of more recent origin. Primary among these is an analysis of the effect of progressive particle size reduction on the nature and yield of the extracts obtained. The results of this analysis have been substantiated by replicate studies of three of the sediments investigated, and by a comparative study of cherts ranging in age from about 0.2×10^9 yr to 1.9×10^9 yr.

Description of Samples

The black, fossiliferous cherts selected for analysis were obtained from: (1) the Gunflint Iron Formation, Middle Precambrian, Ontario, Canada (ca. 1.9×10^9 yr old); (2) the Paradise Creek Formation, Late Precambrian, Queensland, Australia (ca. 1.6×10^9 yr old); (3) the Bitter Springs Formation, Late Precambrian, Northern Territory, Australia (ca. 0.9×10^9 yr old); (4) the Rhynie Cherts Beds, Lower Devonian, Aberdeenshire, Scotland (ca. 0.39×10^9 yr old); and (5) the Serian Volcanic Formation, Upper Triassic, Sarawak, Borneo (ca. 0.2×10^9 yr old).

All five cherts are black, waxy, and somewhat lustrous on freshly broken conchoidal faces, and all are noted for the cellularly preserved, permineralized plant fossils they contain. The cherts are composed predominantly of cryptocrystalline quartz and, except for the Triassic chert, have an organic carbon content of somewhat less than 1%; they appear to be chemical sediments displaying little diagenetic change. The three Precambrian sediments exhibit fine, irregular laminations reflecting the presence of stromatolitic algal mats; the younger cherts are more coarsely bedded and are thought to represent silicified peat deposits. The petrology of these cherts, and of associated sediments, indicates a general absence of metamorphism; the brown-to-amber color of the preserved organic matter, as seen in thin sections, presumably reflects a mild thermal history. The origin, lithology, mode of preservation of organic constituents, as well as the thermal history of the five deposits appear, therefore, to be quite similar.

The specimens of black chert from the Gunflint Iron Formation were collected by J. W. Schopf (June, 1968) from a stromatolitic horizon (the Lower Algal Chert Member) exposed on the northern shore of Lake Superior, about 6.4 km west of Schreiber, Ontario. A well-preserved microbiota, including 12 species of plant fossils, has been described from this locality (Barghoorn and Tyler, 1965). The chert samples from the Paradise Creek Formation were collected by J. W. Schopf and F. DeKeyser (May, 1968) from silicified stromatolites occurring in the upper third of the formation, exposed on low hilltops about 13 km southeast of Lady Agnes Mine (72 km northeast of Camooweal, Queensland); organically preserved unicellular algae recently have been reported from these Late Precambrian stromatolites (LICARI et al., 1969). The bedded carbonaceous cherts of the Bitter Springs Formation were collected by J. W. Schopf, R. Shaw and A. Magee (May, 1968) from the uppermost strata of the formation, exposed on a low ridge about 1·6 km north of the Ross River Tourist Camp (Love's Creek Homestead), 64 km east-northeast of Alice Springs, Northern Territory. Thirty species of algae, bacteria, possible fungi and other microorganisms have been described previously from the Bitter Springs cherts (SCHOPF, 1968).

Specimens of the Rhynie Chert (Old Red Sanstone), secured from the A″1 zone of KIDSTON and LANG (1917–1921) and containing numerous axes of the primitive vascular plants *Rhynia* and *Asteroxylon*, were obtained from E. W. R. Stollery (Portsoy Minerals, 12 Sandend, Portsoy, Banffshire, Scotland). The highly carbonaceous Triassic cherts of the Serian Volcanic Formation were collected by G. E. Wilford in the Penrissen region of West Sarawak (WILFORD and KHO, 1965), and were obtained for our study from E. S. Barghoorn; a variety of plant fossils, including dipteridaceous fern sporangia (GASTONY, 1970), are cellularly preserved in these cherts.

EXPERIMENTAL

Contamination controls

To monitor the level, nature and origin of contamination throughout the analytical technique, samples of powdered, freshly ignited firebrick (880°C; 12 hr) were subjected to the entire separation procedure including hydrofluoric acid treatment. These controls showed quite conclusively that, despite all precautions, some detectable contamination of the final products might be expected. In the case of the saturated hydrocarbons, the total contamination detected never exceeded $0·1 \times 10^{-6}$ g; as much as $0·4 \times 10^{-6}$ g of contaminating saturated and unsaturated fatty acids could be found in the final methyl ester fraction. When firebrick was omitted from these tests, however, the detected level of contamination was reduced by several orders of magnitude, presumably reflecting the effect of large surface areas in concentrating contaminants.

Recognizable laboratory contamination was finally reduced to an acceptable level (<1 ng/ml) by carefully distilling all solvents, by eliminating all organic materials (e.g. plastic, rubber, paper, etc.) from the experimental procedures, and by limiting the access of laboratory air.

The degree of contamination of the cherts in the geological environment prior to collection and laboratory analysis is more difficult to assess. To obtain some indication of the amount of *in situ* contamination, the untreated surface of each sample of chert was repeatedly extracted with a benzene/methanol solution. Details of this procedure are given below.

It should be noted that procedures to reduce the level of surface contamination of the rocks prior to extraction were deliberately avoided. Instead, an effort was made to collect and analyze those organic constituents readily extractable from accessible rock surfaces. Such material presumably contains the major portion of recent contamination, and a comparison of this fraction with materials extracted from the interior of the rock, and from acid-resistant organic residues, provides a basis for assessing the degree and depth of *in situ* contamination and for determining the origins of various extractable components.

Analytical procedure

Samples of the Gunflint, Bitter Springs and Rhynie cherts were examined in duplicate; single analyses were made of the Paradise Creek and Borneo cherts. For each extraction, individual pieces of the black cherts, each weighing at least 100 g, were taken as starting material; total sample weights ranged from 100 to 1300 g.

Solvent extractions

Three separate solvent extractions of organic matter from the cherts, at progressively smaller particle sizes, were made on each of the rock specimens analyzed.

(1) *Chips.* After brushing the surfaces of the specimens under running distilled water to remove loose or friable material, the samples were shattered by hammering, and those pieces 1–2 cm in size were collected. These chips were immersed in a solvent solution (30/70 methanol–benzene) in a covered beaker, boiled gently on a steam bath for 30 min, and then allowed to stand at ambient temperature for 24 hr. After decanting the solvent, the chips were washed three times with a hot solution of the same solvent, which was decanted into the original solvent extract. This fraction constituted the "extract of the chips."

(2) *Powder.* The chips were dried under cover at 60° and then immersed in a chromic/sulphuric acid solution overnight to oxidize any remaining surface contaminants. After repeated washing with distilled water, the chips were etched by immersion in 25% hydrofluoric acid for 2 hr; this treatment resulted in a loss of 6–8 per cent in weight, and effected the removal of previously extracted outer surfaces. Following washing and drying, the etched chips were ground in a shatter box mill (Spex Industries) for 4 min. Prolonged grinding was avoided in order to reduce possible contamination and formation of artifacts. Typically, the ground product consisted of the following powder:

$$88\% < 120 \text{ mesh}; \ 8\% < 60 \text{ mesh} > 120 \text{ mesh}; \text{ and } 4\% > 60 \text{ mesh}$$

This powder was refluxed overnight in the benzene/methanol mixture, and the bulk of the solvent was decanted. After repeating this extraction twice with small quantities of solvent, the powder was transferred to a fritted funnel and washed several times with solvent. The washings and extracts were combined to form the "extract of the powder."

(3) *Acid-resistant residue.* The extracted chert powder was transferred to a covered Teflon beaker and dissolved in a minimum amount of 50% hydrofluoric acid (Baker-Analyzed Reagent). After 4 or 5 days when the reaction was complete, the acid-resistant organic residue was recovered by decantation of the acid followed by centrifugation. The resulting residue was generally composed of about 95% amorphous, brown-to-amber-colored organic matter, 2–4% organic microfossils and plant fragments and minor concentrations of insoluble minerals (e.g. pyrite, fluorite). Extraction of this water-washed and dried residue with the benzene/methanol solvent under reflux gave the third solvent extract.

Chromatography

The solvent extracts from the chips, powder, and acid-resistant residue were treated identically. Most of the solvent from each extract was removed by evaporation on a steam bath; the few remaining drops were removed under a stream of nitrogen at ambient temperature.

Alkanes were separated from the extracts by chromatography on pre-washed columns of silica gel and elution with hexane. The extracts resulting from further elution with benzene and methanol were combined and evaporated to near dryness, redissolved in benzene and the solution shaken with a 2% solution of sodium hydroxide to remove free fatty acids. The free acids were separated from the sodium hydroxide solution and converted to their methyl ester derivatives by treatment with diazomethane.

Hydrocarbons were identified by gas–liquid chromatography on S.E. 30 and P.P.E. (Polyphenyl ether) columns using an instrument equipped with a flame ionization detector. Identification was based on a comparison of the retention times of the individual components with those of authentic pristane, phytane and n-alkanes on the same columns and by coinjecting samples with known standards. Methyl esters of fatty acids were identified similarly, except that the P.P.E. column was replaced by a D.E.G.S. (diethylene glycol succinate) column, and saturated and unsaturated methyl ester derivatives were used for comparison.

For positive identification unsaturated methyl esters in the extracts were converted to saturated esters by reduction with hydrogen in the presence of platinum oxide at atmospheric pressure and ambient temperature. Peak shifts and peak height increases were then determined on the products.

The method of analysis enabled 5 ng of both n-alkanes and methyl ester derivatives of fatty acids to be unquestionably detected above noise level, and quantitatively resolved.

In the case of the Gunflint, Bitter Springs and Rhynie cherts, an additional procedure was introduced to recover bound fatty acids. Following benzene/methanol extraction of the chips, powder, and acid-resistant residue, fresh solvent was added and the solution was stirred while drops of hydrochloric acid were added until a pH of approximately 2·0 persisted for 30 min. After filtration, the bound acids were recovered from the solvent solutions, converted to their methyl ester derivatives, and identified by gas–liquid chromatography as described above.

C^{13}/C^{12} isotope ratios were measured on the CO_2 released by combustion of the acid-resistant organic residues following an adaptation of the method described by CRAIG (1953). δC^{13} values were measured in comparison with a PDB stantard. S^{34}/S^{32} determinations were made on elemental sulfur dissolved in the benzene/methanol extracts of the acid-resistant residues; the sulfur was isolated by its reaction with freshly cleaned copper wire, to produce a black surface layer of copper sulfide. Strands of wire were added sequentially to the solution until the copper no longer turned black, indicating that dissolved sulfur had been completely removed from solution. The isotopic ratio of pyritic sulfur from the Gunflint chert was analyzed following the procedure described by KAPLAN *et al.* (1963). All δS^{34} values are based on the Canyon Diablo meteorite standard.

Permeability and porosity

Permeability and porosity measurements were made on the samples by Chevron Oil Field Research Company on 1-in. dia. cores. Permeability was determined from measurements of the flow rate of air through the core which was exposed to a pressure of 300 psi. Porosity was calculated from the weight of brine trapped within the core after an initial evacuation to 15 μmHg and a final exposure to a known brine at 1000 psi. The brine was forced into accessible voids.

No measurements were made on the Paradise Creek chert, since extensive fracturing within the available samples prevented the cutting of a suitable core.

RESULTS AND DISCUSSION

Alkanes

The yields of alkanes obtained from solvent extraction of the cherts are summarized in Table 1; chromatograms of the extracts from the chips, powder, and acid-resistant residue from each chert are shown in Figs. 1 and 2.

Precambrian cherts

As is shown in Fig. 1, the extracts from chips of the three Precambrian cherts investigated contain very similar alkane suites. These similarities, and other features common to these three extracts, may be summarized as follows:

1. Relatively large extract yields, substantially greater than those detected in subsequent extractions, were obtained by immersing the freshly broken chips of chert in solvent for 24 hr. This distribution indicated that most of the soluble organic matter associated with these rocks is readily accessible and must, therefore, be present on the surface of the chips or in micro-fissures or pore systems of dimensions allowing penetration and extraction by benzene or methanol.

2. The ease with which the organic matter was extracted from the chips suggests that it is of relatively recent, rather than of Precambrian origin, since comparable interchange with mobile, organic materials must also have been possible in the geological environment.

3. In general, the extracts from the chips show a smooth distribution of n-alkanes with the maximum concentration occurring close to the C_{22} member. All three also exhibit a marked predominance of n-alkanes with an odd number of

Table 1. Total alkanes extracted from cherts

Chert	Borneo			Rhynie						Bitter Springs					
Age (yr)	0.2×10^9			0.37×10^9						0.9×10^9					
Sample wt. (g)	146			380			100			1100			200		
Extraction stage*	C	P	R	C	P	R	C	P	R	C	P	R	C	P	R
Total yield of alkanes (ppb)	17	19	7	1180	1640	2570	670	3500	3300	8	1	1	4	1	1
Carbon preference index (C_{27}–C_{33}) Odd/even	1.4	1.3	1.1	1.1	1.0	1.1	1.1	1.0	1.0	2.3	1.3	1.2	1.6	1.4	1.3
Pristane (% Alkanes)	1.1	1.1	—	4.0	3.5	6.7	—	2.5	6.4	3.2	+	—	+	—	—
Phytane (% Alkanes)	3.1	2.3	—	2.1	2.1	2.9	—	1.4	2.7	3.6	+	+	+	—	+
$\dfrac{\text{Pristane}}{\text{Phytane}}$	0.4	0.5	—	1.9	1.7	2.3	—	1.8	2.4	0.9	—	—	—	—	—

Chert	Paradise Creek			Gunflint					
Age (yr)	1.6×10^9			1.9×10^9					
Sample wt. (g)	1020			1300			300		
Extraction stage*	C	P	R	C	P	R	C	P	R
Total yield of alkanes (ppb)	21	1	2	26	2	1	32	6	1
Carbon preference index (C_{27}–C_{33}) Odd/even	1.2	1.2	1.0	1.4	1.2	1.2	2.0	1.2	1.2
Pristane (% Alkanes)	3.3	+	1.3	1.9	0.1	+	0.6	0.4	0.2
Phytane (% Alkanes)	4.7	3.5	2.0	3.8	0.2	+	1.3	0.6	0.4
$\dfrac{\text{Pristane}}{\text{Phytane}}$	0.7	—	0.7	0.5	0.5	—	0.5	0.7	0.5

* C = chips; P = powder, R = residue, + trace, — not detected.

carbon atoms for members of the homologous series above C_{27}. Values for the C.P.I. (Carbon Preference Index) (Bray and Evans, 1961) for C_{27} through C_{33} range from 2.3 to 1.2. In addition, the concentration of pristane is less than that of the phytane, with the ratio of these isoprenoids varying between 0.46 and 0.88 for the three samples.

Organic compounds from a variety of sources and at different stages of diagenesis almost certainly contribute to the extracts of the intact rocks. For example, the low ratio of pristane to phytane, similar to that in many crude oils, seems suggestive of a marine environment and considerable geologic age (Brooks and Smith, 1967). In contrast, the marked predominance of alkanes with odd number of carbon atoms in the range C_{32}–C_{36} in the Bitter Springs extract suggests contamination, possibly from a recent soil.

Disappearance of the odd-carbon number preference during diagenesis (assumed to occur with increasing age of burial) has been demonstrated for a very large number of Phanerozoic crude oils and coals (Bray and Evans, 1961; Brooks and Smith, 1967). The continued persistence of a CPI > 1 in n-alkanes of Precambrian age is not to be expected in view of the age unless the sediment had experienced an extremely mild thermal history or the alkanes had been protected from diagenesis in some very special fashion. The general abundance distribution of the alkanes (C_{20}–C_{30}) in the surficial extracts, however, appears to be similar to that found in solvent extracts of higher rank coals, indicating it is not primarily of recent origin.

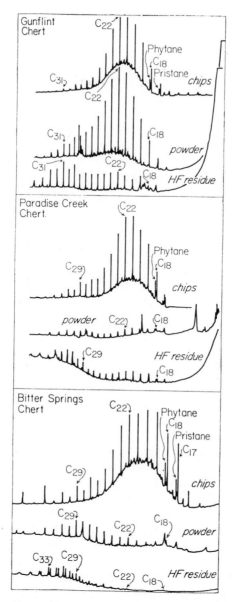

Fig. 1. Distribution of extractable alkanes in chips, powder and hydrofluoric acid residues from three Precambrian cherts.

In view of these considerations and the fact that this material is readily accessible to the exterior as demonstrated by solvent extraction, it seems most unlikely that any significant portion of the extracts from the chips consists of residues of Precambrian age. Although it is conceivable, of course, that some small fraction of these extracts is actually of Precambrian age, the occurrence of such material would be entirely obscured by the predominating constituents of relatively recent origin.

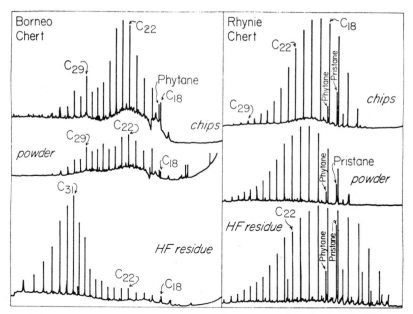

Fig. 2. Distribution of extractable alkanes in chips, powder and hydrofluoric acid residues from a Devonian and a Triassic chert.

Solvent extraction of the powder obtained by grinding the chips yielded small quantities of soluble material. The extracts from both the powdered Bitter Springs and Paradise Creek cherts were qualitatively different from those from the chips; in both cases the high concentrations of saturated hydrocarbons in the C_{22} region, and the associated "hump", were no longer in evidence (Fig. 1). The odd-carbon number preference in the longer chain n-alkanes was decreased noticeably although the relative concentration of higher molecular weight compounds increased. Pristane and phytane occurred in trace quantities only.

The extract from the powdered Gunflint chert, although in many ways resembling the corresponding extracts from the other Precambrian cherts, illustrated several unique features. The general distribution of the n-alkanes did not differ greatly from that in the extract from the chips; but the marked decrease in the concentrations of pristane and phytane relative to the total extract and, more significantly, to the C_{17} and C_{18} n-alkanes, was clearly demonstrated. It seems evident, therefore, that the major portion of these isoprenoid hydrocarbons must be located on surfaces, connected to the exterior, which facilitate their removal by solvents.

In all the Precambrian samples studied, yields of hydrocarbons from the extraction of organic residue remaining after treatment of the powder chert with hydrofluoric acid were very small, the greatest being some 2 ppb. At this level of recovery, the possibility of contamination is so high that the value of further investigation becomes doubtful. However, the yield in itself is of considerable significance since it demonstrates that the organic residues in richly fossiliferous Precambrian cherts, even when exhaustively extracted, yield only trace quantities of soluble organic compounds.

In all three Precambrian sediments, the material extracted from the chips differs significantly from that extracted from the acid-resistant organic residues (Fig. 1), strongly suggesting that the two extracts have different origins; whether the latter is of Precambrian age, however, is uncertain. Although the C.P.I. value approaches unity in the residue extracts, it is nevertheless significantly above unity (see Gunflint residue, Table 1) thus arguing against an age $\geqslant 10^9$ yr. Trace quantities of pristane and phytane were observed in extracts of the acid-resistant residues from the Gunflint and Paradise Creek cherts. In general, their concentration was lowest in these extracts but the pristane/phytane ratio was not significantly different from those of other extracts.

Phanerozoic cherts

Triassic cherts (Borneo). The extracts obtained from this chert (Fig. 2) were generally very similar to those from the Precambrian samples, although the yields of products were more equitably distributed between the three extracts, with a correspondingly larger portion of the total alkanes being recovered from the acid-resistant organic residue.

This latter fraction is of particular interest since a smooth distribution of n-alkanes exhibiting a maximum concentration at C_{31} has not been described in either biological products or their diagenetic derivatives. Presumably, such a distribution might arise from the fractionation of a crude oil with the higher molecular weight n-alkanes remaining behind in the residue. It is also possible that this material represents a laboratory contaminant or artifact. The marked similarity between these Triassic extracts and the corresponding fractions from the Precambrian cherts serves to strengthen the view that most of the latter originated during post-depositional times, and although in part apparently indigenous to cherts, they were primarily not syngenetic with Precambrian deposition. Although the carbon content of the Triassic chert was the highest of all the samples investigated (HF residue, 7·4%), the amount of extractable organic matter was surprisingly low.

Devonian chert (Rhynie). Yields from this sample were at least two orders of magnitude greater than the corresponding extracts from the older cherts; a strict comparison of the products obtained is, therefore, difficult. For example, if the yields and composition of alkanes reported from the Precambrian samples (and from the Triassic chert) were significantly influenced by contaminants, a similar level of contamination in the Rhynie extracts would not be detectable.

The close similarity in composition between the three fractions extracted from the Rhynie chert strongly suggests that the extracted materials have a common origin; the very high yields obtained from the acid-resistant residue seem to indicate that this fraction contains syngenetically emplaced organic matter. If so, diffusion of compounds from the organic material into the surrounding silica has apparently occurred. The concentrations of isoprenoid hydrocarbons relative to total alkanes, shown in Fig. 2 are consistent with this interpretation. The ratio of pristane to n-heptadecane in the extracts from the chips, powder and acid-resistant residue, increases progressively from 0·35 to 0·55 to 0·86; the phytane: n-octadecane ratio similarly increases. It seems apparent that the silicic matrix has acted as a molecular sieve, preferentially retaining the isoprenoid molecules, which have a relatively large cross-sectional area, while allowing the outward diffusion of n-alkanes of corresponding chain length. It further indicates that when significant amounts of soluble materials are present they can be extracted from the kerogen by the procedure used.

The increase in the proportion of the lower molecular weight alkanes in the extract from the acid-resistant residue argues strongly against a significant loss of such hydrocarbons by evaporation or solution during the acid treatment.

The distribution of n-alkanes in the Rhynie chert is generally similar to that of high rank coals and mature crude oils; it seems unlikely, therefore, that the silicic matrix has served to protect the organic constituents from diagenetic alteration. In this respect, the ratio of pristane/phytane as well as concentrations of these isoprenoids, also indicate that the degree of diagenesis is quite advanced since similar values have been observed in coals of the highest rank only (Brooks et al., 1969). The high content of fossil vascular plants in the Rhynie chert, and its origin as a silicified peat deposit suggest that features similar to those of coal might be expected.

Since the Triassic chert from Borneo yielded only trace quantities of soluble organic compounds and is substantially younger than the Devonian Rhynie chert, age alone cannot be the factor controlling the preservation of this material. Furthermore, these two sediments appear to be very similar in lithology and the mode of preservation of their organic constituents, and appear to have had similar origins and mild thermal histories. Differences in the amounts of extractable material obtained from the two cherts are not correlative with total carbon content, which points to differences in the diagenetic mechanisms not yet recognized. A comparison of the distribution of the alkanes extracted from the Triassic chert with those from the much older Precambrian chert does not reveal significant differences in the degree of geochemical maturation, as might be expected from on the great differences in age between these sediments.

Porosity and permeability of the host rock could be of the greatest significance in explaining differences in preservation of organic matter in cherts since they might control the early entrapment of organic compounds, and could certainly influence their subsequent elution and displacement by other materials. It seems possible that the rate of diffusion of isoprenoid hydrocarbons relative to n-alkanes could be controlled by pore size. Diffusion *outward* of these molecules previously associated with the insoluble organic residues of the Rhynie chert, and migration *into* the Gunflint chert from its surroundings, may be of this nature. Similar sieve effects have been suggested previously when it was noted that grinding of rock prior to extraction yielded extracts with an increased content of branched-chain alkanes (Meinschein, 1965).

Free fatty acids

Small quantities of free fatty acids were extracted from all three fractions of each chert investigated. Although the concentration of any individual acid never exceeded 111 ppb, there were marked variations in the concentrations detected.

Hexadecanoic acid was almost invariably the largest single component; the n-saturated C_{14} and C_{18} acids were also prominent, and significant concentrations of the mono-unsaturated n-C_{16} and n-C_{18} acids were usually observed.

The yields and distributions of the free acids (analyzed as their methyl ester derivatives) extracted from the various chert fractions are given in Table 2. In every case, unsaturated fatty acids were prominent in the chromatograms of the

Table 2. Distribution of free acids (ppb of dry sediment)

Carbon No.	11	12	13	14	15	16*	16	17	18*	18	19	20	21	22	23	24
Borneo																
Chips	1·1	1·6	5·4	9·6	5·8	2·6	32·0	1·4	4·6	28·2	—	2·6	—	60·2(2)	—	+
Powder	—	2·1	—	2·6	0·5	1·3	14·1	0·3	4·3	4·0	—	+	—	+	—	+
Residue	—	2·9	—	3·2	1·0	1·0	16·0	—	1·0	7·0	—	—	—	15·6(2)	—	+
Rhynie																
Chips	—	—	—	21·0	18·0	25·5	111·0	6·0	25·5	27·0	—	60·0(1)	+	+	+	+
Powder	—	—	—	—	—	+	+	+	+	+	—	—	—	+	—	—
Residue	—	2·6	1·0	5·3	3·3	1·0	16·0	2·0	2·0	6·0	+	+	+	+	+	+
Bitter Springs																
Chips	—	—	—	+	—	0·2	0·6	+	0·2	+	—	9·0(1)	—	—	—	—
Powder	—	—	—	+	+	+	0·1	—	+	+	—	—	—	—	—	—
Residue	—	1·2	—	5·0	0·9	2·0	20·0	+	5·2	9·6	—	+	—	—	—	+
Paradise Creek																
Chips	0·3	4·0	0·2	11·5	3·0	2·4	21·5	1·5	7·0	10·8	—	+	—	9·1(2)	—	+
Powder	—	+	0·2	0·4	0·3	0·2	2·0	0·1	1·0	0·9	0·2	0·2	0·1	0·1	+	+
Residue	0·3	5·0	0·3	23·7	7·8	5·0	61·3	5·5	10·1	64·0	—	1·5	—	1·0	—	3·0
Gunflint																
Chips	—	—	—	1·2	0·2	0·9	11·5	0·1	1·6	7·2	+	14·6(1)	—	—	—	—
Powder	—	—	—	+	+	+	2·0	+	0·1	0·1	—	—	—	—	—	—
Residue	—	+	—	0·4	0·1	0·4	5·5	+	2·1	2·2	—	0·1(1)	—	—	—	—

(1) and (2) Unidentified acids.
* Trace.
† Unsaturated acids.

freshly collected products; however, the concentration of these acids commonly decreased, or even completely disappeared, within a few days on storage of microgram quantities of these esters in sealed containers.

Reduction of these unsaturated esters with hydrogen in the presence of platinum oxide produced their saturated derivatives and confirmed the identity of the products. The products from the reduction of the unsaturated fatty acids extracted from the hydrofluoric acid-resistant residue of the Gunflint chert are given in Table 3.

Table 3. Hydrogenation of free acids from Gunflint residue

Carbon No.	Distribution as extracted (%)		Distribution after reduction (%)
C_{14}	0·5		0·4
C_{15}	1·6		0·4
C_{16} unsat.	11·4	} 71·6	—
C_{16}	60·2		71·2
C_{17}	1·1		0·4
C_{18} unsat.	14·4	} 25·2	—
C_{18}	10·8		27·7

Relatively high concentrations of unidentified fatty acids were present in extracts from the chips of all five cherts; the retention times of the methyl ester derivatives of these acids were slightly less than those of methylated C_{20} and C_{22} n-saturated acids. In Fig. 3 are shown typical chromatograms of the methyl ester derivatives obtained from extracts of the chert chips. These illustrate the high concentration of unsaturated acids and the distribution of the fatty acids detected in these extracts. This high degree of unsaturation indicates a much more recent origin than the Precambrian, and suggests that these free acids may be contaminants acquired during storage or extraction.

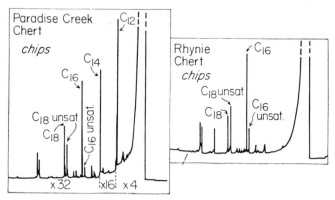

Fig. 3. Examples of free-fatty acid distribution in the chips from one Precambrian and one Devonian chert.

In many cases, the lowest yields of free fatty acids were obtained by extraction of the powdered cherts. Since this extraction was preceded almost immediately by treatment with chromic/sulphuric and hydrofluoric acids, a technique designed to produce a contamination-free surface, a small yield of free acids would be expected if they originated solely from laboratory contamination. Furthermore, the high yields of free acids usually obtained by extraction of the chips were to be expected, since the chips would contain the major portion of contaminants on their surfaces. In some cases, the yield of free fatty acids from extracts of acid-resistant residues was surprisingly high; this may be related to long exposures of the residues to the laboratory atmosphere during the treatment with hydrofluoric acid.

The similarity in composition between the free acids detected in all samples studied suggests that these compounds represent laboratory contamination, possibly of bacterial origin (Tornabene, 1967). If indeed this is the case, their presence may be partly attributable to the large weight of sample extracted. The purely mechanical difficulties in handling large quantities of powdered rock and the relatively long periods of time required to complete standard laboratory procedures (e.g. filtration, extraction, hydrofluoric acid maceration, etc.) in these samples may greatly increase the probability of contamination.

Bound fatty acids

The yields and distributions of fatty acids released from three of the cherts by acidification with hydrochloric acid are summarized in Table 4.

Table 4. Distribution of bound acids (ppb of dry sediment)

Carbon No.	11	12	13	14	15	16*	16	17	18*	18	19	20	21	22	23	24
Rhynie																
Chips	—	—	—	0·5	0·6	1·6	6·3	0·2	3·0	2·6	—	1·5(1)	—	3·6(2)	—	—
Powder	—	—	—	—	—	—	—	—	—	—	—	—	—	—	—	—
Residue	—	—	—	—	—	—	—	—	—	—	—	—	—	—	—	—
Bitter Springs																
Chips	—	—	—	+	—	—	5·0	—	—	—	—	7·5(1)	—	—	—	—
Powder	—	—	—	—	—	0·3	2·8	—	+	+	—	—	—	—	—	—
Residue	—	—	—	+	+	0·1	7·0	+	0·2	0·3	—	—	—	—	—	—
Gunflint																
Chips	—	—	—	+	+	+	44·8	+	+	8·7	—	+	—	—	—	—
Powder	—	—	—	+	+	+	11·0	—	+	+	—	+	—	—	—	—
Residue	—	+	—	0·1	0·3	1·0	16·8	0·3	2·5	1·8	—	+	—	1·2(2)	—	—

(1) and (2) Unidentified acids.
* Unsaturated acids.

Unlike the free fatty acids, the bound acids do not appear to exhibit a regular pattern of distribution. A degree of unsaturation similar to that found in the free acids was apparent in some extracts; in others, particularly those from the Gunflint chips and powdered chert, only saturated compounds were detected. A similar distribution of bound fatty acids from the Gunflint chert has recently been reported by VAN HOEVEN et al. (1969) who interpret the acids as being of probable Precambrian age, preserved by being bound to the chert matrix. Our results neither refute nor confirm this interpretation, but since the greatest yield of saturated acids was obtained from the chips, rather than from the powdered chert, and since unsaturated compounds were detected in the most interior, hydrofluoric acid-resistant residue, the evidence argues against these acids being preserved in the manner suggested.

Particular attention must be drawn to the stability of unsaturated fatty acids. A decrease in concentration of these compounds has definitely been observed on storage of microgram quantities of extracts in the laboratory. Presumably, oxidation reduction or polymerization reactions can effectively alter these unsaturated acids in solution; and the inability to demonstrate their occurrence should not be taken as a firm indication that they were not recently present.

Carbon and sulfur isotopes

Stable isotope data for δC^{13} and δS^{34} are presented in Table 5. The carbon values were measured on hydrofluoric acid-resistant residues, whereas the δS^{34} data were obtained on elemental sulfur extracted from these residues with a methanol/benzene solution. One pyrite sample, freed from the matrix of the Gunflint chert by hydrofluoric acid maceration, was also analyzed.

At first sight it would appear that an inverse correlation may exist between δC^{13} values and the age of the sample analyzed. HOERING (1967) also reports low values of δC^{13} for insoluble organic matter from several Precambrian sediments. Such enrichment in C^{12} is difficult to understand, since diagenetic and metamorphic processes acting on the sediment would result in liberation of isotopically light

Table 5. Ratios of the stable isotopes of carbon and
sulfur in chert fractions

	δC^{13} HF Residue	δS^{34} Free sulfur
Gunflint*	$-33 \cdot 1$	$+19 \cdot 8$
Paradise Creek	$-29 \cdot 3$	$+ 1 \cdot 4$
Bitter Springs	$-24 \cdot 8$	$- 0 \cdot 3$
Rhynie	$-24 \cdot 1$	$-17 \cdot 3$
Borneo	$-23 \cdot 4$	$-10 \cdot 0$

* δS^{34} of pyrite from this fraction was $+14 \cdot 4$.

methane and other short-chained hydrocarbons and the remaining polymer should thus become enriched in C^{13} (Silverman, 1964). One possible explanation is that biosynthesis of the Precambrian organic matter may have occurred under higher partial pressures of CO_2 in acidic or mildly acidic environments. Growth under such conditions may result in a higher enrichment of C^{12} in the organic matter produced (Kaplan and Seckbach, 1970). An alternative explanation is that the insoluble organic fraction represents a high molecular weight polymer, formed by condensation of smaller, unsaturated molecules, that was subsequently metamorphosed during preservation. The δC^{13} values for the Bitter Springs, Rhynie and Borneo cherts are comparable to values characteristic of organic matter derived from both terrigenous and marine sources, and it is not possible to determine their probable origin based on these data alone (Silverman, 1964).

The sulfur isotopic measurements show a wide range in values. The δS^{34} value of sulfur extracted from the Gunflint chert is precisely that of present-day sea water sulfur (dissolved sulfate). The pyrite is $5 \cdot 6\%_0$ lighter, as is usual in co-existing sedimentary sulfur and pyrite (Kaplan et al., 1963). Such a high value of δS^{34} is unusual for biogenic sulfur, and it seems more than a coincidence that it is identical to that of sea water sulfur. The most obvious explanation is that the Gunflint sulfur represents trapped sea water sulfate that was quantitatively reduced to hydrogen sulfide, followed by subsequent oxidation to elemental sulfur. If this interpretation is correct, it would indicate that connate water probably of marine origin or containing dissolved gypsum (presumably of relatively recent origin) has permeated the Gunflint cherts; such solutions might also carry organic contaminants.

The δS^{34} values for the Bitter Springs and Paradise Creek cherts are very similar to that characteristic of meteoritic sulfur. This suggests that the sulfur in these deposits is of igneous origin and that little or no biogenic fractionation occurred during the deposition of these Late Precambrian sediments. The sulfur present in these samples may have been dissolved in the water from which these primary cherts precipitated.

The two Phanerozoic cherts (Rhynie and Borneo) display δS^{34} values ($-17 \cdot 3$ and $-10 \cdot 0$) typical of biologically formed sulfur in present sediments. It seems likely that the sulfur in these two deposits is biogenic; the sulfur isotope ratios of the Precambrian samples provide no evidence for such an origin.

Permeability and porosity measurements

Permeability and porosity data for chert samples collected at the same locations and at the same time as those studied for organic components are given in Table 6.

Table 6. Permeability and porosity measurements on four chert specimens studied

		Permeability (μd)	Porosity (%)
Chert			
	Gunflint	190	0·44
	Gunflint	020	0·55
	Bitter Springs	220	0·62
	Bitter Springs	170	1·15
	Rhynie	660	4·01
	Rhynie	7	4·62
	Borneo	100	0·59
Sandstone		10^3–10^6	5–30
Limestone		10^3–2×10^5	1–15
Shale		$<10^3$	1–15
Granite		<1	

For comparison, ranges of permeability and porosity for sandstones, limestones, shales and granites measured in the same laboratory by identical procedure are also listed. The data show that the bedded cherts possess a finite permeability, not evident from microscopic examination, presumably resulting from microfractures in the matrix. The large variation in permeability measurements obtained for the Rhynie Chert samples is probably due to the presence of larger fissures in the more permeable specimen and may not be entirely representative of other rock samples from this formation. With the exception of the Rhynie chert samples, the porosity of the rocks measured was 1 per cent or less.

It can be seen from Table 6 that measured permeabilities range between 7–650 μd and fall in the general range of many shales. It is not known, of course, whether the microfracturing of these cherts occurred during burial at depth (from tectonic movements) or after exposure at the surface (from diurnal heating and cooling). In either event, if fracturing is present it could permit the entrance of solutions, and under hydrostatic pressure at depth by capillary action at the surface, or the entrance of particle-laden air through "breathing", under the influence of temperature changes.

The porosity is low for all chert samples measured with the exception of the Rhynie chert. The presence of pore spaces may therefore be important in the retention of relatively low-molecular-weight compounds.

CONCLUSIONS

The data presented here raise the same questions that previous studies have posed. Is it feasible to interpret the extractable organic constituents of very ancient sediments as evidence of the biochemical complexity and evolutionary status of the primitive biota? To answer this question a second, more fundamental problem must be considered: Can it be established that the extractable organic compounds in Precambrian sediments are syngenetic with original sedimentation? To approach this problem, we selected three, relatively impermeable, unmetamorphosed, primary Precambrian cherts, known from paleontological studies to contain well-preserved, syngenetically-emplaced, organic microfossils. For comparison, two Phanerozoic cherts of similar origin, lithology, and geologic history were also investigated. From

3

our studies of the extractable hydrocarbons and fatty acids of these sediments, we conclude that only traces of these compounds, in the range of a few ppb or less, are indigenous to the Precambrian cherts. Furthermore, there is no strong evidence to indicate whether this extractable material was emplaced at the time of sedimentation. Some portion of this organic matter may be a product of Precambrian biological activity, derived from the permineralized microorganisms organically preserved in these deposits; the results strongly suggest, however, that the majority of this extractable material is of post-depositional origin. This conclusion seems further supported by S^{34}/S^{32} measurements which, for the Gunflint chert, seem indicative of relatively recent contamination. The paucity of extractable compounds in these sediments may reflect an almost complete diagenetic conversion of the original organic materials to gases of low molecular weight and to insoluble polymers, a process analogous to that observed during coalification.

In view of these considerations and the studies of ABELSON and HARE (1968), who have drawn comparable conclusions regarding indigenous, but not demonstrably syngenetic, amino acids in Precambrian cherts, it appears likely that unless chemical criteria clearly indicative of a "Precambrian origin" can be established, the extractable constituents of very ancient sediments will provide little or no interpretable evidence of early biological processes. As has often been concluded previously, we suggest that the key to the study of early biochemical evolution lies not in the analysis of extractable traces, but rather in the dominant insoluble kerogen-like fraction, and in particular, that material comprising organically preserved microfossils.

Acknowledgments—We thank Dr. ELSO S. BARGHOORN, Department of Biology, Harvard University, for partially defraying costs (through NSF Grant GA-1115) connected with the collection of specimens by J. W. SCHOPF from the Precambrian of Australia, and for providing the samples of Triassic chert used in the present study. We also gratefully acknowledge the field assistance provided in Australia by F. DEKEYSER and R. SHAW of the Bureau of Mineral Resources, and by A. MAGEE, Magellan Petroleum, Ltd. We wish to thank Mr. A. TIMUR, Chevron Oil Field Research Co., La Habra, California, for measurements of permeability and porosity on the cherts. The study was supported by NASA contracts NAS 9-8843 and NAS 9-9941.

REFERENCES

ABELSON P. H. and HARE P. E. (1968) Recent origin of amino acids in the Gunflint chert. *Program, Annual Meeting, Geological Society of America*, Mexico City: 2 (Abs.).

BARGHOORN E. S. and TYLER S. A. (1965) Microorganisms from the Gunflint Chert. *Science* **147**, 563–577.

BRAY E. E. and EVANS E. D. (1961) Distribution of n-paraffins as a clue to recognition of source beds. *Geochim. Cosmochim. Acta* **22**, 2–15.

BROOKS J. D., GOULD K. and SMITH J. W. (1969) Isoprenoid hydrocarbons in coal and petroleum. *Nature* **222**, 257–259.

BROOKS J. D. and SMITH J. W. (1967) The diagenesis of plant lipids during the formation of coal, petroleum and natural gas. I. Changes in the n-paraffin hydrocarbons. *Geochim. Cosmochim. Acta* **31**, 2389–2397.

CRAIG H. (1953) The geochemistry of stable isotopes of carbon. *Geochim. Cosmochim. Acta* **3**, 53–92.

GASTONY G. J. (1970) Sporangial fragments referred to *Dictyophyllum in* Triassic chert from Sarawak. *Amer. J. Botany*. In press.

HOERING T. C. (1967) Organic geochemistry of Precambrian rocks. In *Researches in Geochemistry* (editor P. H. Abelson), Vol. 2, pp. 87–111. John Wiley.

KAPLAN I. R., EMERY K. O. and RITTENBERG S. C. (1963) The distribution and isotopic abundance of sulphur in recent marine sediments off Southern California. *Geochim. Cosmochim. Acta* **27**, 297–331.

KAPLAN I. R. and SECKBACH J. (1970) Growth pattern and C^{13}/G^{13} isotope ratios of an acidophilic hot spring alga cultured under pure CO_2. In press.

KIDSTON R. and LANG W. H. (1917–1921) On old red sandstone plants showing structure, from the Rhynie Chert Bed, Aberdeenshire; Parts I–V. *Trans. Roy. Soc. Edinburgh* **51–52**.

LICARI G. R., CLOUD P. E., JR. and SMITH W. D. (1969) A new chroococcacean alga from the Proterozoic of Queensland. *Proc. Nat. Acad. Sci.* **62**, 56–62.

MEINSCHEIN W. G. (1965) Soudan Formation; Organic extracts of early Precambrian rocks. *Science* **150**, 601–605.

ORO J., NOONER D. W., ZLATKIS A., WIKSTROM S. A. and BARGHOORN E. S. (1965) Hydrocarbons of biological origin in sediments about two billion years old. *Science* **148**, 77–79.

SCHOPF J. W. (1968) Microflora of the Bitter Springs Formation, Late Precambrian, central Australia. *J. Paleontol.* **42**, 651–688.

SCHOPF J. W. (1969) Recent advances in Precambrian Paleobiology. *Grana. Palynologica* **3**. In press.

SCHOPF J. W., KVENVOLDEN K. A. and BARGHOORN E. S. (1968) Amino acids in Precambrian sediments: An assay. *Proc. Nat. Acad. Sci.* **59**, 639–646.

SILVERMAN S. R. (1964) Investigation of petroleum origin and evolution mechanisms by carbon isotope studies. In *Isotopic and Cosmic Chemistry* (editors H. Craig, S. L. Miller and G. J. Wasserburg), pp. 92–102. North-Holland.

TORNABENE T. G. (1967) Distribution and synthesis of hydrocarbons and closely related compounds in microorganisms. Thesis, University of Houston.

VAN HOEVEN W., MAXWELL J. R. and CALVIN M. (1969) Fatty acids and hydrocarbons as evidence of life processes in ancient sediments and crude oils. *Geochim. Cosmochim. Acta* **33**, 877–881.

WILFORD G. E. and KHO C. H. (1965) Penrissen area, West Sarawak, Malaysia. Rept. 2, Geol. Surv. Borneo Region Malaysia.

Reprinted from *Biol. Rev.*, **45**, 319–329, 347–351 (1970)

PRECAMBRIAN MICRO-ORGANISMS AND EVOLUTIONARY EVENTS PRIOR TO THE ORIGIN OF VASCULAR PLANTS

By J. WILLIAM SCHOPF

*Department of Geology, University of California, Los Angeles,
Los Angeles, California 90024*

(*Received* 12 *January* 1970)

* * * * * * *

I. PRECAMBRIAN PALAEOBIOLOGY

The nature of the Precambrian biota—its antiquity, composition and evolution—and the well-known faunal discontinuity near the beginning of the Palaeozoic, have long been recognized as particularly puzzling problems in palaeontology. The evolutionary continuum well-documented in Phanerozoic sediments and the diversity and complexity of the early Palaeozoic biota augur well for a substantial period of Precambrian evolutionary development. Until recently, however, evidence of this development remained largely undeciphered; the nature of Precambrian life was a fertile subject for speculation, essentially unfettered by the poorly known fossil record. The past few years have witnessed a renewed interest in these classic problems and a marked proliferation of available data; this increased activity has resulted in the emergence of a new subdiscipline of palaeontological science, that of Precambrian palaeobiology.

Although widely regarded as a new area of emphasis, Precambrian palaeobiology is firmly rooted in the pioneering studies of the early 1900's by C. D. Walcott and J. W. Gruner, and to a major extent it represents a variation, rather than an innovation, on their original theme. Walcott was an acknowledged leader in the early search for Precambrian fossils and was one of the first to stress the probable algal origin of Precambrian laminated stromatolites (Walcott, 1883, 1899, 1912, 1914). This interpretation, however, and reports by Gruner (1923, 1924a, b, 1925), who claimed to have discovered filamentous microfossils in Precambrian cherts, were viewed with varying degrees of scepticism by contemporary palaeontologists (e.g. Hawley, 1926). Subsequent investigations have shown that in part this scepticism was well founded, and that certain of these early interpretations were erroneous (e.g. Tyler & Barghoorn, 1963). Nevertheless, the fundamental association of cherts, stromatolites and Precambrian microfossils suggested by the studies of Walcott and Gruner has been fully confirmed, and with minor modification has formed the basis of the productive investigations of recent years.

Surprisingly, perhaps, these early studies excited little sustained interest; although several other occurrences of possible microfossils were reported subsequently from Precambrian cherts (e.g. Moore, 1918; Ashley, 1937; Cahen, Jamotte & Mortelmans, 1946), it was not until 1954, with the publication of a short note by Tyler & Barghoorn describing micro-organisms from stromatolitic cherts of the Gunflint Iron-Formation, that the potentialities of the Precambrian began to be widely appreciated. In the 15 years that have followed this report, Precambrian palaeobiology has 'come of age' and the field has developed a distinctive interdisciplinary flavour—a merging of techniques and data from diverse branches of chemistry, geology and biology—to a degree previously unknown in palaeontological science. As the result of this increased activity, numerous sediments containing diverse types of structurally preserved micro-organisms are now known from the Precambrian and organic geochemical studies have yielded putative evidence of the physiology and biochemical complexity of early life. These new data make it possible to outline, in the broadest of terms, major events in Precambrian biological history.

II. LIMITATIONS OF THE PRECAMBRIAN FOSSIL RECORD

At the outset of this discussion it should be stressed that, in spite of the recent progress noted above, the Precambrian biota remains very incompletely known; inferences here drawn from the fragmentary data available should be regarded as being of a most tentative sort.

The deficiencies in the early fossil record are of varied sources. To some degree they reflect a traditional dogma of palaeontology that the 'Precambrian is unfossiliferous'; until recently, this view effectively limited inquiry into the palaeobiology of very ancient sediments. Certain of these gaps, however, may be inherent in the field: The Precambrian, encompassing the earlier seven-eighths of geologic time, presents problems of a rather different sort from those normally encountered in studies of the Phanerozoic 'overburden'.

At present, fewer than three dozen occurrences of cellularly preserved micro-organisms are known from the Precàmbrian, spanning a segment of biologic history more than four times as long as that encompassed by the entire Phanerozoic. The most important of these fossiliferous deposits, and the evidence of early biologic activity that they contain, are listed in Text-fig. 1. Unfortunately, relatively few of these sediments contain communities of micro-organisms preserved *in situ* on which inferences of evolutionary status and palaeoecology might be based most reliably. Furthermore, of the few such assemblages known, all are preserved in inorganically precipitated primary cherts; although these siliceous sediments have provided the geologic setting responsible for cellular preservation of delicate micro-organisms, such cherts reflect unusual ecologic conditions and almost certainly contain a restricted, rather atypical sample of the total biota.

A variety of age effects, essentially inherent in the Precambrian, also complicate the interpretation of the early record. As might be expected, there generally appears to be an inverse correlation between the age of a sedimentary unit studied and the fidelity of organic preservation observed; thus, as the record of biologic activity is traced back into the earliest Precambrian, the morphological fossil record becomes increasingly difficult to decipher. This trend is paralleled by the geochemical degradation of chemical fossils; many biogenic organic compounds detected in Tertiary sediments (e.g. proteins, optically active amino acids, unsaturated fatty acids, predominance of n-alkanes with an odd number of carbon atoms, etc.) are less frequent in Palaeozoic deposits and may be completely absent from sediments of greater geologic age. This loss of information is further compounded by the effects of biologic evolution, so that not only are the earliest organisms relatively poorly preserved, but they are also of limited diversity and of such simple morphology that their biological affinities are difficult to determine. As a result of these age effects, the oldest fossil-like microstructures now known (Engel, Nagy, Nagy, Engel, Kremp & Drew, 1968; Nagy & Nagy, 1969) are of uncertain biogenicity.

Younger Precambrian micro-organisms, although generally better preserved and morphologically more complex than those of the Early Precambrian, rather commonly lack modern and fossil morphological counterparts; the phylogenetic position of such forms is highly conjectural. As Glaessner (1968) has suggested, such organisms might represent phyletic side branches, only remotely related to members of well-known systematic categories. In addition, certain evolutionary transitions (e.g. the development of the eucaryotic cell and the origin of chemosynthetic bacteria) were apparently the result of intracellular changes in ultrastructure and biochemistry, not readily preservable in the fossil record and not originally evinced by changes in organismal morphology. Other transitions (e.g. from unicellular algae to primitive protozoa) may have been reflected initially in behavioural patterns, rather than in obvious differences of form. It is doubtful if the earliest appearance of such evolutionary innovations would be recognized in the geologic record unless chemical fossils were detected indicating the evolution of new, phylogenetically restricted, biosynthetic pathways. A systematic search for chemical fossils of this type (e.g. sterol derivatives perhaps indicating the presence of eucaryotic organisms; derivatives of polyunsaturated fatty acids suggesting the occurrence of oxygen-producing photosynthesizers) has yet to be made, however, and at present the newly derived stock might not be recognized until relatively advanced forms, morphologically comparable to their modern descendants, appeared in the record.

Finally, there are several restrictions imposed by the geologic record itself. In general, Precambrian sediments are of rather limited areal distribution and of moderate to high-grade metamorphism. Moreover, the oldest known rocks ($ca. 3\cdot 5 \times 10^9$ years) are comparable in age to the earliest known fossils (more than $3\cdot 1 \times 10^9$ years). A variety of evidence suggests that these early organisms were physiologically rather advanced; if so, their biochemical complexity would seem to imply the existence of a substantial period of prior evolutionary development. Direct evidence of the beginnings and earliest evolution of living systems may not be detectable unless very ancient sediments, perhaps $3\cdot 5$ to $4\cdot 25 \times 10^9$ years in age, are discovered.

III. CHEMICAL FOSSILS

As an outgrowth of early interest in the chemistry of coals and crude oils and the biologic and geologic processes producing these materials, and gaining general acceptance among palaeontologists with the studies of Abelson (1954, 1959) on amino acids in fossil shells and bones, organic geochemistry has come to play an increasingly significant role in palaeobiological investigations. Nowhere is this more apparent than in studies of Precambrian sediments from which a diverse suite of chemical fossils have been reported during the past decade (e.g. n-alkanes, isoprenoids, steranes, fatty acids, amino acids, porphyrins, sugars, and ratios of the stable isotopes of carbon and of sulphur).

The techniques used to isolate and characterize chemical fossils have been reviewed by Hoering (1967a), who also discussed Precambrian studies carried out before 1967; reviews by Calvin (1969) and Van Hoeven (1969) include certain of the more recent data. Text-fig. 1 summarizes the geologic distribution of the major categories of chemical fossils reported from the Precambrian. Before considering the implications of these distributions, however, the definition of such fossils and the problems inherent to their interpretation must first be examined.

(1) *Definitions*

In its broadest sense, the term 'fossil' may embrace *any* evidence of early life (cf. Barghoorn, Meinschein & Schopf, 1965). The distinction here posed, viz. between 'morphological' and 'chemical' fossils, is used to differentiate the morphological evidence of previously existent organisms (the palaeontologist's 'fossil') from the organic compounds and/or isotopic fractionation effects derived from such organisms. In a sense, this distinction may be somewhat superficial since some morphological fossils are predominantly or entirely composed of organic matter, and since chemical fossils exhibit specific, albeit molecular, morphology. Although other terms might therefore be proposed, this particular nomenclature seems to be in vogue and is gaining wide acceptance (particularly among organic geochemists). In any case, such a distinction is generally useful, for the techniques of detection and the problems encountered in interpretation are inherently different for these two categories of biological remnants.

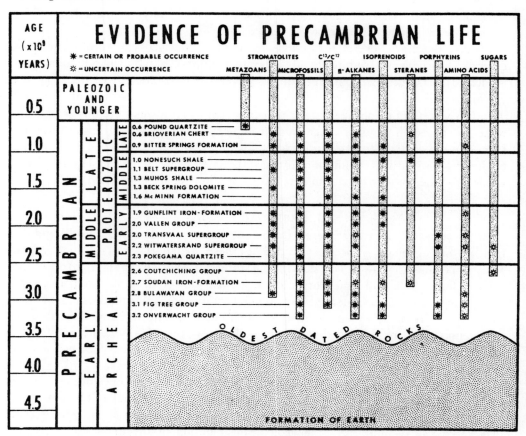

Text-fig. 1. Histogram showing distribution of organic geochemical and morphological evidence of biologic activity reported from Precambrian sediments.

(2) *General limitations*

Unlike analyses of Phanerozoic molluscs (Abelson, 1963; Hare & Mitterer, 1967) and vascular plants (Swain, Bratt & Kirkwood, 1968), chemical studies of Precambrian life have involved the analysis of extracts of whole-rocks, rather than of individual fossil organisms. Since sediments containing essentially monospecific assemblages (e.g. some diatomites and tasmanite coals) are unknown in the Precambrian, such analyses cannot be correlated with particular taxa and are therefore of relatively limited phylogenetic usefulness. Moreover, although it is generally possible to show that these extracted materials are indigenous to a sediment, rather than being the result of contamination introduced in the laboratory, it is substantially more difficult to establish that they are syngenetic with Precambrian sedimentation. This problem arises because the concentrations of organic matter extracted from early sediments are almost always of the order of a few parts per million or less (for exceptions, see Hoering, 1967b) and the deposits are generally permeable to some degree; a minute amount of secondarily emplaced material (e.g. carried by connate water) can greatly influence analytical results. In addition to demonstrating the indigenousness of such chemical fossils, therefore, other criteria (i.e. physical or chemical 'tests') are clearly needed to establish a Precambrian age. If it could be shown, for example, that the $^{13}C:{}^{12}C$ ratios of syngenetically deposited organic matter of differing geologic age vary in a systematic fashion, as some data seem to suggest (Smith, Schopf & Kaplan, 1970), or if the chemistry of organically preserved micro-organisms could be correlated with their micro-morphology (Schopf, 1968b, 1970), this problem might be resolved.

(3) Uncertain occurrences

In the absence of such tests, however, some indication of the probable age of extractable compounds can be obtained from a consideration of their geochemical stabilities. Recent studies seem to indicate that certain amino acids (Abelson, 1959) and sugars (Vallentyne, 1963; Van Hoeven, 1969) are relatively unstable in the geologic environment. To explain the unexpected occurrence of these compounds in extracts of ancient sediments, it has been suggested that they may have been protected from diagenetic alteration by being chemically bound to their encompassing matrix (e.g. Degens, Hunt, Reuter & Reed, 1964; Degens, 1965; Oberlies & Prashnowsky, 1968; Schopf, Kvenvolden & Barghoorn, 1968). Although the possible stabilizing effect of mineral substrates has not been investigated fully, preliminary studies suggest that such effects may be minimal (Abelson & Hare, 1968a, b; Smith et al. 1970).

Certainly, the available data provide insufficient grounds for concluding that some amino acids and sugars cannot survive from the Precambrian. Nevertheless, the syngenetic nature of amino acids (Harrington & Toens, 1963; Prashnowsky & Schidlowski, 1967; Schopf et al. 1968; Oberlies & Prashnowsky, 1968; Pflug, Meinel, Neumann & Meinel, 1969; Kvenvolden, Peterson & Pollock, 1969) and of sugars (Swain, Pakalns & Bratt, 1966; Swain, Rogers, Evans & Wolfe, 1967; Prashnowsky & Schidlowski, 1967; Oberlies & Prashnowsky, 1968; Van Hoeven, 1969) detected in extracts of Precambrian sediments has not been demonstrated; on stability considerations, it seems likely that these compounds are, in part, of secondary and relatively recent origin.

In addition to the geochemically less stable compounds discussed above, a few hydrocarbon occurrences, denoted by the open symbols in Text-fig. 1, are regarded as being of questionable nature. Insufficient data are available to evaluate the report of sterol-like compounds isolated from Late Precambrian Brioverian cherts by Roblot, Chaigneau & Giry (1966). Carbon isotopic data (Hoering, 1967a) suggest that soluble hydrocarbons extracted from sediments of the Middle Precambrian Transvaal Supergroup may not be syngenetic with deposition. Similar reasoning (Hoering, 1967a, b), and mineralogic evidence of an elevated thermal history (J. F. Machamer, cited by Cloud, Gruner & Hagen, 1965), suggest that the n-alkanes, isoprenoids, steranes and fatty acids reported from the Early Precambrian Soudan Iron-Formation (Meinschein,

1965; Burlingame, Haug, Belsky & Calvin, 1965; Oró, Nooner, Zlatkis & Wikström, 1966; Johns, Belsky, McCarthy, Burlingame, Haug, Schnoes, Richter & Calvin, 1966; Van Hoeven, Maxwell & Calvin, 1969; Calvin, 1969) are partially, or entirely, of recent origin.

(4) *Probable occurrences*

Apart from these uncertain occurrences, it is evident from Text-fig. 1 that several types of geochemically stable organic compounds are detectable in Precambrian sediments. Although it should be realized from the arguments presented above that the Precambrian age of these components has not been firmly established, much of the available evidence is consistent with this interpretation. For example: (i) the distributions of *n*-alkanes in these sediments, lacking or exhibiting only slight (Barghoorn *et al.* 1965; Van Hoeven *et al.* 1969) odd-carbon number preference, is suggestive of considerable geologic age; (ii) the carbon isotopic data (Hoering, 1967 *a*) do not seem indicative of secondary emplacement; (iii) palaeontologic studies demonstrate that syngenetically deposited organic matter is present in these sediments; and (iv) stability considerations (Abelson, 1959) indicate that these compounds (denoted by the solid symbols in Text-fig. 1) could reasonably be expected to survive in a geologic setting for several billion years.

On the other hand, it is evident that some portion of these geochemically stable materials might be of post-depositional origin. The difficulties in assessing the degree to which secondary emplacement has occurred are clearly illustrated by variations in analytical results reported from different laboratories in studies of sediments from the same locality. For example, analyses of fossiliferous black cherts from the Gunflint Iron-Formation (from the Schreiber Beach locality of Barghoorn & Tyler, 1965) reported by Smith *et al.* (1970) yielded only 0·029–0·039 p.p.m. of total alkanes (including surface contaminants), as compared with yields of about 5 p.p.m. reported by Oró, Nooner, Zlatkis, Wikström & Barghoorn (1965) and 10–100 p.p.m. reported by Van Hoeven *et al.* (1969); the distributions of *n*-alkanes observed were generally similar, but Oró *et al.* reported no evidence of odd-carbon number preference, Van Hoeven *et al.* detected a predominance of *n*-alkanes with an odd number of carbon atoms in the C_{23}–C_{29} range, and Smith *et al.* showed odd-carbon preference in the C_{27}–C_{33} range. Finally, Oró *et al.* (1965) reported that 'less than 1 per cent of the hydrocarbons recovered from the chert' could reasonably be attributed to surface contamination, as evidenced by extraction of untreated rocks with a benzene-methanol solvent solution; Smith *et al.* (1970), however, report values of 82 and 89% for two separate analyses of similarly extractable material. These wide variations in analytical results presumably reflect local differences in the quality of organic preservation and/or levels of *in situ* contamination, and they present major problems in interpreting the origin(s) of these components.

It is possible that the problems of interpretation illustrated by these analyses are practically, if not theoretically, insuperable and that the extractable components of Precambrian sediments—including those compounds of established geochemical stability—can provide no convincing evidence of early biochemical processes. Whether this conclusion will prove applicable generally remains to be demonstrated; at present, the data suggest that some portion (and perhaps the majority) of these stable compounds, particularly that extracted from the interior of rock specimens, is actually of Precambrian age. In the absence of more definitive criteria to indicate whether these components are syngenetic with Precambrian sedimentation, the only apparent recourse is to use the morphological fossil record as a basis for evaluating the chemical evidence. This solution, of course, is notably imperfect, for it relegates extractable chemical fossils to a subsidiary position, to be accepted only if they are consistent with previous interpretations based on morphological evidence. Nevertheless, such an approach provides a rationale for considering the totality of data available, rather than completely disregarding extractable organic materials as being indigenous, but possibly not syngenetic.

(5) Biogenicity

If it is assumed, therefore, that the occurrences regarded as 'certain or probable' in Text-fig. 1 have been correctly interpreted, four categories of chemical fossils (n-alkanes, isoprenoids, porphyrins and $^{13}C:^{12}C$ values) extend into the Early Precambrian. With the exception of the carbon isotopic values, representatives of each of these categories have been synthesized abiotically in experiments designed to simulate conditions that may have produced organic compounds on the primitive earth. In principle, comparable carbon isotopic ratios might also be abiotically produced, although this has yet to be demonstrated experimentally. The environments inducing these reactions (starting materials, energy sources, etc.) vary widely—some being more plausible in geologic terms than others—and, considering the arsenal of materials available to the synthetic chemist, it should not be surprising that compounds of these types can be produced by solely abiotic processes.

Nevertheless, the distribution of molecular species in the Precambrian sediments listed in Text-fig. 1 seems indicative of biogenicity; while not inconceivable, the possibility of an abiotic origin for these compounds seems relatively remote. It may be noted, for example, that normal alkanes predominate over all other possible isomers in each of these deposits, a feature typical of biological materials, and occur in marked excess over the thermodynamically more stable isoalkanes and cycloalkanes characteristic of many abiotic syntheses. A similar case can be made for the presumed biogenicity of phytane ($C_{20}H_{42}$), an isoprenoid hydrocarbon isolated from the eight sediments indicated. According to Fieser & Fieser (1961, p. 105), there are 366,319 possible structural isomers of this compound (i.e. structural arrangements of 20 carbon atoms and 42 hydrogen atoms, not including stereoisomers); only two or three of these isomers commonly occur as products of living systems (Calvin, 1969) and these appear to be the same isomers detectable in ancient sediments. If these chemical fossils had been synthesized by a random abiotic process, thousands of isomers, many of which are energetically similar, might be expected in these deposits. This specificity, coupled with the detailed similarities of distribution between Precambrian chemical fossils and compounds extracted from extant organisms and Phanerozoic sediments, has persuaded most investigators that an abiotic origin for these compounds is most unlikely (Hoering, 1967a; Calvin, 1969).

(6) *Chemical evidence of photosynthesis*

Assuming that the relatively stable. chemical fossils reported from Precambrian sediments are of biological origin and are of the same age as the sediments from which they have been extracted, what inferences may be drawn regarding the biochemistry of the primitive biota? Based on analogy with extant organisms, and on the apparent universality of biochemical processes in living systems, the most obvious inference is that these compounds, or their geochemical precursors, were produced by a series of enzymically catalysed intracellular reactions. The presence of *n*-alkanes and iso-prenoids (Oró & Nooner, 1967; Meinschein, 1967; Hoering, 1967*a*; MacLeod, 1968; Calvin, 1969) and of porphyrins (Kvenvolden, Hodgson, Peterson & Pollock, 1968; Kvenvolden & Hodgson, 1969) in sediments of the Early Precambrian Swaziland Supergroup (Fig Tree and Onverwacht Groups) would appear to imply, therefore, that organisms exhibiting nucleic acid-directed enzyme synthesis and complex bio-synthetic pathways were extant more than 3000 million years ago.

The probable relationship of these chemical fossils to the physiology of early organisms is more conjectural; but if it is assumed that they played functional roles comparable to those of their modern analogues, biological sources can be suggested that are based on the distribution of these analogues in the extant biota. Thus, since compounds containing isoprene derivatives (e.g. chlorophylls, carotenoids, phyco-bilins, steroids, Vitamin A) are limited in distribution to photosynthetic organisms (including photosynthetic bacteria) and their evolutionary descendants, the presence of isoprenoid hydrocarbons in the Early Precambrian Swaziland Supergroup would seem to indicate that photosynthetic autotrophy had evolved by this point in earth history. The occurrence of porphyrins in these sediments is consistent with this interpretation since porphyrin-containing respiratory pigments (e.g. chlorophylls, cytochromes*) are present in all photosynthetic organisms; moreover, bacterial and algal chlorophylls (as well as those of higher plants) contain a phytol side chain, thought to be the geochemical precursor of pristane and phytane, the predominant isoprenoids extracted from ancient sediments (Oró *et al.* 1965; Calvin, 1969).

The carbon isotopic composition of organic matter in the Swaziland sediments seems similarly suggestive of photosynthetic activity. Studies by Park & Epstein (1960, 1961) have shown that the two stable carbon isotopes, ^{12}C and ^{13}C, undergo kinetic fractionation during the process of green-plant photosynthesis; a series of steps, the most important of which involves the enzyme-mediated carboxylation of ribulose 1-5-diphosphate to yield 3-phosphoglyceric acid, produces organic matter enriched in the lighter isotope, ^{12}C, relative to its concentration in the atmospheric carbon dioxide reservoir. Carbon isotopic ratios have been measured in a wide variety of naturally occurring materials produced by both biologic and inorganic processes, and of both modern and geologic age (for a summary of these values see Degens,

---

* It should be noted that although *Desulfovibrio*, an anaerobic sulphate-reducer, contains cyto-chromes, it seems an unlikely source for Early Precambrian porphyrins since $^{34}S:^{32}S$ measurements (Smith *et al.* 1970) suggest that sulphate-reducing bacteria may not have evolved until much later in geologic time. The cobalt-containing tetrapyrroles (vitamin B_{12}) of anaerobic methane bacteria repre-sent a more plausible source for ancient porphyrins, but such bacteria apparently lack isoprenoids.

1965, p. 282). The $^{13}C:^{12}C$ ratios of apparently syngenetic, non-extractable organic matter in sediments of the Fig Tree Group (Hoering, 1967a) and the Onverwacht Group (D. Oehler, personal communication, 1969) fall well within the range of values characteristic of organic materials of known photosynthetic origin or derivation (e.g. petroleum, coal) and differ markedly from values typical of inorganic (carbonate) carbon compounds. These ratios presumably reflect the occurrence of carbon isotopic fractionation; as such, they constitute suggestive evidence of algal, or possibly bacterial, photosynthesis (measurements of the amount of isotopic fractionation produced by photosynthetic bacteria are apparently unavailable).

In summary, the reported occurrence of isoprenoids and porphyrins in sediments of the Swaziland Supergroup, and the carbon isotopic ratios of organic matter deposited apparently syngenetically with these sediments, are consistent with, and seem suggestive of, the presence of autotrophic photosynthesizers of bacterial and/or algal affinities. For reasons detailed above, this interpretation should be regarded as tentative, to be accepted only if it is consistent with evidence from the morphological fossil record.

IV. THE EARLY PRECAMBRIAN RECORD

It is perhaps a truism that the most interesting and challenging scientific problems involve questions of broadly encompassing import for which solutions have yet to be found, but which have the appearance of being soluble, at least in principle. The origin of biological systems is one such problem. A variety of approaches have been applied to this question, the most promising of which has involved laboratory syntheses of organic 'biologic' compounds in non-biologic systems (for a recent review of these experiments see Ponnamperuma & Gabel, 1968). To provide a putative solution to this problem, however, such experiments must do more than 'merely' spontaneously assemble a group of abiotically produced organic molecules into a complex, self-replicating living system—a feat that may prove attainable within the next few decades—for it is crucial that this end-point be reached through a series of steps that are plausible in terms of the Early Precambrian environment (and even then, only one of the potential pathways to biological complexity will have been traced, with no firm guarantee that living systems originated in precisely this manner).

The plausibility of such experiments (concentrations and types of starting materials, energy sources, etc.) can be determined only by reference to the geologic record, and the limitations it sets on the nature of the primitive environment. In this regard, it should be recognized that Early Precambrian sediments provide no evidence for the existence of the methane-ammonia atmosphere commonly postulated for the primitive earth (Abelson, 1966). On the contrary, certain of the oldest sedimentary sequences now known (e.g. the Swaziland Supergroup with an apparent age of about $3 \cdot 2 \times 10^9$ years) contain inorganically precipitated carbonate units (Ramsay, 1963) indicating that carbon dioxide, and not methane, may have been the dominant form of atmospheric carbon at this point in geologic time. Nevertheless, it is possible, and perhaps likely, that a highly reduced atmosphere was present very early in earth history, before the deposition of the oldest known rocks, some $3 \cdot 5 \times 10^9$ years ago (Donn, Donn &

Valentine, 1965). Thus, the geologic evidence suggests that if living systems originated in the presence of a methane-ammonia atmosphere, an assumption basic to most experimental approaches to the problem (Ponnamperuma & Gabel, 1968), this signal event must have occurred during the earliest third of earth history.

* * * * * * *

Editor's Note: For the sake of brevity the remainder of this paper, which considers the early through late Precambrian microfossil record, has not been reprinted.

IX. REFERENCES

ABELSON, P. H. (1954). Organic constituents of fossils. *Yb. Carnegie Instn Wash.* **53**, 97–101.

ABELSON, P. H. (1959). Geochemistry of organic substances. *Researches in geochemistry* (ed. P. H. Abelson), pp. 79–103. New York: Wiley.

ABELSON, P. H. (1963). Geochemistry of amino acids. *Organic geochemistry* (ed. I. A. Breger), pp. 431–55. New York: Macmillan.

ABELSON, P. H. (1966). Chemical events on the primitive earth. *Proc. natn. Acad. Sci. U.S.* A **55**, 1365–72.

ABELSON, P. H. & HARE, P. E. (1968*a*). Recent origin of amino acids in the Gunflint chert. *Spec. Pap. geol. Soc. Am.* **121**, 2 (Abstr.).

ABELSON, P. H. & HARE, P. E. (1968*b*). Recent amino acids in the Gunflint chert. *Yb. Carnegie Instn Wash.* **67**, 208–10.

ALLSOPP, H. L., ULRYCH, T. J. & NICOLAYSEN, L. O. (1968). Dating some significant events in the history of the Swaziland System by the Rb-Sr isochron method. *Can. J. Earth Sci.* **5**, 605–19.

ASHLEY, B. E. (1937). Fossil algae from the Kundelungu Series of northern Rhodesia. *J. Geol.* **45**, 332–5.

BANKS, H. P. (1968). The early history of land plants. *Evolution and environment* (ed. E. T. Drake), pp. 73–107. New Haven: Yale University Press.

BARGHOORN, E. S., MEINSCHEIN, W. G. & SCHOPF, J. W. (1965). Paleobiology of a Precambrian shale. *Science, N.Y.* **148**, 461–72.

BARGHOORN, E. S. & SCHOPF, J. W. (1965). Microorganisms from the Late Precambrian of central Australia. *Science, N.Y.* **150**, 337–9.

BARGHOORN, E. S. & SCHOPF, J. W. (1966). Microorganisms three billion years old from the Pre-cambrian of South Africa. *Science, N.Y.* **152**, 758–63.

BARGHOORN, E. S. & TYLER, S. A. (1965). Microorganisms from the Gunflint chert. *Science, N.Y.* **147**, 563–77.

BERKNER, L. V. & MARSHALL, L. C. (1965). History of major atmospheric components. *Proc. natn. Acad. Sci. U.S.A.* **53**, 1215–26.

BONDENSEN, E., PEDERSEN, K. R. & JØRGENSEN, O. (1967). Precambrian organisms and the isotopic composition of organic remains in the Ketilidian of south-west Greenland. *Meddr. Grønland*, Bd. 164, nr. 4, 41 pp.

BRINKMANN, R. T. (1969). Dissociation of water vapor and evolution of oxygen in the terrestrial atmo-sphere. *J. geophys. Res.* **74**, 5355–68.

BURLINGAME, A. L., HAUG, P., BELSKY, T. & CALVIN, M. (1965). Occurrence of biogenic steranes and pentacyclic triterpanes in an Eocene shale (52 million years) and in an Early Precambrian shale (2·7 billion years): a preliminary report. *Proc. natn. Acad. Sci. U.S.A.* **54**, 1406–12.

CAHEN, L., JAMOTTE, A. & MORTELMANS, G. (1946). Sur l'existence de microfossiles dans l'horizon des cherts du Kundelungu supérieur. *Annls Soc. géol. Belg.* **70**, B 55–B 65.

CALVIN, M. (1969). *Chemical evolution*, 278 pp. New York: Oxford University Press.

CLOUD, P. E., JR. (1965). Significance of the Gunflint (Precambrian) microflora. *Science, N.Y.* **148**, 27–35.

CLOUD, P. E., JR. (1968). Pre-Metazoan evolution and the origins of the Metazoa. *Evolution and environ-ment* (ed. E. T. Drake), pp. 1–72. New Haven: Yale University Press.

CLOUD, P. E., JR. & HAGEN, H. (1965). Electron microscopy of the Gunflint microflora: Preliminary results. *Proc. natn. Acad. Sci. U.S.A.* **54**, 1–8.

CLOUD, P. E., JR., GRUNER, J. W. & HAGEN, H. (1965). Carbonaceous rocks of the Soudan Iron Forma-tion (Early Precambrian). *Science, N.Y.* **148**, 1713–16.

CLOUD, P. E., JR. & LICARI, G. R. (1968). Microbiotas of the banded iron formations. *Proc. natn. Acad. Sci. U.S.A.* **61**, 779–86.

CLOUD, P. E., JR., LICARI, G. R., WRIGHT, L. A. & TROXEL, B. W. (1969). Proterozoic eucaryotes from eastern California. *Proc. natn. Acad. Sci. U.S.A.* **62**, 623–30.

CROFT, W. N. & GEORGE, E. A. (1959). Blue-green algae from the Middle Devonian of Rhynie, Aber-deenshire. *Bull. Br. Mus. nat. Hist. Geology* **3**, 339–53.

DEGENS, E. T. (1965). *Geochemistry of sediments*, 342 pp. Englewood Cliffs, N.J.: Prentice-Hall.

DEGENS, E. T., HUNT, J. M., REUTER, J. H. & REED, W. E. (1964). Data on the distribution of amino acids and oxygen isotopes in petroleum brine waters of various geologic ages. *Sedimentology* **3**, 199–225.

DONN, W. L., DONN, B. D. & VALENTINE, W. G. (1965). On the early history of the earth. *Bull. geol. Soc. Am.* **76**, 287–306.

ENGEL, A. E. J., NAGY, B., NAGY, L. A., ENGEL, C. G., KREMP, G. O. W. & DREW, C. M. (1968). Alga-like forms in the Onverwacht Series, South Africa: Oldest recognized lifelike forms on earth. *Science, N.Y.* **161**, 1005–8.

FAURE, G. & KOVACH, J. (1969). The age of the Gunflint Iron Formation of the Animikie Series in Ontario, Canada. *Bull. geol. Soc. Am.* **80**, 1725–36.

FIESER, L. F. & FIESER, M. (1961). *Advanced organic chemistry*, 1158 pp. New York: Reinhold.

FRITSCH, F. E. (1945). Studies in the comparative morphology of the algae. IV. Algae and archegoniate plants. *Ann. Bot.* **9**, no. 33, 1–29.

FRITSCH, F. E. (1949). The lines of algal advance. *Biol. Rev.* **24**, 94–124.

GLAESSNER, M. F. (1962). Pre-Cambrian fossils. *Biol. Rev.* **37**, 467–94.

GLAESSNER, M. F. (1966). Precambrian paleontology. *Earth-Sci. Rev.* **1**, 29–50.

GLAESSNER, M. F. (1968). Biological events and the Precambrian time scale. *Can. J. Earth Sci.* **5**, 585–90.

GLAESSNER, M. F. & WADE, M. (1966). The Late Precambrian fossils from Ediacara, South Australia. *Palaeontology* **9**, 599–628.

GRUNER, J. W. (1922). The origin of sedimentary iron formations. The Biwabik Formation of the Mesabi Range. *Econ. Geol.* **17**, 407–60.

GRUNER, J. W. (1924a). Algae believed to be Archaean. *J. Geol.* **31**, 146–8.

GRUNER, J. W. (1924b). Contributions to the geology of the Mesabi Range. *Bull. geol. Surv.* no. 19, 71 pp.

GRUNER, J. W. (1925). Discovery of life in the Archaean. *J. Geol.* **33**, 151–2.

GUTSTADT, A. M. & SCHOPF, J. W. (1969). Possible algal microfossils from the Late Pre-Cambrian of California. *Nature, Lond.* **223**, 165–7.

HARE, P. E. & MITTERER, R. M. (1967) Nonprotein amino acids in fossil shells. *Yb Carnegie Instn Wash.* 65, 362–4.

HARRINGTON, J. W. & TOENS, P. D. (1963). Natural occurrence of amino-acids in dolomitic limestones containing algal growths. *Nature, Lond.* 200, 947–8.

HAWLEY, J. E. (1926). An evaluation of the evidence of life in the Archaean. *J. Geol.* 34, 441–61.

HOERING, T. C. (1967a). The organic geochemistry of Precambrian rocks. *Researches in geochemistry*, vol. II (ed. P. H. Abelson), pp. 87–111. New York: Wiley.

HOERING, T. C. (1967b). Criteria for suitable rocks in Precambrian organic geochemistry. *Yb. Carnegie Instn Wash.* 65, 365–72.

HOFMANN, H. F. & JACKSON, G. D. (1969). Precambrian (Aphebian) microfossils from Belcher Islands, Hudson Bay. *Can. J. Earth Sci.* 6, 1137–44.

HOLMES, A. (1965). *Principles of physical geology*, 1288 pp. New York: Ronald.

HURLEY, P. M., FAIRBAIRN, H. W., PINSON, W. H., JR & HOWE, J. (1962). Unmetamorphosed minerals in the Gunflint Formation used to test the age of the Animikie. *J. Geol.* 70, 489–92.

JOHNS, R. B., BELSKY, T., McCARTHY, E. D., BURLINGAME, A. L., HAUG, P., SCHNOES, H. K., RICHTER, W. & CALVIN, M. (1966). The organic geochemistry of ancient sediments. Part II. *Geochim. cosmochim. Acta* 30, 1191–222.

KUZNETSOV, S. I., IVANOV, M. V. & LYALIKOVA, N. N. (1963). *Introduction to geological microbiology*, translated from Vvedeniye v Geologicheskuyu Mikrobiologiyu, Acad. Sci. U.S.S.R. Press, Moscow, 1962 (C. H. Oppenheimer, editor of English edition, 252 pp. New York: McGraw-Hill).

KVENVOLDEN, K. A., HODGSON, G. W., PETERSON, E. & POLLOCK, G. E. (1968). Organic geochemistry of the Swaziland System, South Africa. *Program, Ann. Meeting, Geol. Soc. Am.* Mexico City, pp. 167–8 (Abstr.).

KVENVOLDEN, K. A. & HODGSON, G. W. (1969). Evidence for porphyrins in Early Precambrian Swaziland System sediments. *Geochim. cosmochim. Acta* 33, 1195–202.

KVENVOLDEN, K. A., PETERSON, E. & POLLOCK, G. E. (1969). Optical configuration of amino-acids in Pre-Cambrian Fig Tree chert. *Nature, Lond.* 221, 141–3.

LABERGE, G. L. (1967). Microfossils and Precambian iron-formations. *Bull. geol. Soc. Am.* 78, 331–42.

LICARI, G. R. & CLOUD, P. E., JR. (1968). Reproductive structures and taxonomic affinities of some nannofossils from the Gunflint Iron Formation. *Proc. natn. Acad. Sci. U.S.A.* 59, 1053–60.

LICARI, G. R., CLOUD, P. E., JR. & SMITH, W. D. (1969). A new chroococcacean alga from the Proterozoic of Queensland. *Proc. natn. Acad. Sci. U.S.A.* 62, 56–62.

MACGREGOR, A. M. (1940). A pre-Cambrian algal limestone in southern Rhodesia. *Trans. geol. Soc. S. Afr.* 43, 9–16.

MACLEOD, W. D., JR. (1968). Combined gas chromatography-mass spectrometry of complex hydrocarbon trace residues in sediments. *J. Gas Chromatogr.* 6, 591–4.

MARGULIS, L. (1968). Evolutionary criteria in thallophytes: A radical alternative. *Science, N.Y.* 161, 1020–2.

MEINSCHEIN, W. G. (1965). Soudan formation: organic extracts of Early Precambrian rocks. *Science, N.Y.* 150, 601–5.

MEINSCHEIN, W. G. (1967). Paleobiochemistry. 1967 *McGraw-Hill Yearbook of Science and Technology*, pp. 283–5.

MEINSCHEIN, W. G., BARGHOORN, E. S. & SCHOPF, J. W. (1964). Biological remnants in a Precambrian sediment. *Science, N.Y.* 145, 262–3.

MOORE, E. S. (1918). The iron formation on Belcher Islands, Hudson Bay, with special reference to its origin and its associated algal limestones. *J. Geol.* 26, 412–38.

NAGY, B. & NAGY, L. A. (1969). Early Pre-Cambrian Onverwacht microstructures: Possibly the oldest fossils on earth? *Nature, Lond.* 223, 1226–9.

NAGY, B. & UREY, H. C. (1969). Organic geochemical investigations in relation to the analyses of returned lunar rock samples. *Life Sci. Space Res.* 7, 31–46.

NICOLAYSEN, L. O. (1962). Stratigraphic interpretation of age measurements in southern Africa. *Petrologic studies: a volume in honor of A. F. Buddington* (ed. A. E. J. Engel et al.), pp. 569–98. New York: Geol. Soc. Am.

OBERLIES, F. & PRASHNOWSKY, A. A. (1968). Biogeochemische und elektronenmikroskopische Untersuchung präkambrischer Gesteine. *Naturwissenschaften* 55, 25–8.

ORÓ, J., NOONER, D. W., ZLATKIS, A., WIKSTRÖM, S. A. & BARGHOORN, E. S. (1965). Hydrocarbons of biological origin in sediments about two billion years old. *Science, N.Y.* 148, 77–9.

ORÓ, J., NOONER, D. W., ZLATKIS, A. & WIKSTRÖM, S. A. (1966). Paraffinic hydrocarbons in the Orgueil, Murray, Mokoia and other meteorites. *Life Sci. Space Res.* 4, 63–100.

ORÓ, J. & NOONER, D. W. (1967). Aliphatic hydrocarbons in Pre-Cambrian rocks. *Nature, Lond.* 213, 1082–5.

PARK, R. & EPSTEIN, S. (1960). Carbon isotopic fractionation during photosynthesis. *Geochim. cosmochim. Acta* **21**, 110–26.

PARK, R. & EPSTEIN, S. (1961). Metabolic fractionation of C^{13} and C^{12} in plants. *Pl. Physiol., Lancaster* **36**, no. 2, 133–8.

PFLUG, H. D. (1964). Niedere Algen und ähnliche Kleinformen aus dem Algonkium des Belt Serie. *Ber. oberhess. Ges. Nat. u. Heilk.* Bd. 33, Heft 4, 403–11.

PFLUG, H. D. (1965). Organische Reste aus der Belt Serie (Algonkium) von Nordamerika. *Paläont. Z.* **39**, 10–25.

PFLUG, H. D. (1966a). Structured organic remains from the Fig Tree Series of the Barberton Mountain Land. *Econ. Geol. Res. Unit, Univ. Witwatersrand, Johannesburg, Inform. Circ.* **28**, 1–14.

PFLUG, H. D. (1966b). Einige Reste niederer Pflanzen aus dem Algonkium. *Palaeontographica* (Abt. B), Bd. 117, 59–74.

PFLUG, H. D. (1967). Structured organic remains from the Fig Tree Series (Precambrian) of the Barberton Mountain Land (South Africa). *Rev. Palaeobotan. Palynol.* **5**, 9–29.

PFLUG, H. D., MEINEL, W., NEUMANN, K. H. & MEINEL, M. (1969). Entwicklungstendenzen des frühen Lebens auf der Erde. *Naturwissenschaften* **56**, 10–14.

PONNAMPERUMA, C. A. & GABEL, N. W. (1968). Current status of chemical studies on the origin of life. *Space Life Sci.* **1**, 64–96.

PRASHNOWSKY, A. A. & SCHIDLOWSKI, M. (1967). Investigation of pre-Cambrian thucolite. *Nature, Lond.* **216**, 560–3.

RAMSAY, J. G. (1963). Structural investigations in the Barberton Mountain Land, eastern Transvaal. *Trans. geol. Soc. S. Afr.* **66**, 353–401.

ROBLOT, M. M. (1963). Découverte de sporomorphes dans des sédiments antérieurs à 550 M.A. (Briovérien). *C. r. hebd. Séanc. Acad. Sci., Paris* **256**, 1557–9.

ROBLOT, M. M. (1964). Sporomorphes du Précambrien Normand. *Revue Micropaléont.* **7**, 153–6.

ROBLOT, M. M., CHAIGNEAU, M. & GIRY, L. (1966). Étude au spectromètre de masse d'extraits d'un phtanite Précambrien. *C. r. hebd. Séanc. Acad. Sci., Paris* **262**, 544–7.

SCHIDLOWSKI, M. (1966). Zellular strukturierte Elemente aus dem Prakambrium des Witwatersrand-Systems (Sudafriça). *Z. dt. geol. Ges.* Bd. **115**, 783–6.

SCHOPF, J. W. (1967). Antiquity and evolution of Precambrian life. 1967 *McGraw-Hill Yearbook of Science and Technology*, pp. 46–55.

SCHOPF, J. W. (1968a). Microflora of the Bitter Springs Formation, Late Precambrian, central Australia. *J. Paleont.* **42**, 651–88.

SCHOPF, J. W. (1968b). Ultrathin section electron microscopy of organically preserved Precambrian microorganisms. *Program, Ann. Meeting, Geol. Soc. Am.*, Mexico City, pp. 267–8 (Abstr.).

SCHOPF, J. W. (1969a). Precambrian microorganisms and evolutionary events prior to the origin of vascular plants. *Program, XI Int. Bot. Congr.*, Seattle, pp. 192 (Abstr.).

SCHOPF, J. W. (1969b). Organically preserved Precambrian microorganisms. *J. Paleont.* **43**, 898 (Abstr.).

SCHOPF, J. W. (1969c). Recent advances in Precambrian paleobiology. *Grana Palynologica* (in the Press).

SCHOPF, J. W. (1970). Electron microscopy of organically preserved Precambrian microorganisms. *J. Paleont.* **44**, 1–6.

SCHOPF, J. W., BARGHOORN, E. S., MASER, M. D. & GORDON, R. O. (1965). Electron microscopy of fossil bacteria two billion years old. *Science, N.Y.* **149**, 1365–7.

SCHOPF, J. W. & BARGHOORN, E. S. (1967). Alga-like fossils from the Early Precambrian of South Africa. *Science, N.Y.* **156**, 508–12.

SCHOPF, J. W. & BARGHOORN, E. S. (1969). Microorganisms from the Late Precambrian of South Australia. *J. Paleont.* **43**, 111–18.

SCHOPF, J. W., KVENVOLDEN, K. A. & BARGHOORN, E. S. (1968). Amino acids in Precambrian sediments: An assay. *Proc. natn. Acad. Sci. U.S.A.* **59**, 639–46.

SIEGEL, S. M. & SIEGEL, B. Z. (1968). A living organism morphologically comparable to the Precambrian genus Kakabekia. *Am. J. Bot.* **55**, 684–7.

SMITH, J. W., SCHOPF, J. W. & KAPLAN, I. R. (1970). Extractable organic matter in Precambrian cherts. *Geochim. cosmochim. Acta* (in the Press).

SWAIN, F. M., PAKALNS, G. V. & BRATT, J. G. (1966). Possible taxonomic interpretation of some Paleozoic and Precambrian carbohydrate residues. *Program, Third Int. Meeting on Organic Geochem.*, London.

SWAIN, F. M., ROGERS, M. A., EVANS, R. D. & WOLFE, R. W. (1967). Distribution of carbohydrate residues in some fossil specimens and associated sedimentary matrix and other geologic samples. *J. sedim. Petrol.* **37**, 12–24.

SWAIN, F. M., BRATT, J. G. & KIRKWOOD, S. (1968). Possible biochemical evolution of carbohydrates of some Paleozoic plants. *J. Paleont.* **42**, 1078–82.

TIMELL, T. E. (1965). Wood and bark polysaccharides. In *Cellular ultrastructure of woody plants* (ed. W. A. Coté, Jr.), pp. 127–56. Syracuse, N.Y. Syracuse Univ. Press.

TYLER, S. A. & BARGHOORN, E. S. (1954). Occurrence of structurally preserved plants in pre-Cambrian rocks of the Canadian Shield. *Science, N. Y.* **119**, 606–8.

TYLER, S. A. & BARGHOORN, E. S. (1963). Ambient pyrite grains in Precambrian cherts. *Am. J. Sci.* **261**, 424–32.

TYNNI, R. & SIIVOLA, J. (1966). On the Precambrian microfossil flora in the siltstone of Muhos, Finland. *C. r. Soc. geol. Finlande* **38**, 127–33.

ULRYCH, T. J., BURGER, A. & NICOLAYSEN, L. O. (1967). Least radiogenic terrestrial leads. *Earth Planetary Sci. Letters* **2**, 179–84.

VALLENTYNE, J. R. (1963). Geochemistry of carbohydrates. *Organic geochemistry* (ed. I. A. Breger), pp. 456–502. New York: Macmillan.

VAN HOEVEN, W. (1969). *Organic geochemistry*. Ph.D. thesis, University of California, Berkeley (January, 1969).

VAN HOEVEN, W., MAXWELL, J. R. & CALVIN, M. (1969). Fatty acids and hydrocarbons as evidence of life processes in ancient sediments and crude oils. *Geochim. cosmochim. Acta* **33**, 877–81.

WALCOTT, C. D. (1883). Pre-Carboniferous strata in the Grand Canyon of the Colorado, Arizona. *Am. J. Sci.* **26**, 437–42.

WALCOTT, C. D. (1899). Pre-Cambrian fossiliferous formations. *Bull. geol. Soc. Am.* **10**, 199–244.

WALCOTT, C. D. (1912). Notes on fossils from limestones of Steeprock Lake, Ontario. *Mem. geol. Surv. Brch Can.* **28**, 16–23.

WALCOTT, C. D. (1914). Pre-Cambrian Algonkian algal flora. *Smithson. misc. Collns* **64**, no. 2, 77–156.

* * * * * * *

Reprinted from *24th International Geological Congress, Sec. 1*, 31–41 (1972)

Organic Geochemistry of Early Precambrian Sediments

KEITH A. KVENVOLDEN,
U.S.A.

ABSTRACT

Portions of the Swaziland Sequence and Bulawayan System rocks in southern Africa are probably time-equivalent with ages exceeding three billion years. These could be the oldest relatively unmetamorphosed sedimentary rocks on the surface of the earth and have particular significance to the geologic record of the origin and evolution of life.

Some of the sediments have been examined for morphological and molecular fossils. Searches for morphological fossils have found evidence of simple microorganisms, including evolutionary forms as complex as filamentous blue-green algae. Organic geochemical studies have shown that a variety of organic compounds are present at very low concentrations. Alkanes (including isoprenoid hydrocarbons), fatty acids, amino acids and porphyrins have been reported. If these compounds are molecular fossils as old as the rocks, then the distribution of these compounds is generally consistent with what might be expected to come from microorganisms as precursors, thus supporting the morphological fossil record. This interpretation requires that the compounds, once trapped in these rocks, have remained essentially unchanged through time. However, the molecular evidence is also generally consistent with the proposition that some of the compounds represent materials that have entered the rocks after lithification. This complication increases the difficulty of interpreting the organic geochemical record with regard to the biochemistry of organisms that lived more than three billion years ago.

INTRODUCTION

MANY of the principles and practices of organic geochemistry during the last decade have been turned toward trying to unravel the chemical evidence in rocks, a task that might lead to an increased understanding of the origin and evolution of life. Finding the oldest rocks containing evidence of life or pre-life processes resulting from chemical evolution became one object of the work. The search has now reached beyond three billion years to the Swaziland Sequence of South Africa and the Bulawayan System of Rhodesia. In these thick sections of volcanic and sedimentary rocks have been found morphological evidence for life as well as organic compounds the significance of which can be debated. These rocks are particularly interesting because they contain the oldest relatively unmetamorphosed sedimentary sequences known and may hold the oldest record of life on earth.

Authors' addresses are given at the back of this book.

GEOLOGY

Swaziland Sequence

The Swaziland Sequence is a thick prism of rocks found in the Eastern Transvaal of South Africa along the border of Swaziland. The Sequence, expressed geographically as the Barberton Mountain Land, contains volcanic and sedimentary rocks with a total thickness which appears to exceed 20 km and is subdivided into three groups (Anhaeusser *et al.*, 1967). The lowest, the Onverwacht Group, is composed of about 15 km of basic and acidic lavas, sedimentary carbonate and chert. The overlying Fig Tree Group is predominantly a sedimentary sequence of about 2 km of graywackes, shale and chert. Moodies Group uncomformably overlies the Fig Tree Group and is made up of about 3.5 km of jasper, quartzites and shales. The Onverwacht Group has recently been subdivided into six formations (listed from oldest to youngest): Sandspruit, Theespruit, Komati, Hooggenoeg, Kromberg and Swartkoppie (Viljoen and Viljoen, 1969); and the Fig Tree Group has been subdivided into three formations (listed in ascending order): Sheba, Bellvue Road and Schoongezicht (Condie *et al.*, 1970). As a result of these subdivisions, some rocks formerly assigned to the Fig Tree Group are now included in the Swartkoppie Formation. The Swaziland Sequence itself is intruded and enveloped by granites.

Bulawayan System

The Bulawayan System is the middle of three Early Precambrian Systems of Rhodesia (Haughton, 1969). It rests unconformably on the Sebakwian System, composed of granulites, banded ironstones, graywackes and lavas. The Bulawayan System constitutes the bulk of the "gold belts" and is composed of as much as 12 km of andesitic lavas, with interbedded sediments of banded ironstones, quartzites, shales and limestones. Predominantly arenaceous sediments, about 3 km thick, of the Shamvian System overlie the Bulawayan rocks. Like the Swaziland Sequence, all three of these systems have been intruded by granites.

AGE AND CORRELATION

Isotopic age determinations have shown that both the Swaziland Sequence and Bulawayan System are probably older than three billion years. Measurements on rocks associated with the Swaziland Sequence indicate that the Sequence is greater than 3.2×10^9 years old (Nicolaysen, 1962). A lava in the Onverwacht Group has recently been dated at 3.36×10^9 years (Van Niekerk and Burger, 1969). Vail and Dodson (1969) have summarized the geochronology of Rhodesia and conclude that the Bulawayan System is 3.0 to 3.3×10^9 years old. Haughton (1969) has discussed the possible contemporaneity of the Swaziland Sequence of South Africa and the three Early Precambrian Systems of Rhodesia by noting the broad similarities between general lithologies in the two regions.

MORPHOLOGICAL FOSSILS

Both the Swaziland Sequence and Bulawayan System contain morphological evidence for very ancient living systems. Samples from the Onverwacht Group (Kromberg Formation?) contain a morphologically variable population of generally cup-shaped microstructures (Engel *et al.*, 1968). The biogenic nature of these microstructures still remains in doubt, however (Nagy and Nagy, 1969). Rod-shaped bacterium-like microfossils and spherical alga-like microfossils have been reported in the Fig Tree Group (actually in the Swartkoppie Formation of the Onverwacht Group) by Barghoorn and Schopf (1966) and by Schopf and Barghoorn (1967). Pflug (1966) found globular and filamentous objects in cherts

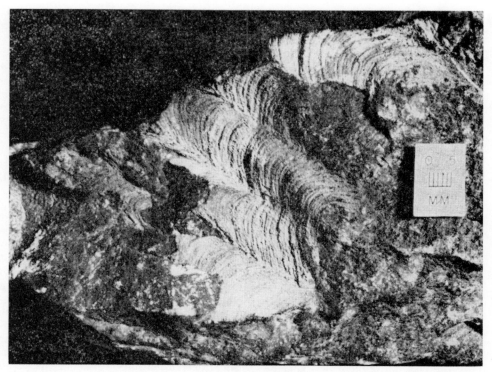

FIGURE 1 — Stromatolitic limestone from the Bulawayan System, Huntsman Limestone Quarries, Rhodesia. Well-defined cresentic laminations resulting from the weathering of stacked hemispheroids composed of laminations of organic material and calcium carbonate.

and shales of the Fig Tree Group (Sheba Formation?). Limestones containing stromatolitic structures were first described by Macgregor (1940) in the Bulawayan System. Oberlies and Prashnowsky (1968) noted spinose organic spheroids associated with the limestones. Recently the author visited Macgregor's locality and found some unusually well-preserved specimens of stromatolitic limestone (Fig. 1). The results of a study of these specimens fully substantiated an algal origin for these stromatolites (Schopf *et al.*, 1971). Apparently autotropic photosynthetic microorganisms were present on earth during the Early Precambrian more than 3×10^9 years ago. The microorganisms of the Swaziland Sequence may have included both heterotrophs and bacterial autotrophs, but not necessarily algal autotrophs. The stromatolites from the Bulawayan System indicate that algal photosynthesis was also an active process more than 3×10^9 years ago.

ORGANIC GEOCHEMISTRY

The search for molecular evidence of early life has paralleled the search for morphological fossils in very ancient rocks. Although organic compounds have been found in Early Precambrian rocks the significance of many of these compounds is uncertain. About 1965, the first warnings were being sounded about the interpretations that could be placed on organic compounds found in small concentrations in very ancient sediments (Hoering, 1965, 1967a, 1967b). A nagging question concerning whether the compounds found were really indigenous to the rocks

24th IGC, 1972 — SECTION 1   33

kept occurring. Do these compounds come from the rocks or do they represent some sort of contamination? Studies were made of the surface and internal portions of various rock samples. Great care was taken in cleaning the rocks prior to extraction. Contamination studies and numerous control experiments characterized many of the organic geochemical analyses applied to the Precambrian rocks. For an example of a particularly detailed investigation, see Meinschein (1965). By the use of careful controls and techniques it generally has been possible to assure that some of the extracted compounds found were indigenous to a given sample and not laboratory contaminations. It is much more difficult to establish that the extracted organic material was deposited contemporaneously with the sediments that later lithified. To be of significance with regard to life at the time of sedimentation, of course, the organic compounds must be as old as the rock itself.

Work on the Swaziland Sequence and Bulawayan System rocks has continued, despite the difficulties of interpretation, mainly because of the optimism of some investigators who generally believe that if there is any chance at all of finding tangible evidence in the geologic record for earliest life or pre-life processes, it will be in these oldest sedimentary units. Many of the early geochemical studies on these rocks were done on individual samples obtained from a few collectors who had visited the areas. More extensive collections have now been made, but the availability of samples has paralleled the increase of questions with regard to what interpretations can be made from the results of the earlier work. At the present time, therefore, few really systematic studies have been carried out because of analytical difficulties in dealing with minute concentrations of organic compounds and data interpretation difficulties once the results have been obtained. This paper will review the organic geochemical results reported thus far for samples of the Swaziland Sequence and Bulawayan System. A summary showing the rock units on which various chemical determinations were made is given in Table 1.

TABLE 1 — Summary of the Organic Geochemical Measurements Made on Rock Units of the Swaziland Sequence and Bulawayan System

Rock Unit	Fossil-like Structures	Organic Carbon	$\delta^{13}C/^{12}C$	n-Alkanes	Iso-prenoid Hydro-carbons	Fatty Acids	Por-phyrins	Amino Acids	Pyrolysis Ozo-nolysis Hydro-genation
Bulawayan System.......	+		+					+	+
Swaziland Sequence Fig Tree Group Sheba Formation...	+	+	+	+	+		?	+	+
Onverwacht Group Swartkoppie Formation...	+	+		+	+			+	
Kromberg Formation .	ᴛ	+		+	+	+	+	+	+
Theespruit Formation...		+	+				+	+	+

TABLE 2 — Carbon Determinations of Early Precambrian Rocks

| Geologic Unit | Total Carbon %[1] | Organic Carbon% | | Carbonate Carbon %[1] |
		Dry Combustion[1]	Wet Combustion[2]	
Fig Tree Shale......	1.24	0.13		1.106
Fig Tree Shale......	1.21	0.02		1.19
Swartkoppie Chert ..	0.13	0.10	0.07	0.032
Kromberg Chert.....	0.12	0.09	0.05	0.030
Theespruit Chert....	1.08	1.02	0.62	0.060
Theespruit Chert....	0.23	0.22	0.15	0.012
Kaap Valley-Type Granite.........	0.011	0.0005		0.006
Gunflint Chert, Canada..........	0.15	0.07	0.03	0.080

[1]Value from Huffman Laboratories, Wheatridge, Colorado. Total carbon determined by dry combustion; carbonate carbon determined manometrically by measurement of CO_2 released by 4N HCl; organic carbon determined by difference.
[2]Values from Soil Control Laboratory, Watsonville, California.

Carbon

The organic carbon content of Swaziland Sequence sediments is variable. Schopf and Barghoorn (1967) point out that carbonaceous cherts and shales from the Fig Tree Group (including samples of the Swartkoppie Formation) contain about 0.5 per cent organic matter. Elemental analyses show that samples from the Theespruit and Kromberg formations contain about 1 per cent total carbon (Scott *et al.*, 1970). Additional carbon determinations have been made by both wet and dry combustion techniques (Table 2). Values for carbon in Kaap Valley-type granite, which intrudes the Swaziland Sequence, and in Gunflint chert from Canada are listed in Table 2 for comparison. For Swaziland Sequence sediments, the total carbon ranges from 0.12 to 1.24 per cent; the organic carbon content of a Bulawayan System sample of stromatolitic limestone is about 0.5 per cent (Hoering, 1964).

Carbon Isotopic Abundances

A limited number of carbon isotopic measurements have been made on Swaziland Sequence and Bulawayan System samples. The $\delta^{13}C$ ratios are listed here relative to the PDB standard. At the present moment (Jan. 1972), it seems that only data for one sample from the Swaziland Sequence have been reported. Hoering (1965) determined that the $\delta^{13}C$ for insoluble organic matter in Fig Tree shale was −28.3 $^o/_{oo}$ and for soluble organic matter −28.9 $^o/_{oo}$. These negative values may indicate that this organic material was fractionated isotopically by photosynthetic processes. The carbon in a sample of Theespruit Formation has a $\delta^{13}C$ of −15$^o/_{oo}$ (Silverman, personal communication). The different isotopic ratios within the Swaziland Sequence may indicate that the isotopic fractionation processes in early Onverwacht time were distinctly different from processes occurring during later Fig Tree time.

Carbon isotopic compositions of coexisting carbonates and reduced organic carbon in samples of stromatolitic limestones from the Bulawayan System were first determined by Hoering (1967b) and later by Schopf *et al.* (1971). The average $\delta^{13}C$ value of the reduced organic carbon (five measurements) was $-32.1^0/_{00}$ and that of the oxidized carbonate carbon (seven measurements) was $0.1^0/_{00}$. The average difference of $-32^0/_{00}$ between coexisting organic and inorganic carbon has been interpreted by Schopf *et al.* (1971) as evidence of photosynthetic isotopic fractionation.

n-Alkanes and Isoprenoid Hydrocarbons

One of the first classes of organic compounds sought in Swaziland Sequence sediments was the *n*-alkanes. These compounds are among the most stable and, consequently, likely to have survived the rigorous conditions imposed by three billion years of geologic time. Hoering (1965, 1967a), Meinschein (1967) and Oro and Nooner (1967) have shown that *n*-alkanes are present in samples from the Fig Tree Group and the Swartkippie Formation. The *n*-alkanes identified ranged from about n-C_{15} to n-C_{35}. The molecular distribution patterns of these compounds showed a dominance of neither odd- nor even-carbon-numbered molecules and are "typical of hydrocarbons formed in association with biologically produced organic matter of all ages" (Hoering, 1965). Isoprenoid hydrocarbons, pristane (C_{19}) and phytane (C_{20}) were identified by Meinschein (1967) in these rocks; Oro and Nooner (1967) found these two compounds as well as the C_{18} isoprenoid hydrocarbon. The concentrations of total alkanes found in Fig Tree samples ranged from 0.003 to 0.15 $\mu g/g$ according to Oro and Nooner (1967), who indicated that the range of variations in concentrations was significant and suggested that the alkanes were indigenous to the rocks. The investigators also reported finding normal C_{18}, C_{20} and C_{22} monoenes, but an independent confirmation of this discovery is still lacking.

MacLeod (1968) and Han and Calvin (1969a) found *n*-alkanes and isoprenoid hydrocarbons in samples on Onverwacht chert (probably from the Kromberg Formation). MacLeod shows *n*-alkanes ranging from n-C_{16} to n-C_{31} having a slight dominance of odd-carbon-numbered molecules and isoprenoid hydrocarbons C_{18} and C_{19}. In contrast, Han and Calvin reported *n*-alkanes n-C_{12} to n-C_{24}, with no significant dominance of odd-carbon-numbered molecules and with isoprenoid hydrocarbons C_{15}, C_{16}, C_{18}, C_{19} and C_{20}. Concentration of total alkanes was 0.05 $\mu g/g$.

The *n*-alkane distribution in Swaziland System sediments are reminiscent of alkanes in some younger Precambrian and Phanerozoic rocks such as in examples described by Van Hoeven *et al.* (1969) and Smith *et al.* (1970). The *n*-alkane distributions are also somewhat like those that have been described for bacteria (Han and Calvin, 1969b).

Fatty Acids

Only one report has appeared concerning the occurrence of fatty acids in either Swaziland Sequence or Bulawayan System sediments. Han and Calvin (1969a) found free and bound fatty acids in a sample of Onverwacht chert (Kromberg Formation?) at a concentration of 0.04 $\mu g/g$. They report the presence of n-C_{14}, n-C_{15}, n-C_{16} and n-C_{18} with the even-carbon-numbered species, particularly n-C_{16}, dominating. This type of distribution is typical of that found in biological systems and is different from that found in extracts of many Phanerozoic sediments where diagenetic processes have led to distributions of saturated *n*-alkanoic acids having equal abundances of both even- and odd-carbon-numbered molecules (Kvenvolden, 1970). The fatty acids in the Onverwacht chert sample are distributed similarly to fatty acids in some younger Precambrian samples (Smith *et al.*, 1970) that are thought to have been contaminated, possibly by bacteria.

Amino Acids

Although the syngenetic nature of amino acids in **Precambrian** sediments has

TABLE 3 — Hydrolyzed Amino Acids in Swaziland Sequence and Bulawayan System Rocks

(Amino acids are listed in order of decreasing abundance)

Sample	Concentration (μg/g)	Amino Acids Detected	Reference
Bulawayan Stromatolite	5.9	ASP, GLY, CYS, TYR, MET, THR, ALA, GLU, VAL, SER	Oberlies and Prashnowsky (1968)
Fig Tree Shale........	0.09	ALA, VAL, GLY	Pflug *et al.* (1969)
Fig Tree Chert.......	0.11	GLY, ALA, and trace of VAL, LEU and ILEU, SER, THR	Pflug *et al.* (1969)
Fig Tree Chert.......	0.19	GLY, SER ,GLU, ALA LEU,VAL, ILEU, THR	Kvenvolden and Peterson[1]
Swartkoppie Formation.........	1.2	GLY, SER, ALA, LEU, VAL, β-ALA, GLU,ILEU, THR, PHE, HIS, PRO, CYSA, TYR, ARG, ASP, MET, β-ABA, ALLOILEU, LYS and ORN, γ-ABA, α-ABA	Schopf *et al.* (1968)
Swartkoppie Formation.........	0.15	GLY, SER, ALA, GLU, LEU, VAL, ILEU, ASP, β-ALA	Kvenvolden and Peterson[1]
Kromberg Formation..	0.24	GLU, GLY, ALA, SER, LEU, VAL, ASP, ILEU	Kvenvolden and Peterson[1]
Theespruit Formation	0.12	GLY, SER, ALA, GLU, ASP, THR, VAL, LEU, ILEU	Kvenvolden and Peterson[1]

Unpublished results

not been demonstrated (Schopf, 1970), amino acids have been found in association with a number of samples from the Swaziland Sequence and Bulawayan System (Oberlies and Prasknowsky, 1968; Schopf *et al.*, 1968; Pflug, *et al.*, 1969; Kvenvolden and Peterson, unpublished results). An abbreviated summary of these findings is shown in Table 3, where the amino acids released by acid hydrolysis are listed. Not only are hydrolyzable amino acids present in Onverwacht samples, but glycine occurs as the principal free amino acid recoverable by leaching the rock with ammonium acetate.

A study of the enantiomeric distribution of amino acids in a sample of Fig Tree chert revealed only the L-isomers (Kvenvolden *et al.*, 1969). This discovery was later confirmed by Oro *et al.* (1971). The fact that the amino acids were not racemic led to conclusions that either the amino acids were protected from diagenesis and racemization by the chert matrix or the amino acids are very modern contaminants.

Porphyrins

Small concentrations of porphyrins were found in Fig Tree shale (5×10^{-6} μg/g) and in samples from the Kromberg (5×10^{-5} μg/g) and Theespruit (5×10^{-6} μg/g) formations of the Onverwacht Group; porphyrins could not be detected in the Fig Tree chert (actually from the Swartkoppie Formation) and in Gunflint chert from Canada (Kvenvolden and Hodgson, 1969). Although the spectral responses clearly indicated the presence of porphyrin-like compounds in a few Swaziland Sequence samples, the significance of such small concentrations may be questioned. Circumstantial evidence seemed to indicate that the porphyrin-like material was indigenous to the sediments; the fact that chlorins were not found was taken as evidence that the pigments found in the rocks were not related to modern contamination. If, indeed, these porphyrin-like pigments are as old as the rocks in which they were found, they possibly could have derived from compounds that participated in photosynthetic functions in Early Precambrian time. Lack of correlation, however, between rocks containing good examples of morphological fossils such as Gunflint chert (Barghoorn and Tyler, 1965) and rocks containing porphyrins reduces the significance that might be attached to finding porphyrins in Swaziland Sequence sediments.

Table 4 presents a summary of the values obtained for various geochemical measurements obtained for the Swaziland Sequence and Bulawayan System sediments.

TABLE 4 — Summary of Quantitative Organic Geochemical Measurements on Swaziland Sequence and Bulawayan System Sediments

	Fig Tree Group	Onverwacht Group	Bulawayan System
Organic Carbon	0.02–0.13%	0.05–1.02%	0.5%
Alkanes	0.003–015 µg/g	0.05 µg/g	not determined
Fatty Acids	not determined	0.04 µg/g	not determined
Amino Acids	0.11–0.19 µg/g	0.12–1.2 µg/g	5.9 µg/g
Porphyrins	5×10^{-6} µg/g	5×10^{-5}-5×10^{-6}µg/g	not determined
$\delta^{13}C_{BDB}$ (organic carbon)	$-28.3\permil$	$-15\permil$	$-32\permil$

INTERPRETATIONS AND CONSEQUENCES

If, and only if, the organic compounds or their precursors, which were extracted from the Swaziland Sequence and Bulawayan System rocks, were deposited contemporaneously with those ancient sediments, then the evidence supports the hypothesis that the compounds originated from photosynthetic and nonphotosynthetic microorganisms and that the compounds have been preserved for three billion years without appreciable alteration. For example, (1) *n*-alkanes showed a broad molecular weight range distribution typical of bacteria; (2) isoprenoid hydrocarbons could have been derived from the isoprenoid hydrocarbons of bacteria or from phytol from chlorophyll of blue-green algae or photosynthetic bacteria or both; (3) fatty acid distributions were typically biological; (4) amino acids having solely L configurations were typical of biological systems; (5) traces of porphyrins could have been derived from chlorophylls of bacteria and blue-green algae; and (6) photosynthetic fractionation of carbon isotopes by autotropic microorganisms could account for the carbon isotopic abundances found. The chemical evidence as thus interpreted would support the morphological fossil record that shows the presence of bacterial and algal microorganisms in these ancient sediments. The evidence would suggest that biochemical pathways three billion years ago were similar to pathways in the modern biosphere.

Unfortunately, it is not possible at present to prove the contemporaneity of the extractable organic molecules and the time of deposition of the ancient sediment. Such proof will be required before the above hypothesis can be even partially accepted. In fact, evidence supporting an alternate hypothesis that most of the extracted organic compounds were not deposited at the time of sedimentation is quite compelling. Several lines of reasoning in support of this alternate contention follow.

A permeability study of a single sample of Onverwacht chert (Kromberg?) by Nagy (1970) showed that this sample had a small (5.7×10^{-7} millidarcy) but finite permeability that would have permitted entry of significant quantities of organic materials over billions of years. Four additional permeability determinations on Swaziland Sequence samples have been made (Sanyal *et al.*, 1971): Swartkoppie Formation, 1.2×10^{-2} millidarcy; Kromberg Formation, 1.2×10^{-3} millidarcy; Hooggenoeg Formation, $\sim 10^{-6}$ millidarcy; and Theespruit Formation, 2.1×10^{-2} millidarcy. These values compare well with those found for later Precambrian cherts (Smith *et al.*, 1970). The additional permeability measurements indicate generally higher values in Onverwacht cherts than the one found previously; these results indicate that fluid flow through these rocks may have been even higher than suggested by Nagy (1970). Because the hydrocarbon distributions found are typical of those present in sediments of almost all geologic ages, these hydrocarbons may have originated in and migrated from younger sediments. The distribution of fatty acids in the Onverwacht chert is more typical of that found in biological systems than that which would be predicted on the basis of Phanerozoic samples. The distribution of fatty acid requires either that the chert matrix protected the biologically derived fatty acids through geologic time from diagenetic changes or that the fatty acids represent contributions from the modern biosphere. The latter alternative seems more likely. Amino acids would be expected to be racemic if they were indeed three billion years old unless, of course, the chert matrix protected and stabilized them throughout geologic time. Evidence presented by Abelson and Hare (1969) suggests that this sort of stabilization is unlikely. Therefore, the presence of only the L-isomer of various amino acids seems to indicate that these compounds are geologically young.

Because of the difficulties in ascertaining the significance of organic molecules recovered in small concentrations from Swaziland Sequence and Bulawayan System sediments by extraction techniques, a few studies have been directed toward trying to discover the nature of the insoluble polymeric organic material called kerogen. With studies of kerogen, one important interpretational problem is reduced because it is quite likely that much, if not all, of the kerogen is indigenous to the samples, and the kerogen precursors were likely deposited at the same time as the sediments.

Three basic lines of attack — hydrogenation, ozonolysis and pyrolysis — have been attempted. Hoering (1964) hydrogenated the kerogen from an algal limestone sample from the Bulawayan System. Low-molecular-weight hydrocarbons ($<C_9$) were generated, but these results did not permit a detailed interpretation concerning the original kerogen structure. Ozonolysis of an Onverwacht chert (Kromberg Formation?) yielded a predominance of aromatic compounds, suggesting that the kerogen polymer consisted of aromatic nuclei connected with short aliphatic chains (Nagy and Nagy, 1969). Pyrolysis experiments to 500°C by Scott *et al.* (1970) yielded mainly aromatic degradation products for samples from the Theespruit and Kromberg formations; a sample from the Fig Tree Group produced mainly *n*-alkanes. These results suggest that there is a basic structural difference in organic matter between Onverwacht and Fig Tree sediments. How-

ever, pyrolysis experiments to 300°C by Hoering (1967a) and pyrolysis-gas chromatography-mass spectrometry by Simmonds *et al.* (1969) on Fig Tree samples produced mainly aromatic compounds.

The future of organic geochemical studies of the Swaziland Sequence and Bulawayan System seems to lie in making systematic determinations on many samples. The parameters to be measured should be those that can reflect the nature of the kerogen and its precursors. The one set of pyrolysis experiments on four samples by Scott *et al.* (1970) apparently showed two different structural types of organic materials in Onverwacht and Fig Tree sediments. If a larger number of samples can be examined and a clear distinction made in all cases between samples from the two groups of rocks, a sufficient basis may be established for discovering the reasons for the differences. A systematic study of carbon isotopic abundances may show variations with regard to stratigraphic position. These variations may in turn reflect real differences in the evolution of Early Precambrian organic materials. Besides systematic studies of kerogen in these samples, a continued effort should be made to explain the origin of molecules recovered by extraction. For example, the amino acids that have been reported quite likely are not indigenous; these compounds may be from such sources as microorganisms, rain water and ground water. Studies to determine the source of nonindigenous compounds will not lead directly to new insights with regard to the origin and evolution of life, but could provide additional understanding of the geochemical processes taking place at the biosphere-lithosphere interface.

REFERENCES

Abelson, P. H., and Hare, P. E., 1969. Recent amino acids in the Gunflint Chert. Carnegie Inst. Yearb. 67, p. 208-210.

Anhaeusser, C. R., Roering, C., Viljoen, M. J., and Viljoen, R. P., 1967. The Barberton Mountain Land: A model of the elements and evolution of an Archean fold belt. Inf. Circ. 38, Econ. Geol. Res. Unit, Univ. Witwatersrand, Johannesburg, South Africa, 31 p.

Barghoorn, E. S., and Schopf, J. W., 1966. Microorganisms three billion years old from the Precambrian of South Africa. Science, 152, p. 758-763.

Barghoorn, E. S., and Tyler, S. A., 1965. Microorganisms from the Gunflint Chert. Science, 147, p. 563-577.

Condie, K. C., Macke, J. E., and Reimer, T. O., 1970. Petrology and geochemistry of Early Precambrian graywackes from the Fig Tree Group, South Africa. Geol. Soc. Am. Bull., 81, p. 2759-2776.

Engel, A. E. J., Nagy, B., Nagy, L. A., Engel, C. G., Kremp, G. O. W., and Drew, C. M., 1968. Alga-like forms in Onverwacht series, South Africa: Oldest recognized lifelike forms on Earth. Science, 161, p. 1005-1008.

Han, J., and Calvin, M., 1969a. Occurrence of fatty acids and aliphatic hydrocarbons in a 3.4-billion-year-old sediment. Nature, 224, p. 576-577.

Han, J., and Calvin, M., 1969b. Hydrocarbon distribution of algae and bacteria and microbiological activity in sediments. Proc. Nat. Acad. of Sci., 64, p. 436-443.

Haughton, S. H., 1969. Geological History of Southern Africa. Geol. Soc. S. Afr., Cape and Transvaal Printers Ltd., Cape Town, 535 p.

Hoering, T. C., 1964. The hydrogenation of kerogen from sedimentary rocks with phosphorus and anhydrous hydrogen iodide. Carnegie Inst. Yearb., 63, p. 258-262.

Hoering, T. C., 1965. The extractable organic matter in Precambrian rocks and the problem of contamination. Carnegie Inst. Yearb. 64, p. 215-218.

Hoering, T. C., 1967a. Criteria for suitable rocks in Precambrian organic geochemistry. Carnegie Inst. Yearb. 65, p. 365-372.

Hoering, T. C., 1967b. The organic geochemistry of Precambrian rocks. *In* Abelson, P. H. (*Editor*) Researches in Geochemistry 2, John Wiley and Sons Inc., New York, p. 87-111.

Kvenvolden, K. A., 1970. Evidence for transformations of normal fatty acids in sediments. *In* Hobson, G. D., and Speers, G. C. (*Editors*). Advances in Organic Geochemistry, 1966. Pergamon Press, Oxford, p. 335-366.

Kvenvolden, K. A., and Hodgson, G. W., 1969. Evidence for porphyrins in Early Precambrian Swaziland System sediments. Geochim. Cosmochim. Acta, 33, p. 1195-1202.

Kvenvolden, K. A., Peterson, E., and Pollock, G. E., 1969. Optical configuration of amino acids in Precambrian Fig Tree Chert. Nature, 221, p. 141-143.

Macgregor, A. M., 1940. A Precambrian alga limestone in Southern Rhodesia. Trans. Geol. Soc., S. Afr. 43, p. 9-15.

MacLeod, W. D., 1968. Combined gas chromatography - mass spectrometry of complex hydrocarbon trace residues in sediments. J. Gas Chrom., 6, p. 591-594.

Meinschein, W. G., 1965. Soudan Formation: Organic extracts of Early Precambrian rocks. Science, 150, p. 601-605.

Meinschein, W. G., 1967. Paleobiochemistry. 1967 McGraw-Hill Yearb. Sci. and Tech., p. 283-285.

Nagy, B., 1970. Porosity and permeability of the Early Precambrian Onverwacht chert: Origin of the hydrocarbon content. Geochim. Cosmochim. Acta, 34, p. 525-527.

Nagy, B., and Nagy, L. A., 1969. Early Precambrian Onverwacht micro-structures: Possibly the oldest fossils on Earth. Nature, 223, p. 1226-1227.

Nicolaysen, L. O., 1962. Stratigraphic interpretation of age measurements in Southern Africa. *In* Engel A. E. J., *et al.*, (*Editors*). Petrologic Studies. A volume to honor A. F. Buddington, Geol. Soc. Am., N.Y., p. 569-598.

Oberlies, F., and Prashnowsky, A. A., 1968. Biogeochemische and elektronenmikroskopische Untersuchung präkambrischer Gesteine. Naturwissenschaften, 55, p. 25-28.

Oro, J., Nakaparksin, S., Lichtenstein, H., and Gil-Av, E., 1971. Configuration of amino acids in carbonaceous chondrites and a Precambrian chert. Nature, 230, p. 107-108.

Oro, J., and Nooner, D. W., 1967. Aliphatic hydrocarbons in Precambrian rocks. Nature, 213, p. 1082-1085.

Pflug, H. D., 1966. Structured organic remains from the Fig Tree Series of the Barberton Mountain Land. Infor. Circ. 28, Econ. Geol. Res. Unit, Univ. Witwatersrand, Johannesburg, p. 1-14.

Pflug, H. D., Meinel, W., Neumann, K. H., and Meinel, M., 1969. Entwicklungstendenzen des Frühenlebens auf der Erde. Naturwissenschaften, 56, p. 10-14.

Sanyal, S., Marsden, S. S., and Kvenvolden, K. A., 1971. Permeabilities of Precambrian Onverwacht cherts and other low-permeability rocks. Nature, 232, p. 325-327.

Schopf, J. W., 1970. Precambrian microorganisms and evolutionary events prior to the origin of vascular plants. Biol. Rev., 45, p. 319-352.

Schopf, J. W., and Barghoorn, E. S., 1967. Alga-like fossils from the Early Precambrian of South Africa. Science, 156, p. 508-511.

Schopf, J. W., Kvenvolden, K. A., and Barghoorn, E. S., 1968. Amino acids in Precambrian sediments: An Assay. Proc. Nat. Acad. Sci., 59, p. 639-646.

Schopf, J. W., Oehler, D. Z., Horodyski, R. J., and Kvenvolden, K. A., 1971. Biogenicity and significance of the oldest known stromatolites. J. Paleontol., 45, p. 477-485.

Scott, W. M., Modzeleski, V. E., and Nagy, B., 1970. Pyrolysis of Early Precambrian Onverwacht organic matter (>3x10⁹ year old). Nature, 225, p. 1129-1130.

Simmonds, P. G., Shulman, G. P., and Stembridge, C. H., 1969. Organic analysis by pyrolysis — gas chromatography-mass spectrometry: a candidate experiment for the biological exploration of Mars. J. Chromatographic Sci., 7, p. 36-41.

Smith, J. W., Schopf, J. W., and Kaplan, I. R., 1970. Extractable organic matter in Precambrian cherts. Geochim. Cosmochim. Acta, 34, p. 659-675.

Vail, J. R., and Dodson, M. H., 1969. Geochronology of Rhodesia. Trans. Geol. Soc. S. Afr., 72, pt. 2, p. 79-113.

Van Hoeven, W., Maxwell, J. R., and Calvin, M., 1969. Fatty acids and hydrocarbons as evidence of life processes in ancient sediments and crude oils. Geochim. Cosmochim. Acta, 33, p. 877-881.

Van Niekerk, C. B., and Burger, A. J., 1969. A note on the minimum age of acid lava of the Onverwacht Series of the Swaziland System. Trans. Geol. Soc. S. Afr., 72, pt. 1, p. 9-21.

Viljoen, M. J., and Viljoen, R. P., 1969. Archaean volcanicity and continental evolution in the Barberton Region, Transvaal. *In* Clifford, T. N., and Gass, I. G., (*Editors*). African Magmatism and Tectonics, Oliver and Boyd, Edinburgh, p. 27-49.

41

38

Reprinted from *Science*, **175**, 1246–1248 (Mar. 17, 1972)

Carbon Isotopic Studies of Organic Matter in Precambrian Rocks

Abstract. *Reduced carbon in early Precambrian cherts of the Fig Tree and upper and middle Onverwacht groups of South Africa is isotopically similar (the average value of $\delta^{13}C_{PDB}$ is −28.7 per mil) to photosynthetically produced organic matter of younger geological age. Reduced carbon in lower Onverwacht cherts (Theespruit formation) is anomalously heavy (the average value of $\delta^{13}C_{PDB}$ is −16.5 per mil). This discontinuity may reflect a major event in biological evolution.*

Since 1967, an impressive array of chemical fossils (such as *n*-alkanes, isoprenoids, porphyrins, and isotopically light carbonaceous material) has been detected in early Precambrian sediments (*1*). These reported occurrences, although subject to question because of possible contamination by younger organic matter (*1, 2*), are consistent with morphological fossil evidence indicating that photoautotrophs probably were extant more than 2.8 × 10⁹ years ago (*1, 3*). Among the chemical evidences suggestive of early autotrophic activity, the carbon isotopic composition of particulate, insoluble organic matter (kerogen) appears to have been least subject to postdepositional contamination (*4–6*). However, $^{13}C/^{12}C$ ratios have been reported for organic components of fewer than three dozen Precambrian deposits (*1–5, 7, 8*). We have therefore undertaken a survey of the carbon isotopic composition of the total organic fraction (composed of more than 95 percent kerogen) of a suite of Precambrian sediments to detect isotopic trends possibly correlative with early evolutionary events (*9*).

Park and Epstein (*10*) have shown that photosynthetic fixation of carbon results in fractionation of the two stable carbon isotopes, ^{12}C and ^{13}C; the organic matter produced is enriched in the lighter isotope relative to the inorganic source carbon. Biogenic organic matter isolated from sedimentary rocks has $\delta^{13}C_{PDB}$ values (*11*) ranging from about −20 to −40 per mil (*12*); $\delta^{13}C_{PDB}$ values of limestones range from about +10 to −14 per mil (*13*). Because of possible variations in the carbon cycle over geologic time and complexities introduced by diagenetic and metamorphic factors, interpretation of Precambrian carbon isotopic data is difficult. Nevertheless, biological fixation of inorganic carbon is the major process on Earth that produces an isotopic fractionation between inorganic and organic carbon of the magnitude and direction found in ancient sediments. Since the isotopic content of marine carbonates of all ages appears to vary little (*3, 12*), fossil organic matter having $\delta^{13}C_{PDB}$ values between −20 and −40 per mil seems reasonably interpreted as being of photosynthetic (*1, 3, 4, 7, 8*) or chemosynthetic (*12*) autotrophic derivation.

The $^{13}C/^{12}C$ ratios of reduced carbon were determined according to a method adapted from Craig (*13*). Each sample was ground to a powder and refluxed with redistilled benzene : methanol (70 : 30 by volume) for 24 hours. Fifteen grams of each sample were then reacted with hydrochloric acid to remove carbonates and with hydrofluoric acid to remove silicates. The resulting acid-resistant organic residue, the "total organic fraction," was combusted in oxygen at 1000°C; the carbon dioxide generated was analyzed on a dual-collecting mass spectrometer. The total organic fractions of three samples were further extracted by refluxing for 24 hours with the benzene : methanol solution; the extracted "soluble organic fraction" and the insoluble "kerogen fraction" were analyzed.

The $\delta^{13}C_{PDB}$ values of the total organic fractions of 39 Precambrian cherts and limestones representing 32 localities have been determined (*14*). These data (Fig. 1) substantially increase the total number of reported analyses of Precambrian organic matter.

Reduced carbon in Phanerozoic sediments tends to exhibit increasing ^{12}C content with increasing geological age (*12*). In the Precambrian, this trend is not marked (Fig. 1), although ^{12}C content does appear to approach a maximum in sediments about 2.1 × 10⁹ years old. With the exception of the very oldest samples analyzed, Precambrian organic carbon generally has $\delta^{13}C_{PDB}$ values between −25 and −35 per mil, well within the isotopic range typical of preserved biogenic organic matter.

Anomalously heavy organic carbon (having a $\delta^{13}C_{PDB}$ value greater than −20 per mil) was detected only in cherts from the lower Swaziland sequence of South Africa; these units are among the oldest sedimentary horizons known. To investigate the discrepancy between these samples and all other cherts, isotopic analyses were made of carbonaceous matter in 16 cherts representing nine stratigraphic horizons in the sequence (*15*).

The Swaziland sequence (Fig. 2), 64,000 feet (19.5 km) thick and well exposed in the Barberton Mountain Land of the southeastern Transvaal, is divided into three units (*16*). The oldest unit, the Onverwacht group, consists of lavas, cherts, and sedimentary carbonates. The Fig Tree group, conformably overlying the Onverwacht, includes shales, graywackes, banded cherts, and lavas. The Moodies group overlies the Fig Tree. The Onverwacht and Fig Tree cherts are thought to have been deposited in a subaqueous, basin-like environment and to have experienced little subsequent alteration (*16*). Rubidium and strontium measurements on intrusives in the Fig Tree group indicate an

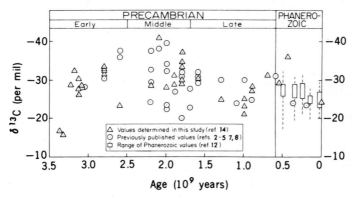

Fig. 1. The $\delta^{13}C_{PDB}$ values of organic carbon in sedimentary rocks.

age in excess of 3×10^9 years (*17*); a maximum age for the sequence of about 3.4×10^9 years is suggested by rubidium-strontium and lead-lead analyses (*18*).

In Fig. 2 the $\delta^{13}C_{PDB}$ values of the analyzed Swaziland samples are shown. The values fall into two distinct categories. One category is comprised of 13 cherts stratigraphically ranging from the middle third of the Onverwacht group (Hooggenoeg formation) to the middle of the Fig Tree group; the $\delta^{13}C_{PDB}$ values range from -25.0 to -33.0 per mil with an average of -28.7 per mil. Hoering's $\delta^{13}C_{PDB}$ value of -28.3 per mil (*4*) for organic matter in a Fig Tree shale falls within this range. The second category is composed of three samples from the lowest third of the Onverwacht group (Theespruit formation); the $\delta^{13}C_{PDB}$ values of these samples range from -14.7 to -19.5 per mil with an average of -16.5 per mil. The $\delta^{13}C_{PDB}$ values of the soluble organic fractions of these three samples range from -25.5 to -26.2 per mil, presumably reflecting a relatively recent origin [see (*4, 5*)]. In contrast, values of the kerogen fractions, apparently syngenetic with Theespruit deposition, range from -14.3 to -18.9 per mil.

Thus, there is an isotopic discontinuity of about 12 parts per thousand between Theespruit and younger organic matter in the Swaziland sequence. Since all Swaziland cherts were analyzed similarly and since triplicate analyses of one Theespruit sample gave highly consistent results (sample 3-T in Fig. 2; $\delta^{13}C_{PDB} = -15.5, -14.8$, and -14.7 per mil), it is unlikely that the unusually heavy isotopic composition of Theespruit carbon is an artifact of experimental procedure.

It might be suggested that the isotopically heavy nature of Theespruit reduced carbon has resulted from metamorphic loss of light carbon species (*19*). In this regard, chert from the middle Hooggenoeg formation, collected immediately adjacent to an intrusive body, provides an indication of possible isotopic effects of contact metamorphism on Swaziland organic carbon. The $\delta^{13}C_{PDB}$ value of reduced carbon in this chert (sample 10-T, -32.5 ± 0.4 per mil) does not differ significantly from values typical of organic matter in other cherts of similar stratigraphic position (Fig. 2), most of which are far removed from zones of contact metamorphism. Since the Theespruit samples were collected more

than 0.4 km away from known intrusives, it seems unlikely that their atypical isotopic composition has resulted from contact metamorphism. Moreover, the Swaziland samples analyzed are from a conformable succession and have been subject to similar depositional and tectonic environments, so that isotopic alteration by diagenesis or regional metamorphism would be expected to have occurred similarly throughout the sequence. The difference in isotopic composition between Theespruit and younger Swaziland carbon is probably not, therefore, a result of postdepositional alteration, a conclusion consistent with the results of other studies (*8, 12, 20*). Thus, the two categories of reduced carbon detected in Swaziland sediments may be of differing origins.

In conclusion, the similarities in the isotopic content of reduced carbon in all but the oldest Precambrian samples analyzed (Fig. 1), together with the essentially constant isotopic composition of oxidized carbon in marine carbonates deposited during the past 2.8 (*3, 12*) or perhaps 3.3×10^9 years (*21*), seem consistent with the existence of autotrophic organisms since the deposition of the middle third of the Onverwacht group, approximately 3.3×10^9 years ago. Kerogen in lower Onverwacht Theespruit cherts has $\delta^{13}C_{PDB}$ values which fall outside the normal range for preserved reduced carbon of established biological origin but which are comparable to values of about -16 per mil characteristic of primordial organic matter in carbonaceous chondrites (*12, 22*). The isotopic discontinuity between Theespruit and younger cherts is diffi-

cult to interpret because of uncertainties regarding the nature of the early carbon cycle and the primitive environment. Possibly this break reflects a geological event that altered the distribution of isotopes in the carbon reservoirs; or, it may mark the time of origin of biochemical mechanisms capable of fractionating carbon isotopes in a manner similar to that of modern autotrophs. The Theespruit organic carbon may even, in part, be a remnant of abiologically produced organic material, isotopically similar to that present in carbonaceous chondrites. Although such speculations are consistent with the known early Precambrian fossil record (*1*) and find marginal support in reported chemical analyses of Onverwacht organic matter (*6, 23*), the data are so few and their potential implication so far-reaching that no firm conclusion should be drawn at the present time.

DOROTHY Z. OEHLER
J. WILLIAM SCHOPF
Department of Geology, University of California, Los Angeles 90024
KEITH A. KVENVOLDEN
*Exobiology Division,
Ames Research Center,
Moffett Field, California 94035*

References and Notes

1. J. W. Schopf, *Biol. Rev. Cambridge Phil. Soc.* **45**, 319 (1970).
2. J. W. Smith, J. W. Schopf, I. R. Kaplan, *Geochim. Cosmochim. Acta* **34**, 659 (1970).
3. J. W. Schopf, D. Z. Oehler, R. H. Horodyski, K. A. Kvenvolden, *J. Paleontol.* **45**, 477 (1971).
4. T. C. Hoering, in *Researches in Geochemistry*, P. H. Abelson, Ed. (Wiley, New York, 1967), vol. 2, pp. 88–111.
5. ———. *Carnegie Inst. Washington Yearb.* **64**, 215 (1965); *ibid.* **65**, 365 (1967).
6. W. M. Scott, V. E. Modzeleski, B. Nagy, *Nature* **225**, 1129 (1970).
7. M. M. Roblot, M. Chaigneau, M. Majzoub,

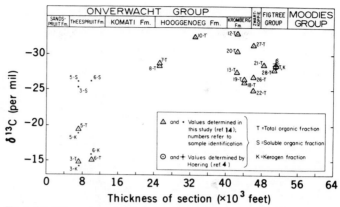

Fig. 2. The $\delta^{13}C_{PDB}$ values of organic carbon in sediments of the Swaziland sequence.

C. R. Acad. Sci. **258**, 253 (1964); J. Hoefs and M. Schidlowski, *Science* **155**, 1096 (1967); E. S. Barghoorn and S. A. Tyler, *ibid.* **147**, 563 (1965); E. S. Barghoorn, W. G. Meinschein, J. W. Schopf, *ibid.* **148**, 461 (1965).
8. E. Bondensen, K. R. Pedersen, O. Jørgensen, *Medd. Groenland* **164**, No. 4 (1967).
9. D. Z. Oehler and J. W. Schopf, *Amer. J. Bot.* **58**, 471 (1971).
10. R. Park and S. Epstein, *Geochim. Cosmochim. Acta* **21**, 110 (1960); *Plant Physiol.* **36**, 133 (1961).
11. The $\delta^{13}C_{PDB}$ (per mil) value is defined as

$$\frac{(^{13}C/^{12}C)_{sample} - (^{13}C/^{12}C)_{PDB}}{(^{13}C/^{12}C)_{PDB}} \times 1000$$

where PDB refers to the Pedee belemnite standard from the Pedee formation, Upper Cretaceous of South Carolina.
12. E. T. Degens, in *Organic Geochemistry*, G. Eglinton and M. T. J. Murphy, Eds. (Springer-Verlag, New York, 1969) pp. 304–329.
13. H. Craig, *Geochim. Cosmochim. Acta* **3**, 53 (1953).
14. The following $\delta^{13}C_{PDB}$ values were determined. The abbreviations are T (total organic fraction), S (soluble organic fraction), K (kerogen fraction), BY (approximate age $\times$ 10^9 years), sample (sample number shown on Fig. 2).
    1) Swaziland samples (cherts, 3.4 to 3.0 BY). Lower third Theespruit formation, 32 km southwest of Barberton: sample 5, T = −19.5, S = −26.1, K = −18.9 per mil; sample 3, T = −15.5, T = −14.8, T = −14.7, S = −25.5, K = −14.3 per mil. Upper third Theespruit formation, 32 km southwest of Barberton: sample 6, T = −15.1, S = −26.2, K = −15.8 per mil. Lower third Hooggenoeg formation, 27 km southwest of Barberton: sample 7, T = −28.8 per mil; sample 8, T = −28.4 per mil. Middle third Hooggenoeg formation, 38 km south-southwest of Barberton: sample 10, T = −32.1, T = −32.9 per mil. Middle third Kromberg formation, 29 km south-southwest of Barberton: sample 12, T = −33.0 per mil; sample 13, T = −27.5 per mil;

sample 20, T = −30.6 per mil. Upper third Kromberg formation, 29 km south of Barberton: sample 18, T = −26.1 per mil; sample 19, T = −26.2 per mil. Zwartkoppie formation, 11 km northeast of Barberton: sample 22, T = −24.9 per mil. Zwartkoppie formation, 24 km east-northeast of Barberton: sample 26, T = −26.9 per mil. Zwartkoppie formation, 23 km east-northeast of Barberton: sample 27, T = −31.4 per mil. Lower third Fig Tree group, 10 km northeast of Barberton: sample 21, T = −28.7 per mil. Middle third Fig Tree group, 13 km south-southeast of Barberton: sample 28, T = −28.0 per mil.
    2) Other Precambrian samples ($\delta^{13}C_{PDB}$ total organic fractions). Limestone: 3.3 to 2.8 BY, Bulawayan group, Rhodesia (three different samples), −33.5, −32.1, −32.5 per mil. Cherts: 3.3 to 2.8 BY, Bulawayan group, Rhodesia, −31.8 per mil; 2.6 BY, Keewatin, Schreiber, Ontario, −33.2 per mil; 2.2 BY, middle Fortescue group, Western Australia, −28.4 per mil; 2.2 BY, upper Fortescue group, Western Australia, −40.8 per mil; 2.0 BY, Transvaal System, South Africa, −28.0 per mil; 1.9 BY, Wittenoom Dolomite, Hamersley group, Western Australia, −29.0 per mil; 1.9 BY, Wittenoom Dolomite, Hamersley group, Western Australia, −31.2 per mil; 1.9 BY, Wittenoom Dolomite/basal Brockman Iron formation, Hamersley group, Western Australia, −30.2 per mil; 1.8 BY, middle to lower Brockman Iron formation, Hamersley group, Western Australia, −34.4 per mil; 1.8 BY, upper Brockman Iron formation, Hamersley group, Western Australia, −29.3 per mil; 1.8 BY, Gunflint Iron formation, Schreiber Beach, Ontario, −37.2 per mil; 1.8 BY, Gunflint Iron formation, Nolalu, Ontario, −34.1 per mil; 1.6 BY, Koolpin Chert, South Alligator group, Northern Territory, Australia, −31.0 per mil; 1.3 BY, Beck Spring Dolomite, California, −25.0 per mil; 1.0 BY, Skillogalee Dolomite, Yatina, South Australia, −21.0 per mil; 1.0 BY, Skillogalee Dolomite, Mundallio Creek, South Australia, −23.0 per mil; 1.0 BY, Skillogalee Dolomite, Depot Creek, South Australia, −25.2 per mil; 0.9 BY, Bitter Springs formation, Jay Creek, central Australia, −28.4 per mil; 0.9 BY, Bitter Springs formation, Ellery Creek, central Australia,

−27.2 per mil; 0.7 BY, Conception group, East Newfoundland, −31.2 per mil.
    3) Phanerozoic samples (cherts, $\delta^{13}C_{PDB}$ total organic fractions). Upper Cambrian, Catlin member, Windfall formation, Nevada, −29.4 per mil; Upper Ordovician, Maravillas Chert, W. Texas, −36.1 per mil; Upper Tertiary, Lost Chicken Creek formation, Alaska, −24.2 per mil.
15. These samples were collected in May 1968 by K.A.K.
16. C. R. Anhaeusser, C. Roering, M. J. Viljoen, R. P. Viljoen, *Information Circular No. 38* (Economic Geology Research Unit, University of Witwatersrand, Johannesburg, 1967); M. J. Viljoen and R. P. Viljoen, *ibid.*, No. 36 (1967).
17. H. L. Allsopp, T. J. Ulrych, L. O. Nicolaysen, *Can. J. Earth Sci.* **5**, 605 (1968).
18. T. J. Ulrych, A. Burger, L. O. Nicolaysen, *Earth Planet. Sci. Lett.* **2**, 179 (1967).
19. D. R. Baker and G. E. Claypool, *Amer. Ass. Petrol. Geol. Bull.* **54**, 456 (1970); F. Barker and I. Friedman, *Geol. Soc. Amer. Bull.* **80**, 1403 (1969).
20. F. E. Wickman, *Geochim. Cosmochim. Acta* **3**, 244 (1953); *ibid.* **9**, 136 (1956); S. Gavelin, *ibid.* **12**, 297 (1957).
21. J. R. Vail and M. H. Dodson, *Trans. Geol. Soc. S. Afr.* **72**, 79 (1969).
22. J. W. Smith and I. R. Kaplan, *Science* **167**, 1367 (1970).
23. M. Calvin, *Chemical Evolution* (Oxford Univ. Press, New York, 1969), pp. 85–88; W. D. MacLeod, Jr., *J. Gas Chromatogr.* **6**, 591 (1968).
24. We thank E. S. Barghoorn, A. R. Palmer, M. E. Taylor, A. M. Gutstadt, and V. M. Page for providing samples; D. A. Pretorius, M. J. Viljoen, R. P. Viljoen, C. R. Anhaeusser, and T. Reimer for assistance in collecting Swaziland samples; I. R. Kaplan for use of laboratory facilities; J. W. Smith, E. Ruth, C. Petrowski, and M. J. Baedecker for technical assistance; and I. R. Kaplan and J. H. Oehler for critical comments and suggestions. Supported by NSF grant GA 23741 and NASA grant NGR 05-007-292 to the University of California at Los Angeles.

13 September 1971; revised 2 December 1971

V
Early Life on Earth—
Paleobotany

Editor's Comments on Papers 39 Through 43

39 **Macgregor:** *A Pre-Cambrian Algal Limestone in Southern Rhodesia*

40 **Schopf, Oehler, Horodyski, and Kvenvolden:** *Biogenicity and Signifcance of the Oldest Known Stromatolites*

41 **Barghoorn and Schopf:** *Microorganisms Three Billion Years Old from the Precambrian of South Africa*

42 **Schopf and Barghoorn:** *Alga-like Fossils from the Early Precambrian of South Africa*

43 **Engel, Nagy, Nagy, Engel, Kremp, and Drew:** *Alga-like Forms in Onverwacht Series, South Africa: Oldest Recognized Lifelike Forms on Earth*

The geological record of earliest life is vague, but paleobotanical studies coupled with geochemical investigations have clearly shown that life has been present on earth for at least 3 billion years. Five papers have been chosen to illustrate the kind of observations that have been made concerning the earliest occurrences of life. Much of the evidence comes from morphological fossils which have been observed, but each of these papers includes, or at least mentions, the application of geochemical measurements. All the papers deal with rocks from the southern part of the African continent, where the Greenstone Belts contain the oldest, relatively unmetamorphosed sedimentary rocks yet known; it is in these rocks that the oldest record of life has been discovered.

Two rock units of importance are the Bulawayan Group and the Swaziland Supergroup. The Bulawayan Group is the middle of three early Precambrian groups of Rhodesia and may have been formed contemporaneously with part of the Swaziland Supergroup of South Africa (Haughton, 1969). The Bulawayan Group is composed of about 12 km of andesitic lavas with interbedded sediments of banded ironstones, quartzites, cherts, shales, and limestones. In the Zwankendaba limestones of the Bulawayan Group, Macgregor (Paper 39) made the important discovery in 1935 of concretionary structures which he considered to be of algal origin after the dark lamentations in the rocks were determined to be composed of graphite. At the time of Macgregor's discovery the very great age of the Bulawayan Group was not fully appreciated. A number of isotopic age determinations and correlations have been made (Holmes, 1954; Nicolaysen, 1962; Haughton, 1969, Vail and Dodson, 1969; Bond et al., 1973), and it is now believed that the Bulawayan Group is 3.0 to 3.3 billion years old.

A reinvestigation by Schopf et al. (Paper 40) of the structures first discovered by Macgregor reaffirmed that the structures were of biologic origin. Thus the limestones of the Bulawayan Group contain the oldest known stromatolites and the oldest evidence of life that can be observed without the aid of microscopy. In establishing the biologic origin of the stromatolites, Schopf et al. made detailed examinations of the morphological relationships of laminations. Also, geochemical measurements were made of the carbon isotopic abundances of the carbonate carbon and the coexisting reduced, organic carbon. The isotopic relationships observed seemed indicative of carbon isotopic fractionation through photosynthesis. These fossils, if indeed constructed by blue-green algae, place a minimum age on the time of origin of oxygen-producing photosynthesis during the evolution of life on earth.

The Swaziland Supergroup is found in the eastern Transvaal of South Africa along the border of Swaziland. The unit exceeds 20 km in thickness and is subdivided into three groups (from oldest to youngest): Onverwacht, Fig Tree, and Moodies (Anhaeusser et al., 1967). The age of the Swaziland Supergroup exceeds 3.2 billion years (Nicolaysen, 1962; Van Niekerk and Burger, 1969; Sinha, 1972; Hurley et al., 1972). In the lower part of the Fig Tree Group (now assigned to the uppermost Onverwacht Group) Barghoon and Schopf (Paper 41) and Schopf and Barghoorn (Paper 42) first discovered evidence for bacteria-like and alga- like microstructures. In support of the morphological evidence for bacteria-like fossils, Barghoorn and Schopf (Paper 41) cite the unpublished work by Meinshein and the paper by Hoering (Paper 31) on organic geochemical analyses showing the presence of small amounts of high-molecular-weight alkanes. Schopf and Barghoorn (Paper 42) discuss in some detail both organic and inorganic geochemical measurements which support the finding of alga-like fossils in Fig Tree cherts. This paper offers a good example of the convergence of lines of both paleontological and geochemical evidence, which together indicate that life was present during the time of formation of these rocks. Independently, Pflug (1966) and Pflug et al. (1969) found globular and filamentous objects in cherts and shales of the Fig Tree Group, and they supported their observations with geochemical measurements of amino acids. It should be remembered, however, that at the present time the significance of simple organic compounds found in these early Precambrian rocks is uncertain, and their presence may have little to do with early Precambrian life.

In the Onverwacht Group, the oldest rocks of the Swaziland Supergroup, Engel et al. (Paper 43) first reported finding spheroidal and cup-shaped structures which could represent fossils of the earliest life yet discovered on earth. With techniques of organic geochemistry the authors reported their initial results on the insoluble kerogen fraction from these rocks. Whereas Onverwacht kerogen appeared to be largely aromatic in composition, Fig Tree kerogen seemed to be essentially aliphatic. The significance of these results with regard to early life has never been made clear. Doubt still remains regarding the possible biogenic nature of the microstructures in the Onverwacht Group (Nagy and Nagy, 1969; Nagy, 1971). On the other hand, Brooks and Muir (1971) and Brooks et al. (1973) have argued that the microstructures should definitely be considered microfossils representing an ancient biota. The evidence obtained thus far is hardly clear cut. Further investigations are needed to develop a truly convincing story that the record of life extends through the time period represented by all the sedimentary rocks of the Onverwacht Group.

Two names very often associated with the paleobotany of the early Precambrian record are Elso S. Barghoorn and J. William Schopf.

Barghoorn is Professor of Botany in the Biology Department at Harvard University. He was born in New York in 1915 and was educated at Miami University and Harvard. His major contributions have been in paleobotany, particularly in the evolution of vascular plants. His interest in paleobotany led him into investigations of the fossil record of the Precambrian. His discovery, along with Stanley Tyler, of microfossils in Gunflint cherts opened up interest in the record of life in the Precambrian and stimulated research efforts which have shown that assembleges of simple organisms can be traced back in time to about 3 billion years.

Schopf was a graduate student with Barghoorn at Harvard. Together they made impressive additions to the growing body of knowledge of Precambrian life. Schopf was born in 1941 in Illinois and attended Oberlin College before entering Harvard. With interest in both biology and geology, he has helped to clarify the paleobiology of the Precambrian. Using optical and electron microscopy combined with studies of organic geochemistry, he has been able to delimit important stages in early biological evolution. Schopf is currently Professor in the Geology Department at the University of California at Los Angeles.

References

Anhaeusser, C. R., C. Roering, M. J. Viljoen, and R. R. Viljoen (1967) The Barberton Mountain-land: a model of the elements and evolution of an Archean fold belt: *Info. Circ. 38, Econ. Geol. Res. Unit., Univ. Witwatersrand,* 31 p.

Bond, G., J. F. Wilson, and N. J. Winnal (1973) Age of the Huntsman limestone (Bulawayan) stromatolites: *Nature,* **244,** 275–276.

Brooks, J., and M. D. Muir (1971) Morphology and chemistry of the organic insoluble matter from Onverwacht Series Precambrian chert and the Orgueil and Murray carbonaceous meteorites: *Grana,* **11,** 9–14.

Brooks, J., M. D. Muir, and G. Shaw (1973) Chemistry and morphology of Precambrian microorganisms: *Nature,* **244,** 215–217.

Haughton, S. H. (1969) *Geological History of Southern Africa*: Geol. Soc. South Africa, Cape and Transvaal Printers Ltd., Capetown, 535p.

Holmes, A. (1954) The oldest dated minerals of the Rhodesian shield: *Nature,* **173,** 612

Hurley, P. M., W. H. Pinson, Jr., B. Nagy, and T. M. Teska (1972) Ancient age of the Middle Marker Horizon, Onverwacht Group, Swaziland Sequence, South Africa: *Earth Planet. Sci. Lett.,* **14,** 360–366.

Nagy, L. A. (1971) Ellipsoidal microstructures of narrow size range in the oldest known sediments on Earth: *Grana,* **11,** 91–94.

Nagy, B., and L. A. Nagy (1969) Early Precambrian Onverwacht microstructures: possibly the oldest fossils on earth: *Nature,* **223,** 1226–1227.

Nicolaysen, L. O. (1962) Stratigraphic interpretation of age measurements in Africa: in *Petrologic Studies: A Volume to Honor A. F. Buddington*: Geological Society of America, Washington, pp. 569–598.

Pflug, H. D. (1966) Structured organic remains from the Fig Tree Series of the Barberton Mountain Land: *Info. Circ. 28, Econ. Geol. Res. Unit, Univ. Witwatersrand,* 1–14.

Pflug, H. D., W. Meinel, K. H. Neumann, and M. Meinel (1969) Entwicklungstendenzen des Fruhenlebens auf der Erde: *Naturwiss.,* **56,** 10–14.

Sinha, A. K. (1972) U-Th-Pb systematics and the age of the Onverwacht Series, South Africa: *Earth Planet. Sci. Lett.,* **16,** 219–227.

Vail, J. R., and Dodson, M. H. (1969) Geochronology of Rhodesia: *Trans. Geol. Soc. South Africa,* **72,** pt. 2, 79–113.

Van Niekerk, C. B., and Burger, A. J. (1969) A note on the minimum age of acid lava of the Onverwacht Series of the Swaziland System: *Trans. Geol. Soc. South Africa,* **72,** pt. 1, 9–21.

Additional Suggested Readings

Cloud, P. E. Jr. (1968) Pre-metazoan evolution and the origins of the Metazoa: in *Evolution and Environment*, edited by E. T. Drake: Yale University Press, New Haven, pp. 1–72.

Echlin, P. (1970) Primitive photosynthetic organisms: in *Advan. Org. Geochem. 1966*, edited by G. D. Hobson and G. C. Speers: Pergamon Press, Elmsford, N.Y., pp. 523–537.

Glaessner, M. F. (1966) Precambrian paleontology: *Earth Sci. Rev.* **1**, 29–50.

Grabert, H. (1973) Die Biologie de Präkambrium: *Zbl. Geol. Paläont. Teil I, 1972*, **5/6**, 316–346.

Nagy, B., and L. A. Nagy (1969) Investigation of the early Precambrian Onverwacht sedimentary rocks in South Africa: in *Advan. Org. Geochem. 1968*, edited by P. A. Schenck and I. Havanaar: Pergamon Press, Elmsford, N.Y., pp. 209–215.

Rutten, M. G. (1957) Origin of life on earth—its evolution and actualism: *Evolution*, **11**, 56–59.

Schopf, J. W. (1967) Antiquity and evolution of Precambrian life: *McGraw-Hill Yearbook of Science and Technology*, pp. 47–55.

Schopf, J. W. (1969) Recent advances in Precambrian paleobiology: *Grana Palynologia*, **9**, 147–168.

39

Reprinted from *Trans. Geol. Soc. South Africa*, **43**, 9–15 (1941)

A Pre-Cambrian Algal Limestone in Southern Rhodesia.

(Read 19th February, 1940.)

By A. M. Macgregor.

[Plates II–V.]

Communicated by permission of the Director, Geological Survey of Southern Rhodesia.

ABSTRACT.

The occurrence is recorded of a graphitic limestone with concretionary structure believed to be of algal origin. The rock occurs at an horizon low down in the " Basement Schists " of the Bembesi gold belt. The principal structures observed are briefly described and illustrated.

Introduction and Stratigraphical Relations.

The Limestone.

Algal Structures.

Conclusion.

Introduction and Stratigraphical Relations.

The Pre-Cambrian surface-formed rocks composing the gold belts of Southern Rhodesia, contain among the preponderating lavas and undifferentiated arkoses and greywackes, ferruginous cherty quartzites and relatively thin bodies of limestone.

Original concretionary structures have been observed in these limestones in a few places, and are particularly well developed at the Huntsman Lime Works near the Turk Mine, thirty-three miles north-north-east of Bulawayo, where they form a large part of the rock which is being quarried. The rock forms part of a generally dolomitic calcareous bed, which can be traced along the strike for several miles associated with banded ironstone, arkoses and conglomerate. These sediments have been defined as the Zwankendaba group, and are bedded between very considerable thicknesses of volcanic greenstones. The geology of the neighbourhood has been described in a bulletin of the Geological Survey of Southern Rhodesia (Macgregor, Ferguson and Amm, 1937), but a brief statement of the stratigraphy may be useful for comparison with other regions.

Owing to the scattered distribution of the older Pre-Cambrian rocks or " Basement Schists " of Southern Rhodesia among wide tracts of intrusive granite, no complete stratigraphical classification has yet been published, but the following provisional table gives a general summary of the determined succession.

TABLE A.

PROVISIONAL TABLE OF PRE-CAMBRIAN FORMATIONS OF SOUTHERN RHODESIA.

Lomagundi System
> *Great unconformity with intrusion of granites*

Upper Sedimentary Series
> *Unconformity*

Lower Sedimentary Series
> *Unconformity ; orogeny and intrusions*

Greenstone System ⎰ Upper Series
⎱ *Unconformity*
Lower Series
> *Great unconformity with intrusion of granites*

Ancient " Magnesian Series."

The stratigraphical succession is discussed in several bulletins of the Geological Survey, notably Nos. 20, 22, 30 and 31.

In no single area is the whole sequence developed. A great thickness of volcanic rocks, basaltic and andesitic lavas and breccias generally with pillow structures forms a large proportion of all the gold belts. The original characters and the succession of these rocks are so uniform throughout the Colony, that one is justified in accepting the volcanic or Greenstone system as a marker horizon, applicable to all the belts. The underlying " Magnesian Series " has been recognized in certain areas only, but one or both of the overlying sedimentary series are found in all the larger belts. All these rocks were folded and intruded by granites, and had suffered denudation to the extent of thousands of feet before the deposition of the Lomagundi system, which is correlated on lithogical grounds with the Transvaal system. The Zwankendaba sedimentary group is bedded among the lavas in the Lower series of the Greenstone system, and is thus among the oldest of the rocks recognized in the majority of the gold belts.

The supposed algal structures are therefore very much older than the somewhat similar structures so carefully described by Professor R. B. Young (Young, 1932 and 1934) and still older than those of the late Pre-Cambrian Beltian of North America.

In the gold belts of the central plateau of Southern Rhodesia, as distinct from the border country of the Limpopo and Zambezi valleys, the regional metamorphism is generally low, with carbonates, sericite and chlorite as characteristic products. It is this low grade of metamorphism which has permitted the original structures in the limestone to be so well preserved.

THE LIMESTONE.

The limestone in the quarry is coloured in various shades of grey ranging almost to black. In places white carbonate veins are very numerous. These appear generally to be formed of dolomite, and other outcrops observed along the line of strike are mainly composed of this mineral. Mr. Williams, the owner of the quarry, regards a dark colour and fine grain as indications of the best rock for burning. A ready sale is obtained for the lime both for building and cyanide purposes. In the burned lime the dolomitic veins assume a darker colour than the limestone matrix which is partially bleached in the kilns.

The chemical characters of the limestone are indicated by the following analysis " A " of a selected specimen of black limestone. For comparison an analysis is given of a similar rock of about the same age from the Que Que Lime Works. The analyses were made by Mr. E. Golding in the Geological Survey Laboratory.

TABLE B.

ANALYSES OF GREY LIMESTONES.

		A Per cent.	B Per cent.
Silica and insoluble silicates	... SiO_2, etc.	1·54	0·76
Alumina and iron oxide ...	... Al_2O_3, Fe_2O_3 ...	3·42	0·80
Magnesia	... MgO	2·00	1·03
Lime	... CaO	50·46	53·22
Phosphorous pentoxide ...	... P_2O_5	0·02	—
Manganous oxide	... MnO	0·03	—
Carbon dioxide	... CO_2	41·80	42·99
Carbon	... C	0·56	1·08
Water	... H_2O	0·39	0·50
	TOTALS	100·22	100·38

In published analyses of limestones the presence of graphite or other free carbon is rarely mentioned. As the graphitic dust floats very conspicuously when the rock is dissolved in acid, this probably indicates that as a rule the quantities are too small to be worth recording. It is, however, mentioned (Clarke, 1924) that the Laurentian limestones of Canada contain large quantities of graphite. This tempts one to enquire to what extent the presence of free graphite is characteristic of the most ancient limestones.

ALGAL STRUCTURES.

On the writer's first visit to the quarry in 1935 the concretionary structure of the limestone was observed and the two diagrammatic sketches illustrated below (Figs. 1 and 2) were made, but he was not convinced that the algal origin could be considered established until laboratory study proved that the dark colouration of the rock was due principally to films of graphite. A later opportunity to visit the quarry occurred in 1938 when a more careful study of the structures was made and more material was collected for examination. The face, however, from which the sketches were made had in the meantime been removed by quarrying.

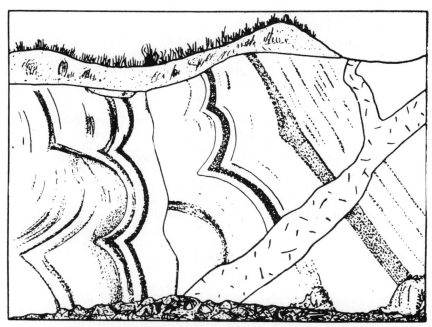

Supposed algal structures, Huntsman Lime Works, thirty-three miles
north-north-east of Bulawayo.

Fig. 1.—West face of quarry, 1935.   Length about fifteen feet.

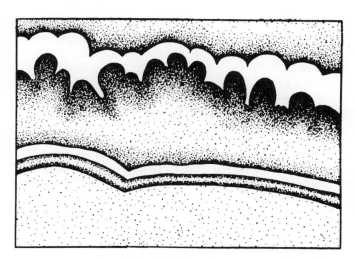

Fig. 2.—Concretionary limestone.   Length about twenty inches.

The limestone may be mottled or massive, but is frequently banded. In its simplest form the banding has the nature of regular bedding. Separate bands vary from fine streaks to beds a few inches thick, and appear to depend upon the varying grain of the calcite and the distribution of the graphite. In other parts the banding shows extraordinary convolutions which are clearly formed, as may be seen in Fig. 1, by concretionary deposition, and not by folding.

Three principal forms of this structure can be recognized, and may be described loosely as domical, dentate and columnar.

The domical form illustrated in the sketches may be seen in many parts of the quarry. The large size of these structures is striking. Plate II, Fig. 2, is a photograph of a weathered surface of rock from the part of the quarry sketched. The banding in this form is widely spaced ; the average width between the darker bands may be as much as 3 millimetres. The spaces are widest in the crest of the curves and narrow downwards towards the congested areas separating the domes. A thin section is illustrated in Plate III, Fig. 1. The structure resembles forms figured and described as Collenia by C. D. Walcott (Walcott, 1914) and Professor R. B. Young (Young, 1932, Plate IV, Fig. 2, and Plate V, Fig. 1). It is probable that the domes contain structures of different kinds, as the dentate form seems to occur as forming part of a large dome in Plate II, Fig. 2.

The dentate form is typically represented by the black bands shown in the photograph of the quarry face (Plate II, Fig. 1). This shows clearly two bands with dentate upper surface resulting from the development of conical or, rather, parabolic protuberances a quarter of an inch to an inch in basal diameter. In the time available it was not possible to collect a specimen from the face, but a piece of similar rock was found on the dump. The dark band is very fine grained and contrasts strongly with the paler mottled colour of the coarser limestone above. A photograph of this specimen, Plate IV, shows the appearance after etching with dilute acid. This treatment brings out the fine lamellar structure of the dark band. Minute films of graphite and apparently silica project from the surface lying parallel with the dentate surface of the band. Under a lens the films are seen to be rather regularly spaced, three to five to the millimetre. Some of the films are rather stronger than others, but these are not spaced at regular intervals. In order to determine the shape of the " teeth " in plan, one end of the specimen was sawn off and horizontal sections cut. These prove that the cone concerned has an elliptical form in plan. A thin section of another specimen of a similar structure is shown in Plate III, Fig. 2.

The third or columnar structure occurs at the northern end of the quarry, where it forms a continuous bed through two pits. In this rock dense black columns two to three inches wide and about a foot long rise from a rather narrow dark band. The photograph (Plate V, Fig. 1) is of an oblique surface showing the columns partly in section. A thin section of the columnar rock is shown in Plate V, Fig. 2. The graphite films in this section are stronger than in the dentate form, are not so regulary disposed, and are more widely spaced.

In the examination of several microscope slides of limestone no definitely cellular structure has been recognized, although rings of dots in places might be so interpreted. Although the grain size is fine for a crystalline limestone, it is still too coarse for minute or delicate organic structures to be retained. The carbon is in the form of minute specks, not as continuous films as megascopic examination suggests.

Bands of chert, which are almost black in the hand specimen, are sometimes present in the limestone. They form layers in the spheroidal growths. Under the microscope the grain is seen to be extremely fine and the rock is therefore a promising one for detailed study. Specks of carbon are present in fine bands, which may be regularly spaced about 0·02 mm. apart. Elsewhere the banding is less definite, but there are numerous circular or elliptical spaces formed of clear chert of slightly coarser grain than the matrix, and surrounded by a thin dark line. In several cases there are secondary incomplete coats on one side which in the section appears to be always the same, and is presumably the top. These structures are comparable with the *oncolites* described by Dr. Young.

CONCLUSION.

In writing this note it has been my intention to record the occurrence, rather than to describe in detail the various algal structures in the Zwankendaba limestone of the Huntsman Lime Works. The literature on the subject is growing rapidly, and many new names are being introduced for forms which are not always easy to distinguish by the descriptions.

The fossils are really little more than middens formed of precipitated calcium carbonate, silica and carbon, deposited beneath and around the tangled threads of simple plants.

Some general forms have been repeated throughout the geological record, but the variety of different forms occurring at this and other localities, suggests that several species and genera took part in the deposition. In the earliest geological times when the algae were undergoing their great evolutionary diversification it is probable that distinct structures were developed which have not been repeated. Is it too much to hope that the study of the widespread algal deposits will be made to throw a light on the stratigraphical relations of the middle and later Pre-Cambrian rocks of the world ?

In ending I wish to express my thanks to Mr. N. E. Barlow for his help in the preparation, and to Mr. F. L. Amm for photographs of the thin sections. I am particularly grateful to Dr R. B. Young for his help and interest.

LIST OF REFERENCES.

CLARKE, F. W. 1924. The Data of Geochemistry. *United States Geological Survey Bulletin* 770, p. 331.

MACGREGOR, A. M. 1932. The Geology of the Country around Que Que, Gwelo District. *Southern Rhodesia Geological Survey Bulletin* 20.

MACGREGOR, A. M. 1937. The Geology of the Country around Hunters Road, Gwelo District. *Southern Rhodesia Geological Survey Bulletin* 31.

MACGREGOR, A.M., FERGUSON, J. C., and AMM, F. L. 1939. The Geology of the Country around the Queen's Mine, Bulawayo District. *Southern Rhodesia Geological Survey Bulletin* 30.

TYNDALE–BISCOE, R. McI. 1933. The Geology of the Central Part of the Mazoe Valley Gold Belt. *Southern Rhodesia Geological Survey Bulletin* 22.

WALCOTT, C. D. 1914. Pre-Cambrian Algonkian Algal Flora. Cambrian Geology and Palaeontology III, No. 2. *Smithsonian Miscellaneous Collection.* Vol. 64, No. 2.

YOUNG, R. B. 1932. The Occurrence of Stromatolitic or Algal Limestone in the Campbell Rand Series of Griqualand West. *Transactions of the Geological Society of South Africa.* Vol. XXXV, pp. 29–36.

YOUNG, R. B. 1934. A comparison of Certain Stromatolitic Rocks in the Dolomite Series of South Africa with Modern Algal Sediments in the Bahamas. *Transactions of the Geological Society of South Africa.* Vol. XXXVII, pp. 153–62.

Fig. 2.—Massive domed limestone, dolomitic vein on left.

Structures in limestone, Huntsman Lime Works.

Fig. 1.—Dentate bands in west face of quarry.

PLATE III.

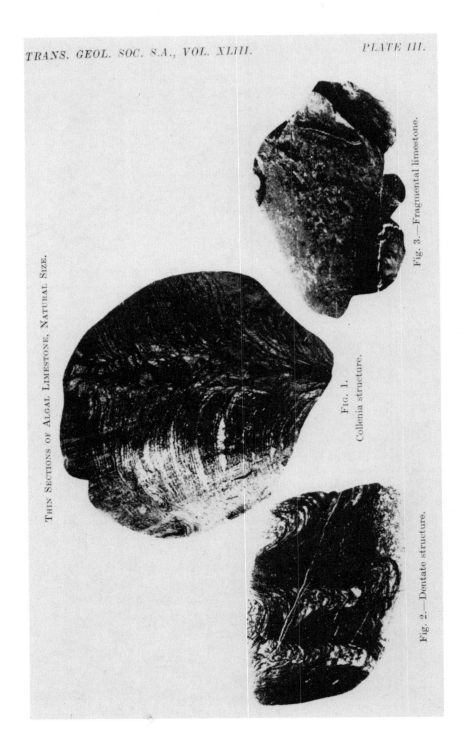

Thin Sections of Algal Limestone, Natural Size.

Fig. 1.
Collenia structure.

Fig. 2.—Dentate structure.

Fig. 3.—Fragmental limestone.

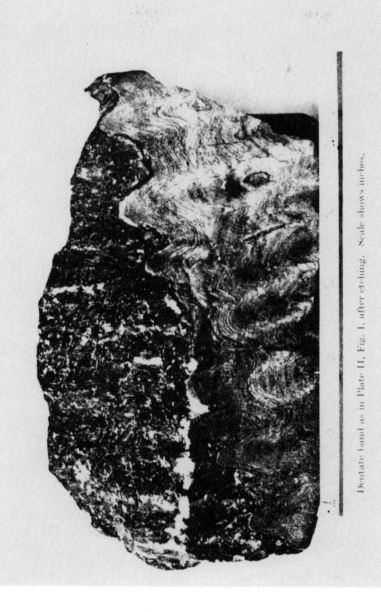

Dentate band as in Plate II, Fig. 1, after etching. Scale shows inches.

COLLUMNAR STRUCTURES,

Fig. 1.— Photograph of inclined surface, north face of quarry.

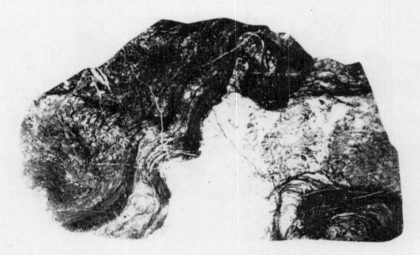

Fig. 2.— Thin section of the same. Nine-tenths natural size.

40

Reprinted from *Jour. Paleon.*, **45**(3), 477–485 (1971)

BIOGENICITY AND SIGNIFICANCE OF THE OLDEST
KNOWN STROMATOLITES

J. WILLIAM SCHOPF, DOROTHY Z. OEHLER, ROBERT J. HORODYSKI, AND
KEITH A. KVENVOLDEN
Department of Geology, University of California, Los Angeles
Exobiology Division, Ames Research Center, Moffett Field, California

ABSTRACT—Optical microscopic and carbon isotopic studies have demonstrated the biologic origin of the oldest stromatolitic structures known, occurring in limestones of the Bulawayan Group of Rhodesia, more than 2.6 billion years in age; this conclusion substantiates an interpretation first suggested by A. M. Macgregor in 1941. In part, the laminae of the specimens analyzed are stratiform, comprising short columns of laterally contiguous, stacked hemispheroids and representing a stromatolitic form not previously described from the Bulawayan sediments. A difference in δ C^{13} values of about 32 parts per thousand between carbonate carbon and reduced organic carbon coexisting in these structures seems indicative of photosynthetic, carbon isotopic fractionation. The gross morphology, second-order organization and carbon isotopic composition of the Bulawayan structures support their interpretation as remnants of laminated biocoenoses composed of filamentous blue-green algae and associated microorganisms. These Early Precambrian stromatolites appear to place a minimum age on the time of origin of cyanophycean algae, oxygen-producing photosynthesis, the filamentous habit, and integrated biological communities.

INTRODUCTION

STROMATOLITES are finely laminated organo-sedimentary structures, commonly calcareous in composition and columnar to hemispherical in gross morphology, resulting from accretion of detrital and precipitated minerals on successive sheet-like mats formed by communities of microorganisms, predominated by filamentous blue-green algae. Beginning as early as 1858, with studies of *Eozoon canadense* by W. E. Logan (summarized by Dawson, 1875), finely layered stromatolite-like deposits have been postulated as evidence of Precambrian life, an interpretation that has been subject to considerable criticism (e.g., Seward, 1941, p. 79–92). Recently, however, with increased understanding of modern algal biocoenoses (Ginsburg et al., 1954; Logan, Rezak and Ginsburg, 1964; Monty, 1967; Sharp, 1969; Gebelein, 1969) and based on the discovery of cellularly preserved cyanophytes in fossil stromatolitic deposits (Barghoorn and Tyler, 1965; Licari, Cloud and Smith, 1969; Gustadt and Schopf, 1969; Cloud et al., 1969), the algal origin of several Precambrian stromatolites has been firmly established. It seems reasonable to infer, therefore, that most Precambrian stromatolitic sediments, and particularly those occurring in mound-shaped masses and exhibiting alternating carbonaceous and inorganic laminations, are similarly biogenic.

The gross morphology and microstructure of

EXPLANATION OF PLATE 56 (see following page)

Stromatolitic limestone (sample Bul-7) from the Early Precambrian Bulawayan Group, Huntsman Limestone Quarries, Rhodesia. Line for scale in each figure represents 1.0 cm.

FIG. 1—Weathered surface showing crescentic laminations.
2—Petrographic thin section showing three zones of different microstructure: *i*) A lower zone, composed of convex-upward lamellae and interspersed calcite lenses; *ii*) a middle zone predominantly composed of sparry calcite; and *iii*) an upper zone characterized by planar laminations.
3—Thin section showing laterally contiguous columnar aggregates of closely spaced, stacked, hemispheroidal lamellae in the lower zone.
4—Thin section showing lower, middle and upper zones; note the weak tendency toward columnar organization of organic lamellae in the middle and upper zones.
5—Thin section showing weakly columnar organization of carbonaceous layers in the middle zone.

477

Schopf, Oehler, Horodyski
& Kvenvolden

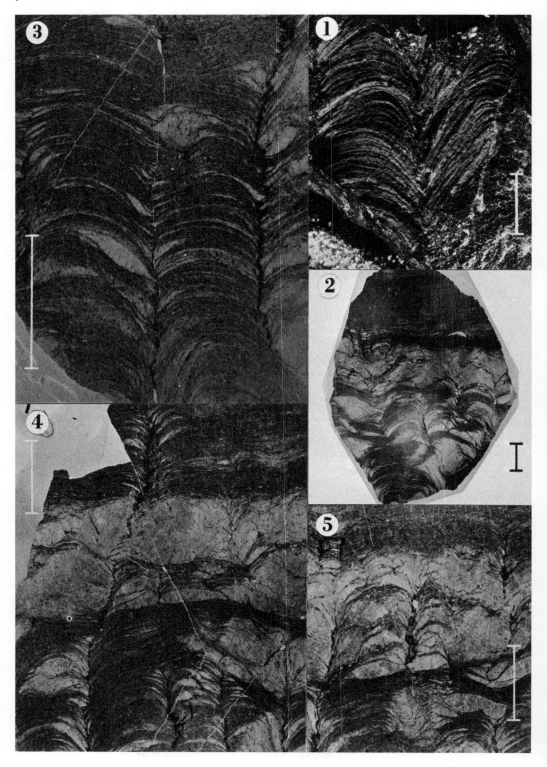

stromatolites result from complex, and as yet poorly understood, biotic-environmental inter-actions. Despite difficulties of identification aris-ing from the use of diverse taxonomic systems by stromatolite specialists (summarized by Hof-mann, 1969), some form-genera appear to have limited time-ranges and wide geographic distri-butions; such forms have been proposed as "in-dex fossils" for intra- and inter-continental bio-stratigraphic correlation of Late Precambrian deposits (Raaben, 1969; Glaessner, Preiss and Walter, 1969; Cloud and Semikhatov, 1969). Correlations based on the occurrence of stroma-tolitic zones in strata of differing lithologies have been corroborated by radiometric data (Raaben, 1969); it seems likely, therefore, that the morphology of at least some form-genera may be influenced primarily by biotic, rather than environmental factors. Although the mi-croorganisms responsible for stromatolite deposi-tion are rarely preserved, the few known micro-biotas associated with Precambrian stromato-lites (Barghoorn and Tyler, 1965; Schopf, 1968; Cloud et al., 1969; Hofmann and Jackson, 1969) apparently comprise integrated and perhaps partially symbiotic communities (Schopf and Barghoorn, 1969). Thus, changes in stromatoli-tic morphology, resulting in limited time-ranges of form-genera, might reflect the rise to domi-nance of a previously subordinate member of the biota, or the introduction of a new thallo-phytic stock that resulted in readjustment of symbiotic relationships and a concomitant change in growth habit.

The oldest stromatolitic structures known are apparently those discovered in 1935 by A. M. Macgregor at the Hunstman Limestone Quar-ries in southwestern Rhodesia, in limestones of the Early Precambrian Bulawayan Group, thought to be more than 2.6 billion years in age (Nicolaysen, 1962; Haughton, 1969, p. 33). Be-tween the time of Macgregor's discovery of these finely layered structures and the publica-tion of his report (1941), a quarry face show-ing the structures exposed *in situ* had been re-moved; subsequently, the quarry was abandoned and partially flooded. The very great age of the Bulawayan Group was not fully appreciated un-til much later (Holmes, 1954), by which time the stromatolitic locality was no longer accessi-ble and only small chunks of the distinctively laminated limestone, scattered among talus at the quarry site, were available for examination. In recent years it has become increasingly evi-dent that not only are the Bulawayan structures apparently the oldest stromatolitic deposits known, but in addition, that they seem to be of an anomalously great age; stromatolites are not widespread in the geologic record until the Mid-dle Precambrian (being particularly abundant in sediments younger than 2.3 billion years) and only one other presumably Early Precambrian occurrence, in calcareous deposits at Steep Rock Lake, Ontario, has been noted (Jolliffe, 1955).*
For these reasons, and despite Macgregor's well-illustrated and soundly reasoned report, his interpretation of an algal origin for the Bula-wayan structures has been doubted, accepted only with reservation, or regarded as probably correct but incompletely demonstrated (e.g., Andrews, 1961, p. 49; Cloud and Abelson, 1961, p. 1706; Murray, 1965, p. 10; Glaessner, 1966a, p. 36 and 1966b, p. 309; Cloud, 1965, p. 32 and 1968, p. 11; Oparin, 1968, p. 196; Houghton, 1969, pp. 29, 500; for contrasting views see Schopf and Barghoorn, 1967, p. 508; Hoering, 1967, p. 91; Schopf, 1969, p. 153 and 1970; Cloud and Semikhatov, 1969, p. 1030).

The present study of finely laminated, unusu-ally well-preserved specimens of stromatolitic limestone from the Huntsman Quarries pro-vides additional morphological and chemical data with which to assess Macgregor's interpreta-tion. Although the systematic description and taxonomic designation of these forms, needed for purposes of reference, await the availability of additional material, the results here reported fully substantiate an algal origin for the Bula-wayan structures; it seems evident that stroma-tolitic communities of filamentous, microscopic, photosynthetic blue-green algae and associated microorganisms were extant during the Early Precambrian.

MATERIAL AND METHODS

Stratigraphy.—Samples of stromatolitic lime-stone and laminated carbonaceous chert were collected in 1968 by one of us (K.A.K.) from ·talus piles, mine dumps and exposed quarry walls at the Huntsman Limestone Quar-ries near Turk Mine, about 55 km north-north-east of Bulawayo, Rhodesia (text-fig. 1). Ac-cording to the stratigraphic sequence established by Macgregor (1947, 1951), these sediments (the "Zwankendaba limestone" of Macgregor, 1941) are referred to the lower portion of the Bulawayan Group (the "Greenstone System" of earlier publications), a thick volcanic sequence with prominent pillow lavas containing inter-bedded units of banded ironstone and associated graywackes, conglomerates, quartzites, cherts

* The deformed "oolitic" objects of suggested algal origin reported by Ramsay (1963) from the Early Precambrian Fig Tree Group of South Africa are neither stromatolitic nor, as Cloud (1968) has noted, convincingly biogenic.

and limestones that comprises the bulk of the Rhodesian gold belts (Macgregor, 1951, pp. xxxiii–xxxv).

Age.—Isotopic age determinations, using the strontium, argon and lead techniques, on pegmatites intruding rocks of the Bulawayan Group near Bikita (265 km east of Turk Mine) and Salisbury (330 km northeast of Turk Mine) indicate a minimum age of about 2,600 million years for the Bulawayan sediments (summarized by Holmes and Cahen, 1957, p. 86 and by Nicolaysen, 1962); a maximum age of about 3,100 million years is suggested by potassium/argon ratios measured in a granitic boulder obtained from a conglomeratic unit near the base of the Bulawayan Group at the Sebakwe River, 130 km northeast of Turk Mine (Wasserburg, Hayden and Jensen, 1956). Based on these data, and on correlations with the Nyanzian Group of Kenya, the Bulawayan Group is generally regarded as being 2,800 to 3,100 million years in age (Holmes and Cahen, 1957, pp. 91, 103; Furon, 1963, p. 329; Holmes, 1965, pp. 369, 371); Haughton (1969, p. 60) has discussed the possible contemporaneity of the Bulawayan Group and members of the Swaziland Supergroup of the eastern Transvaal, a correlation suggestive of an age in excess of 3,000 million years.

Morphology and mineralogy.—Twelve specimens of the Bulawayan limestone, each having a maximum dimension of 10 to 15 cm and exhibiting one or more areas of iron-stained, crescentic laminations on weathered surfaces (text-

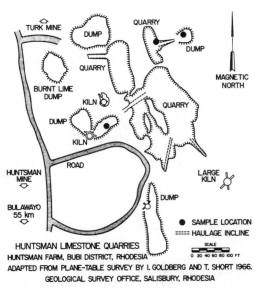

TEXT-FIG. 1—Map of Huntsman Limestone Quarries, southwestern Rhodesia, showing sample localities.

fig. 2), were selected for study. The most promising of these (sample Bul-7; approximately 10 × 9 × 7 cm in size), exhibiting seven areas of lamination on external surfaces, (Pl. 56, fig. 1), was gently broken along planes of weakness perpendicular to the laminations exposing five additional laminated areas; these interior areas were similarly iron-stained, presumably by percolating ground water. The broken fragments were photographed and a plaster cast of each was prepared; the pieces were then cemented together with epoxy resin in their original configuration, and a plaster cast was prepared of the total specimen. The plaster casts and photographs were used for reference in the study of both sides of a series of polished thick sections of the specimen, serially cut at intervals of 2.5 mm, oriented perpendicular to the laminations; petrographic thin sections were subsequently prepared from these serial thick sections.

Optical microscopic studies were also made of cellulose acetate peels of polished and acid-etched rock fragments and thin sections, of hydrochloric acid-resistant organic residues, and of petrographic thin sections and serial sections of five other stromatolitic specimens. Hydrofluoric acid-resistant organic residues and thin sections of chert samples were similarly investigated. Optical microscopic identification of sparry calcite in specimens of stromatolitic limestone was confirmed by electron probe X-ray microanalysis in comparison with calcite and dolomite standards, and by staining with alizarin red S (Friedman, 1959).

Carbon isotopes.—C^{13}/C^{12} ratios of coexisting organic and inorganic (carbonate) material in several specimens of the Bulawayan limestone were measured following an adaptation of a method described by Craig (1953); δ C^{13} values were determined in comparison with a PDB standard*. Each sample was ground to a fine powder in a "shatter box" disc pulverizer, and was extracted to remove surficial organic contamination with a hot refluxing solution of redistilled benzene methanol (70:30 by volume) for 24 hours. Approximately 0.05 gm of the extracted sample was reacted under vacuum with 1N HCl; the CO_2 released (from carbonate phases) was analyzed on a dual collecting, Nuclide Corporation Mass Spectrometer. The remaining portion

* $$\delta C^{13}(°/_{oo}) = \frac{(C^{13}/C^{12} \text{ sample} - C^{13}/C^{12} \text{ standard})}{C^{13}/C^{12} \text{ standard}} \times 1000$$

The PDB standard (Peedee belemnite) here used is from the Peedee Formation, Upper Cretaceous of South Carolina. The National Bureau of Standards (NBS) reference sample No. 20 (Solenhofen limestone) used by other workers (e.g., Hoering, 1967) equals + 1.37 ‰ relative to PDB.

TEXT-FIG. 2—Stromatolitic limestone from the Early Precambrian Bulawayan Group, Huntsman Limestone Quarries, Rhodesia, showing well-defined crescentic laminations on a weathered surface.

of each sample was reacted with NCl to remove carbonates and with HF to remove silicates; the resulting acid-resistant organic residue was combusted in an oxygen atmosphere at 1000°C and the CO_2 generated was analyzed for carbon isotopic content.

RESULTS

Stromatolitic morphology.—Macgregor (1941) provisionally recognized three categories of "algal structures" in the Bulawayan limestones: 1) Forms columnar in gross morphology, 5 to 7.5 cm in diameter and about 30 cm in length, exhibiting rather widely spaced (0.25 to 2.5 mm apart), irregularly distributed, graphitic laminations; 2) domical structures, about 1.2 m in diameter and 80 cm in height, having widely spaced (approximately 3 mm apart) graphitic lamellae that exhibit both strongly arched and dentate organization; and 3) extensive laminated beds, 2.5 to 8 cm in thickness, having an undulatory upper surface and exhibiting a dentate second-order organization composed of upwardly-convex, stacked, parabolic cones (about 2 cm in basal diameter) of alternating calcareous and graphitic laminations.

Because the specimens here studied represent only fragments of stromatolites, their description is based solely on second-order morphology; their relationship to Macgregor's three categories of "algal structures" is therefore somewhat uncertain. On cursory examination, the crescentic laminations evident on weathered surfaces (text-fig. 2) appear roughly comparable to the dentate organization exhibited by categories two and three, above. Studies of serial sections demonstrate, however, that the lamellar geometry of these specimens varies from highly disordered (cf. category 1) to relatively regular. Where most regularly organized, the carbonaceous laminae are gently convex-upward and are closely and rather uniformly spaced (pl. 56, fig. 3); they comprise short vertical columns of roughly constant diameter, composed of laterally contiguous stacked hemispheroids. This stratiform second-order morphology markedly differs from the conical dentate organization noted by Macgregor, and apparently represents a stromatolitic form not previously described from the Bulawayan sediments.

As is shown in plate 56, figure 2, the stromatolitic specimen studied in detail by means of serial polished sections and serial thin sections (sample Bul-7) is composed of three zones having different microstructure: *i*) A lower zone composed of convex-upward carbonaceous and calcareous laminae and interspersed concavo-convex (less commonly plano-convex and biconvex) lenses of sparry calcite; *ii*) a middle zone predominantly composed of sparry calcite lenses and containing relatively few, commonly highly disordered, organic layers; and *iii*) an upper zone composed almost entirely of planar to slightly undulatory, closely spaced, thin carbonaceous and calcareous laminations.

Although incompletely developed in the thin section shown in Plate 56, figure 2, the convex laminae of the lower zone comprise short regular columns of stacked hemispheroids (pl. 56, fig. 3). The laminae are smooth, diffuse to distinct, 0.05 to 0.2 mm thick, and are composed of alternating layers of micro-granular (0.01 to 0.02 mm in diameter) calcite and particulate organic matter; they vary from slightly to strongly convex-upward with a ratio of width to height commonly between 3:1 and 5:1. Where relatively well-defined, the columns are cylindrical and 1.5 to 5 cm in length; column diameters generally vary from 0.8 to 1.3 cm, but margins with adjacent columns are commonly indistinct due to lateral fusion resulting in columnar structures of somewhat larger diameter (up to 2.5 cm). Weathering and iron-staining of these laterally contiguous columns produced the crescentic laminations evident on external surfaces (text-fig. 2; pl. 56, fig. 1).

As seen in Plate 56, figures 2 through 4, discrete lenses of sparry calcite, as much as 10 mm in width and 2 mm in thickness, disrupt the symmetrical distribution of lamellae in the lower zone; these pods may span an entire column, but more commonly they appear to be concentrated at the margins of hemispheroids. They are composed of microgranular sparry calcite (0.01 to 0.06 mm in diameter) similar to that comprising calcareous laminae, and although they are generally devoid of organic matter and have sharp contacts with adjacent laminations, they are in some places transected by wispy carbonaceous layers. These lenses seem generally comparable to the calcitic "birdseye" open spaced structure reported from other stromatolitic deposits and interpreted as evidence of a supratidal to intertidal environment (Ham, 1952; Wolf, 1965; Tebbutt, Conley and Boyd, 1965; Schenk, 1967). Their distribution, morphology and mineralogy suggest that they have resulted from carbonate precipitation within originally gas-filled pockets in the laminated sediments.

The middle zone (pl. 56, fig. 2) consists almost entirely of sparry calcite lenses, mineralogically similar to those of the lower zone; these lenses are separated by a few carbonaceous layers that are highly disordered or are subparallel and exhibit a weak tendency toward columnar organization (pl. 56, figs. 4, 5).

The upper zone (pl. 56, figs. 2, 4) is composed of thin alternating calcareous and organic lamellae comparable to those of the lower zone except that they are somewhat more closely spaced, are much less undulatory, and generally are not interrupted by lenses of calcite spar. A very weak tendency toward columnar organization may be seen near the bottom of this zone in Plate I, figure 4: An irregularly serrated, dark vertical line, marking the plane of juncture of adjacent columns in the lower zone, passes through the middle zone (where it represents the margin of two poorly defined columns) and in the upper zone marks a plane of flexure of organic lamellae and the occurrence of sparry calcite lenses comparable to those associated with the margins of stacked hemispheroids in the lower zone.

Micropaleontologic studies.—In an attempt to detect microfossils in the Bulawayan sediments, optical microscopic studies were made of petrographic thin sections and acid-resistant organic residues of laminated black cherts and stromatolitic limestones. Thirty thin sections were examined in detail; with the possible exception of minute organic particles marginally suggestive of fossil bacteria, no microfossils were detected. This apparent absence of preserved microorganisms is not surprising; the grain size of the calcareous sediments, although relatively fine for limestones, is generally much coarser than that typical of deposits containing well-preserved, organic-walled microfossils. In addition, much of the calcareous matrix has been recrystallized, a process that commonly obliterates organically preserved microorganisms. Comparative studies of Recent calcareous stromatolites containing partially degraded algal filaments (collected by A. T. Groves on the shores of pluvial Lake Stephanie in southern Ethiopia and thought to be 5,000 to 10,000 years old) suggest that recrystallization and the resulting physical destruction of microorganisms is a relatively rapid geologic process. It is probably for this reason that, with but few exceptions (e.g., Walcott, 1914; Schopf and Barghoorn, 1969), Precambrian calcareous stromatolites and laminated sediments appear devoid of microfossils.

As noted by Macgregor (1941), the carbonaceous Bulawayan cherts are finer grained than the associated limestones and might therefore be expected to be more promising for micropaleontologic studies. That microfossils were not detected is probably attributable to the relatively high degree of chemical alteration of the included organic matter. Recent studies have shown that certain Precambrian black cherts are profusely fossiliferous (Barghoorn and

Tyler, 1965; Schopf, 1968; Cloud et al., 1969); in all such cases the organic-walled microfossils are reddish-amber or brown in color, a character reflecting the absence of marked geochemical degradation (Staplin, 1969). In contrast, the particulate carbonaceous components of the Bulawayan limestones and cherts are black, opaque and apparently graphitic in composition. This degree of chemical alteration is typical of the organic matter preserved in very ancient sediments and provides a partial explanation for the paucity of known Early Precambrian microfossils (Schopf, 1970).

Although the samples here studied do not contain recognizable microorganisms, Oberlies and Prashnowsky (1968) detected spinose spheroids ("Mikrosporen") in surface replicas of specimens of Bulawayan limestone studied by electron microscopy; based on published photomicrographs, there can be little doubt that these organic structures are biogenic. However, despite precautions taken to avoid introduction of modern contaminants, the suggested Early Precambrian age of these spheroids (and of amino acids reported in the same paper; for discussion, see Schopf, 1970) appears to be less than fully substantiated. Additional electron microscopic evidence of the relationship of these sporelike bodies to mineralogic features of the Bulawayan sediments (e.g., Schopf et al., 1965; Barghoorn and Schopf, 1966), or optical microscopic evidence of their occurrence in petrographic thin sections, seems needed to firmly demonstrate their presumed indigenous nature.

Carbon isotopic analyses.—Nier and Gulbrandsen (1939) were apparently the first to note that the C^{13}/C^{12} ratio of plant material is significantly lower than carbon isotopic ratios typical of carbonate minerals and atmospheric carbon dioxide. Park and Epstein (1960, 1961) have suggested that this difference arises from isotopic fractionation during the first steps in the biochemical reduction of CO_2 and that photosynthetically produced organic matter is thereby enriched in C^{12} relative to the atmospheric carbon dioxide reservoir and to inorganic carbonate rocks. Carbon isotopic ratios (expressed as δC^{13} values) have been determined for several hundred samples of inorganic and biological origin, and of both modern and Phanerozoic age (summarized by Degens, 1965, 1969, and by Welte, 1965); relatively few analyses have been made of Precambrian carbon (Hoering, 1967; Smith, Schopf and Kaplan, 1970). In all of these determinations, including those of Precambrian samples, δC^{13} values of carbonaceous material of demonstrable photosynthetic origin or derivation are significantly

lower than values exhibited by coexisting carbonate minerals; fossil organic matter has a δC^{13} value between $-24\%o$ and $-34\%o$, whereas the isotopic composition of marine limestones, such as those of the Bulawayan Group (Reynolds, 1965), is typically in the range from $+5\%o$ to $-5\%o$. Carbon isotopic distributions of this type in unmetamorphosed Precambrian sediments have been interpreted as evidence of photosynthetic fractionation (Barghoorn, Meinschein and Schopf, 1965; Hoering, 1967; Schopf, 1970).

TABLE 1—Carbon isotopic composition of coexisting carbonates and reduced organic carbon in samples of stromatolitic limestone from the Bulawayan Group, Huntsman Limestone Quarries, Rhodesia

Sample	δC^{13} reduced organic	δC^{13} oxidized carbonate	δC^{13} difference ($\%o$)
Bul-5	−33.5	+0.8	34.3
Bul-9	−32.5	+1.2	33.7
Bul-8	−32.1	+0.9	33.0
Bul-3	−31.8	−0.8	31.0
"stromatolite"[a]	−30.5[a]	−2.1[a]	28.4
Bul-1	not determined	−0.7	—
Bul-2	not determined	0.0	—

[a] Measurements by Hoering (1967) using NBS standard converted relative to PDB standard for comparison with values here reported.

In Table I are shown the δC^{13} values of coexisting carbonates and organic carbon measured in samples of the stromatolitic Bulawayan limestone; for comparison, a pair of values reported by Hoering (1967) is also listed. The average δC^{13} value of the reduced organic carbon (5 measurements) is $-32.1\%o$ and that of the oxidized carbonate carbon (7 measurements) is $-0.1\%o$; the average difference in δC^{13} values measured between coexisting organic and inorganic carbon fractions (5 measurements) is about 32 parts per thousand.

DISCUSSION AND SIGNIFICANCE

The data here presented seem clearly to indicate that the Bulawayan structures are of algal origin and, as such, they fully substantiate Macgregor's (1941) original interpretation. Evidence supporting this conclusion may be summarized as follows:

1) The three categories of stromatolitic gross morphology recognized by Macgregor are exhibited by modern and fossil algal biocoenoses.

2) The second-order organization reported by Macgregor, and the columnar aggregation of stacked hemispheroids here described, are simi-

larly exhibited by stromatolitic structures of established algal origin; they primarily result from a phototrophic growth habit characteristic of microorganisms comprising stromatolitic algal communities.

3) The rather uniform spacing of the carbonaceous laminae comprising short vertical columns (Pl. 56, fig. 3), their symmetrical geometry and their regular alternation with calcareous layers are characteristic of stromatolitic algal mats, both fossil and modern.

4) The occurrence of calcite lenses at columnar margins, interpreted as infilled gas bubbles, is a feature exhibited by younger stromatolites and is consistent with a biological origin; these bubbles may have resulted from trapped CH_4, H_2 and CO_2 generated by bacterial decay of dead algal filaments, or they may have initially contained photosynthetically produced oxygen and afforded limited buoyancy to the biological community.

5) The large difference in δ C^{13} values between coexisting organic and inorganic carbon compounds in the unmetamorphosed Bulawayan sediments is most reasonably interpreted as having resulted from carbon isotopic fractionation during the biological fixation of carbon dioxide by photosynthetic microorganisms.

6) The concatenation of stromatolitic gross morphology, second-order organization of carbonaceous lamellae, and carbon isotopic composition and distribution exhibited by the Bulawayan structures cannot be explained by known nonbiological processes (Donaldson, 1963); these features are typical only of biological communities chiefly composed of phototrophic microorganisms exhibiting oxygen-producing (i.e., non-bacterial) photosynthesis.

7) The apparent occurrence of spheroidal microfossils in the Bulawayan sediments reported by other workers is consistent with a biological origin for the stromatolitic structures; the presumed indigenous nature of these spinous spheroids, however, needs additional confirmation.

Because only second-order morphology is exhibited by the specimens here studied, their relationship to named stromatolitic groups is uncertain. Based on the data available, however, the range of possible affinities can be narrowed. Thus, form-genera characterized by highly branched second-order organization (e.g., *Anabaria, Gymnosolen*), encapsulating laminae (e.g., *Osagia, Ottonosia*) and conical laminae (e.g., *Conophyton, Jacutophyton*) appear to be excluded. The stratiform organization of the specimens examined seems generally comparable to that exhibited by *Weedia, Stratifera* and,

to a lesser degree, *"Collenia" kona* (personal communication, H. J. Hofmann, 1970). However, studies of intact stromatolites, showing the relationship between microstructure and gross morphology are needed before proper taxonomic treatment can be given the Bulawayan structures.

These Early Precambrian algal stromatolites represent an important benchmark in early biological evolution. The existence of photosynthetic microorganisms at the time of deposition of the Swaziland Supergroup, perhaps 300 million years earlier than Bulawayan sedimentation, has been inferred on the basis of micropaleontological and geochemical evidence (Pflug, 1966, 1967; Schopf and Barghoorn, 1967); this evidence, however, seems equally compatible with the presence of either algal or bacterial photosynthesizers (Schopf, 1970). Thus, although photoautotrophy (e.g., bacterial photosynthesis) was apparently an early biological innovation, algal photosynthesis, producing gaseous oxygen as a byproduct, may not have evolved until somewhat later. The Bulawayan algal stromatolites, with an apparent minimum age of 2.6 billion years and a probable age of between 2.8 and 3.1 billion years, apparently place a minimum age on the time of origin of this most important biochemical process, and of the appearance of biologically generated atmospheric oxygen.

By analogy with younger stromatolitic algal communities, both fossil and modern, the Bulawayan stromatolites appear to represent remnants of laminated biocoenoses composed of blue-green algae and associated microorganisms; the predominance of filamentous cyanophytes (Oscillatoriaceae, Nostocaceae) in the organic lamellae of such communities is well-established (Sharp, 1969; Barghoorn and Tyler, 1965) with successive organic-rich layers being produced as the phototactic algal trichomes and reproductive hormogones migrate upward (by gliding motility) through accumulated detritus, leaving behind layers of empty mucilaginous sheaths and dead algal filaments. Studies of modern algal biocoenoses at Shark Bay, Western Australia, have shown that laminated stromatolites are formed only when the community is dominated by filamentous forms; assemblages dominated by coccoid cyanophytes result in deposition of unlaminated "thrombolites" (personal communication, P. F. Hoffman, 1970). Thus, the Bulawayan deposit is also interpreted as placing a minimum age on the time of origin of cyanophycean algae, of the filamentous habit, and of integrated biological communities of procaryotic microorganisms presumably including producers (blue-green algae), reducers (aerobic and anaerobic bacteria) and consumers (bacteria, predatory by absorption). This interpretation is consistent with the known geologic distribution of cyanophycean families (Schopf, 1970) and is supported by the reported occurrence of filamentous and unicellular alga-like and bacterium-like microfossils in several Early Precambrian sediments (Gruner, 1925, p. 152; Pflug, 1966, 1967; Barghoorn and Schopf, 1966; Schopf and Barghoorn, 1967; LaBerge ,1967; Cloud and Licari, 1968).

ACKNOWLEDGEMENTS

We thank Mr. Euen Morrison, Rhodesian Geological Survey Office, Bulawayo, for his guidance and help in collecting the samples here described. In addition, we thank the following individuals of the Department of Geology, University of California, Los Angeles: Dr. Ian R. Kaplan, for providing facilities for isotopic analyses; Mr. Ed Ruth, for mass spectrometric determinations of carbon isotopes; Mr. Robert Jones, for electron probe analyses; and Mrs. Carol Lewis and Mr. John Oehler, for suggestions and assistance. This work was supported in part by NASA contract NAS 9-9941 to U.C.L.A.

REFERENCES

Andrews, H. N., Jr. 1961. Studies in Paleobotany. John Wiley & Sons, Inc., New York. 487 p.
Barghoorn, E. S., and S. A. Tyler. 1965. Microorganisms from the Gunflint chert. Science 147: 563–577.
Barghoorn, E. S., W. G. Meinschein, and J. W. Schopf. 1965. Paleobiology of a Precambrian shale. Science 148. 461–472.
Barghoorn, E. S., and J. W. Schopf. 1966. Microorganisms three billion years old from the Precambrian of South Africa. Science 152:758–763.
Cloud, P. E., Jr. 1965. Significance of the Gunflint (Precambrian) microflora. Science 148:27–35.
———. 1968. Pre-metazoan evolution and the origins of the Metazoa, p. 1–72. In E. T. Drake [ed.] Evolution and Environment. Yale Univ. Press, New Haven.
———, and P. H. Abelson. 1961. Woodring conference on major biologic innovations and the geologic record. Nat. Acad. Sci. (U.S.), Proc. 47:1705–1712.
———, and G. R. Licari. 1968. Microbiotas of the banded iron formations. Nat. Acad. Sci. (U.S.), Proc. 61:779–786.
———, G. R. Licari, L. A. Wright, and B. W. Troxel. 1969. Proterozoic eucaryotes from eastern California. Nat. Acad. Sci. (U.S.), Proc. 62:623–630.
———, and M. A. Semikhatov. 1969. Proterozoic stromatolite zonation. Amer. J. Sci. 267:1017–1061.
Craig, H. 1953. The geochemistry of stable isotopes of carbon. Geochim. Cosmochim. Acta 3:53–92.
Dawson, J. W. 1875. The Dawn of Life. Hodder & Stoughton, London. 239 p.
Degens, E. T. 1965. Geochemistry of Sediments. Prentice-Hall, Englewood Cliffs, New Jersey. 342 p.

———, 1969. Biogeochemistry of stable carbon isotopes, p. 304–330. In G. Eglinton, and M. T. J. Murphy [ed.] Organic Geochemistry. Springer-Verlag, New York.
Donaldson, J. A. 1963. Stromatolites in the Denault Formation, Marion Lake, coast of Labrador, Newfoundland. Geol. Surv. Canada Bull. 102. 33 p.
Friedman, G. M. 1959. Identification of carbonate minerals by staining methods. J. Sed. Petrol 29: 87–97.
Furon, R. 1963. Geology of Africa. Oliver & Boyd, London. 377 p.
Gabelein, C. D. 1969. Distribution, morphology, and accretion rate of Recent subtidal algal stromatolites. J. Sed. Petrol. 39:49–69.
Ginsburg, R. N., L. B. Isham, S. J. Bein, and J. Kuperberg. 1954. Laminated algal sediments of south Florida and their recognition in the fossil record. Marine Laboratory, Univ. Miami, Coral Gables, Florida, Rept. No. 54–21. 33 p.
Glaessner, M. F. 1966a. Precambrian paleontology. Earth-Sci. Rev. 1:29–50.
———, 1966b. The first three billion years of life on earth. J. Geography 75:307–315.
———, W. V. Preiss, and M. R. Walter. 1969. Precambrian columnar stromatolites in Australia: Morphological and stratigraphic analysis. Science 164: 1056–1058.
Gruner, J. W. 1925. Discovery of life in the Archaean. J. Geol. 33:151–152.
Gutstadt, A. M., and J. W. Schopf. 1969. Possible algal microfossils from the Late Pre-Cambrian of California. Nature 223:165–167.
Ham, W. E. 1952. Algal origin of the "Birdseye" Limestone in the McLish Formation. Oklahoma Acad. Sci., Proc. 33:200–203.
Haughton, S. H. 1969. Geological History of Southern Africa. Geol. Soc. South Africa, Cape Town. 535 p.
Hoering, T. C. 1967. The organic geochemistry of Precambrian rocks, p. 87–111. In P. H. Abelson [ed.] Researches in Geochemistry, vol. 2, John Wiley & Sons, Inc., New York.
Hofmann, H. J. 1969. Attributes of stromatolites. Geol. Surv. Canada Paper 69–39. 58 p.
———, and G. D. Jackson. 1969. Precambrian (Aphebian) microfossils from Belcher Islands, Hudson Bay. Canadian J. Earth Sci. 6:1137–1144.
Holmes, A. 1954. The oldest dated minerals of the Rhodesian Shield. Nature 173:612.
———, 1965. Principles of Physical Geology. Ronald Press, New York. 1288 p.
———, and L. Cahen. 1957. Géochronologie africaine 1956. Acad. Roy. Sci. Colon (Bruxelles), Cl. Sci. Nat., Mem. B, n. s., vol. 5, fasc. 1. 169 p.
Jolliffe, A. W. 1955. Geology and iron ores of Steep Rock Lake. Econ. Geol. 50:373–398.
LaBerge, G. L. 1967. Microfossils and Precambrian Iron-Formations. Geol. Soc. Amer. Bull. 78: 331–342.
Licari, G. R., P. E. Cloud, Jr., and W. D. Smith. 1969. A new chroococcacean alga from the Proterozoic of Queensland. Nat. Acad. Sci. (U.S.), Proc. 62:56–62.
Logan, B. W., R. Rezak, and R. N. Ginsburg. 1964. Classification and environmental significance of algal stromatolites. J. Geol. 72:68–83.
Macgregor, A. M. 1941 (1940). A pre-Cambrian algal limestone in Southern Rhodesia. Trans. Geol. Soc. South Africa 43:9–16.
———, 1947. An outline of the geological history of Southern Rhodesia. Southern Rhodesia Geol. Surv. Bull. 38. 73 p.

——, 1951. Some milestones in the Precambrian of Southern Rhodesia. Geol. Soc. South Africa, Proc. 54:xxvii–lxxi.

Monty, C. L. V. 1967. Distribution and structure of Recent stromatolitic algal mats, eastern Andros Island, Bahamas. Soc. Géol. Belg., Ann. 90(1–3): 55–102.

Murray, G. E. 1965. Indigenous Precambrian petroleum? Amer. Assoc. Petroleum Geol. Bull. 49: 3–21.

Nicolaysen, L. O. 1962. Stratigraphic interpretation of age measurements in southern Africa, p. 569–598. *In* A. E. J. Engel et al. [ed.] Petrologic Studies: A Volume in Honor of A. F. Buddington. Geol. Soc. Amer., New York.

Nier, A. O., and E. A. Gulbrandsen. 1939. Variations in the relative abundances of the carbon isotopes. J. Amer. Chem. Soc. 61:697–699.

Oberlies, F., and A. A. Prashnowsky. 1968. Biogeochemische und elektronenmikroskopische Untersuchung präkambrischer Gesteine. Naturwissenschaften 55:25–28.

Oparin, A. I. 1968. Genesis and Evolutionary Development of Life [Transl. from Russian by E. Maass]. Academic Press, New York. 203 p.

Park, R., and S. Epstein, 1960. Carbon isotopic fractionation during photosynthesis. Geochim. Cosmochim. Acta 21:110–126.

——. 1961. Metabolic fractionation of C^{13} and C^{12} in plants. Plant Physiol. 36(2):133–138.

Pflug, H. D. 1966. Structured organic remains from the Fig Tree Series of the Barberton Mountain Land. Econ. Geol. Res. Unit, Univ. Witwatersrand, Johannesburg, Inform. Circ. 28:1–14.

——, 1967. Structured organic remains from the Fig Tree Series (Precambrian) of the Barberton Mountain Land (South Africa). Rev. Palaeobot. Palynol. 5:9–29.

Raaben, M. E. 1969. Columnar stromatolites and Late Precambrian stratigraphy. Amer. J. Sci. 267: 1–18.

Ramsay, J. G. 1963. Structural investigations in the Barberton Mountain Land, eastern Transvaal. Geol. Soc. South Africa Trans. 66:353–398.

Reynolds, R. C., Jr. 1965. The concentration of boron in Precambrian seas. Geochim. Cosmochim. Acta. 29:1–16.

Schenk, P. E. 1967. The Macumber Formation of the Maritime Provinces, Canada: A Mississippian analogue to Recent strand-line carbonates of the Persian Gulf. J. Sed. Petrol. 37:365–376.

Schopf, J. W. 1968. Microflora of the Bitter Springs Formation, Late Precambrian, central Australia. J. Paleontology 42:651–688.

——, 1969. Recent advances in Precambrian paleobiology. Grana Palynologica 9:147–168.

——, 1970. Precambrian micro-organisms and evolutionary events prior to the origin of vascular plants. Cambridge Phil. Soc., Biol. Rev. 45(3): 319–352.

——, E. S. Barghoorn, M. D. Maser, and R. O. Gordon. 1965. Electron microscopy of fossil bacteria two billion years old. Science 149:1365–1367.

——, and E. S. Barghoorn. 1967. Alga-like fossils from the Early Precambrian of South Africa. Science 156:508–512.

——, 1969. Microorganisms from the Late Precambrian of South Australia. J. Paleontology 43: 111–118.

Seward, A. C. 1941. Plant Life Through the Ages. 2nd ed. Cambridge Univ. Press. 607 p.

Sharp, J. H. 1969. Blue-green algae and carbonates—*Schizothrix calcicola* and algal stromatolites from Bermuda. Limnol. Oceanography 14:568–578.

Smith J. W., J. W. Schopf, and I. R. Kaplan. 1970. Extractable organic matter in Precambrian cherts. Geochim. Cosmochim. Acta 34:659–675.

Staplin, F. L. 1969. Sedimentary organic matter, organic metamorphism, and oil and gas occurrence. Canadian Petroleum Geology Bull. 17:47–66.

Tebbutt, G. E., C. D. Conley, and D. W. Boyd. 1965. Lithogenesis of a distinctive carbonate rock fabric. Wyoming Geol. Surv., Contributions to Geology 4: 1–13.

Walcott, C. D. 1914. Pre-Cambrian Algonkian algal flora. Smithsonian Inst., Misc. Coll. 64(2): 77–156.

Wasserburg, G. J., R. J. Hayden, and K. J. Jensen. 1956. A^{40}-K^{40} dating of igneous rocks and sediments. Geochim. Cosmochim. Acta 10:153–165.

Welte, D. H. 1965. Relation between petroleum and source rock. Amer. Assoc. Petroleum Geol. Bull. 49:2246–2268.

Wolf, K. H. 1965. Littoral environment indicated by open-space structures in algal limestones. Palaeogeog., Palaeoclimatol., Palaeoecol. 1:183–223.

MANUSCRIPT RECEIVED JULY 30, 1970

41

Reprinted from *Science*, **152**(3723), 758–763 (1966)

Elso S. Barghoorn and J. William Schopf:

Microorganisms Three Billion Years Old from the Precambrian of South Africa

Abstract. *A minute, bacterium-like, rod-shaped organism, Eobacterium isolatum, has been found organically and structurally preserved in black chert from the Fig Tree Series (3.1 × 10⁹ years old) of South Africa. Filamentous organic structures of probable biological origin, and complex alkanes, which apparently contain small amounts of the isoprenoid hydrocarbons pristane and phytane, are also indigenous to this Early Precambrian sediment. These organic remnants comprise the oldest known evidence of biological organization in the geologic record.*

During the past decade, and particularly within the past few years, there has developed a marked revival of interest in the classical problem of the antiquity of life on earth. Although parallel to a current expansion of interest in theoretical and experimental approaches to the origin of living systems, paleobiological research on the antiquity of life is developing within a geological and geochronological framework which is unique in the history of paleontology. To a great extent this derives from a firmly based and fast-growing body of data on the radiogenic age of the earth and, in particular, on the age of its Precambrian sedimentary rock systems. The application of electron microscopy to organic sediments, moreover, has made it possible to transcend the limits of optical microscopy in observing minute objects of possible biological origin. In addition, development of analytical instruments for the detection and characterization of organic compounds in sediments provides refined techniques for the detection of correlative evidence of past biological systems.

We now report results of the application of optical and electron microscopy to sediments exceeding 3000 million (3 × 10⁹) years in age. The sediments are organic-rich black cherts, interpreted as primary chemical precipitates, in which are found discrete microfossils of bacterium-like size and form, as well as larger remnants of partially organized organic matter. Unpublished results of organic geochemical analyses establish that small concentrations of complex, high-molecular-weight alkanes are also present in this Early Precambrian sediment (1). Chromatographic analyses indicate that the isoprenoid hydrocarbons, pristane and phytane, are minor constituents of the indigenous alkanes (1). These hydrocarbons, thought to be "of definitely biological origin" (2), have been identified in other fossiliferous Precambrian sediments (2, 3) and are commonly suggested to have been derived by the chemical alteration of carotenoid or chlorophyll pigments (2, 3, 4).

The black cherts under consideration were collected by E. S. Barghoorn in February 1965 from rocks of the Fig Tree Series of the Swaziland System, eastern Transvaal, South Africa. The exact locality of collection is an outcrop excellently exposed by road cutting, close to the entrance to the Daylight Mine (Barbrook Mining Co.), 28 km east-northeast of Barberton, South Africa. The Barberton district has long been an area of gold mining, and extensive geological study has been made of the complex geology of the Swaziland System and the immediately overlying Moodies System of sediments. The structural relations of the "Barberton Mountain Land" have been discussed by Ramsay (5), from whose extensive discussion of the Fig Tree Series we have drawn in part (6).

The Fig Tree sediments comprise a thick series of graywackes, slates, and shales, with interbedded well-developed horizons of banded chert, jasper, and ironstone. The cherts are the only wide-ranging reliable stratigraphic markers in little-deformed as well as in folded and contorted areas of the sedimentary sequence. Although it is probable that the thickest of the green and gray chert units (up to 175 m thick) are secondary replacements, it is our interpretation that the thinner black fossiliferous cherts of the Fig Tree strata represent primary chemical precipitates. In this respect they appear to be of the same or comparable chemical mode of origin as those of the Middle Precambrian Gunflint Iron Formation of Ontario, Canada (7) and the biohermal chalcedonic cherts of the Late Precambrian Bitter Springs Limestone of central Australia (8), both of which contain remarkably well-preserved three-dimensional microorganisms of considerable morphological diversity. In the field, the black cherts of the Fig Tree Series show a striking megascopic resemblance to certain of the massive blocky and waxy cherts of the Gunflint Iron Formation. Unlike the Gunflint cherts, however, they are not associated with

stromatolitic gross algal structure. Ramsay (5) notes the presence of "oolites" incorporated as sedimentary constituents in one unit of the graywackes of the Fig Tree Series. These siliceous bodies appear to contain organic matter surrounding a finely crystalline silica core, and Ramsay suggests that they may be of algal origin (9). These "oolites" are stratigraphically and lithologically unrelated to the black cherts described here.

The age of the Swaziland System and that of the mineral belts in both South Africa and Southern Rhodesia has been the object of considerable geochronological study (10). The results have consistently indicated a very great age, certainly among the oldest determined from Early Precambrian rocks currently exposed on the earth's surface. Recent unpublished measurements on shales and graywackes from the Fig Tree Series, with the rubidium-strontium whole-rock method, indicate an age greater than 3100 million years for the last period of strontium isotopic homogenization for the sediments (11). These results confirm studies indicating an age for the Fig Tree sediments of at least 3 billion years (10).

The physical organization of organic matter in chert from the Fig Tree Series has been studied in thin sections and in macerations by optical microscopy, and in surface replicas and in macerations by electron microscopy. Optical examination of thin sections of the chert shows that the abundant dark-colored organic material is arranged in irregular laminations approximately parallel to the bedding planes, a distribution indicative of its original sedimentary deposition. The fact that much of the organic material transgresses chalcedony grain boundaries, as seen in thin sections in polarized light, indicates that it was emplaced prior to the crystallization of the surrounding silicic matrix, and is consistent with a primary sedimentary origin for the chert.

Although the laminar organization of the organic matter is evident, individual constituents of the layers are generally too small to be optically well resolvable. The organic lamellae appear to be comprised of discrete, minute, apparently branching, threadlike structures and isolated, spheroidal or rod-shaped particles, but neither precise morphology nor definite biological organization is apparent. In chert residues resistant to hydrofluoric acid these minute organic objects appear to retain their individual filiform or particulate character, a fact that suggests they possess distinct organization and are not random aggregates of organic material.

We have used electron microscopy to determine the morphology of organic structures in the Fig Tree chert. Application of this technique is based on our similar studies of chert from the approximately 2-billion-year-old Gunflint Iron Formation. Electron microscopy of surface replicas of the Gunflint chert has demonstrated the occurrence of organically well-preserved rod-shaped and coccoid bacteria (*12*). Other investigators have used electron microscopy to detect remarkably well-defined imprints of fossil bacteria in pyrite 300 million years old (*13, 14*), and micrographs of minute structures of possible biological origin have been figured from a sediment at least 2.5 billion years old (*15*).

Although similar organic objects were micrographically observed in surface replicas and in hydrofluoric acid macerations of the Fig Tree chert, the maceration technique appears to be somewhat unreliable (*16*). Original organic microstructure is often altered or destroyed when fragile carbonaceous fossils are freed by the acid digestion of their encompassing silicic matrix. In addition, any contaminants introduced in the maceration technique are often very difficult to differentiate from indigenous organic particles.

Electron microscopy of surface replicas of the rock is not subject to the disadvantages inherent in the process of maceration. In this technique, minute organic fossils can be separated intact from their encompassing mineral matrix and micrographically examined in their original state of preservation. Moreover, evident physical relationships between the microfossils, their imprints, and associated mineralogical structures of the replicated matrix establish the indigenous nature of the organic entities.

Examination of surface replicas of thin sections of the Fig Tree chert proved the most rewarding method of studying the morphology of indigenous organic structures. Replicas of each of the several thin sections investigated were prepared in the following manner. A polished thin section was ground according to standard procedures with 0.05-micron γ-alumina on a rayon lap for the final polish (*17*). The thin section was optically examined at high magnification, and areas where organic

matter was exposed at the rock surface were noted on a traced outline map of the section. After washing with distilled water, the section was etched for 2 minutes by immersion in a 4.88-percent (by volume) solution of reagent-grade hydrofluoric acid and distilled water. After the section had dried under cover, the etched surface was flooded with a 5-percent (g/ml) solution of parlodion in amyl acetate (reagent grade) and allowed to dry.

The outline map was turned over to show an inverted image of the thin-section outline; transparent tape (sticky-side-up) was placed on top of the map, and half-centimeter squares of one-ply paper were attached to the tape at positions corresponding to areas of organic concentration noted on the map. The outline of the section was traced from the map onto the tape, and the tape was turned sticky-side-down and superimposed on the parlodion-coated thin section. The tape and the adhering parlodion surface replica were then stripped free from the etched rock surface.

In a vacuum evaporator the parlodion replica was shadowed with platinum at about a 2:1 angle and was replicated

with an evaporated carbon film. The parlodion-platinum-carbon sandwiches overlying each of the small squares of paper were easily separated from the remainder of the replica and were immersed in amyl acetate (reagent grade). After a few hours the parlodion film dissolved and the remaining platinum-carbon replicas were picked up on microscope grids. Micrographs were taken with an RCA-EMU-3F electron microscope.

This double replica technique allows the correlation and morphological comparison of objects visible in the light microscope with those revealed by electron microscopy, and enables the investigator to ignore relatively unpromising areas of the rock surface. In addition, the outline map of the thin section serves as a record of the areas investigated and can be used repeatedly as a guide in the repeated replication of specific areas, with or without additional etching.

Minute organic structures, exposed on the rock surface by dissolution of their surrounding mineral matrix, often adhere to and are physically supported by the original parlodion replicas. Some of these structurally intact bodies, those

Figs. 1–9. Negative prints of electron micrographs of platinum-carbon surface replicas of chert from the Fig Tree Series showing *Eobacterium isolatum*, n. gen., n. sp., preserved both organically and as imprints in the rock surface: line in each figure represents one micron.

Fig. 1. Electron dense, organically preserved rod-shaped cell about 0.6 μ long (white, below) and its imprint in the chert surface (above). During preparation of the sample the bacterium-like fossil was displaced from its original position in the chert matrix. The presence of imprints and the fact that they transgress structures of the mineralogical matrix, such as grain boundaries and polishing scratches, indicate that the minute organism is indigenous to the rock.

Fig. 2. Organically preserved cell about 0.5 μ long (white, below) displaced from its imprint in the rock surface (above). Note the irregular, granular texture of the cellular imprint. Subparallel, horizontally oriented lineations in the rock surface are polishing scratches; a prominent grain boundary is present to the right of the imprint.

Fig. 3. Rod-shaped imprint in chert surface. About 0.75 μ long, this is one of the longest imprints of *E. isolatum* observed.

Fig. 4. Organic, somewhat flattened bacterium-like cell transgressing chalcedony grain boundary. The continuity of the grain boundary, passing through the fossil organism, demonstrates that *E. isolatum* is indigenous to the rock and is consistent with a primary origin for the chert.

Fig. 5. Organically preserved cell of *E. isolatum* showing the short, broad, rod-shaped morphology of the fossil organism. This well-preserved cell demonstrates the morphological similarity of the ancient organism to modern bacillar bacteria.

Fig. 6. Thin organic film, probably representing the remnants of a bacterium-like cell which was deformed during preservation.

Fig. 7. Circular structure interpreted as a transverse section through a cell of *E. isolatum*. Although the cellular contents appear to have been replaced by silica, the cell wall is organically preserved. The cell wall, about 0.015 μ thick, has a two-layered organization (shown at point of arrow) and is comparable in thickness and structure to cell walls of many modern bacteria.

Fig. 8. Circular imprint about 0.28 μ in diameter interpreted as a transverse section through the bacterium-like organism. The morphological uniformity of these structures (Figs. 7 and 9) and their size range (Fig. 10) support their interpretation as transverse sections of *E. isolatum*.

Fig. 9. Poorly preserved organic structure thought to be a transverse section of *E. isolatum*. The arrows point to a chalcedony grain boundary which passes through the circular structure and establishes the indigenous nature of these ancient organic remnants.

2

394

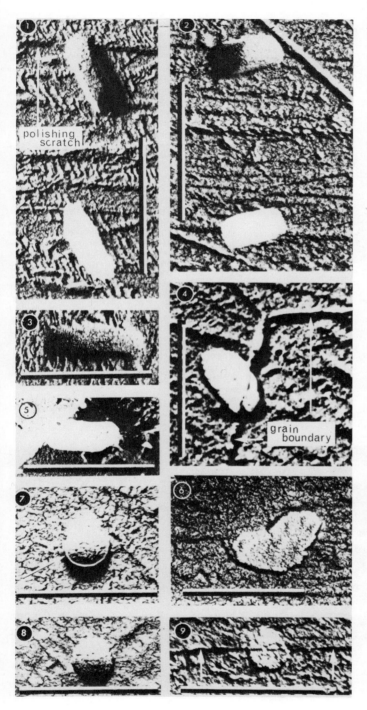

which are transferred to the final platinum-carbon replicas, can be micrographically examined in their original state of organic preservation. In this way, organically preserved rod-shaped cells and filamentous structures were separated from the chert and studied in their unaltered relationships to mineralogical structures (Figs. 4, 11, and 12).

In those cases in which the rod-shaped cells were not transferred from the parlodion film to the platinum-carbon replicas, imprints, shadowed as if they represented raised structures, were obtained in the final platinum-carbon replicas (Fig. 3). In some cases, upon dissolution of the parlodion film, the rod-shaped cells became displaced from their original positions in the replicated surface but adhered nearby to the platinum-carbon film (Figs. 1 and 2); these unusual occurrences show clearly the relationships between the organically preserved microfossils, their imprints in the rock surface, and the mineralogical structures of the replicated matrix.

Several types of organic structures having a characteristic and consistent morphology have been observed in surface replicas of the Fig Tree chert. Additional studies are necessary to fully characterize many of these structures and we here limit our discussion to two of the more interesting types— short, broad, bacterium-like rods (Figs. 1–9) and fibrous, branching threads (Figs. 11 and 12).

Isolated, rod-shaped bacterium-like cells occur as constituents of the organic lamellae in the chert and are similar in morphology and in distribution to spheroidal or rod-shaped organic particles seen optically in thin sections. Forty-five of the cells, preserved either organically (Figs. 1, 2, and 4–6) or as imprints in the rock surface (Figs. 1–3) have been observed in surface replicas. Both intact, relatively undistorted cells (Figs. 1–3, and 5) and deformed or greatly altered cellular remnants (Figs. 4 and 6) are present. Well-preserved cells typically have rounded ends (Figs. 1, 4, and 5) and are longer than twice their breadth, characteristics exhibited by modern bacillar bacteria. The twelve organic, relatively undistorted cells plotted in Fig. 10 have an average length of 0.56 μ and an average width of 0.24 μ. The rod-shaped imprints appear to be slightly longer and broader than are the organic cells they originally contained (Figs. 1, 2, and 10).

3

395

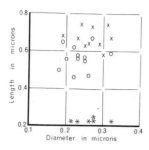

Fig. 10. Scatter diagram showing size distribution for 28 well-preserved cells of *Eobacterium isolatum* n. gen., n. sp., observed in surface replicas of chert from the Fig Tree Series. "O's" show diameters and lengths of organically preserved cells; "X's" show diameters and lengths of rod-shaped imprints in rock surface. Asterisks show diameters of circular structures interpreted as transverse sections of the bacterium-like organism (note that lengths cannot be measured for these bodies). The rod-shaped imprints are generally slightly broader and longer than are the organic cells they originally contained.

Fourteen circular structures, preserved both organically and as imprints, have been observed in replicas of the chert (Figs. 7–9). These structures, interpreted as transverse sections of the rod-shaped cells, are approximately 0.26 μ in diameter and typically have an outer wall about 0.015 μ thick (Figs. 7 and 8). In the best preserved structures the wall is seen to be composed of two layers (Fig. 7) and is comparable in thickness and in organization to cell walls of many modern bacteria (*18*).

Our interpretation that these structures represent transverse sections of the bacterium-like cells is supported by the following observations: (i) the circular structures are morphologically quite uniform (Figs. 7–9); (ii) their diameters are very similar to the widths of the rod-shaped cells (Fig. 10); (iii) they have been observed only in areas of thin sections in which the rod-shaped forms are also present; (iv) the occurrence of transverse sections is consistent with the apparently random longitudinal orientation exhibited by the rod-shaped cells. This interpretation is further strengthened by the bacterium-like morphology of both the elongate cells and the circular structures. The fact that no sheath or sheath-like residue has been observed in either transverse or longitudinal views of the microorganism is consistent with its apparently isolated, noncolonial growth habit. Although the apparent lack of flagella suggests that the organism may have been nonmotile, this lack may be the result of loss of such structures at death.

That these minute fossils are indigenous to the rock rather than laboratory contaminants is supported by the following six considerations: (i) the forms occur in replicas both organically preserved and as imprints in the rock surface and show variation in completeness of preservation; (ii) they are orient-

ed in a variety of positions, not only parallel with but passing into the prepared rock surface; (iii) they transgress mineralogical structures of the chert matrix such as polishing scratches (Figs. 1 and 3) and chalcedony grain boundaries (Figs. 4 and 9); (iv) they are present in repeated replicas of specific areas of the same thin section, and in several areas of several thin sections· (v) they are similar in morphology and in distribution to organic structures seen optically in thin sections, and (vi) they are absent from glass microscope slides replicated side-by-side with thin sections of the chert, and are absent from preparations of Lakeside Cement, the mounting medium for the thin sections.

That these microorganisms are both organically preserved and often relatively undistorted is consistent with the occurrence of complex organic molecules within the rock, and is not surprising in view of the occurrence of similar bacterial fossils in other Precambrian cherts (*12*). The Fig Tree organisms are comparable in size, shape, complexity of structure, and isolated habit to many modern bacillar bacteria. Although they may have had a nonphotosynthetic metabolism, there is insufficient information available about their paleobiochemistry to realistically evaluate such a suggestion. In view of the fact that more than 3 billion years of evolutionary history separate these forms from possible modern counterparts, and more than 1 billion years separate them from the oldest previous-

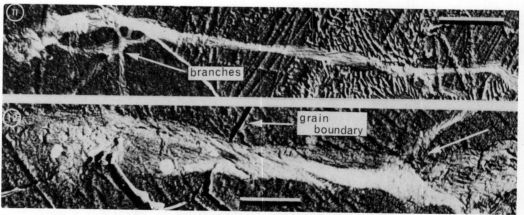
branches

grain boundary

Figs. 11 and 12. Filamentous organic material of probable biological origin shown in negative prints of micrographs of platinum-carbon replicas of chert from the Fig Tree Series; line in each figure represents one micron. Fig. 11. Threadlike organic structure approximately 8.5 μ long. The branching, fibrillar nature of this organic residue, comparable in appearance to degraded plant material, is suggestive of biological origin. Fig. 12. Linear organic residue similar to that shown in Fig. 11. The texture and varying thickness of the structure and the fact that it transgresses chalcedony grain boundaries demonstrate that it is indigenous to the rock. Arrow at right points to the origin of a lateral branch.

4

ly reported bacterium-like microfossils (*12*), physiological and environmental conclusions based upon a morphological comparison between them and more recent microorganisms would be of questionable significance.

For purposes of reference, it seems desirable to propose a formal taxonomic designation for this ancient bacterium-like microorganism. The photographic record must serve as "type material" for this taxon inasmuch as the original replicas are perishable and unique (*19*). In the absence of definitive information regarding the physiology of this organism, the binomial here proposed designates a morphological genus based solely on form.

Eobacterium, new genus.

Diagnosis: Short, broad rods, usually with well-rounded ends and approximately circular in transverse section. Length usually between two and three times greater than diameter. Thickness of cell wall usually between 0.05 and 0.10 times the cell diameter. Cell wall may have two distinct layers; external surface may appear granular.

Etymology: With reference to Early Precambrian age and rod-shaped form of type species.

Type Species: *Eobacterium isolatum*, new species.

Diagnosis: Unicellular, isolated rods. Length usually between 0.45 and 0.70 μ, with average length of about 0.55 μ; diameter usually between 0.18 and 0.32 μ, with average diameter of about 0.25 μ. Cell wall approximately 0.015 μ thick, composed of two layers of approximately equal thickness. Cells lack sheath or sheath-like residue. External surface often appears granular.

Etymology: With reference to non-colonial, unicellular growth habit.

Type Locality: Black chert facies in upper third of Early Precambrian Fig Tree Series, exposed by road cutting 100 m northwest of surface opening to Daylight Mine (Barbrook Mining Co.), 28 km east-northeast of Barberton, South Africa.

Type Material: Figures 1–9 show representative members of species; Fig. 10 shows ranges of size and shape of species; Figs. 1, 2, 5, and 7 are cited as primary "type material."

In addition to *E. isolatum*, electron microscopy of surface replicas has revealed the presence of filamentous organic structures in the Fig Tree chert (Figs. 11 and 12). Although these organic residues lack the regularity and clearly biological character of the bacterium-like organism, they are significant inasmuch as they appear to show a greater degree of structural complexi-

ty. Their evident physical association with mineralogical structures of the matrix (Figs. 11 and 12) and their similarity in morphology and distribution to filiform structures seen optically in thin sections establish that these elongate threadlike forms are indigenous to the chert. Their origin, however, is difficult to determine.

The fibrillar (Fig. 11), branching (Figs. 11 and 12) morphology of these complex structures is suggestive of the high degree of molecular and polymeric order characteristic of living systems; their general appearance is not dissimilar from degraded plant material. However, the fact that they appear to lack such structures as cell walls or transverse septae, and are generally quite irregular in form, suggests that although their molecular components are probably biogenic the branching morphology may be an artifact of inorganic processes operating during crystallization of the matrix. Additionally, these forms might represent ordered remnants of organic material produced abiotically in the early stages of organic evolution (*20*). Although we regard these threadlike forms as almost certainly biogenic, additional investigation is necessary to clarify their mode of origin.

The Middle Precambrian Gunflint chert from southern Ontario contains the oldest known structurally preserved evidence of multicellular plant life (*2, 7*). The diversity and complexity of this approximately 2-billion-year-old microfossil assemblage, and the occurrence of possible biogenic remnants in sediments thought to be older than 2.5 billion years (*15, 21*), have constituted putative, yet somewhat equivocal, evidence suggesting that biological systems originated early in Precambrian time. The occurrence of bacterium-like microfossils, presumably biogenic organic filaments, and complex biologically important hydrocarbons in the Early Precambrian Fig Tree chert establishes that organisms were in existence at least 3.1 billion years ago, and indicates that life on earth must have originated during the preceding 30 percent of the earth's history.

References and Notes

1. W. G. Meinschein, private communication, 20 December 1965; alkanes from Fig Tree sediments have also been reported by T. C. Hoering in *Carnegie Institution Year Book 64* (Washington, D.C., 1965), p. 215.

2. J. Oró, D. W. Nooner, A. Zlatkis, S. A. Wikström, E. S. Barghoorn, *Science* **148**, 77 (1965).
3. W. G. Meinschein, E. S. Barghoorn, J. W. Schopf, *ibid.* **145**, 262 (1964); E. S. Barghoorn, W. G. Meinschein, J. W. Schopf, *ibid.* **148**, 461 (1965); G. Eglinton, P. M. Scott, T. Belsky, A. L. Burlingame, M. Calvin, *ibid.* **145**, 263 (1964).
4. J. C. Bendoraitis, B. L. Brown, L. S. Hepner, *Anal. Chem.* **34**, 49 (1962); M. Blumer and W. D. Snyder, *Science* **150**, 1588 (1965).
5. J. G. Ramsay, *Trans. Geol. Soc. So. Africa*, in press.
6. ———, permission by private communication, 14 December 1965.
7. S. A. Tyler and E. S. Barghoorn, *Science* **119**, 606 (1954); E. S. Barghoorn and S. A. Tyler, *ibid.* **147**, 563 (1965); P. E. Cloud, Jr., *ibid.* **148**, 27 (1965); E. S. Barghoorn and S. A. Tyler, *Ann. N.Y. Acad. Sci.* **108**, 451 (1963); E. S. Barghoorn and S. A. Tyler, in *Current Aspects of Exobiology*, G. Mamikunian and M. H. Briggs, Eds. (Jet Propulsion Laboratory, Pasadena, 1965), p. 93; W. W. Moorhouse and F. W. Beales, *Trans. Roy. Soc. Can. Sect. III Ser. 3* **56**, 97 (1962).
8. E. S. Barghoorn and J. W. Schopf, *Science* **150**, 337 (1965).
9. Other supposed biological structures from Swaziland rocks have been referred to by H. Pflug in the newspaper "Die Welt" 30 October 1965.
10. L. O. Nicolaysen, in *Petrographic Studies: A Volume in Honor of A. F. Buddington*, A. E. J. Engel *et al.*, Eds. (Geological Society of America, New York, 1962), p. 569; H. L. Allsopp, H. R. Roberts, G. D. L. Schreiner, D. R. Hunter, *J. Geophys. Res.* **67**, 5307 (1962).
11. T. J. Ulrych, private communication, 7 Dec. 1965; information pertaining to these measurements will appear in an article in preparation by T. J. Ulrych and H. L. Allsopp.
12. J. W. Schopf, E. S. Barghoorn, M. D. Maser, R. O. Gordon, *Science* **149**, 1365 (1965); J. W. Schopf and E. S. Barghoorn, Abstr. Ann. Meeting, Geol. Soc. Amer., Kansas City (1965), p. 147.
13. J. M. Schopf, E. G. Ehlers, D. V. Stiles, J. D. Birle, *Proc. Amer. Phil. Soc.* **109**, 288 (1965).
14. E. G. Ehlers, D. V. Stiles, J. D. Birle, *Science* **148**, 1719 (1965).
15. P. E. Cloud, Jr., J. W. Gruner, H. Hagen, *ibid.* **148**, 1713 (1965).
16. The maceration technique has been described by Schopf *et al.* (*12*).
17. AB Gamma Micropolish and AB Microcloth were obtained from Buehler, Ltd., Evanston, Ill.
18. K. V. Thimann, *The Life of Bacteria* (Macmillan, New York, ed. 2, 1963), p. 109.
19. As is noted by J. M. Schopf *et al.* (*13*) in an analogous circumstance: "This is permissible according to regular exemptions given in the Botanical Code of Nomenclature (Art. 9, Note 3) and the International Code of Nomenclature of Bacteria and Viruses (Rule 9d, Note)."
20. We here refer to the abiotically produced organic compounds suggested by A. I. Oparin as the precursors of biological systems (A. I. Oparin, *The Origin of Life on the Earth* (Oliver and Boyd, London, ed. 3, 1957); also *The Origin of Life on the Earth*, Proc. 1st Symposium Moscow, 1957, F. Clark and R. Synge, Eds. (Pergamon, London, 1959). If this speculation were substantiated, *E. isolatum* might have been heterotrophic, metabolizing this organic matter.
21. A. M. Macgregor, *Trans. Geol. Soc. So. Africa* **43**, 9 (1941); T. Belsky, R. B. Johns, E. D. McCarthy, A. L. Burlingame, W. Richter, M. Calvin, *Nature* **206**, 446 (1965); W. G. Meinschein, *Science* **150**, 601 (1965); K. M. Madison, *Trans. Illinois State Acad. Sci.* **50**, 287 (1957).
22. We thank Dr. W. G. Meinschein for access to unpublished data on the organic geochemistry of Fig Tree sediments, Drs. J. G. Ramsay, T. J. Ulrych, L. O. Nicolaysen, H. L. Allsopp, and D. Macauly for assistance in the field and/or for allowing access to unpublished data on the geology of the Fig Tree Series, Dr. M. D. Maser for helpful suggestions on techniques of electron microscopy, and Mrs. Dorothy D. Barghoorn for assistance in the field. This work was supported by NSF grants GP-2794 and G-19727 and Public Health Service grant GM-06637 to Harvard University. J.W.S. is a NSF graduate fellow.

27 January 1966

5

397

42

Reprinted from *Science*, **156**(3774), 508–512 (1967)

Alga-Like Fossils from the Early Precambrian of South Africa

J. WILLIAM SCHOPF and
ELSO S. BARGHOORN

Abstract. *Micropaleontological studies of carbonaceous chert from the Fig Tree Series of South Africa (> 3.1 × 10⁹ years old) revealed the presence of spheroidal microfossils, here designated* Archaeosphaeroides barbertonensis, *interpreted as probably representing the remnants of unicellular alga-like organisms. The presumed photosynthetic nature of these primitive microorganisms seems corroborated by organic geochemical and carbon isotopic studies of the Fig Tree organic matter, and is consistent with the geologically and mineralogically indicated Early Precambrian environment. These alga-like spheroids, together with a bacterium-like organism previously described from the Fig Tree chert, are the oldest fossil organisms now known.*

Laminated stromatolitic structures, comparable in gross morphology to the biohermal deposits of modern blue-green and red algae, have long been known from sediments of Precambrian age. Although generally devoid of cellularly preserved microfossils, these structures have been regarded as presumptive evidence of early algal activity. Firm evidence for this supposition comes from the recent investigations of siliceous stromatolites of Middle Precambrian age from the Gunflint Iron Formation (1.9 × 10⁹ years old) along the northern shore of Lake Superior in Ontario, Canada. Various microorganisms, some morphologically similar to modern blue-green algae, are structurally preserved in the organic lamellae of these black chert stromatolites (*1, 2*). Cellular microfossils are similarly preserved in laminated, primary black cherts associated with stromatolites of the genus *Collenia* in the Late Precambrian Bitter Springs Formation (about 1.0 × 10⁹ years old) of central Australia (*3*). Morphologically, several of these billion-year-old organisms are referable to modern families of blue-green and green algae. These two microfossil assemblages and the many deposits of calcareous stromatolites and carbonaceous sediments of comparable age establish that diverse types of primitive algae were present relatively early in geologic time.

The oldest known stromatolites are apparently those described and figured by A. M. Macgregor (*4*) from Early Precambrian limestones near Turk Mine, 53 km north-north-east of Bulawayo, Rhodesia. Columnar, dentate, and domical structures have been noted in the deposit; the domical forms have been compared with *Collenia* (*4*). Comparison of the stable isotopes C¹² and C¹³ in reduced organic carbon, present as finely bedded graphitic laminations,

and in oxidized inorganic carbon, present in the carbonate matrix, demonstrated an isotopic fractionation probably of photosynthetic origin (*5*). Microfossils have not been reported in the calcareous Bulawayan stromatolites, but the gross morphology, carbon isotopic composition, and presence of organic matter are consistent with an algal origin. The Bulawayan limestones have been correlated, by lithologic similarity and stratigraphic setting (*6*), with sediments occurring in the Bikita tin field, about 300 km east of Bulawayo (*6, 7*). A pegmatite dike, intrusive in these sediments near Bikita, has been dated at 2640 ± 40 million years (*7–9*). The Bulawayan stromatolites have therefore been regarded as older than about 2.6

× 10⁹ years; some authors suggest an age approaching 3 billion years (*10*).

Several investigators have examined other Early Precambrian sediments for evidence of biological activity (*11*). Of particular interest are shales and cherts from the Fig Tree Series of the Upper Swaziland System, eastern Transvaal, South Africa (*12–14, 21*). The Fig Tree sediments comprise a thick sequence of graywackes, slates, and shales with interbedded well-developed horizons of chert, jasper, and ironstone. These sediments overlie pyroclastic deposits and dolomitic limestones of the Lower Swaziland System, and they immediately underlie sediments of the Moodies System (*15*). This Early Precambrian sedimentary sequence occurs in an area actively mined for gold; the geology, stratigraphy, and structural relationships of these deposits have been extensively studied (*15, 16*).

The age of the Swaziland System has been the object of considerable geochronological investigation (*9, 17*). Results have consistently indicated an age of deposition greater than 3 billion years, and a granite intruding the Swaziland System has a tentative rubidium-strontium age of 3440 ± 300 million years (*17*). Recent measurements on shales and graywackes from the Fig Tree Series, with the rubidium-

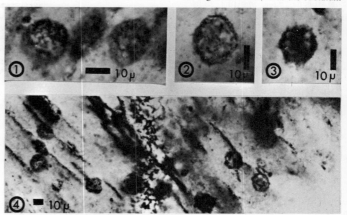

Figs. 1–4. Spheroidal, organic alga-like microfossils (*Archaeosphaeroides barbertonensis* n. gen., n. sp.) in black chert of the Early Precambrian Fig Tree Series (> 3.1 × 10⁹ years old), near Barberton, South Africa. All structures shown in thin sections photographed in transmitted light. Fig. 1. Spheroidal and somewhat distorted alga-like fossils showing the typical, irregularly reticulate surface texture. Fig. 2. Median optical section of the type specimen of *A. barbertonensis*, showing the continuity and somewhat variable thickness of the cell wall. Fig. 3. Alga-like spheroid containing "coalified" organic material interpreted as representing the coalesced remnants of the original cytoplasmic cellular contents. Fig. 4. Organic spheroidal microfossils exhibiting varying degrees of completeness of preservation. The diagonally oriented lamellae are composed of organic particles aligned parallel to the bedding planes; the vertically oriented irregular dark material is anthraxolitic.

strontium whole-rock method, indicate a minimum age of 3.1×10^9 years for the last period of strontium isotopic homogenization of these sediments (*18*).

Biohermal stromatolites have not been reported from the Fig Tree sediments. Ramsey (*15*), however, has figured apparently deformed, concentrically laminated "oolites" which he suggests may be of algal origin. These curious structures occur in graywackes of the sequence, and are stratigraphically and lithologically unrelated to the carbonaceous cherts which we investigated for possible microfossil content.

Chalcedonic black cherts, interpreted as primary in origin (*12*), are particularly abundant in the upper third of the Fig Tree Series. The physical organization of the organic matter indigenous to these cherts was studied in thin sections and in hydrofluoric acid macerations by optical microscopy, and in surface replicas and in macerations by electron microscopy. Much of the carbonaceous material occurs as irregularly shaped, minute, dispersed organic bodies predominantly aligned parallel to bedding planes, and, as seen in the optical microscope, is generally devoid of evident biogenic morphology. However, electron microscopy revealed a minute, rod-shaped bacterium-like organism, *Eobacterium isolatum*, structurally and organically preserved in this Early Precambrian sediment (*12*). Also present in thin sections and in surface replicas of the chert are irregular filamentous organic structures morphologically comparable to degraded plant material (*12*); their probable biological origin, however, cannot be established by morphology alone.

Spheroidal, dark-colored organic bodies, comparable in general morphology to certain modern unicellular algae, have now been discovered in the same locality and black chert facies of the Fig Tree Series as *E. isolatum* (*19*). Figures 1–4 show several of these bodies photographed in thin sections in transmitted light. Both well-defined spheroids (Figs. 1–4) and distorted, partially flattened cell-like remnants (Figs. 1 and 4) are present. Measurements of 28 spheroids show that they range in diameter between about 15.6 μ and 23.3 μ, with an average of 18.7 μ (Fig. 5). The cell-like spheroidal bodies, which often exhibit a reticulate surface texture (Fig. 1), are three-dimensionally preserved in the chalcedonic matrix. Although sufficient organic matter is present to outline the shape of these spheroids, the walls are often rather

poorly preserved. In better-preserved spheroids the wall is about 1 μ thick (Fig. 2). Irregular masses of organic material occasionally present within the spheroids (Fig. 3) appear to represent the coalesced and "coalified" remnants of the original internal contents, a feature rather commonly observed in fossil algae preserved in cherts (for example, *20*, plate 44). The organic bodies are two to six times larger than the chalcedony grains of the chert matrix in which they are embedded, and the boundaries of the chalcedony grains pass through the organic structures without deforming their spheroidal shape. They are demonstrably not organic coatings of mineral grains, and clearly not mineral artifacts. Similar organic spheroids, interpreted as being of biological origin, from the Fig Tree sediments were reported by H. Pflug (*21*).

In the paleontological assessment of minute structures of Precambrian age and of possible biological origin, it is often difficult to differentiate between inorganically produced pseudofossils and partially degraded remnants of primitive microorganisms. This is particularly true in the interpretation of structures from sediments such as those of the approximately 3 billion-year-old Fig Tree Series, for which there are no known fossils of equivalent age for morphological comparison. Nevertheless, the Fig Tree spheroids are almost certainly of biological origin, probably representing the remnants of single-celled alga-like microorganisms. Their

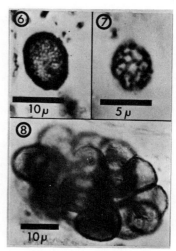

Figs. 6–8. Spheroidal, organic alga-like microfossils in black cherts of Middle and Late Precambrian age. All structures shown in thin sections photographed in transmitted light. Fig. 6. Type specimen of *Huroniospora microreticulata* Barghoorn (*1*), from chert of the Middle Precambrian Gunflint Iron Formation (1.9 × 10⁹ years old), Ontario, Canada. Fig. 7. Type specimen of *H. macroreticulata* Barghoorn (*1*), from the Gunflint chert, showing the morphological similarity between this organism and *Archaeosphaeroides barbertonensis*. Fig. 8. Well-preserved, spheroidal blue-green algae from chert of the Late Precambrian Bitter Springs Formation (*3*), about 1 billion years old, from central Australia.

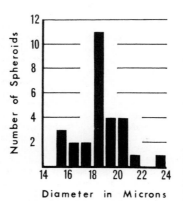

Fig. 5. Size distribution of 28 well-defined representatives of *Archaeosphaeroides barbertonensis* n. gen., n. sp., in chert from Fig Tree Series. The diameters of the alga-like bodies were measured from photomicrographs showing the organisms in thin sections of the chert.

organic composition, morphological consistency, limited size range, mode of preservation, and morphological similarity to known spheroidal algae, both modern and fossil, support this interpretation. Their physical relationship to grains of the chalcedonic matrix seems to render an inorganic origin untenable, and they are demonstrably indigenous to the sediment and are not laboratory contaminants.

In size, shape, and general organization, the Fig Tree spheroids are comparable to certain members of the modern blue-green algal group Chroococcales. They are morphologically dissimilar to known bacteria. In general appearance they seem quite similar to reticulate, spheroidal alga-like microorganisms from the Middle Precambrian Gunflint chert (Figs. 6 and 7), and they are morphologically comparable to well-preserved blue-green algae from the Late Precambrian Bitter Springs Formation (Fig. 8). However, perhaps because of their relatively poor preservation, the Fig Tree organisms cannot be assigned with certainty to any extant algal group. Moreover, in view of the extraordinary antiquity of these organisms, and the intervening 3 billion years of evolutionary history, there is no reason to expect that modern morphological counterparts exist.

For purposes of reference, it seems desirable to propose a taxonomic designation for these spheroidal microfossils and, in view of their somewhat uncertain biological affinities, to designate them by a binomial having morphological and geological, rather than phyletic, significance or implication.

Archaeosphaeroides, new genus.

Diagnosis: Cells small, spherical, spheroidal, or ellipsoidal, not angular; solitary, unicellular, noncolonial; more or less circular in cross section. External wall texture varies from psilate or nearly psilate to coarsely reticulate. Cell walls membranous, thin, may be ruptured and distorted. Internal organic matter may be present.

Etymology: With reference to Early Precambrian age and spheroidal form of type species.

Type species: *Archaeosphaeroides barbertonensis*, new species.

Diagnosis: Unicellular, isolated organic spheroids. Cross-sectional diameter varies between about 15 and 24 μ, with an average diameter of approximately 19 μ. Cell wall delicate, thin, commonly of somewhat variable thickness, often coarsely and irregularly reticulate. Internal organic matter occasionally present, clumped in an irregular mass near the center of the cell.

Etymology: With reference to type locality in the Barberton Mountain Land, near Barberton, South Africa.

Type locality: Black chert facies in upper third of Early Precambrian Fig Tree series, Upper Swaziland System, exposed by road cutting 100 m northwest of surface opening to Daylight Mine (Barbrook Mining Co.), 28 km east-northeast of Barberton, eastern Transvaal, South Africa.

Type specimen: Figures 1–4 show representative members of the species; Fig. 5 shows the range of measured diameters for 28 well-defined individuals; the organism shown in Fig. 2 is cited as the type specimen (thin section D/FT-1, Paleobotanical Collection, Harvard University, No. 58436).

Recent studies of the organic chemistry and carbon isotopic composition of the Fig Tree organic matter, and a consideration of the geology and mineralogy of the Fig Tree sediments, appear to provide additional evidence for the existence of photosynthetic organisms in Fig Tree time. The reactants in green-plant photosynthesis, carbon dioxide and water, were apparently present in the Fig Tree environment. The precipitation of calcite and of dolomite seems to require the availability of free carbon dioxide (22). Thus, the occurrence of sedimentary dolomitic limestones in the Lower Swaziland System (15), immediately underlying the Fig Tree sediments, may be interpreted to indicate that atmospheric CO_2 was present during this period of geologic time. Such an interpretation is consistent with the occurrence of limestones of comparable age in the Bulawayan Group of Rhodesia. The presence of liquid water is indicated by the occurrence of water-laid deposits, containing such depositional features as pillow lavas, current-bedding, ripple marks, and mud cracks, in the Swaziland and overlying Moodies System (15). The existence of hematite deposits in the form of banded iron-stones and ferruginous shales in the Fig Tree Series has been interpreted as indicating the presence of free oxygen (23), possibly produced by photosynthetic organisms. However, as Holland pointed out (22), ferrous iron can be oxidized in the presence of very low concentrations of atmospheric oxygen; this oxygen may conceivably reflect the abiotic, ultraviolet-induced photolysis of water, rather than the presence of photosynthetic organisms. Thus, the geological and mineralogical evidence from the Fig Tree sediments seems consistent with, but probably does not necessarily require, the presence of a photosynthetic biota.

Carbonaceous cherts and shales from the Fig Tree Series contain about 0.5 percent of organic matter, of which somewhat less than 1 percent is soluble in a 3 : 1 benzene : methanol solvent solution. In this soluble fraction, W. Meinschein (24) and T. Hoering (13) identified alkane hydrocarbons, ranging from about 17 to 35 carbon atoms in length and "typical of hydrocarbons formed in association with biologically produced organic matter of all ages" (13). The isoprenoid hydrocarbons pristane and phytane have been reported as minor constituents of the indigenous alkanes (12, 14, 24). Pristane and phytanes have been identified in other Precambrian deposits containing organically preserved plant microfossils (2, 25), and they are widely distributed in photosynthetically produced organic matter of younger geologic age (26). These isoprenoids are commonly regarded as geochemical or biochemical derivatives of the phytyl alcohol moiety of chlorophyll (2, 14, 25, 27), and for this reason their presence seems suggestive of photosynthetic activity. In the modern biota, isoprene derivatives are apparently represented in all organisms with the exception of nonphotosynthetic anaerobic bacteria (28). Thus, judging from extant organisms, the occurrence of fossil isoprenoids seems to require the existence of organisms different from, and probably more advanced than, nonphotosynthetic anaerobes. The reported occurrence of pristane and phytane in the Fig Tree sediments is consistent with, and seems to suggest, the presence of photosynthetic microorganisms.

In the process of photosynthesis, modern plants tend to selectively metabolize carbon dioxide containing the lighter carbon isotope C^{12} in preference to CO_2 containing the heavier isotope C^{13}. This partial fractionation results in an enrichment of C^{12} in photosynthetically produced organic matter, as compared with its concentration in the atmosphere and in precipitated carbonate sediments. Park and Epstein (29) have shown that in tomato plants this fractionation is biochemically complex, apparently involving both kinetic and enzymatic processes favoring the preferential incorporation of the lighter isotope in the organic matter produced. Ratios of C^{12} to C^{13} have been measured in many types of materials, of both modern and geological age, and of both biological and inorganic origin (30).

Hoering (13) determined the carbon isotopic composition of both the benzene-soluble and the insoluble kerogen fractions of organic matter indigenous to the Fig Tree sediments. The measured isotopic ratios fall well within the range exhibited by many crude oils and other organic materials of known photosynthetic origin or derivation, and they are distinctly different from values characteristic of inorganically produced carbon compounds. The carbon isotopic composition of the Fig Tree organic matter is comparable to that measured on reduced carbon from the Middle Precambrian Gunflint chert (1, 2), and to the isotopic composition of organic compounds from the Late Precambrian Nonesuch Shale (25), both of which contain pristane and phytane, as well as cellularly preserved plant microfossils.

The chemical pathways from the

3

sedimentary deposition of organic matter to kerogen and the extractable organic components in rocks are not known. In addition, the complexities of the carbon-cycle in nature and its possible variations over geologic time are not well understood. Nevertheless, green-plant photosynthesis is the sole means of carbon fixation and isotopic fractionation of quantitative importance operating in modern environment, and the observed $C^{12} : C^{13}$ ratios seem indicative of its existence in Fig Tree time.

Together with *Eobacterium isolatum*, a bacterium-like microorganism previously described from the Fig Tree chert, the organic spheroids, here designated *Archaeosphaeroides barbertonensis*, constitute the oldest fossil organisms now known. These spheroidal microfossils probably represent the remnants of unicellular, noncolonial alga-like organisms. Possibly they are related to, or were perhaps the evolutionary precursors of, modern coccoid blue-green algae. The presumed photosynthetic nature of these primitive microorganisms seems corroborated by geological and organic geochemical considerations. The carbon isotopic composition of the Fig Tree organic matter and the reported occurrence of organic compounds related to chlorophyll suggest the presence of photosynthetic activity; the geology and mineralogy of these Early Precambrian sediments are consistent with this interpretation. The apparent existence of photosynthetic microorganisms in Fig Tree time, more than 3100 million years ago, is similarly in accord with the occurrence of laminated stromatolites of probable algal origin in Early Precambrian limestones near Bulawayo, Rhodesia. Moreover, the early appearance of the photosynthetic process provides a credible and logical explanation for the extensive occurrence of reduced carbon and organic compounds in many Early Precambrian sediments. Photosynthetic algae or alga-like organisms

probably originated quite early in the evolution of biological systems, apparently during the 30 percent of the earth's history preceding the deposition of the Fig Tree Series.

References and Notes

1. S. A. Tyler and E. S. Barghoorn, *Science* 119, 606 (1954); E. S. Barghoorn and S. A. Tyler, *ibid.* 147, 563 (1965); J. W. Schopf, E. S. Barghoorn, M. D. Maser, R. O. Gordon, *ibid.* 149, 1365 (1965); P. E. Cloud, Jr., *ibid.* 148, 27 (1965); E. S. Barghoorn and S. A. Tyler, *Ann. N.Y. Acad. Sci.* 108, 451 (1963); W. W. Moorhouse and F. W. Beales, *Trans. Roy. Soc. Can. Sect. III* [3] 56, 97 (1962).
2. J. Oró, D. W. Nooner, A. Zlatkis, S. A. Wikström, E. S. Barghoorn, *Science* 148, 77 (1965).
3. E. S. Barghoorn and J. W. Schopf, *ibid.* 150, 337 (1965); J. W. Schopf and E. S. Barghoorn, in *Abstracts*, Geol. Soc. Amer. Ann. Meeting, San Francisco, Calif. (1966), p. 193; J. W. Schopf, in *1967 McGraw-Hill Yearbook of Science and Technology*, pp. 46–55.
4. A. M. Macgregor, *Trans. Geol. Soc. S. Africa* 43, 9 (1940).
5. T. C. Hoering, *Carnegie Inst. Washington Year Book* 61, 190 (1962).
6. R. Tyndale-Biscoe, *Trans. Geol. Soc. S. Africa* 54, 11 (1951).
7. A. Holmes, *Nature* 173, 4405 (1954).
8. L. H. Ahrens, in *Crust of the Earth*, A. Poldervaart, Ed. [*Geol. Soc. Amer. Spec. Papers* 62, 155 (1955)].
9. L. O. Nicolaysen, in *Petrological Studies: A Volume to Honor A. F. Buddington*, A. E. J. Engel *et al.*, Eds. (Geological Soc of America, New York, 1962), pp. 569–598.
10. A. Holmes, *Principles of Physical Geology* (Ronald Press, New York, ed. 2, 1965), p. 371; A. M. Macgregor, *Trans. Geol. Soc. S. Africa* 54, 25 (1951).
11. P. E. Cloud, Jr., J. W. Gruner, H. Hagen, *Science* 148, 1713 (1965); T. Belsky, R. B. Johns, E. D. McCarthy, A. L. Burlingame, W. Richter, M. Calvin, *Nature* 206, 446 (1965); W. G. Meinschein, *Science* 150, 601 (1965); K. M. Madison, *Trans. Illinois State Acad. Sci.* 50, 287 (1957).
12. E. S. Barghoorn and J. W. Schopf, *Science* 152, 758 (1966).
13. T. C. Hoering, *Carnegie Inst. Washington Year Book* 64, 215 (1965); *ibid.* 65, 365 (1966).
14. G. Eglinton and M. Calvin, *Sci. Amer.* 216, 32 (1967).
15. J. G. Ramsay, *Trans. Geol. Soc. S. Africa*, in press; *Information Circ. No. 14* (Econ. Geol. Res. Unit, Univ. Witwatersrand, Johannesburg, 1963).
16. O. R. van Eeden *et al.*, *Geol. Surv. S. Africa Spec. Publ. No. 15* (1956); A. L. Hall, *Mem. Geol. Surv. S. Africa No. 16* (1918).
17. H. L. Allsopp, H. R. Roberts, G. D. L. Schreiner, D. R. Hunter, *J. Geophys. Res.* 67, 5307 (1962).
18. T. J. Ulrych, personal communication, 7 December 1965; information pertaining to these measurements will appear in an article in preparation by T. J. Ulrych, H. L. Allsopp, and L. O. Nicolaysen.
19. Black chert facies in upper third of Fig Tree Series, exposed by road cutting 100 m northwest of surface opening to Daylight Mine (Barbrook Mining Co.), 28 km east-northeast of Barberton, South Africa. Samples on which these studies are based were collected by E.S.B. in February 1965.
20. W. N. Croft and E. A. George, *Bull. Brit. Museum Geol.* 3, 339 (1959).
21. H. D. Pflug, in *Die Welt* (Berlin, Germany), 30 October 1965.
22. H. D. Holland, in *Petrological Studies: A Volume to Honor A. F. Buddington*, A. E. J. Engel *et al.*, Eds. (Geol. Soc. Amer., New York, 1962), pp. 447–477.
23. C. F. Davidson, *Proc. Nat. Acad. Sci. U.S.* 53, 1194 (1965).
24. W. G. Meinschein, personal communication, 20 December 1965. However, see W. G. Meinschein, in *1967 McGraw-Hill Yearbook of Science and Technology*, p. 283.
25. W. G. Meinschein, E. S. Barghoorn, J. W. Schopf, *Science* 145, 262 (1964); E. S. Barghoorn, W. G. Meinschein, J. W. Schopf, *ibid.* 148, 461 (1965); G. Eglinton, P. M. Scott, T. Belsky, A. L. Burlingame, M. Calvin, *ibid.* 145, 263 (1964).
26. R. B. Johns, T. Belsky, E. D. McCarthy, A. L. Burlingame, P. Haug, H. K. Schnoes, W. Richter, M. Calvin, *Geochim. Cosmochim. Acta* 30, 1191 (1967); N. A. Sorenson and J. Mehlum, *Acta. Chem. Scand.* 2, 140 (1948); J. Cason and D. W. Graham, *Tetrahedron* 21, 471 (1965); J. D. Mold, R. K. Stevens, R. E. Means, H. M. Ruth, *Nature* 199, 283 (1963); M. Blumer and D. W. Thomas, *Science* 148, 370 (1965).
27. J. C. Bendoraitis, B. L. Brown, R. S. Hepner, *Anal. Chem.* 34, 49 (1963); M. Blumer and D. W. Thomas, *Science* 147, 1148 (1965); M. Blumer and W. D. Snyder, *ibid.* 150, 1588 (1965).
28. K. Bloch in *Taxonomic Biochemistry and Serology*, C. Leone, Ed. (Ronald Press, New York, 1964), p. 385.
29. R. Park and S. Epstein, *Geochim. Cosmochim. Acta* 21, 110 (1960); *Plant Physiol.* 36, 133 (1961).
30. F. E. Wickman, *Geochim. Cosmochim. Acta* 2, 243 (1952); H. Craig, *ibid.* 3, 53 (1953); W. A. Hodgson, *ibid.* 30, 1223 (1967); S. R. Silverman and S. Epstein, *Bull. Amer. Assoc. Petrol. Geologists* 42, 988 (1958); W. R. Eckelmann *et al.*, *ibid.* 46, 699 (1962); S. R. Silverman, in *Isotopic and Cosmic Chemistry*, H. Craig *et al.*, Eds. (North-Holland, Amsterdam, 1964), pp. 92–102; B. Nagy, *Proc. Nat. Acad. Sci. U.S.* 56, 389 (1966).
31. We thank Drs. W. G. Meinschein, J. G. Ramsay, T. J. Ulrych, L. O. Nicolaysen, H. L. Allsopp, D. S. Macauly, and Mrs. Dorothy O. Barghoorn for assistance. Supported by NSF grants GP-2794 and GA-650 to Harvard University and an NSF fellowship to J.W.S.

23 January 1967

43

Reprinted from *Science*, **161**, 1005–1008 (Sept. 6, 1968)

Alga-like Forms in
Onverwacht Series, South Africa:
Oldest Recognized Lifelike Forms on Earth

ALBERT E. J. ENGEL, BARTHOLOMEW NAGY,
LOIS ANNE NAGY, CELESTE G. ENGEL,
GERHARD O. W. KREMP, and CHARLES M. DREW

Abstract. *Spheroidal and cupshaped, carbonaceous alga-like bodies, as well as filamentous structures and amorphous carbonaceous matter occur in sedimentary rocks of the Onverwacht Series (Swaziland System) in South Africa. The Onverwacht sediments are older than 3.2 eons, and they are probably the oldest, little-altered sedimentary rocks on Earth. The basal Onverwacht sediments lie approximately 10,000 meters stratigraphically below the Fig Tree sedimentary rocks, from which similar organic microstructures have been interpreted as alga-like micro-fossils. The Onverwacht spheroids and filaments are best preserved in black, carbon-rich cherts and siliceous argillites interlayered with thick sequences of lavas. These lifelike forms and the associated carbonaceous substances are probably biological in origin. If so, the origins of unicellular life on Earth are buried in older rocks now obliterated by igneous and metamorphic events.*

The search for evidence of early terrestrial life in the better preserved, old Precambrian sedimentary rocks has revealed a wide variety of unequivocal fossils as well as problematical structures and carbonaceous materials of uncertain origin (*1*). Some of the microstructures referred to as exhibiting "alga-like" and "filamentous" morphologies occur in carbonaceous argillites, siltstones, and cherts from South Africa that are more than 3 eons (3×10^9 years) old (*2, 3*).

The oldest unequivocal fossils are

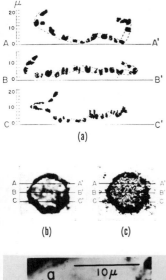

(a)

(b)   (c)

Fig. 1 (left top). (a–c) Photomicrograph and optical cross sections of one of the carbonaceous spheroids freed from the mineral matrix of the lower Onverwacht chert zone SF-77 by acid maceration; largest diameter, 74 μ. (a) Three cross sections constructed along lines A–A', B–B' and C–C' from a total of 38 photomicrographs taken partly at 1-μ and partly at ½-μ intervals vertically through the object. (b) The same object at 4 μ above the base.

probably the laminated algal stromatolites found in carbonate rocks of the Bulawayan Series near Bulawayo, in southern Rhodesia (4). Although the age of the Bulawayan stromatolites is arguable and is based on a correlation of the Bulawayan type series with similar rocks approximately 300 km east-northeast, the geological relations in this region strongly suggest that these algal remains are at least 2.7 eons old,

and possibly more than 3 eons old (5).

Recently, however, very old microfossil-like forms and carbonaceous matter have been found in the sedimentary rocks of the Fig Tree Series (2, 3). The Fig Tree is the middle member of the Precambrian Swaziland System, which is well exposed some 650 km south-southeast of Bulawayo, in the Barberton-Badplaace region of the eastern Transvaal, South Africa. In this region, (the Barberton Mountain Land) the Swaziland System is demonstrably older than 3 eons (6). Spheroidal organic bodies (that range from 5 to 25 μm in diameter) and filamentous wisps of carbonaceous matter are present in the Fig Tree Series. Rodlike to filamentous forms having the appearance of bacteria (7) and concentrically laminated "oolites" possibly of algal origin (8), have also been reported from the Fig Tree sediments. The Fig Tree spheroids have been interpreted as fossil algae by Pflug (3) and as alga-like microfossils by Schopf and Barghoorn (2); the carbonaceous filaments also may be algal in origin (2, 3). The spheroidal forms that Schopf and Barghoorn regard as remnants of unicellular, noncolonial, alga-like organisms, have been named *Archaeosphaeroides barbertonensis* (2). We observed similar organic spheroids and filaments in the Fig Tree beds and also in successively lower stratigraphic zones in the very thick underlying Onverwacht Series. The sedimentary rocks in the Onverwacht are probably the oldest exposed, well-preserved beds in South Africa and are perhaps the oldest exposed beds on the earth.

We note herein (i) the stratigraphic features and interrelations of the several rock series in the Swaziland System, (ii) the occurrence and morphology of abundant carbonaceous alga-like and filamentous forms in the Onverwacht sediments, and (iii) the presence of inorganic, siliceous spheroids that are morphologically suggestive of alga-like microfossils but definitely not of organic origin in associated basaltic lavas. These inorganic spheroids are illustrated to indicate the extreme caution necessary in the interpretation of the origin and occurrence of cell-like microscopic structures that occur in many rock types of diverse ages and origins throughout the Precambrian rock systems.

The Swaziland System is a remarkably complete, well-preserved, oceanic to island-arc and continental borderland sequence of volcanic and sedimentary rocks in the Rhodesian or South African

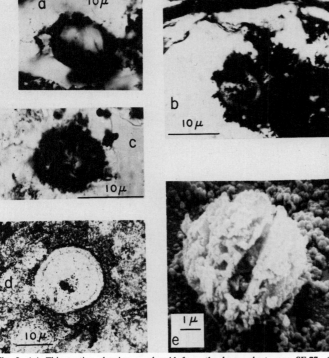

Fig. 2. (a) Thin section showing a spheroid from the lower chert zone SF-77. (b and c) Thin sections of carbonaceous argillite with spheroidal and filamentous forms from the upper Onverwacht sedimentary zone ACS-14; detailed microscopic examination of the round particle in (b) suggests it is cupshaped. (d) Thin section of Onverwacht pillow lava showing an inorganic spheroidal microstructure with a pseudo-"double wall" and central body. This is definitely nonbiological in origin. (e) An "organic" body from the lowest chert zone SF-77. This particle was freed from a preparation of powdered, carbonaceous chert by treatment with hot 6N HCl and hot 48 percent HF for 1 hour and was then coated with a thin layer of gold-palladium.

shield (*9*). Rocks of this system are in part sheared and faulted and in part undeformed. Most of the system is metamorphosed, commonly to the greenschist facies, although some segments, especially the marginal contacts with granitic rocks, are reconstituted into the amphibolite facies. In a few scattered areas, constituent igneous and sedimentary rocks are relatively free of metamorphic overprint. The Swaziland System is subdivided into three rock series. In order of decreasing geologic age these are: (i) Onverwacht Series, (ii) Fig Tree Series, and (iii) Moodies Series (*9*).

During the past 5 years the Onverwacht Series has been carefully studied and mapped by M. and R. Viljoen at and near the type locality (*9*). One of us (A.E.J.E.) has examined in detail many of the better exposed parts of the Onverwacht Series. In and near the Komati River Valley, 15 to 30 km east of Badplaace, many of the initial sedimentary and volcanic structures and textures are well preserved. The series consists largely of successions of ultramafic and mafic lavas with quite subordinate layers of dacitic agglomerate, tuff, chert, and clastic sediments. Beds of carbonaceous argillite and carbonate occur as interlayers with the cherts in the lower, middle, and upper Onverwacht.

The amounts of carbonate sediment are crudely proportional to the amount of chert and argillite. The lowermost sedimentary zone is largely tuff, with thin lenses of laminated, carbonaceous and white chert. Most of the sedimentary beds and the lavas were deposited in water. This is clearly indicated by graded- and cross-bedding, detailed and uniform laminations, and scour and fill structures in the sediments, and by numerous and widespread pillows in the associated lavas. The thickness of the Onverwacht at and near the type locality varies up to about 11,000 m.

The Onverwacht Series is overlain, in part unconformably, by the Fig Tree Series. The Fig Tree varies in thickness up to approximately 3600 m and consists largely of graywackes, carbonaceous argillites, arkoses, and argillaceous sandstones with subordinate amounts of banded, iron-bearing and carbonaceous cherts, tuffs, and conglomerates (*9*). Volcanic rocks are rare, except for the tuff components in the finer-grained clastic sediments.

The Fig Tree Series is, in turn, overlain unconformably by the Moodies Series, which consists mostly of arkosic

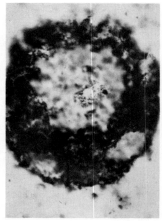

Fig. 3. Lifelike form, 106 μ in diameter, typical of many forms found in carbonrich, siliceous sedimentary rocks of the Onverwacht series in South Africa.

sandstones, shales, and thick orthoquartzites which are in part recycled (*9*). The thickness of the Moodies also appears to vary up to approximately 3600 m.

In conjunction with detailed petrographic and chemical studies of sedimentary and igneous rocks of the Swaziland System, one of us (C.G.E.) isolated both siliceous and carbonaceous particles and carbonaceous filaments from cherts, argillites and carbonate

beds. In the acid-resistant residues, filamentous and spheroidal alga-like forms of the types described from the Fig Tree (*2, 3*) were abundant. The Fig Tree and Onverwacht microstructures are similar in morphology, but the Onverwacht spheroids are commonly larger in size than the spheroidal bodies in the Fig Tree Series (*2*).

From studies to date, gradations in both size and form in the organic spheroids from the lowest Onverwacht beds to the lower Fig Tree rocks are suggested. The carbon-bearing filamentous forms are of diverse morphology.

Our morphological studies have been concentrated largely in two stratigraphic zones: (i) the lowermost sedimentary rocks in the Onverwacht (hereafter referred to as lower chert zone SF-77) which occurs 350 to 600 m above the contact of granitic gneisses and granodioritic plutons with the oldest recognizable Onverwacht lavas, and some 10,000 m below the base of the Fig Tree, and (ii) carbonaceous chert and argillite in the upper Onverwacht (referred to as ACS-14) roughly 2500 m below the base of the Fig Tree Series and approximately 7500 m stratigraphically above the lower chert zone SF-77. Several studies have been made of the organic geochemistry of the Fig Tree sediments (*10*). Our analysis of the Onverwacht carbonaceous matter is incomplete. They are involved mainly with the insoluble kerogen fraction because

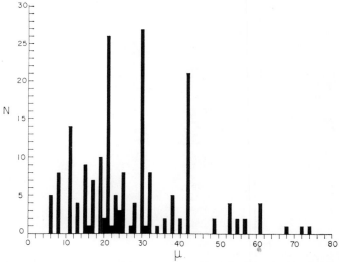

Fig. 4. Size distributions of 190 spheroidal particles in the Onverwacht lower chert zone SF–77. Note the large range in diameter of the 190 spheroids measured and what may be a polymodal size-frequency distribution.

solutions percolating through the beds during geoiogic time may contaminate these sediments with soluble organic material of younger geologic age. Initial analyses indicate a kerogen which is largely aromatic in composition. In contrast, much of the Fig Tree kerogen is essentially aliphatic (*11*).

In both the carbonaceous chert zones (SF-77 and ACS-14) spheroidal forms are more common than the filamentous forms. The microstructures were studied in thin sections of the rocks and in powdered preparations. Some of the powdered rock was treated with hot 6*N* HCl and then with hot 48 percent HF for 1 hour, and then the acid-resistant residues were examined. Most of the spheroidal structures found in powdered preparations and in thin sections are cupshaped (Fig. 1, a–c). The Fig Tree spheroids reported by Schopf and Barghoorn have a maximum size of 24 μm and a mean of about 19 (*2*). The largest Fig Tree spheroids we have seen are about 30 μm in diameter. Many spheroids in the Onverwacht are much larger (Fig. 3). The spheroidal particles in both Fig Tree and Onverwacht are definitely indigenous to the sedimentary beds as is indicated by their presence in thin sections of the rocks [Figs. 2 (a–c) and 3].

Thin sections of the Onverwacht lavas also contain inorganic spheroids that approach in morphology the alga-like forms in the associated sediment (Fig. 4). Many of the spheroids in the lavas tend to be precisely shaped and they possess distinctive, if pseudo, double walls. Several microbiologists who examined the thin sections, but who were not told that the enclosing rock is of igneous origin, interpreted these forms found in the pillow lavas as definite remains of unicellular life. A photomicrograph of one of these pseudomicrofossils with a double wall is shown in Fig. 2d. Two of us (C.G.E. and A.E.J.E.) have observed a wide variety of equally lifelike mineral artifacts in diverse rock types. Bramlette (*12*) has noted the problem, and most petrographers and some micropaleontologists are aware that fossillike forms are commonly produced by inorganic processes. It is impossible that any of these forms in the altered igneous rocks are fossil remnants of life. It could be inferred that some of the lifelike microstructures in the Onverwacht and Fig Tree sedimentary rocks may have had their source in weathered or tectonically disintegrated and eroded igneous rocks, or as products of other nonbiological processes.

However, the spheroids within the carbonaceous Onverwacht sediments not only have the morphologies of fossils, but also are intimately associated with the kerogen-bearing carbonaceous substances which appear to form parts of their walls and interiors. They are also closely associated with kerogen-bearing, filamentous forms which have the appearance of microfossils (*2*). These features and relations suggest that the enclosed carbonaceous, spheroidal, and filamentous forms may be microfossils.

Difficulties of interpretation of the carbonaceous filaments are, however, apparent from inspection of Fig. 2b. The filamentous layers in this carbonaceous argillic chert are of diverse morphology, many remarkably lifelike, similar to those found in the Fig Tree (*2*). Some of these filaments are probably carbonaceous material of accidental form, distributed unevenly and also in regular geometric patterns along the bedding planes by sedimentary and diagenetic processes. But many appear to be true fossils, although less well preserved than those found in younger Precambrian sediments (*1*).

Establishing the presence of biological activity during the very early Precambrian clearly poses difficult problems. Although the Fig Tree and Onverwacht organic spheroids and filaments are probably of biological origin, skepticism about this sort of evidence of early Precambrian life is appropriate. Unfortunately, little-altered sedimentary rocks as old or older than the Onverwacht chert SF-77 are unknown on Earth. If the carbonaceous forms in the Onverwacht are fossils, the origin of unicellular life presumably has occurred in still older rocks destroyed by superimposed igneous and metamorphic episodes in the evolving earth.

References and Notes

1. P. E. Cloud, *Memoirs, Centennial Symposium Volume* (Peabody Museum Yale Univ. Press, New Haven, Conn., in press); J. W. Schopf and E. S. Barghoorn, *Science* **156**, 508 (1967); E. S. Barghoorn and S. A. Tyler, *ibid.* **147**, 563 (1965); E. S. Barghoorn and J. W. Schopf, *ibid.* **152**, 758 (1966); E. S. Barghoorn, *ibid.*, p. 758; H. D. Pflug, *Economic Geology Research Unit, Information Circular 28* (University of Witwatersrand, Johannesburg, South Africa, 1966); P. E. Cloud, *Science* **148**, 27 (1965); P. E. Cloud, J. W. Gruner, H. Hagen, *ibid.*, p. 1713; M. F. Glaessner and B. Daily, *South Austr. Rec.* **13**, 3, 369 (1959); M. F. Glaessner, *Earth Sci. Rev.* **1**, 29 (1966); T. C. Hoering, *Annual Report Director Geophysical Laboratory* (Carnegie Institution of Washington, Washington, D.C., 1961), p. 190; H. Winter, *Economic Geology Research Unit, Information Circular 16* (Univ. of Witwatersrand, Johannesburg, 1963); J. S. Harington and J. J. Cilliers, *Geochim. Cosmochim. Acta* **27**, 412 (1963); P. Hoffman, *Science* **157**, 1043 (1967); C. Marshall, J. May, C. Perret, *ibid.* **144**, 290 (1964); R. Rezak, *U.S. Geol. Survey Professional Paper 294-D* (U.S. Government Printing Office, Washington, D.C., 1957); J. Hoefs and M. Schidlowski, *Science* **155**, 1096 (1967); J. A. Donaldson, *Geological Survey of Canada Bulletin* (Geol. Survey Canada, Ottawa, 1963), p. 102.
2. E. S. Barghoorn and J. W. Schopf, *Science* **152**, 758 (1966); J. W. Schopf and E. S. Barghoorn, *ibid.* **156**, 508 (1967).
3. H. D. Pflug, *Econ. Geol. Res. Unit, Infor. Circ. 28* (University of Witwatersrand, Johannesburg, South Africa, 1966).
4. A. A. MacGregor, *Trans. Geol. Soc. S. Africa* **54**, 35 (1951).
5. Dr. J. Wiles, director of the Geological Survey of Rhodesia, and his colleagues conclude that the remarkable similarity in rock successions and interrelations of the carbonate-bearing greenstones at and near Bulawayo as compared with those to the northeast toward and in the Bikita district indicate similar ages. Near Bikita, pegmatite dikes cutting Bulawayon meta-sedimentary rocks are at least 2.6 $\times$ 10⁹ years old. In other areas of southern Rhodesia Bulawayon-like meta-sedimentary rocks are cut by granitic rocks as old as 3 $\times$ 10⁹ years. A. Holmes, *Nature* **173**, 4405 (1954); L. O. Nicolaysen, in *Petrologic Studies: A Volume in Honor of A. F. Buddington*, A. E. J. Engel, B. F. Leonard, H. L. James, Eds. (Geological Society of America, New York, 1962), p. 569; H. D. Pflug, *Econ. Geol. Res. Unit, Infor. Circ.* **28** (Univ. of the Witwatersrand, Johannesburg, South Africa, 1966); J. W. Schopf and E. S. Barghoorn, *Science* **156**, 508 (1967).
6. C. E. Hedge, private communication, May 1967; information pertaining to these measurements will appear in an article by C. E. Hedge, M. Tatsumoto, and A. E. J. Engel (in preparation); H. L. Allsop, H. R. Roberts, G. Schreiner, D. R. Hunter, *J. Geophys. Res.* **67**, 5307 (1962).
7. E. S. Barghoorn and J. W. Schopf, *Science* **152**, 758 (1966).
8. I. G. Ramsay, *Trans. Geol. Soc. S. Africa*, in press.
9. D. J. L. Visser, O. R. Van Eeden *et al.*, *South African Geol. Survey, Spec. Paper 15* (1956); C. R. Anhaeusser, C. Roering, M. J. Viljoen, R. P. Viljoen, *Econ. Geol. Res. Unit, Infor. Circ.* **38** (University of the Witwatersrand, Johannesburg, South Africa, 1967); A. E. J. Engel, *Econ. Geol. Res. Unit, Infor. Circ.* **27** (University of the Witwatersrand, Johannesburg, South Africa, 1966).
10. T. C. Hoering, in *Researches in Geochemistry*, P. H. Abelson, Ed. (Wiley, New York, 1967), vol. 2, p. 87; J. Oró and D. W. Nooner, *Nature* **213**, 1082 (1967); J. W. Schopf, K. A. Kvenvolden, E. S. Barghoorn, *Proc. Nat. Acad. Sci. U.S.* **59**, 639 (1968).
11. L. A. Nagy, M. C. Bitz, C. G. Engel, A. E. J. Engel, *Int. Geol. Congress 23rd Prague* (1968), abstr.
12. M. N. Bramlette, *Science* **158**, 673 (1967).
13. Supported by NASA grants NGR-05-009-043, NsG-541 and NsG-321, and by NSF grant GA-800, and by the U.S. Geological Survey. The petrographic thin sections were prepared by Mrs. G. S. Rev at Columbia University in New York and by R. Dehaven of the Scripps Institution of Oceanography, University of California at San Diego. The cooperation of the Economic Geology Research unit, University of the Witwatersrand, and the complementary field studies of M. Viljoen and R. Viljoen are acknowledged.

1 March 1968

Author Citation Index

Abelson, P. H., 32, 34, 38, 59, 65, 155, 173, 241, 254, 274, 281, 293, 308, 312, 336, 348, 362, 391
Adelberg, E. A., 312
Ahrens, L. H., 401
Akabori, S., 32, 45
Alberty, R. A., 173
Aldrich, H. R., 206
Aldrich, L. T., 238
Alevandrov, E. A., 206
Allen, H. A., 65, 179
Allen, J. R. L., 267
Allen, R. O., 136
Allen, W. V., 43, 58, 159
Alling, H. L., 206
Allsopp, H. L., 180, 348, 366, 397, 401, 405
Amm, F. L., 378
Anders, E., 9, 10, 62, 65, 71, 133, 134, 135, 136, 151, 155, 159
Anderson, W., 209
Andrews, H. N., Jr., 391
Anhaeusser, C. R., 362, 366, 370, 405
Armstrong, H. S., 208, 238
Arnon, D. I., 55
Ashley, B. E., 348
Attaway, D. H., 293, 296
Ault, W. V., 90, 238, 293
Axon, H. J., 133

Baadsgaard, H., 180, 206, 207, 255
Bada, J. L., 59, 173
Bahadur, K., 32, 38
Bajor, M., 133
Baker, B. L., 43, 71, 133, 134, 293
Baker, D. R., 366
Baker, E. G., 293

Baldwin, R. B., 23
Ball, J. S., 293
Banks, H. P., 348
Bardwell, D. C., 42
Barghoorn, E. S., 34, 173, 179, 209, 254, 274, 281, 293, 296, 308, 312, 336, 337, 348, 349, 350, 351, 352, 362, 363, 365, 391, 392, 397, 401, 405
Barker, D. S., 173
Barker, F., 366
Barker, H. A., 173
Barnes, I., 65
Bar-Nun, A., 59
Bar-Nun, N., 59
Barrer, R. M., 173
Barrett, L. P., 209
Basu, S. K., 238
Bateman, A. M., 194
Bates, R. G., 173
Bauer, S. H., 59
Bauman, A. J., 151, 274
Bayley, R. W., 206
Bayley, W. S., 209
Baylor, E. R., 175
Beales, F. W., 208, 397, 401
Beck, A., 42
Becker, R. H., 167
Bedford, R. L., 65
Behrendt, M., 155
Bein, S. J., 391
Belsky, T., 150, 155, 274, 275, 293, 296, 308, 349, 350, 397, 401
Bender, M. L., 173
Bendoraitis, J. G., 150, 281, 293, 296, 397, 401
Benoiton, N. L., 59, 62

407

Subject Index

419